RECLAMATION, TREATMENT AND UTILIZATION OF COAL MINING WASTES

PROCEEDINGS OF THE THIRD INTERNATIONAL SYMPOSIUM ON THE
RECLAMATION, TREATMENT AND UTILIZATION OF COAL MINING WASTES
GLASGOW / UNITED KINGDOM / 3 - 7 SEPTEMBER 1990

Reclamation, Treatment and Utilization of Coal Mining Wastes

Edited by
A.K.M. RAINBOW
British Coal Corporation, Minestone Services, Hebburn, Tyne and Wear

A.A.BALKEMA / ROTTERDAM / BROOKFIELD / 1990

The texts of the various papers in this volume were set individually by typists under the supervision of each of the authors concerned.

Published by

A.A. Balkema, P.O. Box 1675, 3000 BR Rotterdam, Netherlands
A.A. Balkema Publishers, Old Post Road, Brookfield, VT 05036, USA

ISBN 90 6191 154 0

Reclamation, Treatment and Utilization of Coal Mining Wastes, Rainbow (ed.) © 1990 P

Foreword

N 90 6191 154 0

Eur. Ing. Dr. A. K. M. Rainbow
Head, British Coal Corporation
Minestone Services

Bachelor of Science (Hons) Civil Engineer
Doctor of Philosophy (Mining Eng.)
Chartered Engineer
Fellow, Institution of Civil Engineers
Fellow, Institution of Mining & Metallurgy
Fellow, Institution of Highways & Trans.
Fellow, Institution of Water & Env. Management
Fellow, Geological Society
Member, Geotechnical Society

Organiser & Editor

First, Second and Third International
Symposium on the Reclamation, Treatment
and Utilization of Coal Mining Wastes

I believe that many of the delegates attending this, the 3rd International Symposium on the Reclamation, Treatment and Utilization of Coal Mining Wastes will have recognised that we could not have chosen a worse title – 'Coal Mining Wastes' – waste suggests rubbish – rubbish is regarded as a worthless, useless or unwanted matter. I now believe that it would have been more prudent to have substituted MINESTONE for Coal Mining Wastes.

Minestone is a generic term for unburnt coal mining waste and was first introduced by the then National Coal Board to distinguish between well burnt and unburnt material but which is now in international usage.

It is indisputable that the Coal Mining Industries are faced with very high cost in disposing of their wastes and as a result a number of specialised groups have been formed within the mining organisations. These groups are actively engaged in developing ways of reducing those costs. An obvious way of reducing the costs of disposal is to find ways of utilizing the non-coal fraction before it incurs disposal cost – this means finding ways of utilizing 'fresh wrought' material. Fresh

...s different from material which has been placed on a tip and which will have ...degrees of physical, chemical and mineralogical weathering.

...coal producing nations have recognised the economic sense of utilizing their coal ...in construction and civil engineering projects as well as the rehabilitation of derelict ...ict land – often caused by mining activities. Many countries have also recognised the ...s benefit of utilizing cheap coal-mining waste which can conserve the more valuable ...sources for more appropriate use.

...st savings on civil engineering projects, especially where large volumes of fill are required ...e substantial. Unfortunately, there is a reluctance by those responsible for approving ...al selection to use waste materials solely on the basis that they are classified as 'wastes'.

...thusiasm and dedication by research workers and entrepreneurs has inspired other organisa-...ns, including some Governments to sponsor even more research particularly into the area of ...valuation into the properties and potential further utilization of minestone. The United Nations are sponsoring a task force to further these ambitions – thus creating an international understanding of what you can and equally important what you cannot or should not do with minestone.

Experience shows that once civil engineers have been made aware of the suitability of coal mining wastes – such as the suitability of minestone in waterway engineering they have little trouble in extending that use to projects which require the construction of islands in the sea. These islands can be used to provide numerous off-shore facilities and in many cases be an extension of a shore line installation such as a berth or for coastal protection works.

It has been found that by adding certain additives (usually cement) the load bearing properties of minestone can be improved to the extent that it can replace more conventional materials for sub-base and full depth construction for hardstandings, car parks, road sub-bases and other facilities.

Some nationally controlled and larger coal mining enterprises have, in addition to establishing units whose main objective is to reduce coal mining costs have also established units concerned with environmental, ecological, social and psychological issues.

As a result of changing trends in coal utilization, coal quality, or particle size it is reasonable to presume that coal which could not be marketed in certain periods were discarded onto the coal waste tips.

For various reasons throughout the years some of the coal ignited resulting in some form of nuisance to the surrounding communities. There was a need to control and minimise this nuisance and techniques were introduced with those aims in mind.

In circumstances where there was no heating – but evidence that the waste heaps contained substantial quantities of coal, techniques were established to re-work the heaps to recover the coal.

Certain researchers have identified that coal mining wastes contain some base elements and have set about and established methods for separating or extracting these substances.

This Symposium was designed to afford an opportunity to researchers in all these fields to express their aspirations, views and to be able to present their findings. We, at British Coal, Minestone Services are sure that this opportunity has been grasped firmly and I am equally sure that as with the previous two Symposia you will not be disappointed with the technical content of this the Third.

Our social programme having both a Scottish and Czechoslovakian flavour is, I trust, accept-able.

My message to you all is have a successful Symposium, meet old friends, make new ones and come again in three years time.

Reclamation, Treatment and Utilization of Coal Mining Wastes, Rainbow (ed.) © 1990 Balkema, Rotterdam. ISBN 90 6191 154 0

Lord Haslam – Chairman, British Coal Corporation

Lord Haslam
Chairman, British Coal Corporation

Lord Haslam was appointed Chairman of the British Coal Corporation (formerly the National Coal Board) on September 1, 1986. He had been Deputy Chairman and Chairman-Designate since November 1985.

He was formerly Chairman of British Steel Corporation and Chairman of Tate and Lyle plc. He was knighted in 1985 and made a Life Peer in the Queen's Birthday Honours in June 1990.

OTHER CURRENT APPOINTMENTS INCLUDE:

August 1983	Member of the Council, Confederation of British Industry (CBI)
July 1983	Member of Nationalised Industries Chairmen's Group and Chairman in 1985-1986
March 1985	Director of the Bank of England
March 1985	Governor of the National Institute of Economic & Social Research
May 1985	Governor of Henley Management College
November 1985	Chairman of Manchester Business School
September 1986	Advisory Director of Unilever
January 1989	President, Institution of Mining Engineers

Lord Haslam, married with two sons and four grandchildren, was born on February 4 1923 at Bolton, Lancashire. His hobbies are golf and travel.

He was educated at Bolton School and Birmingham University, where he graduated as a BSc in Coal Mining, in 1944. Having worked underground during every holiday at University, he joined Manchester Collieries Ltd, who operated the former Chanters, Gibfield, Bedford and Mosley Common collieries, in 1944. With experience of a full range of mining work he qualified for his Colliery Manager's Certificate in 1947 and he is one of the longest-serving members of the Institution of Mining Engineers. In January 1947 the colliery company became part of the North Western Division of the newly-formed National Coal Board.

Later in 1947 he joined the Nobel Division of Imperial Chemical Industries as a mining engineer specialising in explosives in the mining, quarrying and oil prospecting industries. For the next 10 years he travelled the world, visiting and working in mines and quarries in every continent. Subsequently, he had wide general management experience in three ICI divisions, becoming Chairman of ICI Fibres, a Main Board Director, ICI's Personnel Director for four years and Deputy Chairman.

APPOINTMENTS – IMPERIAL CHEMICAL INDUSTRIES

1947	Technical Service Engineer, Nobel Division
1955	Technical Service Assistant Manager, Nobel Division
1957	Staff Manager, Nobel Division
1960	Personnel Director, Nobel Division
1962	Deputy Regional Manager, Southern Region Sales Office
1963	Commercial Director, Plastics Division
1964	Director and General Manager, Films Group
1966	Deputy Chairman, Plastics Division and General Manager, Films
	Chairman of British Visqueen Ltd and Bexford Ltd
1969	Deputy Chairman, ICI Fibres – Commercial Sales.
	Marketing Chairman of ICI committee proposing Staff Job Assessment Scheme
1971	Chairman of ICI Fibres
	Director of Fiber Industries Inc. (USA)
1974	Main Board Director
1975 - 1977	Director of Imperial Metal Industries Ltd (ICI sold its shareholding in IMI in November 1977)
1978 - 1979	Director of AECI
1980	Deputy Chairman
1979 - 1981	Director/Deputy Chairman resident in USA
	Chairman, ICI Americas
	Director, C-I-L Inc.
	Chairman, ICI Canada

Lord Haslam was appointed non-executive Director of Tate and Lyle plc in 1978, Deputy Chairman in 1982 and was Chairman from 1983 to March 1986.

Appointed to the board of British Steel Corporation in July 1983, he was Chairman from September 1983 to March 1986.

OTHER APPOINTMENTS INCLUDED:

1985 - 1988	Member of the National Economic Development Council
1982 - 1983	Non-Executive Director, Cable & Wireless
1982 - 1983	Chairman of North American Advisory Group
1981 - 1985	Member of British Overseas Trade Board
1971 - 1974	Chairman of Man-Made Fibre Producers Committee
1972 - 1974	Chairman of European Synthetic Fibre Manufacturers (CIRFS)
1972 - 1974	Vice Chairman, British Textile Confederation

Lord Haslam was made a Freeman of the City of London in March 1985 and received honorary degrees of Doctor of Technology from Brunel University in May 1987, and Doctor of Engineering from Birmingham University in July 1987.

Reclamation, Treatment and Utilization of Coal Mining Wastes, Rainbow (ed.) © 1990 Balkema, Rotterdam. ISBN 90 6191 154 0

Table of contents

Reclamation, Treatment and Utilization of Coal Mining Wastes, Rainbow (ed.) © 1990 Balkema, Rotterdam. ISBN 90 6191 154 0

The role of the mixing moisture content in determining the strength and durability of cement stabilised minestone

M. D. A. Thomas
Department of Materials, Imperial College of Science and Technology, London, UK

R. J. Kettle
Department of Civil Engineering, Aston University, Birmingham, UK

J. A. Morton
Department of Geological Sciences, University of Birmingham, UK

ABSTRACT: Cement stabilised minestone (CSM) specimens were produced at a range of mixing moisture contents from 2% below to 3% above the optimum for compaction. Mixing "dry" of optimum produced a CSM of considerably reduced strength and resistance to immersion. CSM compacted at higher than optimum moisture contents also had reduced 7 day strength compared to mixes produced at optimum although these "wet" mixes gained strength beyond 7 days at a faster rate than the optimum mixes especially at higher cement contents. In addition, specimens produced at higher moisture contents exhibited less expansion when immersed and showed a greater resistance to cyclic freeze-thaw and wet-dry testing. The expansion mechanisms are discussed. It is concluded that provided short-term strength requirements are met, the performance of CSM can be enhanced by increasing the moisture content at the mixer.

1 INTRODUCTION

Cement stabilised minestone (CSM) has been successfully utilised as a structural layer in roads, hardstandings, coal stacking areas and car parks (Tanfield, 1978; Sleeman, 1984). However, not all minestones are suitable for stabilisation and previous studies have indicated that the coarser, less plastic minestones have the greatest potential for cement stabilisation (Kettle and Williams, 1977; Kettle and McNulty, 1981; McNulty; 1985). More recent study has shown that whereas the short-term durability of CSM may be controlled largely by raw minestone properties such as: grading, plasticity, slaking resistance, water soluble sulphate content (Thomas et al; 1987) the long term durability is greatly influenced by the sulphur bearing mineralogy of the raw minestone (Thomas et al; 1989). However, despite the importance of the properties of the raw minestone, other factors such as mix design and construction practice have a major effect on the performance of the complete product (Minestone Services, 1983). This was emphasised in 1980 when problems of deformation of a pavement constructed with CSM occurred shortly after it was placed (Byrd, 1980). The deformation was attributed to the expansion of the CSM due to moisture uptake which was exacerbated both by the material being placed 'dry' of the optimum moisture content and by poorly finished construction joints.

This study investigates the role of mixing moisture content on the strength and durability of cement stabilised minestone. Two minestones were selected for this study, Snowdon and Wardley, based on their apparent suitability for utilization as cement stablised materials. Both of these materials have been used extensively in the construction of cement stabilised pavement layers.

2 EXPERIMENTAL PROCEDURE

2.1 Materials

The minestones used were sampled from Wardley and Snowdon collieries and grading, specific gravity and plasticity characteristics of the materials are given in Table 1. The compaction characteristics were determined for a mixture of minestone with 7.5% Ordinary Portland Cement (by weight of dry minestone) in accordance with Test 1 of BS 1924 (British Standards Institution, 1975) and the results are given in Table 1.

Ordinary Portland Cement (OPC) was used throughout the study for the manufacture of the stabilised samples.

2.2 Specimen preparation

The minestones were air dried in the laboratory and screened at 28mm, the coarse fraction being discarded. Seven days prior to stabilisation, the minestones were pre-wetted to a moisture content 2% below the target mixing moisture content for stabilisation and stored in sealed containers. Final mixing of the pre-wetted minestone, OPC and additional water was carried out using a twin 2-bladed, pug-mill mixer as this type of mixer is known to produce an intimate mix (Kettle and McNulty, 1981).

The cement stabilised minestone was compacted into 100mm cube moulds in two equal layers using a vibrating hammer under a load of 400N for 60s per layer. The cube moulds were sealed from the atmosphere and stored at 20°C for 24 hours. After this period the CSM specimens were demoulded and stored in sealed polythene bags at 20°C until test.

2.3 Testing of CSM specimens

Compressive strength determinations were carried out on CSM specimens that had been standard cured (sealed in polythene at constant moisture content) for 7, 14 and 17 days.

The resistance to immersion was determined by placing CSM specimens that had been standard cured for 7 days in water at 20°C for periods of either 7 or 140 days. The immersion resistance is expressed in terms of the Immersion Ratio (IR) which is the ratio of the strength of immersed specimens (7 day immersion) to the strength of specimens standard cured for 14 days. In addition, dimensional changes were monitored throughout the immersion period.

Following standard curing for 7 days, CSM specimens were subjected to either the Freeze-thaw test (ASTM, 1980a), or the Wet-dry test (ASTM, 1980b). Resistance to freeze-thaw and wet-dry cycles was assessed by determining residual strength and by monitoring dimensional changes.

3. RESULTS

The strength and resistance to immersion results for Wardley and Snowdon CSM mixed at a range of moisture contents are given in Table 2.

The mixing moisture contents for this series of mixes were selected to produce a range from 2% below to 3.5% above the optimum moisture contents of 8.6% and 6.8% for Wardley and Snowdon CSM respectively. However, it was observed that the optimum moisture content for the compaction of CSM in the cube moulds was approximately 1% higher than that determined by the compaction test using a cylindrical mould. This effect is shown in Figure 1, and is possibly the result of differences in wall friction of the moulds or the ratios of the maximum particle size to mould volume (higher for the cube specimens).

Figures 2 and 3 show the effect of the mixing moisture content on the strength of Wardley and Snowdon CSM. For both materials, the maximum 7 day strength and the maximum dry density occur at approximately the same moisture content. A higher than optimum moisture content is required to achieve the maximum 14 day strength and this is especially noticeable for the mixes at the higher cement content.

The relationship between the strength and dry density of cement stabilised materials is well established (Grimer, 1958) with greater dry densities obviously producing higher mechanical strength. However, the strength of CSM is also dependent on cementing action and the moisture content required for maximum compaction does not necessarily coincide with that for maximum compressive strength (Davidson et al, 1962). The absorption of moisture by the minestone may result in an internal

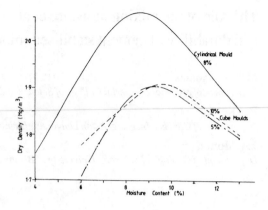

FIGURE 1 Compaction characteristics for Wardley CSM (Cement contents given on the graphs).

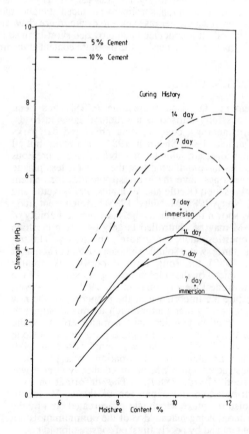

FIGURE 2 Strength vs Moisture Content for Wardley CSM.

2

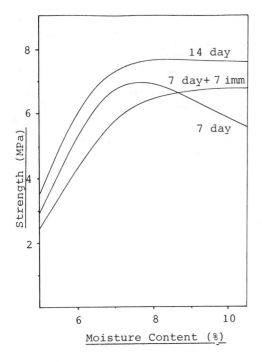

FIGURE 3 Strength vs
Moisture Content
for Snowdon CSM

Table 3 gives the strength results for a second series of Wardley and Snowdon CSM with a cement content of 8% (by mass of dry shale). The optimum moisture contents used for this series was adjusted to allow for the increase in the optimum moisture content to 9.5% and 8.0% observed for the Wardley and Snowdon cubes prepared for first mix series (Table 1). The range of mixing moisture contents was selected to cover the range from optimum to 3% above optimum. Figures 4 and 5 show the effect of the initial mixing moisture content on the strength of standard cured specimens and the residual strength of specimens subjected to freeze-thaw and wet-dry cycles and 140 day immersion. As with the first series of tests, the maximum 7 day strength coincides with the maximum dry density (i.e. when the mixing moisture content is at the optimum), whereas the later-age strengths reach a maximum when the mixing moisture content is above optimum.

The residual strengths of the specimens for both CSM's subjected to the durability tests are improved when the mixing moisture content is increased above optimum. However, the performance of Wardley CSM reaches a maximum when the moisture content is between 1 and 2% above optimum with further increases in mixing water producing a reduced residual strength. The performance of the Snowdon CSM continues to improve for mixing moisture contents up to 3% above optimum and the data suggest that greater increases in the residual strength of specimens subjected to immersion or freeze-thaw testing may be expected if the moisture content at mixing were increased still further.

depletion of moisture thereby inhibiting the hydration of cement. These effects would become more noticeable with increasing age and cement content.

At lower than optimum moisture content there is a marked reduction in compressive strength. The 7 day strength of these mixes is reduced on average by more than 50% due to decreasing the moisture content from optimum to 2% below optimum.

The results in Table 2 and Figures 2 and 3 clearly show that the resistance to immersion is improved by increasing the mixing moisture within the range studied. The increased moisture not only leads to increased residual strength following immersion but also results in an increase in the Immersion Ratio (IR).

For CSM mixed at 3.5% above optimum the specimens retain their 7 day strength after 7 days immersion, with the Snowdon CSM showing significant strength increase during immersion at the higher mixing moisture content. The results in Table 2 also show that increasing mixing moisture content of CSM produces a considerable reduction in the amount of expansion of specimens during immersion. The effect of increasing the cement content is to increase the strength of standard cured immersed specimens and to reduce the expansion of specimens during immersion. However, the increased cement content does not result in an increase in the IR.

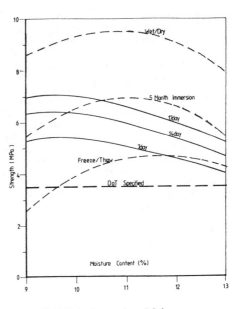

FIGURE 4 Strength vs Moisture Content for Wardley CSM.

3

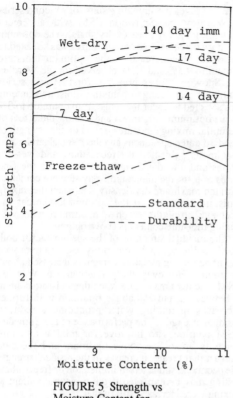

FIGURE 5 Strength vs
Moisture Content for
Snowdon CSM

Figure 6 shows the expansion of specimens of Wardley CSM during immersion. The expansion in the direction of compaction (Z direction) is on average 40% greater than the expansion in the direction perpendicular to compact (X-Y direction). Nearly 60% of the 140 day expansion occurs during the first 7 days immersion, but the data suggest that expansion would have continued beyond the duration of this study albeit at a much reduced rate. The effect of increasing the mixing moisture content is clearly evident with the amount of expansion of Wardley CSM being reduced by more than 50% with an increase in the mixing moisture content of just 3%.

Figure 7 shows the dimensional changes in the direction perpendicular to compaction (X-Y direction) for specimens of Wardley CSM during cyclic freeze-thaw testing. The effect of mixing moisture content is clearly visible, increases in moisture resulting in a reduction in the ultimate expansion. The behaviour of these specimens is similar to that observed by McNulty (1985) for CSM, but it is not consistent with the behaviour observed by other workers (Packard and Chapman, 1963) for soil-cement. The latter workers suggested that deterioration or 'failure' of specimens in the freeze-thaw test is indicated by a expansion on thawing above the original length and that this often coincides with an expansion during freezing. From

Figure 7 it can be seen that all three Wardley CSM mixes expand above their original length during the first thawing period. However, only the material compacted at optimum moisture content exhibited a length increase during freezing and this occurred during the eleventh cycle. Only the specimens from this mix and the Snowdon mix also compacted at optimum moisture content, exhibited any visual signs of deterioration, with cracking occurring after three and seven cycles respectively. Packard and Chapman (1963) also suggested that during this test specimens could be expected to gain strength due to further cement hydration and that this gain would be approximately equivalent to a further 10 days standard cure. Consequently, specimens suffering no damage should have a strength equal to the 17 day standard cure strength. None of the Wardley or Snowdon CSM specimens achieve this target and only the mixes with the highest moisture contents actually gain strength above their initial 7 day standard cure strength.

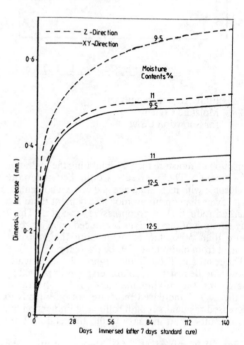

FIGURE 6 Swelling Rates of Wardley CSM

Dimensional changes for specimens of Wardley CSM during wet-dry cycling are shown in Figure 8. The overall expansion of the CSM is reduced at higher mixing moisture contents. Packard (1962) suggested criteria for this test where specimen 'failure' is indicated by a full recovery of the initial shrinkage. As with the freeze-thaw test all the CSM mixes tested failed to meet this criteria, with CSM's mixed at the optimum moisture content failing during the fourth wetting cycle and CSM's at higher mixing

4

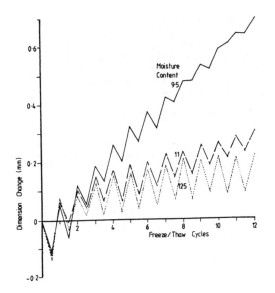

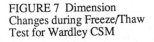

FIGURE 7 Dimension
Changes during Freeze/Thaw
Test for Wardley CSM

FIGURE 8 Dimension
Changes during Wet-Dry Test
for Wardley CSM

moisture contents failing during the fifth wetting cycle. Once the initial shrinkage was reduced, CSM specimens from all mixes continued to show increasing expansion during subsequent wetting cycles. Specimens mixed at optimum moisture content showed a net expansion of more than 0.4% after completing twelve cycles, whereas specimens mixed at optimum +3% moisture content exhibited approximately half of this expansion. Little evaporable water remained in these specimens after the final drying period, the maximum being 6% for Snowdon CSM mixed at the highest moisture content.

Packard (1962) suggested that the strength gain for specimens subjected to this test is equivalent to 39 days standard-cured strength. No determinations were made of the 39 day standard-cured strength but extrapolation of data in Figures 4 and 5 would suggest that the Wardley specimens would have certainly met this criteria and the Snowdon specimens would probably be borderline.

4. DISCUSSION

The results for both Wardley and Snowdon CSM show that increasing the mixing moisture content of these materials resulted in increased resistance to immersion, freeze-thaw cycles and wet-dry cycles. In addition, despite a reduction in the 7 day strength, later-age strengths of standard-cured specimens were increased for specimens with higher mixing moisture content.

The maximum 7 day strength occurs at approximately the same moisture content as the maximum dry density. The relationship between the strength and dry density of cement stabilised materials is well-established (Grimer, 1958) and is due to the greater degree of mechanical strength in mixes with high dry density. However, the strength of CSM is also dependent on the cementing action of the Portland cement and the moisture content required for maximum compaction does not necessarily coincide with that for maximum compressive strength (Davidson et al, 1962). The absorption of mixing water by the minestone may result in an internal depletion of moisture, thereby inhibiting the hydration of cement and reducing compressive strength at later ages. These effects can be seen in Figures 2 and 3 where little strength increase between 7 and 14 days is observed for CSM mixed at optimum moisture content or below. However, at higher mixing moisture content, despite the reduction in 7 day strength, there is greater strength gain between 7 and 14 days due to cement hydration and this can lead to higher long-term strength in these mixes.

At lower than optimum moisture content there is a reduction in dry density which leads to lower mechanical strength and this is accompanied by severe depletion in the moisture available for cement hydration and this combination leads to significantly reduced compressive strength at all ages.

The performance of the CSM specimens during the 7 day immersion test and the longer 140 day immersion test is clearly dependent on the mixing moisture content of the CSM. The expansion of cement stabilised minestone during immersion may be due to both the attack on cement hydrates from sulphates in the minestone and the expansion of the minestone itself. Sulphate attack will depend upon the quantity and availability of sulphates within the raw minestone and will be unaffected by the mixing moisture content. When the availability of moisture is increased, expansion of the shales and mudrocks in minestone may occur due to either (i) interparticle swelling mechanisms such as air-breakage, ionic dispersion or ion adsorption (Taylor and Spears, 1970); Badger et al, 1956; McNulty, 1985), or (ii) interparticle swelling of expansive clay minerals (Taylor and Spears, 1970). All these mechanisms are accompanied by an uptake of moisture.

The expansion and strength loss of hardened CSM during immersion can be reduced by increasing the moisture available during mixing. This allows much of the potentially damaging expansion to occur prior to compaction and reduces the hygroscopic requirements of the moisture and the capacity for future moisture absorption and swelling.

Although the CSM with higher mixing moisture contents show reduced expansion during immersion, the increased moisture makes these materials more susceptible to the initial freeze cycle of the freeze-thaw test. However, despite the increased contraction exhibited by CSM with high moisture contents during freezing, these CSM's showed reduced final expansion and increased residual strength following both freeze-thaw testing compared with the CSM compacted at optimum moisture content. Earlier studies on CSM (McNulty, 1985) have also shown the initial moisture content has a major influence on the performance during freeze-thaw testing and that the presence of moisture during thawing influences the rate of strength development and degree of expansion.

Although all the mixes tested failed to meet the severe criteria suggested by Packard and Chapman (1964), both the Snowdon and the Wardley CSM compacted at the highest moisture contents have residual strengths following freeze-thaw testing in excess of their 7 day standard cure strength and therefore exhibit no strength loss.

Overall, the results for this study emphasise the importance of the initial mixing moisture content on the strength and durability of cement stabilised minestone. Providing early-age strength requirements can be met, the quality of the material can be improved considerably by increasing the mixing moisture content to about 2% above the optimum for compaction. The results also indicate the severe consequences of producing CSM at moisture contents below the optimum.

5. CONCLUSIONS

The strength and durability of cement stabilised minestone is greatly influenced by the mixing moisture content of the material. Increasing the moisture content above optimum reduces the short-term strength but leads to increased long-term strength and increased resistance to the effects of immersion, freeze-thaw and wet-dry cycles. These improvements were observed for moisture contents up to 2 or 3% above optimum. Reducing the mixing moisture water below optimum results in considerable reduction in strength and durability.

6. ACKNOWLEDGEMENTS

The authors are indebted to British Coal for funding the research project. The views expressed are those of the authors and not necessarily of British Coal.

7. REFERENCES

ASTM, 1980a Standard methods for freezing and thawing tests of compacted soil-cement mixtures. American Society for Testing of Materials, Designation D560.

ASTM, 1980b Standard methods for wetting and drying tests of compacted soil-cement mixtures. American Society for Testing of Materials, Designation D559.

Badger C W, Cummings A D and Whitmore P L 1956 The disintegration of shales in water, Journal Institute Fuel, Oct. pp417-423.

British Standards Institution, 1975 Methods of test f for stabilised soils BS1924, BSI, London.

Byrd T V 1980 Soil Canterbury Tale, New Civil Engineer, 1980.

Davidson D T, Pitre G L, Mateos M and George K P 1962, Moisture Density, Moisture-Strength and Compaction Characteristics of Cement-Treated Soil Mixtures, Highway Research Board, Botteton No. 353, pp42-63.

Grimer F J, 1958 A laboratory investigation into some of the factors affecting the strength of soil-cement. Road Research Laboratory, Research Note No. 3288.

Kettle R J & McNulty A T, 1981. The improvement of colliery shale Proc. 2nd Australian Conference on Engineering Materials, University of New South Wales, pp273-284.

Kettle R J & Williams R I T, 1977. Frost action in cement stabilised colliery shale, Transportation Research Record, No. 641, pp41-48.

McNulty A T, 1985, The durability of cement bound minestone PhD Thesis, University of Aston, UK.

Minestone Services, 1983. Cement bound minestone users guide for pavement construction:, British Coal Corporation, London.

Packard R G, 1962 , Alternative methods for measuring freeze-thaw and wet-dry resistance of soil-cement mixtures. Highway Research Board, Bulletin 353, pp8-41.

Packard R G and Chapman G A, 1963, Developments and durability testing of soil-cement mixtures, Highway Research Board, Record No.36, pp97-123.

Sleeman W, 194 Practical application of cement bound minestone within the British Coal Mining Industry, Proc. Int. Sym. Reclamation, Treatment and Utilisation of Coal Mining Wastes, National

Coal Board, London, pp 53.1 - 53.19.

Tanfield D A, 1978, The use of cement stabilised colliery spoils in pavement construction. Proc. Int. Conf. Use of By-Products and Wastes in Civil Engineering, Vol III, Ecole Nationale des Ponts et Chaussess, Paris, pp151-160.

Taylor K and Spears D A, 1970. The breakdown of British Coal measure rocks, Int. Journal of Rock Mech. and Mining Sci., Vol. 7, pp 481-501.

Thomas M D A, Kettle R J and Morton J A, 1987, Short-term durability of cement stabilised minestone, in reclamation, Treatment and Utilization of Coal Mining Wastes (edited by A K M Rainbow), Elsevier Science Publications B V, Amsterdam.

Thomas M D A, Kettle R J and Morton J A, 1989, The oxidation of pyrite in cement stabilised colliery shale. Q J Eng Geol., Vol. 22, pp207-218.

TABLE 1 - PHYSICAL PROPERTIES OF 2 MINESTONES

	ATTERBERG PLASTIC	LIMITS LIQUID	SPECIFIC GRAVITY	COMPACTION CHARACTERISTICS		GRADING CHARACTERISTICS PERCENTAGE FINER THAN:-				
				OPTIMUM MOISTURE CONTENT	MAXIMUM DRY DENSITY	- mm -			microns	
	%	%		(%)	kg/m^3	37.5	20	5	600	63
WARDLEY	24	33	2.36	8.6	2060	94	82	50	20	9
SNOWDOWN	19	29	2.55	6.8	2170	90	65	26	6	4

TABLE 2   PERFORMANCE OF CSM IN IMMERSION TEST

Minestone	Moulding Moisture Content	Cement Content	Dry Density	Unconfined Compressive Strength			Immersion Ratio	Length (%)	
				7d	14d	7d + 7imm	I.R.	Direction X-Y	Z
	W'	S	d'						
	%	%	kg/m^3	MPa	MPa	MPa	%		
Wardley	6.5	5	1730	1.4	1.5	0.8	50	0.31	0.55
	8.5		1890	3.4	3.6	2.0	57	0.35	0.55
	10.0		1890	3.9	4.4	2.7	60	0.35	0.47
	12.0		1830	2.7	3.4	2.7	79	0.19	0.27
	6.5	10	1800	2.5	3.5	1.9	54	0.33	0.45
	8.5		1890	5.3	5.9	3.4	57	0.30	0.40
	10.0		1900	6.8	7.2	4.5	62	0.29	0.38
	12.0		1840	5.9	7.7	5.9	77	0.22	0.33
Snowdon	5.0	10	1900	2.9	3.5	2.4	68	0.34	0.49
	7.0		2010	6.8	7.3	5.8	79	0.27	0.40
	8.5		2000	6.6	7.6	6.6	87	0.24	0.33
	10.5		1940	5.6	7.5	6.8	90	0.15	0.21

TABLE 3 - PERFORMANCE OF CSM IN FREEZE/THAW, WET/DRY TESTS AND LONG-TERM IMMERSION TESTS.

| Minestone | Moulding Moisture Content | Dry Density | Unconfined Compressive Strength (MPa) | | | | | |
			7 day	14 day	17 day	Freeze/Thaw	Wet/Dry	5 month Immersion
	%	kg/m^3						
Wardley	9.5	1,900	5.4	6.4	7.1	3.3	8.9	6.0
	11.0	1,850	4.9	5.9	6.7	4.5	9.5	6.9
	12.5	1,800	4.5	5.3	5.7	4.6	8.6	6.2
Snowdown	8.0	2020	6.8	7.3	7.6	3.9	7.7	7.5
	9.5	1990	6.4	7.7	8.5	5.2	8.8	8.6
	11.0	1900	5.4	7.5	8.0	6.0	8.5	9.0

Cement content 8% for all mixes

Reclamation, Treatment and Utilization of Coal Mining Wastes, Rainbow (ed.) © 1990 Balkema, Rotterdam. ISBN 90 6191 154 0

Influence of geometry on strength and elasticity of cement and coal washery refuse material models

J. K. Hii, N. I. Aziz, S. Zhang & Y. H. Wu
Department of Civil and Mining Engineering, The University of Wollongong, NSW, Australia

ABSTRACT: This paper presents the results of laboratory tests conducted to investigate the influence of geometry on the strength and elasticity of cement and coal washery refuse (CCWR) material models. CCWR models were tested, simulating underground monolithic pack support with different widths in a coal seam of uniform height. The study also investigated the load deformation characteristics of CCWR models under the test conditions. Results indicate that geometry has a significant effect on both the mechanical properties and load deformation characteristics of CCWR models.

1 INTRODUCTION

The support of gateroads for Longwall development panels is becoming increasingly important in strata control in underground coal mining operations. Laboratory tests were conducted to investigate the effect of geometry on the mechanical properties and behaviour of CCWR models, and to ascertain the suitability and effectiveness of the material which is to be used as a means of underground roof support.

An outline of a set of standard quantities and conditions used in the preparations and testings of CCWR models is given. A computerised strength testing system was developed to facilitate accurate experimental data logging and processing using Hewlett Packard 3054A Automatic Data Acquisition and Control Systems. The main purposes of this study are to investigate the effects of geometry and end constraints on the strength and load deformation behaviour of CCWR models. The experimental data of the mechanical properties and behaviour of CCWR models are analysed and discussed in the paper.

It is envisaged that the results obtained can serve as a realistic basis for numerical analysis of underground pack under variable geological and stress conditions. The experimental data generated from this study can also be used for pack designs in underground coal mining operations. Suggestions for further work are highlighted.

2 TEST PROGRAM AND PROCEDURE

The test program was conducted in a series of tests which is listed as series I in Table 1. A special set of steel and PVC plastic moulds were designed and fabricated to complete these tests.

Table 1. Summary of tests.

Test Series	Description of Tests	Experimental Measurements	No of Tests
I	Unconfined compression strength tests.	Force, axial and lateral displacement.	19

2.1 Material studied

Type A ordinary Portland cement, minus 10mm coal washery refuse (CWR) and calcium chloride set accelerator were used in making the test specimens.

2.2 West Cliff Colliery coal preparation plant

The coal washery (West Cliff Colliery, Appin, NSW, Australia) throughput is 600 tonnes dry feed per hour and it produces a hard coking coal and CWR. The coal washery feed consists of coarse -127mm x 12.7mm and fine -12.7mm x 0 coal travelling in separate conveyor belts. A 424 McNally jig is used to treat the coal sized at -127mm x 12.7mm. The -12.7mm x 0.5mm coal is processed by 4 x 610mm D.S.M cyclones, and the 0.5mm x 0 coal is recovered by froth flotation.

2.3 Sampling and collection of CWR

Sampling of CWR took place on seven consecutive

weekdays. Samples were taken three to four times a day by stopping the refuse conveyor belt and taking the CWR from a suitable length by the full width of the conveyor belt. Care was taken to retain any fine CWR by scraping or otherwise removing it from the conveyor belt and adding it to the sample. The sampling operations were conducted with the aim to ensure that the samples obtained represent, as far as possible, the true nature and condition of the bulk of CWR from which they were drawn.

2.4 Processing of CWR

The sample-increments collected were re-combined to form bulk samples, then reduced by sample division to form the samples which were sieved through a standard 10mm sieve. The oversize CWR was passed through a small jaw crusher and the crushed CWR was re-screened. The oversize material was rejected (this constituted some 15.0% of the feed but could be reduced by more extensive crushing in practice). The processed CWR was then used for making the test specimens. The main purpose of processing the CWR was to produce a minus 10mm product which was necessary for the ultimate placement by pneumatic and high-density hydraulic conveying.

2.5 Grain size analysis of CWR using mechanical method

The grain size analysis was performed in accordance with AS 1289 C6.1-1977 standard method of sieving analysis to determine the relative proportions of different grain sizes which made up a given mass of CWR. It is realized that the sample is actually a statistical representative of the mass of CWR. However, it is difficult to determine the individual sizes of CWR. It is, therefore, cautioned that the test can only bracket the various ranges of sizes. Figure 1 shows typical grain shapes of West Cliff CWR.

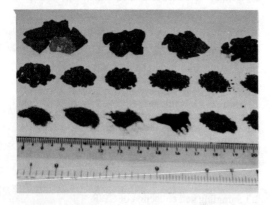

Figure 1. Typical grain shapes of West Cliff CWR.

Figure 2 shows the grain size distribution curves of the processed CWR. The relative density of the processed CWR was measured using Beckman Model 930 air comparison pycnometer and found to be 2.31. This value is comparable to that obtained by Thomas (1986) for a different sample where the relative density was found to be 2.54, also measured using Beckman Model 930 air comparison pycnometer.

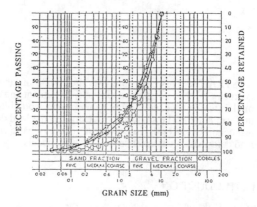

Figure 2. Grain size distribution curves of processed CWR.

2.6 X-ray diffraction analysis

Chemical composition analyses of West Cliff CWR were carried out. The results of x-ray diffraction analyses of West Cliff CWR are given in Table 2. Silicon oxide, aluminium oxide and iron oxide compounds are the three major compounds found in West Cliff CWR. They are on average 60.0% of silicon oxide, 21.0% aluminium oxide and 5.0% iron oxide which constitute some 86.0% of the total composition of West Cliff CWR.

Table 2. Chemical compositions of West Cliff CWR.

SAMPLE	A	B	C	D	E	F	G
Silicon as SiO_2	67.17	64.69	65.81	65.93	67.62	43.3	43.2
Aluminium as Al_2O_3	24.54	22.65	22.70	22.43	23.87	14.7	14.9
Iron as Fe_2O_3	4.19	6.51	5.68	5.75	3.56	3.7	3.9
Potassium as K_2O	2.93	3.08	3.09	2.97	3.23	1.9	1.9
Calcium as CaO	1.09	1.74	1.45	1.98	1.28	2.2	2.3
Magnesium as MgO	0.76	0.92	1.24	1.34	1.26	1.0	1.1
Titanium as TiO_2	1.01	1.05	1.07	1.02	1.02	0.67	0.67
Manganese as MnO	0.10	0.14	0.10	0.12	0.06	0.06	0.07
Phosphorus as P_2O_5	0.05	0.26	0.19	0.28	0.15	0.13	0.14
Sodium as Na_2O	-	-	-	-	-	0.1	0.08
Ignition loss	21.9	25.3	25.3	-	-	32.1	31.7
Total	101.9	101.0	101.3	101.8	102.0	99.9	100.0

Note that all values given in Table 2 are expressed in percentage.

10

The results indicate that West Cliff CWR may be classified as a siliceous or pozzolanic material. It is very obvious by comparing Tables 2 and 3 that the chemical composition of West Cliff CWR is very similar to that of UK fine discards. However, it is also apparent that West Cliff CWR (60.0% SiO_2) contains a comparatively higher percentage of silicon oxide than that of UK fine discards (maximum 46.0% SiO_2). It is indicative that West Cliff CWR which contains silicon oxide, aluminium oxide, calcium oxide and magnesium oxide exhibits pozzolanic characteristics.

Table 3. Average major geomechanical compounds in fine discards (UK). (after Taylor, 1984).

	Minimum	Maximum
SiO_2	10.74	46.19
Al_2O_3	5.45	29.91
Fe_2O_3	2.14	8.70
MnO	N.D.	N.D.
MgO	0.05	2.38
CaO	0.19	7.21
Na_2O	0.10	0.82
K_2O	0.74	4.40
TiO_2	0.46	1.06
S	0.44	7.85
P_2O_5	0.03	0.18

N.D. denotes not determined.
All values presented in Table 3 are expressed in percentage.

Chemical composition analyses of coal preparation wastes conducted in the USSR have indicated that coal preparation wastes generated from different coalfields exhibit a significant variation in their chemical compositions. Chemical compositions of coal preparation wastes from four main coalfields of the Soviet Union are presented in Table 4. It is obvious from Table 4 that the chemical composition of coal preparation wastes differs markedly from one location to the next and even within the limit of any

Table 4. Chemical compositions of coal preparation wastes for four main coalfields of the Soviet Union. (after Ruban and Shpirt, 1987).

Content	Coalfields (method of preparation)							
	Donbass		Kuzbass		Karaganda		Pechora	Ekibastuz
%	grav.	float.	grav.	float.	grav.	float.	grav.	grav.[®]
SiO_2	50-62	51-58	57-78	50-67	53-63	54-56	61-65	53-64
Al_2O_3	17-31	18-32	14-26	14-31	23-35	23-28	20-24	27-39
Fe_2O_3	3-16	4-12	2-10	2-9	4-7	6-13	6-8	0.2-5
CaO	0.3-5	2-4	1-7	1-10	1-6	1-5	1-2	0.4-2
MgO	0.8-2	1-2	0.3-3	0.7-3	0.3-1	1-2	2-4	0.4-1

[®]Wastes of open cut mining operations and coal preparation pilot plant.

particular coalfields. According to Ruban and Shpirt (1987) for each individual coal washery plant the variation in chemical composition between samples collected on any particular day is rather small. It can be noted from Table 2 that the variation in chemical composition between samples of West Cliff CWR is also small.

The most important clay mineral constituents of West Cliff CWR are 50% to 60% of kaolinite, 35% to 40% of illite and 5% to 10% of montmorillonite (montmorillonite is of course the well known troublesome clay). The major minerals found in West Cliff CWR include well crystallized quartz, degraded muscovite and calcite.

The following metals (done by compton scatter) also exist in West Cliff CWR:
Sample D - 130 ppm of rubidium, 222 ppm of strontium, 276 ppm of zirconium, 17 ppm of niobium, 17 ppm of lead, 23 ppm of thorium, 35 ppm of yttrium, 37 ppm of nickel, 63 ppm of copper, 52 ppm of zinc and 35 ppm of gallium.
Sample E - 148 ppm of rubidium, 182 ppm of strontium, 289 ppm of zirconium, 17 ppm of niobium, 18 ppm of lead, 21 ppm of thorium, 32 ppm of yttrium, 43 ppm of nickel, 70 ppm of copper, 86 ppm of zinc and 36 ppm of gallium.

2.7 Procedures

CCWR models (scale 1:15) with a height of 150mm and diameters of 55.5mm, 75.5mm, 102.8mm, 149.5mm, 206mm, 242mm and 314.5mm were tested, simulating underground monolithic pack support with different widths in a coal seam of uniform height. After they were cast in accordance with AS1012 Part II, method of mixing concrete in the laboratory, the CCWR models were cured at 20°C in the humidity room, the condition of which was the closest approximation to real mine situation.

The unconfined compressive strength tests (in accordance with ISRM suggested methods) were conducted on CCWR models 28 days after casting. Unconfined compressive strength, modulus of elasticity and Poisson's ratio were the primary properties investigated. The study also investigated the load deformation characteristics of CCWR models under the test condition.

2.8 Testing equipment and technique

The servo-controlled strength testing system consists of five units:
   1. a reaction loading frame,
   2. an electronic control console,
   3. a hydraulic power pack,
   4. a LVDT strain measurement rig and/or strain gauges, and
   5. a HP 3054A automatic data acquisition and control system.
An Instron 8033 servo-controlled stiff testing machine is used in this testing system. This machine

is microprocessor controlled and a simple keyboard serves as the interface with the operator. Magnetic tape cassettes are used to load the test programmes. The Instron 8033 is capable of applying compressive and tensile loads of 500KN over a working stroke from minus 75mm to plus 75mm with an overall system stiffness greater than 1060 KN/mm.

The load cell, type 2518-110, which is employed in this machine has been tested for accuracy and linearity on a suitable calibration device, which in turn has been certified by the National Physical Laboratory to an accuracy better than 0.05%. The accuracy of each range of the cell has been found to equal or exceed 0.2% of cell rated output, or 0.5% of indicated load, whichever is greater.

2.9 Data acquisition system components

The data acquisition and control system for the monitoring of both axial and diametric strains of CCWR material specimens consists of:
  1. the HP 9826 microcomputer,
  2. the HP 3054A data acquisition and control system, and
  3. the HP 7DCDT series of displacement transducers (linear variable differential transformers (LVDT) and/or strain gauges).

A loading rate of 0.0015mm/s was used in all tests. The strength testing system is capable of measuring both axial and diametral strains to an accuracy of $\pm 0.5\%$ of reading. The details of this strength testing system has been reported elsewhere (Hii and Aziz, 1989).

3 TEST RESULTS

The summary of results of unconfined compressive strength tests is presented in Table 5. Figures 3 through 7 show the effect of geometry on the properties and behaviour of CCWR models. The results indicate that the geometry has a significant effect on both the mechanical properties and load deformation behaviour of CCWR models.

3.1 Effect of geometry on strength

Figure 3 shows the effect of geometry on the unconfined compressive strength of CCWR models. It can be noted from Figure 3 that generally the unconfined compressive strength of the model increases with the increase in the width to height (w/h) ratio. The increase in model diameters from 55.5mm through to 206mm results in a small increase in the strength of the model. However, the increase in model diameters from 206mm through to 242mm gives a marked increase in the strength of the model. In this series of experiments the optimum w/h ratio for maximum model strength appears to be 1.63.

3.2 Effect of geometry on axial modulus

Figure 4 illustrates the effect of geometry on the axial modulus of CCWR models. The increase from 0.34

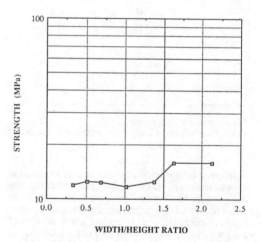

Figure 3. Effect of geometry on the unconfined compressive strength of CCWR models.

Table 5. Summary of results of compressive strength tests for CCWR models.

Specimen Diameter	Max. Load (KN)	Max. Strength (MPa)	Axial Modulus (GPa)	Diametric Modulus (GPa)	Poisson's Ratio	W/H	Moist Density (Kg/m^3)	Stiffness (KN/mm)
55.5mm	23.800	11.883	6.582	30.663	0.24	0.34	1900.6	40.29
75.5mm	56.667	12.408	8.002	35.673	0.24	0.51	1873.6	95.16
102.8mm	100.500	12.335	8.513	36.860	0.23	0.69	1914.5	121.53
149.5mm	204.380	11.643	7.467	33.798	0.21	1.02	1890.3	180.80
206.0mm	414.000	12.422	5.979	38.609	0.16	1.38	1910.8	246.85
242.0mm	730.000	15.871	4.745	43.127	0.11	1.63	1919.0	284.83
314.5mm	1232.500	15.866	3.407	43.394	0.08	2.13	1896.0	334.15

w/h through to 0.69 w/h ratios of models gives a significant increase in the axial modulus of the models. A further increase (from 0.69 w/h through to 2.13 w/h) in the w/h ratio of the model results in a decrease in the axial modulus of the model. The optimum geometry for maximum axial modulus is the model with a w/h ratio of 0.69.

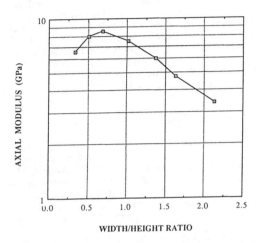

Figure 4. Effect of geometry on the axial modulus of CCWR models.

3.3 Effect of geometry on diametric modulus

The effect of geometry on the diametric modulus of CCWR models is shown in Figure 5. Generally, the diametric modulus of the model increases with the increase in its w/h ratio except for the model with a w/h ratio of 1.02 (where a lower value than expected has been measured).

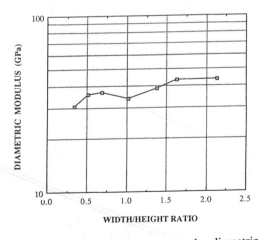

Figure 5. Effect of geometry on the diametric modulus of CCWR models.

3.4 Effect of geometry on Poisson's ratio

The effect of geometry on the Poisson's ratio of CCWR models is illustrated in Figure 6. It is obvious from Figure 6 that as the w/h ratio of the model increases its Poisson's ratio decreases.

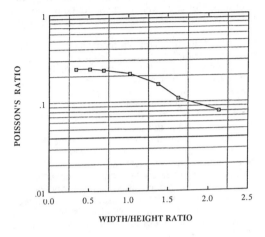

Figure 6. Effect of geometry on the Poisson's ratio of CCWR models.

3.5 Effect of geometry on stiffness

The effect of geometry on the stiffness of CCWR models is illustrated in Figure 7. Generally, the stiffness of CCWR material model increases with the increase in its w/h ratio.

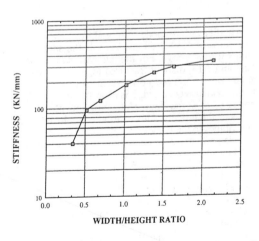

Figure 7. Effect of geometry on the stiffness of CCWR models.

4 DISCUSSION OF TEST RESULTS

Generally, the measured strength of the model increases with the increase in its diameter. It is indicative that the model with a higher w/h ratio exhibits a corresponding greater end constraint effect. Due to the end constraint phenomenon, confining pressures were created around the core of the model and hence this increased the strength of the model.

Figure 8 illustrates the observed shapes of CCWR models at complete failures. It clearly shows the effect of geometry on the load deformation behaviour of CCWR models. Complete collapse failures were observed for models with low w/h ratios of 0.34. Failure due to unconfined compressive testing of CCWR models with w/h ratios of 0.51, 0.69, 1.02 can be described in two phases:

1. Small axial cracks appeared just before the maximum compressive strength was reached. These cracks indicated tensile cleavage.

2. As strain increased these cracks widened and large slabs of the model surface broke away leaving a typical hour glass failure shape. This indicated a concentration of shear stress along some diagonal plane within the model and this caused it to fail in shear.

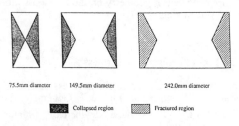

75.5mm diameter   149.5mm diameter   242.0mm diameter

▓ Collapsed region   ▨ Fractured region

Figure 8. Observed shapes of CCWR models at complete failures.

Another failure mode observed, for models with w/h ratios of 1.38, 1.63 and 2.13, is the shear cone with splitting above. For models with w/h ratios of 1.63 and 2.13, distinct fractured regions or yielding zones were observed at the surfaces of the models but no collapse occurred. It is clear that these yielded zones offer some degree of confinement to inner cores of the models. This confinement phenomenon contributes to the higher strength of the model.

It is particularly interesting to note that there is a marked increase in the strength of CCWR models with w/h of 1.63. This increase in the strength of CCWR models becomes comparatively small for models with even higher w/h ratios than 1.63. The practical significance of this singular aspect is that, in CCWR pack design for roof support in a real mine situation, the w/h ratio of the pack can neither be too low nor too high and there exists an optimal w/h ratio of the pack.

For roof support in a real mine situation, if the w/h ratio of CCWR pack is lower than 1.63, there may be a danger of inadequate roof support which can lead to complete failure of the pack support and the roof. On the other hand, if the w/h ratio of CCWR pack is too high, resulting in a very stiff pack which can cause floor heave and roadway closure problems, and this may involve unacceptable cost penalties. Of course, a more detailed study on this aspect is a vast subject in its own, therefore, more comprehensive investigations on the effect of end constraint on insitu CCWR pack and triaxial strength properties of CCWR materials are essential. Some studies on these aspects have been carried out by the authors. The results of these findings have been reported separately in Wu et al. (1989), Hii et al. (1990) and Zhang and Hii (1990a, 1990b).

5 CONCLUSIONS

It is shown that the geometry has a significant effect on both the mechanical properties and load deformation characteristics of CCWR models. The measured strength of the model increases with the increase in its diameter. It is envisaged that the results obtained can serve as a realistic basis for numerical analysis of underground pack under variable geological and stress conditions. The experimental data of the mechanical properties of CCWR models generated from this study can also be used in the pack design in underground coal mining operations.

6 ACKNOWLEDGEMENTS

The authors thank The University of Wollongong for providing facilities for the work to be carried out. Technical staff of the Department of Civil and Mining, The University of Wollongong and Australian Coal Industry Research Laboratories, Bellambi are particularly acknowledged for their great help throughout the course of this work.

7 REFERENCES

AS 1012 Part II, Method for Mixing Concrete in the Laboratory.

Atkins, A.S., Aziz, N.I., Singh, R.N. and Bridgewood, E.W. 1986. Integrated Methods of Colliery Waste Disposal for Ground Control and Safety, The AusIMM Illawarra Branch, Ground Movement and Control Related to Coal Mining Symposium, Wollongong, Australia, August, pp. 263-268.

Hii, J.K. and Aziz, N.I. 1989. Microcomputer Automation of the Strength Testing System for Use in a Rock Mechanics Laboratory, Australasian Instrumentation and Measurement Conference, The Institution of Engineers Australia, Adelaide, South Australia, 14th-16th November, pp.51-55.

Hii, J.K., Zhang, S. and Wu, Y.H. 1990. Evaluation of the Triaxial Strength of Cement and Coal Washery Refuse Concrete, The Twelfth

Australasian Conference on the Mechanics of Structures and Materials, 24th-26th September, Organised by School of Civil Engineering, Queensland University of Technology, Australia, in Press.

Rock Characterization Testing and Monitoring, ISRM Suggested Methods, Edited by Brown, E T, Pergamon Press, 1981.

Ruban, V.A. and Shpirt, M.Y. 1987. Utilization of Mining Operations and Coal Preparation Processes Wastes in the USSR and the principles of Their Classification, Second International Symposium on the Reclamation, Treatment and Utilization of Coal Mining Wastes, Edited by Rainbow, A.K.B., Elsevier, 7th-11th September, Nottingham University, pp. 45-54.

Wu, Y.H., Hii, J.K. and Aziz, N.I. 1989. A New Yield Function for CCWR Concretes, First Conference on Concrete and Structures, Kuala Lumpur, Malaysia, 3rd-4th October, pp.173-176.

Zhang, S. and Hii, J.K. 1990a. An Analytical Approach for the Estimation of Pillar Strength, Ninth Conference, Ground Control in Mining, Edited by Peng, S.S., 4th-6th June, Sheraton Lakeview Resort and Conference Center Morgantown, WV 26505 USA, in Press.

Zhang, S. and Hii, J.K. 1990b. A Computer Model for the Prediction of Fully Mobilized Pillar Strength, 3rd Conference on Ground Control Problems in the Illinois Coal Basin, Edited by Chugh, Y.P., Southern Illinois University at Carbondale, USA, 8th-10th August, in Press.

Reclamation, Treatment and Utilization of Coal Mining Wastes, Rainbow (ed.) © 1990 Balkema, Rotterdam. ISBN 90 6191 154 0

Construction of a road base using coal wastes stabilized with cement

J.González Cañibano, M.Garcia & J.A.F.Valcarce
HUNOSA, Dirección Técnica, Oviedo, Spain

ABSTRACT: This article sets out the results obtained in laboratory together with a detailed description of the results of trials carried out on a full-scale roadbase made with coal wastes stabilized with cement.

1 INTRODUCTION

It is well known that one of the most interesting applications of coal wastes, from the point of view of material use, is in earthworks: infills and embankments for roads, highways, dams, etc., as these allow large amounts of material to be placed, without, in most cases, need for any kind of prior treatment.

Although it seems likely that the quantities of wastes to be used in earthworks and infilling will peak and then decline owing to the fact that fewer highways and motorways are to be built, they may well still be required for repair work.

The present article details the results obtained in laboratory tests and tests carried out in the construction of a full-scale trial of roadbase to demonstrate technical viability of coal wastes stabilized with cement as roadbase material. This could suppose a further application which would help solve the problems which these wastes present, especially in areas where it is difficult to find base and subbase material at a reasonable price, but which at the same time have available wastes deriving from coal exploitation.

2 LABORATORY TESTS

Bearing in mind the site at which the road was to be built, the Reicastro spoil heap was selected. It contains some 2.5×10^6 m3 of coal wastes, mainly washery, and is situated at some 12 km from the works.

Samples of the materials were taken (M-1 and M-2) present in the spoil heap, and laboratory tests were performed: Normal and Modified Proctor (Normal Proctor test: using a cylindrical mould 102 cm in diameter, 1000 cubic centimeter, a rammer of 2.5 kg; Modified Proctor test: using a cylindrical mould 152.4 mm in diameter, 2320 cubic centimeter, a rammer of 4.535 kg), Normal and Modified C.B.R. (C.B.R.: California Bearing Ratio according to ASTM.:1883 or BS.:1377), sand equivalent and

Los Angeles wastage, according to PG3/75 (Portfolio of General Norms for Road and Bridge Works) issued by the Ministry for Public Works of Spain regarding materials for use granular layers.

As can be seen from Fig.1, en which the particle size distribution range (as required by the above mentioned PG3/75) and the particle size distribution of both samples are set out, M-1 is within this range except in the 10 mm screen, while M-2 falls outside the specifications in 10 and 5 mm screens.

However, the results obtained from the other tests on the two samples meet the requirements of PG3/75 (though M-2 has a slighty lower C.B.R. index) for roadbase materials, as can be seen from Table I.

Sample material suffers degradation both when compacted with Normal Proctor and Modified Proctor energy, as can be seen from Fig. 2, which represents the original particle size distribution and after the compacting tests of the M-2 sample.

At the same time, as was to be expected, sand equivalent values also decreased, especially when Modified Proctor energy was applied, as can be seen in Table II.

Swelling determined in the C.B.R. tests is nil, and the average permeability coefficients vary between 1.14 x 10^{-4} cm/sec for sample M-2 and 1.91 x 10^{-6} cm/sec for sample M-1, values equivalent to those which rock fractured with clay of fine sand-filled fissures would present. It is, therefore a kind of material which varies from slightly-draining to draining, according to the sample.

Material degradation is an important subject, as it advises against the use of the same as subbase or granular base, since as part of the base and having to bear a significant percentage of the stress to which this is subjected, there must of needs be a degradation once the road is open to traffic, lowering the quality and leading to, among other phenomena, a reduction in permeability with a consequent loss in drainage capacity.

Despite their degradability, the coal wastes in the samples tested retained sufficient support capacity, as shown in the C.B.R. indices, which are derived from the degraded material, and absence of swelling in the presence of water (again shown in the C.B.R. test), suggesting that they might be employed, stabilized with cement, as a material to roadbases.

To this end, tests were carried out on the soluble sulfates content, percentages of between 0.03 and 0.07 being found, below the 0.5% maximum required by PG3/75.

This low sulfate content means that any of the following types of cements may be used for their stabilization: Portland, Portland with active additions, puzzolanic and cements with additional properties. Portland with active additions was chosen, with a compresion strength of 350 kg/cm2.

The M-2 waste sample was chosen because the results obtained were lower than those with M-1.

The wastes were mixed with 3 different percentages of cement: 4, 6 and 8%, and these mixes were compacted with Normal Proctor energy as required by PG3/75. Figures 3 and 4 show the variations in density and pore index, optimum humidities being those between 9.5 and 9.8%. Percentage of cement had hardly any influence on the variation.

These mixes were aslo tested under simple compression. Fig. 5, representing evolution as a function of cement added, shows that with 5-6% of cement the 15 kg/cm^2 required by PG3/75 for use in foundation layers is already surpassed, and that with 7-8% it surpasses the 20 kg/cm^2 required for roadbases.

In the light of the results obtained, it was thought that it might be possible to achieve even better performance if the wastes were compacted using Modified Proctor energy, but the only effect apparent was an increase in density, as can be seen in Fig. 3, accompanied by a drop in the pore index, Fig. 4, with no corresponding increase in resistance to compression (Fig. 5), which remained at the same value as that obtained when using Normal Proctor energy. For this reason, we did not proceed to test for 8% cement with Modified Proctor energy.

3 CONSTRUCTION OF A TRIAL BASE OF COAL WASTES STABILIZED WITH CEMENT.

This was carried out taking advantage of the building of a new road from the old N-630 to the Olloniego Pit.

The road in question was 1625 m in length, to be built over an already-existing path the surface of which would be used, after suitable treatment, as a foundation layer on which would be set a 15 cm thick granular subbase, a 15 cm granular base and 5 cm surface layer of asphalt agglomerate.

The foundation layer is designed to support a traffic equivalent to an accumulated number of 13 t axles of $10^4 - 8.10^4$.

Fig. 6 represents a cross-section of the trial area in which coal wastes stabilized with 8% cement were employed.

Once the artificial ballast subbase had been laid and compacted with a percentage compacting of 101% of the maximum density corresponding to the Modified Proctor test, 40 m were selected as the trial area.

The coal wastes were deposited on the subbase and evenly spread by a power grader. The trial area was then divided up into squares, in which the quantity of cement calculated to give 8% uniform spread was distributed.

The real humidity of the coal waste and the cement was determined and water necessary to attain the optimum humidity of 9.5% which had been obtained in laboratory tests was added. This water was distributed by means of a tank fitted with a sprinkler device duly calibrated.

The mixing of wastes, cement and water was done by a power grader, and the mix was then levelled.

Compacting was carried out using a 10.2 t static weight roller; the roller was passed over first twice without vibrating, and then another two times vibrating. In situ density was then determined by means of radioactive isotopes using Model 2401 Troxler apparatus previously calibrated. 104% compactation was achieved with respect to the maximum density of the Normal Proctor test, above the 100% required by PG3/75.

Two further passes were performed using the vibrating roller to determine the effect according to the number of times the roller was used. The density was checked using the same method and 103% compactation was achieved, which is to say that density had not increased, and therefore no further rollings were made.

7 days later, load tests were carried out with a 30 cm diameter base, in accordance with DIN 18134 which gave the following deformation modules:

$$E_1 = 2348 \ kg/cm^2$$

$$E_2 = 3600 \ kg/cm^2$$

and giving a relationship between the modules of

$$\frac{E_2}{E_1} = 1.53$$

Set as optimum values for bases layers

$$E_1 = 700 \ kg/cm^2$$

$$E_2 = 1400 \ kg/cm^2$$

The results obtained may be considered optimum.

Six years after construction of the trial roadbase, test carried out revealed no difference with regard to the rest of the road (whose base had not been built using coal wastes stabilized with cement) nor were crachs, crushing or fissures due to faults in the layers in the road observed.

4 CONCLUSIONS

It may be deduced from the above that coal wastes may be employed in base and subbase layers as long as they are stabilized with cement.

The optimum percentage of cement is between 6 and 8% depending on the quality of the wastes and the kind of road in which they are to be used: foundation or base layers.

As for the economic viability of cement-stabilized coal wastes in road bases or subbases, this will depend on the close availability of traditional materials and the proximity of the coal wastes to the point of use.

These results coincide with those resulting from earlier studies carried out in other countries such as the United Kingdom (1) and the Federal Republic of Germany (2).

ACKNOWLEDGEMENTS

The authors wish to express their thanks to the Management of HUNOSA for the facilities afforded them in the preparation of this article, to Doña Josefina Grela for the typing and to D. Ceferino Fdez. Cuetos for the drawings.

REFERENCES

(1) Kettle, R.J.; Rainbow, A.K.M. 1983. The stabilisation of colliery spoil. Symposium on the Utilization of Waste from Coal Mining and Preparation, Tatabanya, Hungary, vol. III, 17-22 October.
(2) Leininger, D.; Erdmann, W.; Köhling, R.; Petry, R.; Schieder, Th. 1983. Recent development in the utilization of preparation refuse in the Federal Republic of Germany. Symposium on the Utilization of Waste from Coal Mining and Preparation, Tatabanya, Hungary, vol IV, 17-22 October.

Table I. Results of laboratory tests

	test	PG3/75 Specification	Sample M-1	Sample M-1
Plasticity	Liquid limit	< 25	No	No
	Plastic limit	–	No	No
	Plasticity index	< 6	No	No
Compaction — Normal Proctor	Maximum dry density (g/cm³)	–	2.04	1.81
	Optimum moisture (%)	–	7.6	11.6
Compaction — Modified Proctor	Maximum dry density (g/cm³)	–	2.12	1.96
	Optimum moisture (%)	–	5.8	7.7
C.B.R. — Normal	CBR index	–	23	7
	Swelling (%)	–	0	0
C.B.R. — Modified	CBR index	> 20	28	15
	Swelling (%)	–	0	0
Sand equivalent		> 25	58	84
Los Angeles wastage		< 50	35	36
Organic material (%)		–	0.21	0.31
Soluble sulfates (%)		< 0.5	0.03	0.07

Table II. Sand equivalent

Sample	Before compacting	After	
		Normal Proctor test	Modified Proctor test
M-1	58	54	28
M-2	84	61	51

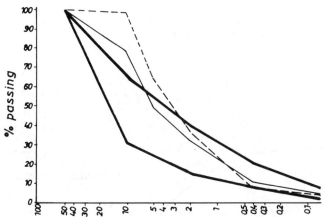

Fig. 1   Required particle size distribution and size distribution of M-1 and M-2 coal wastes samples

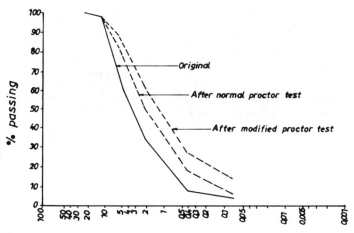

Fig. 2   Effect of the compaction on coal waste particle size distribution

21

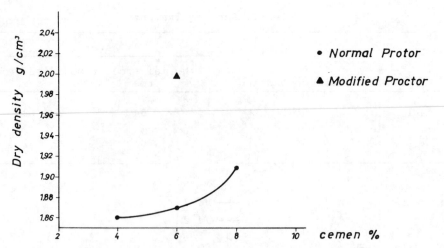

Fig. 3. *Variation of density with respect to*
percentage cement content.

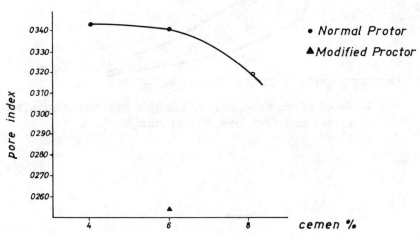

Fig. 4. *Variation of pore index with respect to*
percentage cement content.

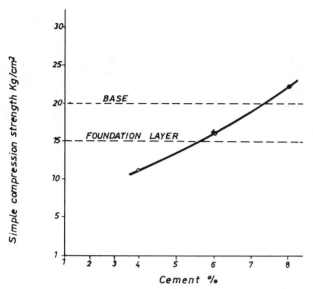

Fig. 5. Variation in compression strength after 7 days with respect to percentage of cement added.

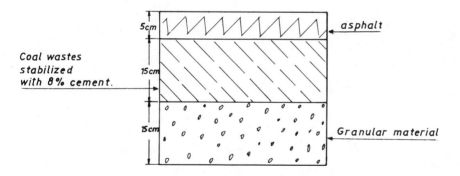

Fig 6. Cross-section of the trial area roadbase

Reclamation, Treatment and Utilization of Coal Mining Wastes, Rainbow (ed.) © 1990 Balkema, Rotterdam. ISBN 90 6191 154 0

Construction and performance evaluation of a cement-stabilized coal ash base course in a highway pavement

D.H.Gray & E.Tons
University of Michigan, Ann Arbor, Mich., USA

W.H.Berry, Jr.
Consumers Power Company, Jackson, Mich., USA

ABSTRACT: A base course beneath a highway shoulder pavement was constructed using a compacted, aggregate-free, cement stabilized coal fly ash. A high carbon, Class "F" fly ash stabilized with 12% by weight Portland cement was used. Tests and analyses employed to evaluate the performance of the base are described

1.0 INTRODUCTION

A compacted, aggregate free, cement-stabilized base course was constructed in the late Spring of 1987, under a test section of road shoulder along Michigan Route M-54. This project entailed new construction of several miles of four-lane highway with a nine-foot shoulder on either side. The site selected was a four-lane state highway connecting I-75 directly south of the city of Flint. This realignment of M-54, also known as Dort Highway, by-passes the town of Grand Blanc. The trial road shoulder base course was placed on both sides of a 1500-foot long section of the four-lane highway. The combined length of the fly ash shoulder on both sides totaled 3,000 feet. An additional 3,000-foot section of conventional shoulder adjacent to the fly ash test section was established as a control section. The finished M-54 highway was opened to traffic in June 1987.

Fly ash used for this project was a high-carbon, Class "F", dry hopper ash from a coal fired electric power plant located at Essexville, Michigan, 50 miles north of the demonstration site. The cement stabilized fly ash base course was placed and compacted in two lifts using conventional equipment. The final thickness of the base course was 10 inches. An asphalt-concrete levelling course was placed over the fly ash shoulder seven days after construction.

2.0 LABORATORY TESTING PROGRAM AND MIX DESIGN

2.1 Scope and Objectives

The main objective of the laboratory testing program was to determine the mix design (i.e., relative proportions of ash/cement/ water) and compaction speci-

fications in order to meet strength and durability requirements. Recommended strength/ durability criteria (DiGioia *et al.*, 1986) for cement-stabilized fly ash base courses require the following:

1. The 7-day unconfined compressive strength of strength of the mix, cured under moist conditions, must be 400-450 psi;

2. The strength must increase with time; and,

3. The minimum strength after vacuum saturation should not drop below 400 psi.

The fly ash used in the base course was a Class "F" hopper ash obtained from the D.E. Karn plant, Consumers Power Company, Essexville, Michigan. The cement used was Type I, supplied by the Dundee Cement Company, Dundee, Michigan.

The following tests were conducted on the ash:

1.	Specific Gravity	ASTM D854
2.	Grain Size Analysis	ASTM D442
3.	Moisture-Density	ASTM D1557
4.	Unconfined Compression .	ASTM D2166
5.	Vacuum Saturation	ASTM C593
6.	Frost Heave	BRRL LR90

Standard ASTM testing procedure was followed in all cases except for the frost heave test for which there is presently no ASTM standard. A frost heave test developed by the British Road Research Laboratory (Croney and Jacobs, 1967) was adopted instead. Details of the laboratory testing program together with the index and engineering properties of compacted, ash-cement mixtures are described elsewhere (Berry et al., 1989).

2.2 Mix Design and Specifications

On the basis of the laboratory testing program specifications were prepared for a compacted, cement
stabilized fly ash base course. A cement content of
12% by dry weight of solids was specified in order
to meet strength-durability criteria and minimize frost
heaving. A relatively high unburned carbon content
in the ash (7.3% by wt) required the addition of a
fairly high amount of cement.

The required thickness of the fly ash base was estimated using the AASHTO design equation. The
final thickness of the fly ash base course for the M-
54 highway shoulder was calculated to be 10 inches
after compaction and trimming. The minimum
required sand subbase thickness was computed to be
12 inches. The adjacent control section had a standard Michigan Class "A" design with a 5-inch thick
asphalt-concrete base, 4-inch aggregate base, and 2.5
to 3.5 asphalt wearing surface, all supported on a
sand subbase. A stratigraphic profile of both the fly
ash and control section is shown in Figure 1.

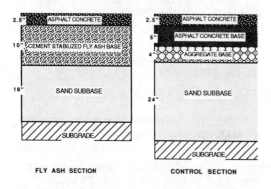

Figure 1. Stratigraphic, vertical profiles of control
and fly ash test sections respectively

Initial compaction of the fly ash base was to be
carried out with a rubber-tired roller followed by a
vibratory roller. The target field density was specified at 98% of the maximum dry density based on the
Modified Proctor test. The maximum allowable time
between mixing and final compaction was set at one
hour to minimize the loss of strength that can occur
as a result of long elapsed times between the two
events. The final compacted layer of cement stabilized fly ash was to be sealed as soon as possible to
prevent loss of moisture and ensure adequate curing.
A bituminous asphalt wearing surface was to be
placed after a minimum 7-day curing period.

The laboratory testing program revealed the importance of mixing on strength and durability. Consequently, close attention was paid to developing a
mixing protocol in order to insure good mix uniformity and dispersion of cement. Specifications called

for initial cement mixing with dry hopper ash at a
40:60 ratio. The purpose of this initial mixing step
was to insure good dispersion of the cement. This
40:60 mix was then combined and mixed with calculated amounts of conditioned (moist) fly ash and
water to achieve the final desired cement-ash and
water-solids ratios.

3.0 FIELD CONSTRUCTION

3.1 Site Selection and Road Shoulder Layout

In cooperation with the Michigan Department of
Transportation (MDOT), a 1,500-foot test section
along Michigan Route M-54 was selected for evaluation of the cement-fly ash base course. The M-54
project involved the relocation and construction of a
four-lane highway with nine-foot wide shoulders on
each side. The experimental fly ash-cement base
was placed under the road shoulder on either side of
the 1500-foot test section. The test section was
selected so that part was in cut and part on fill. A
schematic diagram of the alignment and location of
the test section is shown in Figure 2.

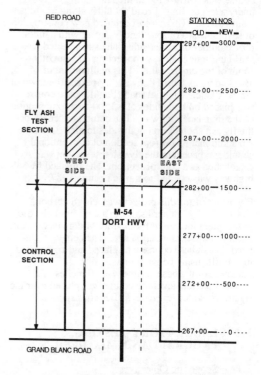

Figure 2. Plan view of Route M-54 showing location of control and fly ash test sections
and station designations.

3.2 Materials and Handling

The demonstration project required enough material for a 10-inch thick, 9-foot wide and 3,000-foot long shoulder base course. This requirement translated into about 690 tons of fly ash. The dry fly ash from the plant hoppers was conditioned (at approximately 25% water content), transported by 50-ton, double-bottom trucks, and stockpiled at the site for each day's production. The remaining fly ash was pre-mixed dry with cement at the power plant, brought to the site by a self-unloading, double-bottom cement transporter, and blown into a closed-feed silo mounted on the mixing unit. The proportions were 40% fly ash and 60% cement for the dry mix. This initial dry mixing step was included in order to insure good dispersion of the cement in the final mix. The dry (40:60) mix, the moist conditioned fly ash, and sufficient makeup water were fed independently into a continuous mixing and discharging pug mill to yield a final mix with the specified water and cement contents.

3.3 Mixing and Spreading

The pug mill mixer was placed near the stockpile of conditioned ash. The conditioned ash was fed to the mixer by a front end loader (see Figure 3). The 40:60 dry mix bin was kept full by the dry cement

Figure 3. Mixing equipment used in field demonstration. A dry ash-cement blend and moist ash were fed into portable pugmill truck-transporter.

A water truck was kept on standby next to the mixer to supply the prescribed moisture to the mix. The mix was discharged continuously from the pugmill, scooped up by another front end loader, and then loaded into five-ton trucks. The trucks in turn transported the ready mix to a spreading machine travelling along the shoulder as shown in Figure 4.

3.4 Compaction

The fly ash-cement mixture was placed and compacted in two lifts, using both a steel-wheel and

Figure 4. Emplacement of fly ash-cement mix in shoulder trench using a conventional travelling spreader box.

rubber-tired roller. After placement from a spreader, fly ash is usually too loose and fluffy for compression by roller. Roller compaction of such material can cause "shoving" or formation of a bow wave ahead of the roller. Accordingly, a small bulldozer or tractor is often used to track the mix once or twice in order to obtain a working surface on which a roller can operate. This step was omitted on this project without causing too much difficulty by controlling the manner and sequence of roller compaction.

An asphalt emulsion seal coat consisting of 0.1 to 0.2 gallons per sq. yd. of emulsified asphalt was applied to the surface immediately after compaction. As asphalt cap or wearing surface with a minimum thickness of 2 1/2 inches was placed over the compacted fly ash shoulder after a minimum 7-day curing period.

4.0 PERFORMANCE EVALUATION

4.1 Scope and Objectives

One of the of the main objectives of the project was to evaluate the engineering performance of the base course following its construction. The base course has been subjected to freezing temperatures during three winters (as of this writing)--the most critical period with regard to potential frost heave problems.

Performance was gaged periodically by a number of post construction tests and surveys. Both nondestructive and destructive methods were employed for this purpose. These tests included the following:

1. Clegg impact testing
2. Moisture-density tests on base course core samples
3. Unconfined compression tests on base course core samples

4. Edge break surveys
5. Vertical displacement measurements
6. Crack pattern survey
7. Seismic analyses

Testing procedures and results of each of these tests and/or surveys are described in subsequent sections of this report.

3.2   Clegg Impact Tests

Impact tests were run on the compacted surface of the fly ash base course using a Clegg impact tester as shown in Figure 5. This device records the deceleration of a standard weight dropped from a fixed distance onto the base course surface. The Clegg "impact" reading (CIR) obtained in this manner can be correlated (Clegg, 1976) with the modulus and strength of a compacted base. This test was run every 50 feet along the surface and at different elapsed times up to four days following compaction. The test is non-destructive, fast, and provides a good record of the spatial variation and development of strength/stiffness with time.

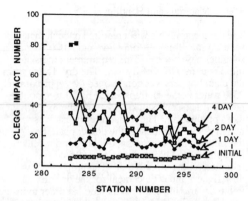

Figure 6. Clegg impact readings at different elapsed times and at different locations along the west side shoulder.

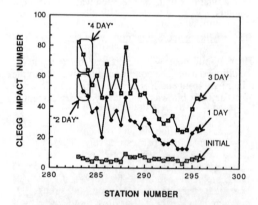

Figure 7. Clegg impact readings at different elapsed times and at different locations along the east side shoulder.

The reason for the low values at Stns. 293 & 294 is not readily apparent. Low cement contents are a possible explanation, however, the coincidence of low values at exactly the same location, but on opposite sides of the highway, suggest other explanations as well. The highway traverses a topographic low at this location. A small amount of standing water was observed in the road shoulder trench before placement of the fly ash base. This moisture may have adversely affected compaction of the fly ash and the cement set, thus explaining much lower rate of strength increase relative to other locations.

Figure 5. Clegg impact tester used to monitor strength and stiffness of compacted, cement stabilized fly ash base.

The results of the Clegg impact test along the length of the fly ash test section on both the east and west sides are shown in Figures 6 and 7 respectively. Strength gains tended to be higher on the south end of the project. In addition, east side readings were generally higher than the corresponding west side readings on the opposite side of the highway. A pronounced dip in Clegg impact values was recorded in the vicinity of Stns. 293 to 294 on both sides of the highway. These low values suggest a weaker base at this location and help to identify an area which bears close scrutiny as time goes by.

3.3   Moisture-Density Tests

The compacted, cement stabilized fly ash base course was constructed during the period May 14-19, 1987. The Michigan Dept. of Transportation drilled 4-inch diameter core samples from the fly ash test section on

28

26 August '87, 3 December '87, and 18 April '88. These dates correspond approximately to 90-, 180-, and 270-day test samples. The cores were sealed in plastic bags and transferred to the laboratory for determination of their moisture content, density, and compressive strength.

Samples were obtained from 9-10 coring locations on either side of the highway. The locations were selected randomly with the exception of two locations which were cored each time in the same vicinity of stations with known high and low Clegg impact readings respectively, viz., Stns. 282+00 (east) and 292+00 (west). Test specimens were recovered from both the top and bottom por tions of the fly ash field cores. The total length of the asphalt cap and fly ash base portions of the core at each location was recorded as well.

Moisture contents for the 90-day cores samples ranged from 26 to 44% (dry weight basis) with an average value of 33.5 %. With one or two possible exceptions, all of the cores samples exhibited moisture contents well in excess of the Modified Proctor optimum moisture determined in the laboratory investigation to lie between 24 to 26 % of dry weight for cements contents ranging from 0 to 15 % dry weight. Moisture contents measured on the 180-day (6-month) cores were even higher, ranging from 29 to 53 % with an average value of 36.5 %.

Dry densities for the 90- and 180-day samples averaged 69 and 72 pcf respectively. These values lie below the target density of 78 pcf, and correspond to a relative compaction of slightly under 90 percent based on the the Modified AASHTO test. The average density, moisture content, (and strength) computed for the the 270-day test samples are skewed because many of the samples were either shattered or too soft to handle and hence, unusable.

Correlations between strength and moisture/density are discussed in the next section along with the probable influence of variations in field cement content. Cements contents in the field cores varied from 2.8 to 19.1 % dry weight The target cement content was 12 %. This variation is consistent with cement contents measured on pugmill samples during construction which varied from 5.9 to 18.5 % by weight with an average of 11 %.

.3.4   Unconfined Compression Tests

Unconfined compression tests were run on 2-inch cube samples that were trimmed from the cores by dry sawing. Strength tests were run on samples recovered after 90-, 180-, and 270-days. The average unconfined compressive strength at 90 and 180 days was 489 and 410 psi respectively. These averages exceed the minimum, required, 7-day strength of 400 psi. The average unconfined compressive strength at 270 days is unreliable because the sample population was too small.

Although the average strength met or exceeded the minimum, required value there was considerable variation among samples. Strengths ranged from less than 100 to about 1000 psi. Only about half the samples tested met or exceeded the minimum, target strength of 400 psi. No additional strength gain was observed after 90 days.

Correlations were examined between unconfined compressive strength and other soil properties such as density, moisture content, cement content, and Clegg impact readings. The relationship between unconfined compressive strength at 90 and 180 days versus dry density are shown plotted in Figures 8 and 9 respectively. Fair to good correlations were obtained in both cases with the strength increasing with dry density in an exponential fashion as shown. The coefficient of correlation for 90- and 180-day strength vs. dry density was 0.61 and 0.95.

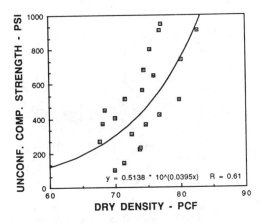

Figure 8. Unconf. compressive strength vs. dry density. 90-day samples

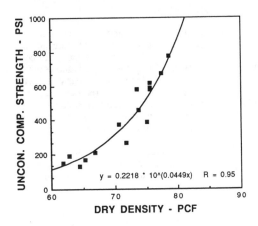

Figure 9. Unconf. compressive strength vs. dry density. 180-day samples.

The result of all strength tests (@ 90 and 180 days) are plotted vs. dry density in a single graph in Figure 10. The trend of increasing strength with density is clear and evident. The width of the data band can be attributed to variations in cement content of the samples. The results of laboratory tests on samples with known cement contents are also superimposed on the graph. A series or family of curves results which roughly parallel the field data band. Cement contents that were measured on field samples are also noted on the graph. The position of these field data points (with their known cement contents) are consistent with the laboratory strength-density-cement content curves. The influence of water content is reflected in the dry density value...the higher the water content, the lower the dry density and the corresponding strength.

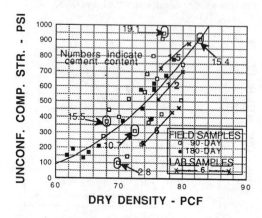

Figure 10. Unconf. compressive strength vs. dry density. Results of 90- and 180-day field cores plus laboratory test samples. Cement contents (where known) also plotted.

3.5 Edge Break Survey

Another indicator of stability and durability of the road shoulder is the extent of "edge breaking" along the outside edge of the pavement. The extent of edge breaking was assessed along the fly ash and control sections respectively. Edge breaking was classified as either major or minor. "Major" refers to crumbling, disintegration, and/or excessive settlement at the edge of the pavement whereas "minor" refers to a slight cracking and/or deflection at the edge.

The amount of edge breaking in these two categories was paced off along the entire length of the control and fly ash sections in June of 1989. The results of the survey are shown in Table 1.

The results of the survey show that edge breaking was not as extensive along the fly ash test section compared to the control section. This finding was true for both categories of severity. It is interesting to note the pronounced contrast in *major* edge

TABLE 1. AMOUNT OF PAVEMENT EDGE BREAKING OBSERVED ALONG ROAD SHOULDER.

SECTION	MINOR BREAKS (no. of paces)			MAJOR BREAKS (no. of paces)		
	East	West	Total	East	West	Total
CONTROL	199	172	371	123	123	246
FLY ASH	159	122	281	15	109	124

breaking along the fly ash test section between the east side (15 paces) vs. west site (109 paces). This difference in resistance to serious disintegration/ crumbling at the edge is also consistent with the findings of the Clegg impact tests on the surface and unconfined compression tests on core samples. In general impact readings were much higher on the east side than the west (compare Figures 6 and 7). The same also holds true for unconfined compression test results.

3.6 Vertical Displacement Measurements

During the winter of 1988-89 a visual inspection of the fly ash road shoulder in March appeared to indicate that some vertical heave had occured at certain locations. The heave was manifest most visibly in the form of a localized, slightly raised crown or mound with an associated cracking pattern on the west side road shoulder in the vicinity of Stn. 292+ 00. This particular response is discussed more fully in the next section.

After construction a decision was made to cut a joint between the road shoulder and pavement...but only along the east side test section. The joint was filled with a bituminous sealant. The west side test section was left alone. The purpose of the this action was to determine the influence, if any, of the presence of a joint between the shoulder and pavement. A slight amount of differential heave appeared to take place along the joint during the winter of 1989 . This vertical heave subsided, however, to a very small residual value by May of 1989.

In order to study and document this suspected vertical heave more thoroughly, two series of vertical displacement measurements were initiated in the fall of 1989. The first consisted of measuring the vertical separation distance or offset every 25 feet along the joint. The second consisted of running a line of levels every 100 feet down the middle of the fly ash test section road shoulder on either side of the the highway. Both these measurements were repeated periodically during the winter of 1990 to see if there any hard evidence exists for frost heaving, and if so, its magnitude and extent. Results of this survey were not available at the time of writing.

3.7   Crack Pattern Survey

Periodic inspection visits were made to the site following construction of the road shoulder base in May of 1987. One of the main purposes of these visits was to look for any crack development in the asphalt pavement cap. The first crack was observed on April 13, 1988, in the vicinity of Stn. 25+15 (new station designation) on the West side shoulder. The cracking consisted of two, narrow, approximately parallel cracks with a combined length of 6 feet at a distance of about 3 feet from the shoulder-travelled way pavement joint (edge of white line). No other structural cracks were observed in the West side shoulder.

The cracks at Stn. 25+15 have been inspected and measured frequently. They have grown slowly and developed branches with time. Their combined or total length (including major branches) were about 18 feet at the beginning of the 1990 winter season. The crack growth or development over time is illustrated in Figure 11. The temporal pattern in Figure 11 shows that crack growth is most rapid during the winter months of January, February, and March. More frequent observations are needed during these months in order to record the changes that are taking place.

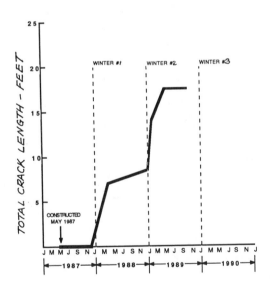

Figure 11. Crack length development at Station 25+15, West side.

Additional study and observations are required to explain fully the reasons for the observed cracking at Stn. 25+15. The density and strength of cores from this location were generally much lower than other locations. The cement content of a core from this location was also quite low (viz., 2.8% by wt). visual evidence during the winter indicated that localized heaving was occuring here as well. Snow plow striations in the pavement surface were clearly visible here as a result of a slightly raised "mound" in the pavement surface.

4.0   DISCUSSION AND CONCLUSIONS

Performance monitoring and evaluation of a compacted, cement stabilized fly ash base showed that in general the base has held up quite well. No widespread nor major problems, e.g., crumbling, disintegration, "potholing", excessive heave and/or settlement, have occurred after an elapsed time of 2 1/2 years following construction (in May of 1987). Problems with surface heave and pavement cracking have so far been restricted to a few local areas. Heaving and cracking in these areas occur primarily during the winter and are associated with frost effects. These localized problems appear to be the result of low density and strength in combination with a low cement content in the fly ash base and a thin asphalt cap or wearing surface.

The following conclusions can be drawn from the performance monitoring to date:

1.   The Clegg impact readings provided an initial indication of the spatial and temporal variation in stiffness of the compacted, cement stabilized fly ash and were good predictors of post construction strength and durability.

2.   The average moisture content/dry density of core samples from the fly ash test section at 90 and 180 days was 33.5 %/68.9 pcf and 36.5 %/71.2 pcf respectively. These average moisture contents were considerably higher than the target moulding water content of 24-26 %. The average densities were somewhat lower than the target density of 78 pcf, and correspond to a relative compaction slightly under 90 percent based on the Modified AASHTO test.

3.   The average unconfined compressive strength of random core samples at 90 and 180 days was 489 and 410 psi respectively. These averages slightly exceeded the target value of 400 psi (7-day strength), however, about half the samples fell under this value. Areas with very low strengths, e.g., Stn. 292 (West), also exhibited the most severe problems with heaving and cracking.

4.   Unconfined compressive strengths appear to correlate strongly with dry density in an exponential relationship. Deviations from this relationship (i.e., higher or lower strengths at a given density) could be ex-

plained by variations in cement content. The cement content of the field cores that were tested varied from 2.8 to 18.1 % dry wt. basis. The target cement content was 12 %.

5. No significant difference was observed between the control and fly ash test section in the extent of edge breaking along the road shoulder pavement.

6. Fairly extensive and progressive cracking was observed in the vicinity of Stn. 292+00 (West). Cores from this location exhibited lower strengths and cement contents; the asphalt cap was also considerably thinner here compared to other locations. The crack in the pavement at this location grew progressively with the passage of each winter season and appeared related to and/or the consequence of associated frost heaving in the underlying fly ash base. This problem appears to be localized to this one area.

5.0   ACKNOWLEDGEMENTS

The work described herein was supported by a research grant from Consumers Power Company. Funding for construction of the experimental base course was provided by the Michigan Department of Transportation.

6.0   REFERENCES

1. Berry, W.H., Gray, D.H., Stoll, U.W., and Tons, E. (1988). Use of Coal Ash in Highway Construction: Michigan Demonstration Project. Electric Power Research Institute, Rept. No. GS-6155, Palo Alto, CA.

2. Clegg, B. (1976). An Impact Testing Device for In-Situ Base Course Evaluation, *Proceedings* 8th ARRB Conf., Vol. 8, No. 8, pp. 1-6.

3. DiGioia, A.M *et al*. (1986). Fly Ash Design Manual for Road and Site Applications: Vol. 1: dry or Conditioned placement. Interim Report prepared for the Electric Power Res. Institute, EPRI CS-4419 (Volure 1)/RP2422-2, Feb. 1986, 243 pp.

4. Croney, D. and Jacobs, J.C. (1967). The Frost Susceptibility of Soils and Road Materials. Laboratory Report RRL #90, British Ministry of Transport, Road Research Laboratory, 1967, 68 pp.

Reclamation, Treatment and Utilization of Coal Mining Wastes, Rainbow (ed.) © 1990 Balkema, Rotterdam. ISBN 90 6191 154 0

The influence of particle size on the performance of cement bound minestone

R.J.Kettle
Department of Civil Engineering, Aston University, Birmingham, UK

ABSTRACT: The paper describes a preliminary investigation of the role of maximum particle size on the performance of cement bound minestone, assessed in terms of the 7-day compressive strength. Mixes were prepared with maximum particle size ranging from 28 to 5mm, using both 50 and 100mm diameter constant volume moulds. The results clearly demonstrate the influence of maximum particle size with the strength increasing as the maximum particle size was reduced. Mechanisms for failure are presented and these are also used to explain the role of both maximum particle size and mould size on the measured strength. Both cement and a mixture of cement and lime were used as stabilisers and the results indicate that the two-stage treatment did produce benefits with the particular minestone, particularly when it was used to stabilise the materials with maximum particle sizes below 14mm.

1 INTRODUCTION

Considerable quantities of waste material are produced in coal mining, and the waste is usually referred to as minestone. These deposits are composed of the debris removed from coal workings during the extraction of coal and the majority of the material comes from the carboniferous series of rocks lying close to the coal measures. Much of it consists of shale containing small amounts of bituminous and carbonaceous matter, mudstone, argillaceous sandstone, fireclay, ironstone and limestone. (Fraser, 1974)

Given the massive quantities of minestone stockpiled in Great Britain, only a relatively small quantity has been utilised in highway construction. The prime utilisation of road construction has principally been as a fill material and a wide range of minestones can be used as bulk fill. More recently interest has centred on its possible inclusion in the pavement structure. Laboratory tests suggest that some form of stabilisation will be essential if the minestone is to satisfy the performance requirements for pavement material (Kettle, 1969; Sherwood and Pocock, 1969). Mixtures of cement or cement and hydrated lime have been found to be satisfactory. Sometimes the mixtures present chemical problems due to the presence of sulphates, but stabilisation reduced the effect of frost and chemical action (Kettle, 1974; McNulty, 1985; Thomas 1986). There is evidence of considerable differences between the various sources of minestone, and each material must be rigorously tested before it can be employed. It is therefore necessary to determine the physical properties of the materials: particle size distribution, compaction characteristics when mixed with stabilisers, strength, frost resistance, and resistance to immersion in water.

Generally, the mechanism of both cement and cement and lime stabilisation is influenced by soil technology. When cement is present in the granular shale, the cementation is probably very similar to that in concrete, except that the cement paste does not fill the voids between the soil particles. In other words, cementation is primarily by means of mechanical bonding of the hydration products to the rough mineral surfaces. In the silt and clay fraction, cementation becomes a combination of mechanical bonding and chemical bonding which involves a reaction between the cement and the surfaces of these particles. The mechanism perhaps can be demonstrated by visualising the grains of cement as a nucleus to which the fine soil particles adhere. As the cement in the soil is increased, the quantity of free silt and clay is progressively reduced and producing a coarse-grained material of a lower water holding capacity and increased volume stability and supporting value. As more and more cement is added, the quantity of coarse-grained material is increased until the point is reached where all the soil grains remain in a solid mass as befits a structural material. Lime as an admixture is mainly to reduce the activity of clay, that is, by reducing the plasticity index.

2 SCOPE AND OBJECTIVES OF INVESTIGATION

The main objective of this project was to investigate the influence of particle size on the unconfined compressive strength of cement stabilised minestone. The effect of mould size in relationship with the particle size was also investigated.

Previous works (McNulty, 1985; Thomas, 1986)

have shown that cement treatment improved both the strength and the durability of Cement Bound Minestone (CBM). However, with the finer grade minestones the benefits of cement stabilisation are not so apparent. It has been shown (Moussas and Kettle, 1979) that two stage treatment involving both cement and lime could improve the strength significantly. The lime is intended to reduce the activity of the clay minerals with the cement bonding together the individual grains, or aggregations within the lime modified spoil. Earlier studies (McNulty, 1985) have shown that favourable results can be obtained with an additive content of 10% (by weight of oven dry shale) comprising 10% cement or 8% cement and 2% lime. Indeed, the response to this level of additive is a good indication of commercial exploitation and so this was adopted throughout the investigation.

The compacted samples of CSM were produced in constant volume cylindrical moulds by static compaction (British Standards 1975a)). These were moulded at the appropriate optimum moisture content, as determined with the vibrating hammer test (British Standards 1975b) and compacted to the corresponding maximum dry density. Specimens were prepared from materials covering a maximum particle size ranging from 28mm, through 20, 14 and 10mm down to 5mm. Two mould sizes were employed 100mm by 50mm (dia) and 200mm by 100mm (dia). The influence of particle sizes was judged primarily by the 7 - day unconfined compressive strength of these cylindrical samples.

3 MATERIALS USED

Two minestones from Kent were used for the study:
  (1) Betteshanger
  (2) Snowdown
It has shown (McNulty, 1985) that, with CSM from Snowdown, there was a noticeable difference in strength between the 'small' mould used in the laboratory and the 'large' mould used in the field for test samples. Samples of CSM from both Snowdown and Betteshanger minestone had performed favourably in laboratory studies with strength levels approaching the DoT requirements (Department of Transport, 1986) for CSM 1 and a strong resistance to immersion in water. On this basis it was therefore decided to use these minestones from Kent for this investigation.

The minestone samples were supplied by British Coal and, before commencing the main investigation the particle sizes distributions were established. The gradings (Thomas, 1986) of the two materials indicate that both samples are within the limits given in the DoT specification (1986) for CSM 1 material. The stabilisation was undertaken with Ordinary Portland Cement and Hydrated Lime.

4 METHODS AND TECHNIQUE AND DEVIATIONS FROM STANDARD PRACTICE

4.1 Drying and Sieving

To minimise degradation and modification the minestones were allowed to dry at room temperature for four to five days. They were then separated on particular BS test sieves, 28, 20, 14 10 and 5mm, to produce the individual test samples. The moisture content of the air-dry material was determined before storage in air tight containers.

4.2 Mixing

A twin Z-bladed pug mill mixer was used throughout the investigation. This was selected because it is known to be capable of producing relatively consistent results (Kettle, 1973) and it has been used in previous studies (McNulty, 1985; Thomas, 1986). In all cases, the minestone and stabilisers were dry mixed for one minute, after which the water was added to raise the mixture to the pre-determined optimum moisture content. The mixing was continued for a further two minutes.

4.3 Determination of Optimum Moisture Content and Maximum Dry Density

When soil cement (CSM 1) is used in highway pavements, the moisture content of the mixed material is usually (Thomas, 1986; Ministry of Transport, 1978) between the optimum value and 2 per cent above, the optimum being determined by the vibrating hammer compaction test (British Standards, 1975b). As all mixes were to be prepared with 10% stabiliser, either 10% cement or 8% cement and 2% lime, it was decided to only reform the compaction tests on the materials prepared with the cement and lime and to use these values of optimum moisture content and maximum dry density for the preparation of all mixes. This was unlikely (Kettle, 1973; McNulty, 1985) to produce a significant error in the moulding parameters. For all cases, the targeted maximum dry density corresponded to 100% of the value determined with the vibrating hammer.

4.4 Preparation of Specimens

Two different sizes of constant volume cylindrical mould were used with the height to diameter ratio of 2:1 in accordance with BS 1924 (British Standards, 1975a) and these are referred to as either 'small' or 'large'. The small specimens refer to samples measuring 100mm x 50mm (dia) and the 'large' samples were 200mm x 100mm (dia).

For all specimens the mix proportions and batching were by weight. All the specimens were prepared to a constant volume by static compaction to a pre-determined maximum dry density and optimum moisture content obtained from the vibrating hammer compaction test following the procedures described in BS 1924 (British Standards, 1975a). The following maximum particle sizes were used for the small specimen: 28 mm, 20 mm, 14 mm, 10 mm and 5mm. Whereas, for the large specimen, only 28 mm and 14 mm were selected owing to the vast amount of material involved in specimen manufacture. Care was taken to ensure that the time between mixing and compacting of specimens were within 20 minutes.

Additional specimens were made at density levels other than the maximum value to examine the effect of density on strength. To fully assess the role of the maximum particle size on compressive it was decided to calculate both the mean strength and the variation. Accordingly, for each mix, five small samples were manufactured following the former requirements (Ministry of Transport, 1975) of the Ministry of Transport. With the large mould only three replicate samples were prepared as it was believed that these would provide a representative value of the variation.

4.5 Curing

The specimens were cured in polythene bags at a temperature of $20^0 \pm 2^0C$, and no significant moisture loss was recorded throughout the investigation. The 7-day curing period adopted for test purposes was chosen for convenience and is purely arbitrary, since it provided that all comparisons would be made at the same age, the actual age being of little importance.

5. STRENGTH TESTING METHODS

As the unconfined compressive strength test is one of the easier parameters to determine, this is the test most frequently employed for studying the properties of stabilised materials. This test was, therefore, adopted for the assessment of the factors affecting the strength of materials.

Initially, all the small specimens were tested on a deformation-controlled loading machine, normally used for measuring the CBR and unconfined compressive strength of natural soils (British Standards, 1975b). But, unfortunately, as the investigation progressed it became clear that this system was not capable of exerting a force corresponding to the failure strength of the 'large' specimens. As access to another proving ring type testing machine of larger loading capacity was not available, an alternative hydraulic compression machine (Denison) was used for testing 'large' specimens instead.

The proving ring applies the load to the specimen at a uniform rate of 1mm per minute whereas with the Denison machine, the specimens are loaded at constant rate of stress, corresponding to $3.5N/mm^2/min$ used for cubes (British Standards, 1975a). A difference in strength could be expected if the 'same' specimens are subjected to two different testing methods and, therefore, a supplementary investigation was carried out to determine an appropriate correlation factor. This is given in the Appendix and it was used to permit valid comparisons between the various results.

6 DISCUSSION OF THE RESULTS

6.1 Compaction

The prime factor influencing compaction is the moisture content, with the maximum dry density being achieved at the appropriate optimum moisture content. With the Snowdown samples this optimum was established on samples with 28mm maximum particle size. However, difficulties were experienced when preparing samples of a maximum particle of 14mm and below. Clearly the composition of the minestone was influencing the compaction characteristics, for the amount of fine material would be proportionately greater when the particle size is reduced, leading to a higher specific surface area of material to be wetted. As a result, the optimum moisture content determined with larger particle sizes is insufficient for compaction. It is also believed that the specific gravity of the minestone is dependent on the maximum particle size since the proportion of organic materials, e.g. coal, are likely to be greater, as the particle size is reduced. Therefore, in order for valid comparison to be made, it is ideal to determine the compaction characteristic for each particular maximum size.

However, to limit the number of compaction tests, it was assumed that the compaction curves for the same material with the same amount of stabilisers would have a similar shape, independent of particle size distribution. A mix was made up to moisture content judged by eye to be on the dry side of the optimum, and a vibrating hammer compaction test carried out immediately. Following the addition of sufficient moisture to raise the material to above the optimum, this test was repeated. On the assumption that the shape of the curve was not affected by the change in maximum particle size, an appropriate curve could be fitted between the two points to provide the optimum moisture content and maximum dry density. Thus compaction characteristics were determined for materials passing 28mm, 14mm and 5 mm. For materials passing 20mm, compaction characteristics for 28mm were used, similarly for 10mm, compaction characteristics for 14mm were used. Although this is not entirely satisfactory, the errors introduced would be small and were judged to be acceptable. Again these characteristics were only carried out on mixtures stabilised with 8% cement and 2% lime. It would be assumed that the compaction characteristics with 10% cement would be very similar.

As no difficulty was experienced in compaction of the Betteshanger samples it was believed that the compaction characteristics did not change appreciably with different particle sizes. This could be due to the nature of the grading of Betteshanger.

Minestones are essentially compacted muds and therefore possess a laminated structure. Variation in moisture content may disrupt this structure leading to deterioration. During compaction weaker particles are also subject to degradation by simply breaking along the laminal planes. The addition of lime and cement to shale may modify the behaviour of the

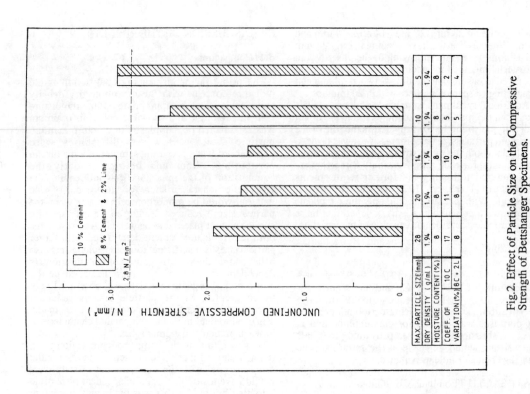

Fig.2. Effect of Particle Size on the Compressive Strength of Bettshanger Specimens.

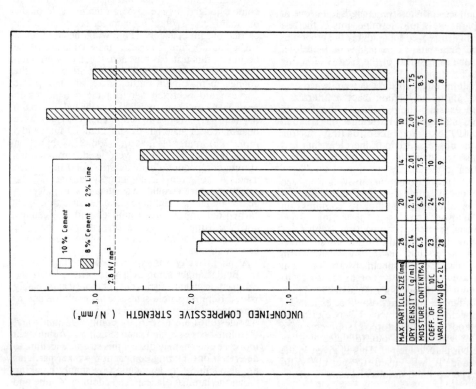

Fig.1. Effect of Particle Size on the Compressive Strength of Snowdown Specimens.

finer fraction by floculation and cation exchange. All these factors complicate the compaction characteristics of stabilised shales and therefore difficult to define.

7.2 Compressive Strength

The results of the 7-day compressive strength are shown in Figures 1 and 2. These show the effect of different particle sizes for small specimens. In earlier studies (Kettle, 1973; McNulty, 1985) it has been indicated that when 7-day strength of statically compacted cylindrical specimens is in excess of $2.8MN/m^2$, the stabilised material is suitable for induction in pavement construction and this value is indicated on the appropriate figures.

The results show a general trend of increasing strength for both Snowdown and Betteshanger specimens as the particle sizes were reduced when 10% cement was used. The exception was for Snowdown with particle size below 5mm, of which the strength is below the expected value. This can be explained by the fact that lumps were formed while preparing these specimens. It was observed that these consisted of clay-cement and not lumps of clay coated with cement. This leads to uneven cement distribution and hence improper bonding.

There is strong evidence that the strength of the stabilised minestones is primarily due to weak shale particles within the minestone. On examination of the failed specimens it was observed that large particles often occupied a disproportionate fraction of the total failure plane and therefore had an exaggerated influence on the failure. When the maximum particle size was reduced, a more uniform and continuous matrix was formed and, as a result, interparticle bonding became more predominant. A possible fracture mechanism is determined by the natural strength of the binder, i.e. the interaction between cement and the finer fraction rather than the natural strength of the particles. Hence, these failure characteristics explained one of the reasons governing the observed trend of increasing strength with decreasing particle sizes. This is explained diagramatically in Figure 3.

Obviously there are many factors which influenced the failure trend based on the effect of particle size. The method of static compaction does not reorientate or redistribute material within the mould as effectively as vibrating compaction (Omar and Brown, 1981) By visual examination, during compaction, voids had been bridged by large particles and these only became partially filled with material on completion of moulding. Such areas of low density are thought to be partially responsible for the lower strength of specimens with larger maximum particle sizes.

As minestone has, essentially, a finely laminated structure, it could be easily broken into parallel sided fragments during compaction without a coating of cement paste. This further contributes to the lower strengths for large particle size, as it is obvious the failure plane will be predominantly along the fracture line of the shale, as shown in Figure 4.

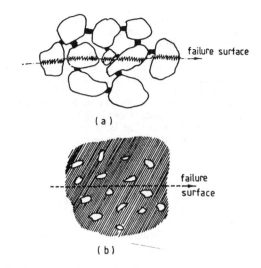

Fig.3. Fracture Mechanism in Stabilised Minestone

(a) strength determined by natural strength of shale

(b) strength determined by natural strength of binder

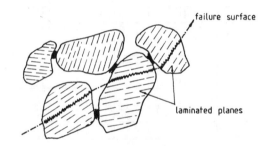

Fig.4. Fracture mechanism

Based upon the above explanations the coefficient of variation also demonstrated the effect of particle size on failure characteristics. Specimens with larger particle sizes have a greater coefficient. This could be explained by the orientation of the partticles which resulted in the fluctuation in compresive strength. For specimens with smaller particle sizes a uniform and continuous matrix can be obtained hence leading to a more uniform compressive strength.

A comparison between 'small' and 'large' specimens further supports the failure characteristics. On account of the particle size in relation to mould size e.g. 28mm in 100mm diameter specimens and 14mm in 50mm diameter specimens are directly proportional. Hence, they would be expected to have similar stengths and this is supported by the results in Table 2, thereby further supporting the suggested failure characteristics.

In earlier studies (McNulty, 1985) it had been assumed that the mixing action produced sufficient breakdown to justify the use of 50mm (diameter) moulds for maximum particle sizes up to 25mm. However, an examination of Snowdown material given standard mixing in the Z-blade mixer indicated that the extent of this breakdown was limited, as can be seen from the particle size distribution curves plotted in Figure 5. It therefore seems likely that many of these large particles would be intact when placed in the appropriate moulds and that this could influence the relationship between particle size, mould size and strength. It is suggested that the large particles compacted in the 50mm (diameter) moulds would undergo more local fracture during compaction than when compacted in the larger mould. This is also clearly demonstrated in Table 2 and so further supports the suggested failure mechanisms. This is in contrast to the behaviour of concrete specimens where large specimens produced lower measured strengths due to the increased probability of the presence of critical flows in the larger specimens (Neville, 1981). With the minestones the extra crushing that occurs in the small moulds leads to the creation of additional critical flaws before loading and, hence a lower strength.

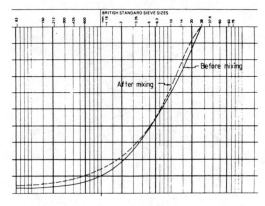

Fig.5. Grading Curves for Snowdown Minestone

Further examination of Figure 1 and 2 indicates that the effects of lime on the stabilised shales is no. apparent as has been suggested (Moussas and Kettle, 1979) and, in actual fact, very different for the two minestones. For Snowdown specimens, a mixture of 8% cement and 2% lime follows the same failure trend as that of 10% cement, whereas the results of the Betteshanger specimens do not demonstrate any noticeable pattern with lime, except that for smaller particle size. There is a strong indication that lime does bring a positive change in the condition of the finer soil as shown by the marked rise in the compressive strength for both 5 mm and 10 mm. This may be explained by the grading and consideration of the clay mineralogy of the materials. It is also noted that the compressive strength of the

stabilised Snowdown shales is greater for 8% cement and 2% lime rather than with 10% cement alone. This demonstrates the effect of lime on the clay particles which superimposed on the effect of cementation leading to the development of greater strength. This was not the case for Betteshanger specimens with maximum particle sizes above 14mm. Although, with smaller particle sizes, a definite improvement was obtained in compressive strength. It is believed that this is due to the effect of lime on the greater proportion of clay and silt particles as indicated by the grading.

A comparison between the strength improvement with particle size of stabilised Betteshanger shales with 10% cement and 8% cement and 2% lime are shown in Table 2 and emphasis the benefits obtained in the laboratory by limiting the maximum particle size.

8   CONCLUSIONS

1. The unconfined compressive strength of minestones from Snowdown and Betteshanger are directly influenced by different particle size in a specimen. The strength increased when the particle size was reduced. With 10 percent cement, both were able to meet the targeted strength requirement for soil cement. These materials have a 7-day cylinder strength in excess of $2.8N/mm^2$ of 100 mm x 50 mm specimen with particle size below 10 mm and 5mm for both Snowdown and Betteshanger. With the 200 mm x 100 mm specimen both are also capable of achieving the strength in access of $2.8N/mm^2$.

The substitution of large particles of shale with gravel was not investigated. It is believed that with an appropriate blending of gravel with the finer fraction of the shales would improve the strength greatly which may enhance its suitability but would increase costs. Sieving the minestone down to finer particles may be a problem, but this form of stabilisation of soil-aggregate with cement and lime may prove to be beneficial and should be further investigated for both minestone and other materials.

2 The effect of mould size was clearly demonstrated with compaction into largest moulds generally producing higher 7-day compressive strengths. It was considered that excessive particle fracture was created when the specimens were produced in small moulds leading to lower measured strengths.

3 The effect of lime was evident. The strength was improved significantly with Snowdown material, exceeding the value when stabilised with 10% cement.

9   ACKNOWLEDGEMENT

The views expressed are those of the author and not necessarily of British Coal, who kindly supplied the test samples.

10.   REFERENCES

British Standards Institution, 1975a Methods of test for stabilised soils BS 1924, BSI London.

British Standards Institution, 1975b Methods of test for soils for civil engineering purposes BS1377 BSI London.

Department of Transport, 1986 Specification for road and bridge works, HMSO, London.

Fraser C K 1974 The use of unburnt colliery shale in road construction. TRRL Supplementary Report 20 UC.

Kettle R J 1969 Cement stabilised unburnt colliery shale. Road and Road Construction, 47, No 559, pp 200-206.

Kettle R J 1973 Freezing behaviour of colliery shale. PhD thesis, University of Surrey, Guildford, UK, 1973.

McNulty, A T 1985 Durability criteria for cement bound minestone. PhD thesis,
University of Aston, UK.

Ministry of Transport 1975 Specification for road and bridge works, HMSO, London.

Moussas J H, Kettle R J 1979 Lime stabilisation of colliery shale. Final year project. Department of Civil Engineering, University of Aston.

Neville A M 1981 Properties of concrete, Third Edition, Pitman, London.

Omar A and Brown S 1981. Investigation into the effect of compaction techniques on unconfined compressive strength of stabilised unburnt colliery shale. Final year report. Civil Engineering Department. University of Aston.

Sherwood P T, Pocock R G 1969 The utilisation of cement-stabilised waste materials in road construction. Road and Road Construction, 47m No 554, pp43-50.

Thomas M D A 196 The performance of cement stabilised minestone. PhD thesis, University of Aston, UK.

11.   APPENDIX

Correlation factor for compressive strength

All the stabilised shale specimens were prepared under the same conditions and were tested as follows:

a) For 100 mm x 50 mm cylindrical specimens, the apparatus used and procedure adopted followed the recommendations given in BS 1924 (British Standards, 1975a)

b) For 200 mm x 100 mm ϕ cylindrical specimens, again the procedure adopted was based on that given in BS 1924, but the specimens were loaded at constant stress rate of 3.5 N/mm^2/min rather than constant rate of deformation of 1 mm/min.

As a result two distinctive modes of failure were observed, with method (a) generally producing a shear failure together with a barrelling effect whereas method (b) produced two cones of undamaged material adjacent to the ends of the specimens. This can be explained by the fact that portions of the specimen nearest to the platen of the testing machine are restrained by friction between the ends of the specimen and the platens. Failure was proceeded by the formation of tensile cracks inclined at small angles to the direction of load, once these cracks had developed, the strength of the specimen was dependent upon its internal friction. It is noted that the platens used in method (b) were much harder than that used in method (a), hence it would be expected that the failure strength of the specimens tested in method (b) would have higher strength.

When determining the correlation factor, specimens were made from Betteshanger with maximum particle size 14 mm and to stabiliser content of 8% cement and 2% lime. This was arbitrarily chosen and the correlation factor is given in Table A1

Table 1: Comparison of 'small' and 'large' specimen

Source	Maximum particle size (mm)	Dry density (g/ml)	Moisture content (%)	'small' specimen 10C	'small' specimen 8C + 2L	'large' specimen 10C	'large' specimen 8C + 2L
SNOW-DOWN	28	2.01	7.5	1.64	2.27	2.63	-
		2.05	7.5	2.67	-	3.00	2.95
		2.14	6.5	1.96	1.91	-	-
	14	2.01	7.5	2.47	2.53	2.82	3.46
		2.14	6.5	2.69	3.17	-	-
BETTES-HANGER	28	1.94	8.0	1.69	1.93	2.26	1.91
	14	1.94	8.0	2.13	1.59	2.85	2.45

The compressive strength headers are $Compressive\ Strength\ (N/mm^2)$.

Note: C = % of cement
L = % of lime

Table A1 Strength Correlation

	Strength (N/mm^2)	Variation (%)	Strength ratio (Correlation factor)
Method (a) (Proving Ring)	2.13	10	$\dfrac{2.13}{2.43} = 0.871$
Method (b)· (Denison)	2.43	13	

Table 2: Strength improvement with decreasing particle size in Betteshanger specimens

Maximum particle size (mm)	Unconfined Compressive strength (N/mm^2) 10% cement	8% C + 2% L	Strength improvement in percentage (%) 10% cement	8%C + 2% L
14	2.13	1.39		
			17.8	50.3
10	2.51	2.39		
			17.1	23.0
5	2.94	2.94		

Reclamation, Treatment and Utilization of Coal Mining Wastes, Rainbow (ed.) © 1990 Balkema, Rotterdam. ISBN 90 6191 154 0

Flocculation of fine coal waste – At discharge

Bill M. Stewart, Ronald R. Backer & Richard A. Busch
Spokane Research Center, US Bureau of Mines, Spokane, Wash., USA

ABSTRACT: An evaluation of dewatering fine coal refuse slurry using chemicals was undertaken by the U.S. Bureau of Mines. Two separate field tests have been completed. In the first test, 18.1 kg of lime and 0.9 kg of anionic polymer per ton of dry solids were manually mixed with the underflow from the preparation plant thickener. Clear, free water was immediately liberated from the slurry. The percent solids by weight of the slurry increased from about 17 to 30 immediately, and to 45 after 5 days. This indicated a 63-pct reduction in total volume after 5 days from 4.86 m^3/mt of solids originally to 1.82 m^3/mt of solids.

In the second test at a different site, automatic flocculant system was used to treat fine coal waste slurry from a single-point discharge. Slurry flowing at rates from 1.51 to 2.65 m^3/min with a specific gravity range of 1.15 to 1.25 was successfully flocculated. The system was designed to maintain the optimum polymer dosage at all times. This was accomplished by controlling the dilute polymer pump rate with a slurry mass flow rate signal. For this test, polymer dilution was automatic and clarified water from the preparation plant was used for dilution. Both anionic and cationic polymers were used. The average untreated slurry density was 30.5 wt pct (2.64 m^3/mt of solids). The slurry density was 64.3 wt pct (1.23 m^3/mt of solids) after treatment and 78.0 wt pct (1.03 m^3/mt of solids) after 60 days.

1 INTRODUCTION

Currently, many coal mining operations are devoting more engineering and other efforts to waste disposal. More and more manpower, money, and equipment are being used for waste disposal. The main reasons are dirtier coal from automated mining systems creating more fine waste, new and more stringent regulations, and a greater concern for the health and safety of the people who work and live near the waste disposal sites. Although the slurry impoundment technique was cost effective in the past, current regulations have greatly increased the cost of this disposal method. Impoundments also require a great deal of disposal area. Because coal waste is not generally marketable and surface areas for disposal are becoming less available, it is understandable that management is open to innovative waste disposal concepts that maintain health and safety standards at less cost.

1.1 Background

A comprehensive research effort was initiated to solve this problem. A small belt press was rented and several samples of coal waste, metal mine mill tailings and uranium tailings were treated. A reasonably good effluent "cake" was produced with various polymer additives. Additional tests using a solid bowl centrifuge, vacuum-disk filter, plate and frame filter press (pressure filters), and thermo drying processes were also performed. Each of these tests demonstrated some degree of success, but each also had some problems. Some could not effectively dewater wastes with high clay and ash content, some had low throughput capacity, and some had high capital, maintenance, and operating costs. The decision therefore was made to try flocculating fine waste and allow it to dewater naturally without mechanical equipment.

A series of "bench" tests was performed in the laboratory utilizing a polymer and a gravity drainage system. For some coal waste products it was also necessary to add lime to increase the pH and enhance the action of the coagulant. The results of these simple laboratory tests were very encouraging, indicating that a field project should be designed to demonstrate this concept.

A cooperative agreement was negotiated with the Washington Irrigation and Development Co. (WIDCO), a coal mine operator in western Washington. The arrangement was to tap a 0.38 m^3/min sample from the thickener underflow, condition it with lime in a 3.79 m^3 baffled mixing tank, add dilute polymer at the discharge end, and deposit the mixture in large earth cells.

The success of this field experiment indicated that the deposition could be made on a naturally sloping site with no peripheral embankments. Additional testing appeared to be warranted. Therefore, the Jewell Smokeless Coal Corp. (JSCC) mine at Vansant, VA, was selected, and a full-scale, automated mixing and dosing scheme was designed to treat up to 2.27 m^3/min. This time the treated material would be dumped on a sloping site just above the existing impoundment.

1.2 Laboratory and Field Tests -- WIDCO

To develop the concept of dewatering using polymers only, laboratory tests were essential. Results of the laboratory tests indicated the need for larger field tests.

Representative samples of coal-refuse slurry were obtained from WIDCO. Based on polymer data for belt press tests at WIDCO, American Cyanamid Superfloc 1202 (reference to specific equipment or trade names does not imply endorsement by the Bureau of Mines) was used. The WIDCO experience also indicated lime was necessary to raise the pH of the slurry to obtain desired flocs. Two laboratory tests were performed. In the first series of tests, lime was mixed with 1,000 mL of slurry and dumped on a sloped 1.2- by 2.4-m piece of plywood. The polymer dosage was varied for each 1,000-mL pour, and the consistency of the flocculated slurry was noted. Moisture content samples were taken at various times after each pour. In the second series of tests, 11.3 to 13.6 kg of slurry was conditioned with lime and mixed with various dosages of polymer. Immediately after the slurry reacted, it was poured into a 0.02-m^3 bucket equipped with a

vertical drain through which the separated water could readily flow. The rate of water drainage and the moisture contents of each sample were determined.

Results of the laboratory tests indicated that dosages of 0.68 to 0.91 kg of polymer and 18.1 kg of lime per ton of solids would result in a well-flocculated slurry that would release water readily. At this point large-scale field tests were considered necessary to validate the small-scale laboratory results.

While the site for the field tests were being prepared at the WIDCO Mine, equipment was gathered at the Spokane Research Center. The equipment consisted of (1) a 3.78-m^3 mixing tank with manifold and baffles for mixing and conditioning, (2) four 0.21-m^3 drums with jet pump system (one pair of drums was used for lime-water mix and the other pair for neat polymer dilution), (3) four 0.38-m^3/min, centrifugal recirculation pumps, and (4) a 10-cm-diam slurry feedline. Based on laboratory results and past WIDCO experience, the slurry-lime mixture had to be retained in the mixing tank for 10 min to thoroughly condition the slurry. The controlled low feed rate of 0.38 m^3/min to the mixing tank provided the time needed to condition the slurry. The dilute polymer was introduced at the tank exit to mix with the treated slurry in the discharge line.

The site for the field test included four 1.8- by 15.3- by 30.5-m earthen dewatering test cells. Each cell was constructed with bottom drains connected to PVC pipe that passed through the test cell embankment. These pipes were used to collect and convey separated water from the test cells. Detailed information on the test cell layout, drainage system, and equipment setup can be found in Bureau of Mines Report of Investigations 8581 (Backer and Busch, 1981).

Two separate treated slurry runs were conducted in test cells 2 and 3. Test cell 4 was filled with untreated slurry and served as a control cell. Untreated slurry in test cell 1 was used for electrokinetic research; therefore, experimental data are not included in this paper (results of electrokinetic treatment in cell 1 are published in Bureau of Mines Report of Investigations 8666). The operational data for both runs in cells 2 and 3 and for the control cell 4 are found in Table 1. The major difference in the two runs is that in the first run the bottom drain pipe was not opened until after filling was completed and in the second run the drain pipe was opened

Table 1. Operational data.

	First run			Second run	
	Cell 2	Cell 3	Cell 4	Cell 2	Cell 3
Fill time.......h..	15.85	16.75	11.00	50.75	51.83
Feed:					
Rate.....m³/min..	0.38	0.38	0.63	0.41	0.41
Total........m³..	360.0	395.5	412.2	1152.5	1177.7
Solids...wt pct..	21.0	17.3	19.0	16.9	17.4
Specific gravity.	1.19	1.10	1.10	1.20	1.20
Solids.........mt..	90.2	72.1	86.5	249.8	261.8
Lime..........kg..	1837.0	1429.0	NA	4876.0	5647.0
.......kg/mt..	20.4	19.8	NA	19.5	21.6
Polymer........kg..	74.4	61.7	NA	235.9	271.2
.....kg/mt..	0.83	0.86	NA	0.95	1.03

NA Not applicable; untreated control cell.

at the beginning of fill in addition to siphoning excess surface water. For this reason, as can be seen in Table 1, cells 2 and 3 accepted three times more slurry in the second run than in the first run.

1.3 Results -- WIDCO Field Test

Chemical treatment of the slurry caused the water to rapidly separate from the flocculated solids and become free surface water that could easily be removed. As the depth of solids increased in the test cells, the water could not permeate the solids to the bottom drain. This was mainly due to a rather low permeability $(5.02 \times 10^{-6}$ cm/s) of the flocculated solids. The average slurry density in cell 3 (second run) was 43.8 wt pct solids after 4 days and 48 pct solids after 47 days. Cell 2 showed slightly better (higher) results but was aided by vertical drains. These limited field tests indicated that, with chemical treatment, slurry density can be increased to about 45 wt pct (1.82 m³/mt of solids) in 5 days. Based on an initial slurry density of 17 wt pct solids (4.86 m³/mt of solids), only about one-third to the volume would be required to retain treated fines with respect to untreated material. By contrast, samples taken in the WIDCO impoundment, after many years of settling, had a total volume reduction of only 43 vol pct. Figure 1 shows the moisture-volume relationship of the WIDCO treated and untreated slurry. A dramatic reduction in volume is realized when the slurry density is increased to about 50 wt pct solids. Further increases in slurry density do not appreciable reduce volume. The reduction in volume achieved

by polymer treatment could double slurry storage capacity and proportionally increase the life of the impoundment.

1.4 Laboratory and Field Tests -- JSCC

Successful WIDCO test results led to a second field test at Jewell Smokeless Coal Corp. near Vansant, VA. The purposes of this test were (1) to update and automate the flocculant dilution and mixing system, (2) to optimize the injection of dilute flocculant, (3) to treat single-point discharge with slurry flow rates up to 2.27 m³/min (six times more than the flow rate treated at WIDCO), and (4) primarily to test the concept of slope deposition.

Representative slurry samples were sent to the Spokane Research Center laboratory. For the laboratory tests two polymers were used, American Cyanimide Superfloc 1202 and Nalco 8873. The latter is the product used at JSCC for pre-belt-press conditioning. Lime conditioning was not required. The 1,000-mL tests and the 0.02-m³ bucket tests previously described were repeated using both polymers. The results of nine bucket tests showed an average slurry density of 55.7 wt pct solids after 18 h. The average initial slurry density was 27.0 wt pct solids. Detailed results of these tests are available at the Spokane Research Center. Two other tests, referred to as trough tests, were completed. In those tests, treated slurry was discharged into a 13.7-m-long trough sloped at 1 pct grade, with a drain at the downstream end. The purpose of these tests was to simulate field flow conditions and to determine if a 20-s retention time (before discharge) was sufficient for the dilute polymer to flocculate the slurry. The bucket tests required 10- to 15-s retention time before reaction with controlled mixing. For the field test, with uncontrolled mixing, a somewhat longer period of 20 s was deemed necessary. Data supplied by the mine showed that slurry was discharged at an average of 2.69 m³/min through a 15.2-cm line (about 2.4 m/s). To simulate this in the laboratory, 50 gal of slurry was pumped through a 2.54-cm line at 0.076 m³/min (about 2.4 m/s). To obtain the 20-s retention time, 12 m of 5.1-cm line was used, which reduced the flow velocity to 0.6 m/s. The dilute polymer was injected (at a controlled rate) into a nipple installed at the beginning of the 5.1-cm line. In both trough tests, the slurry discharged with a thick milkshake consistency with excellent flocculated particles. Clear water immediately ran off to the drain.

1.4.1 Equipment

The equipment for the JSCC field test was planned for automatic dilution of neat polymer and optimized injection of dilute polymer. The equipment consisted of (1) a 1.14-m³ neat polymer tank, (2) a 7.57-m³ dilute polymer tank with high- and low-level probes, (3) a polymer-water mixing system consisting of a centrifugal water booster pump, a variable speed polymer gear pump, and a stata-tube mixer, (4) a variable-speed dilute gear pump, (5) a 5.1-cm flow meter for dilute polymer flow rate, (6) a 15.2-cm flow meter for slurry flow rate, (7) a 15.2-cm nuclear densometer for slurry specific gravity, and (8) a four-pen recorder with a built-in math module. Water was supplied from a 45.4-m³ tank provided by the mine. After calibrating the neat polymer pump, the neat polymer was mixed with water at a 0.75-pct concentration and the dilute polymer was pumped to the 7.57-m³ tank. When the dilute polymer reached the high level probe in the tank, the neat polymer pump and the water booster pump automatically shut off. When the dilute polymer cleared the low-level probe, the neat polymer and water booster pumps automatically came on, refilling the dilute tank. The rate of fill was about 0.15 m³/min. The dilute polymer was pumped into the slurry at an average rate of 0.06 m³/min, providing a continuous slurry treatment. The speed of the dilute polymer pump was automatically adjusted by the slurry mass flow rate signal to the controller of the variable-speed pump. The mass flow rate signal was calculated by the math module in the recorder from signals provided by the 15.2-cm magnetic flow meter and nuclear densometer, which continuously measured the flow rate and specific gravity of the slurry. By this method, as the mass flow rate of the slurry decreased or increased, the speed of the dilute polymer pump decreased or increased accordingly, resulting in efficient polymer use. The four-pen recorder was used to continuously plot the dilute polymer flow rate, slurry flow rate, slurry specific gravity, and mass flow rate. Figure 2 shows a schematic of the equipment setup.

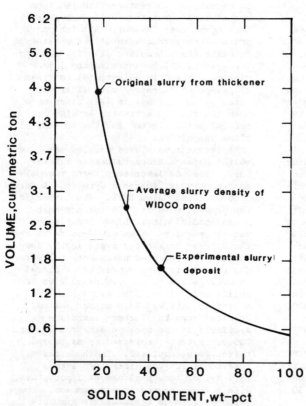

Figure 1. Moisture-volume relationship of WIDCO slurry.

FINE COAL WASTE FLOCCULATION SYSTEM

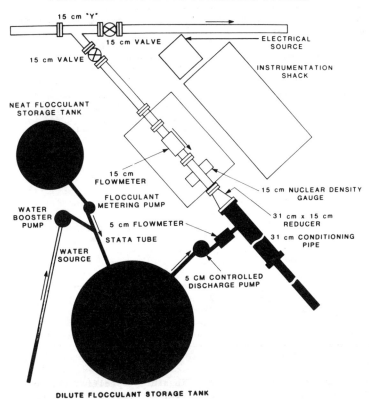

15 cm "Y"

15 cm VALVE

15 cm VALVE

ELECTRICAL
SOURCE

INSTRUMENTATION
SHACK

NEAT FLOCCULANT
STORAGE TANK

15 cm
FLOWMETER

FLOCCULANT
METERING PUMP

WATER
BOOSTER
PUMP

5 cm FLOWMETER

STATA TUBE

WATER
SOURCE

15 cm NUCLEAR DENSITY
GAUGE

31 cm x 15 cm
REDUCER

31 cm CONDITIONING
PIPE

5 CM CONTROLLED
DISCHARGE PUMP

DILUTE FLOCCULANT STORAGE TANK

Figure 2. Schematic of flocculation system.

1.5 Results -- JSCC

After the equipment was set up and field
calibrated, slurry treatment began.
During the first 2 full days of treatment the slurry did not flocculate as
expected. The slurry thickened but not
nearly as well as observed in the laboratory. The finest particles were not being flocculated. Several attempts were
made to improve flocculation. These included (1) manually adjusting the dilute
polymer flow rate, (2) reducing the flow
rate of slurry, (3) circulating the neat
polymer prior to dilution, (4) increasing and decreasing the concentration of
dilute polymer, (5) reducing downstream
flow of treated flow slurry after discharge, and (6) moving the discharge
pipe to flatter ground. None of these
improved the thickening of the slurry.
It was also noted that during manual adjustment of the dilute polymer flow rate,
both 0.75- and 1.0-kg/mt dosage amounts
appeared to overflocculate the slurry,

and at about 0.5-kg/mt
dosage the slurry ap-
peared thicker. This
contradicted laboratory
results. It was clear
that the mineralogy of
the slurry was different.
After discussing the
problem with a represen-
tative of the chemical
supplier and the prepar-
ation plant manager, a
small amount of cationic
polymer was added to the
treated slurry at dis-
charge. This vastly im-
proved slurry floccula-
tion with results similar
to those observed in the
laboratory (without cati-
onic polymer added). A
0.02-m^3 plastic jug was
fitted with a valved cop-
per tubing and used to
manually trickle the cat-
ionic polymer into the
treated slurry at dis-
charge. The amount of
cationic polymer required
was based upon visual ob-
servation of the thicken-
ed slurry, but was opti-
mized automatically as
was the anionic polymer.
    The mean operational
data for the final 5 days
of testing are shown in
Table 2. During that
period, a total of 3661 m^3 of slurry
containing 1365 mt of solids was treated
with 805.5 kg of anionic polymer and
257.2 kg of cationic coagulant. The
average untreated slurry density was 30.5
wt pct solids (2.64 m^3/mt of solids).
Five samples were collected at the surface 18 h after depositing the treated
slurry on day 4. The average slurry
density for these five samples was 64.3
wt pct (1.23 m^3/mt of solids). Another
five samples were collected 65 h after
the deposit on day 5. The average slurry
density for these five samples was 64.6
wt pct (1.22 m^3/mt of solids), indicating only a slight change from the 18-h
samples. The final sampling of the
treated slurry was made 60 days after
deposition. Samples were taken at depths
from 30.5 cm to 122 cm at 15.25-m intervals from discharge to about 45.7 m downstream. The results are presented in
Table 3. The average slurry density
for the 10 samples is 78.0 wt pct (1.03
m^3/mt). The data show that with polymer

45

Table 2. Mean operation data, JSCC.

Day	Slurry feed, m³/min	Slurry, specific gravity	Total slurry, m³	Solids, wt pct	Solids, mt	Polymer, kg/mt solids	Coagulant, kg/mg solids
1....	2.19	1.222	603.4	32.7	240.7	0.58	0.20
2....	2.30	1.202	681.6	26.5	261.91	0.63	0.20
3....	2.25	1.218	909.2	29.4	325.7	0.58	0.20
4....	2.17	1.247	575.9	32.8	235.7	0.59	0.22
5....	2.30	1.233	891.0	31.2	342.9	0.60	0.15

Table 3. Slurry density after 60 days.

Sample No.	Depth, cm	Slurry density, wt pct	Volume, m³/mt	Average slurry, specific gravity
1......	91.4	76.0	1.07	1.224
2......	61.0	85.3	0.95	1.224
3......	30.5	81.7	1.00	1.224
4......	122.0	74.3	1.09	1.224
5......	91.4	75.9	1.07	1.224
6......	61.0	82.9	0.98	1.224
7......	30.5	85.5	0.95	1.224
8......	30.5	77.8	1.04	1.224
9......	91.4	68.9	1.18	1.224
10.....	30.5	72.0	1.13	1.244

treatment a decrease in volume requirements for fine coal waste disposal of a factor of 2.1 and 2.6 is obtainable after 18 h and 60 days, respectively.

2 CONCLUSIONS

The laboratory and field experiments indicate that the addition of polymer in the proper dosage has a dramatic and beneficial effect on the dewatering process. Some coal slurries require pH adjustment to achieve this effect. As the treated slurry was discharged, an immediate separation of water and solids occurred. The flocculated solids readily settled and relatively clear water was liberated. Untreated slurry would take much longer periods of time to settle with more suspended solids remaining in the surface water. Polymer treatment could double the capacity for stored slurry.

It is necessary to add the polymer near discharge, and to thoroughly, mix the polymer into the slurry. Usually, natural flow turbulation will achieve the mixing. Proper dilution of neat polymer is also important. Relatively inexpensive equipment is available for polymer dilution. In both field tests, extreme

fluctuation in slurry specific gravity was encountered. Automatic control of the dilute polymer pump as a functioning of slurry mass flow rate prevented over or under treatment.

To aid water separation, deposition on a slightly sloping area is required. Retention of the water is necessary. However, if a closed system is used, the retention dam would be small because water would continually be pumped back to the preparation plant for coal cleaning and to the polymer system for dilution.

With the JSCC material, two laboratory events were different than field events: (1) The polymer dosage requirement in the field was less than in the laboratory and (2) a cationic coagulant was needed in the field but not in the laboratory. Approximately 6 months lapsed between the time the laboratory sample was collected and the field test was performed. Although mineralogical tests were not conducted, the differences between laboratory and field results are believed to be due to mineralogical changes of the fine coal refuse slurry during this time lapse. For this reason, laboratory investigations to determine polymer type, dosages, and mixing requirements, have to be performed prior to large scale field application, and polymer treatment may also have to be changed if the slurry properties change with time.

The equipment used to complete the field test was relatively inexpensive and capable for optimizing polymer dosage. The polymers added were similar to that required by a belt press or other mechanical dewatering systems. All mines, but especially those having a limited waste disposal area, could benefit from polymer treatment. To complete the disposal procedures, coarse waste could be dumped over the dewatered fines to aid in reclamation and overall safety and maintenance of the waste site. Continuing research will investigate this concept.

3 ACKNOWLEDGEMENTS

The authors wish to thank Jewell Smokeless Coal Corp. and Washington Irrigation and Development Co. for providing sites for field tests and the people of these companies who so graciously helped and cooperated.

REFERENCE

Backer, R.R., and R.A. Busch. 1981. Fine Coal-Refuse Slurry Dewatering. BuMines RI 8581, pp 1-18.

Reclamation, Treatment and Utilization of Coal Mining Wastes, Rainbow (ed.) © 1990 Balkema, Rotterdam. ISBN 90 6191 154 0

Engineering properties of Australian coal mine tailings relevant to their disposal and rehabilitation

D.J.Williams & P.H.Morris
The University of Queensland, St.Lucia, Qld, Australia

ABSTRACT: It has only been with the dramatic escalation of coal mining activity in Australia over the last decade and the increasing community awareness of environmental issues that the disposal of coal mine tailings and the rehabilitation of tailings deposits have been seriously addressed in that country. As a first step in researching this area, it has been necessary to gather data on the engineering properties and behaviour of coal mine tailings. The main properties of interest are strength, compressibility, and permeability. Strength controls the capacity of the tailings deposit to support the loading imposed by a cover of fill. The compressibility of the tailings controls settlement under load, and their permeability controls the rate of settlement under load. These properties are in turn a function of other more fundamental properties of the tailings, including the in situ moisture content and suction level, Atterberg limits, particle size distribution, and specific gravity of the tailings. This paper presents the results of laboratory and field testing to define the engineering properties of particular Australian coal mine tailings relevant to their disposal and rehabilitation.

1 INTRODUCTION

Coal mine tailings constitute the fine grained waste from the washing of run-of-mine coal. This waste material leaves the washery at a solids concentration of only a few percent by weight ([weight of solids]/[weight of solids + weight of water] expressed as a percentage). The tailings are generally piped from the washery to a thickening tank where flocculants are added at a dosage of about 0.05 % by weight. The solids concentration of the thickener underflow is typically 30 to 35 %. The underflow slurry is conventionally piped to a tailings dam for disposal. Discharge rates from the tailings pipe are typically in the range 30 to 50 L.s^{-1}. At the few mines still without a thickener circuit in their washery, the tailings may be mechanically thickened to no more than 10 to 15 % solids before disposal. Where the tailings dam is distant from the washery, it may be necessary to limit the solids concentration of the tailings slurry to between 10 % and 15 % to facilitate pumping.

At mines where available tailings slurry storage capacity is severely limited, it may be necessary to dewater the thickener underflow further by mechanical means. This is done either in solid bowl centrifuges or by band press filters. In both cases, further flocculant (up to 10 times the earlier dosage) is required, and the process (particularly for band press filters) may be sensitive to the acid balance of the tailings. Apart from the expense of the extra flocculant required, the dewtering equipment is expensive to install and maintain. Solid bowl centrifuges and band press filters can increase the solids concentration of the tailings to between 60 % and 65 % and between 65 % and 70 %, respectively. The output from a centrifuge is just transportable by conveyor, while the filter cake from a properly operating band press filter can readily be transported by conveyor and is suitable for mixing with coarse reject for disposal in waste dumps.

Following discharge from a pipeline into a conventional storage, coal tailings undergo hydraulic sorting on a delta with an average slope of about 1 in

100. Within the decant pond beyond the delta the tailings undergo sedimentation, achieving a solids concentration of typically 40 to 50 %. Self-weight consolidation of the deposit eventually produces a solids concentration of around 70 %. A falling water-table within the tailings deposit allows the uppermost tailings to desiccate. The solids concentration of the desiccated crust varies from between 70 % and 75 % at the water-table increasing to about 85 % at the surface (where the tailings become significantly desaturated).

2 PHYSICAL PROCESSES INVOLVED IN TAILINGS DISPOSAL AND REHABILITATION

2.1 Sub-aerial deposition

The discharge of tailings slurry, typically at a solids concentration of about 35 %, from a pipe or launder onto a tailings delta is known as sub-aerial deposition. The slope, particle size, strength and permeability of the deposited tailings decrease with increasing distance from the point of discharge. Williams and Morris (1990) compared two non-dimensionalised profiles suitable for fitting to the resulting field profile. A knowledge of the shape of the profile enables optimum use to be made of the available tailings storage. Blight et al (1985) showed that the non-dimensionalised profile may also be obtained with reasonable reliability from the results of small scale laboratory flume tests. In addition, Williams and Morris obtained a reasonable fit to the hydraulic sorting of particles which takes place down the profile, based on river transport equations. A knowledge of particle sorting is of use in predicting the changing engineering properties of the tailings down the profile, which are strongly related to particle size.

2.2 Sedimentation

Sedimentation or sub-aqueous deposition requires quiescent conditions and hence is limited to the decant pond beyond the tailings delta. At low initial solids concentrations unhindered (Stokesian) sedimentation occurs. At higher initial solids concentrations or as the concentration increases, particles interact giving hindered sedimentation. The resulting sedimentation rates are therefore strongly inversely proportional to the solids concentration of the slurry. A two-fold increase in the initial solids concentration gives rise to about a ten-fold increase in the sedimentation time. There may therefore be little time saved by increasing the concentration of the tailings slurry prior to deposition. However, this costly exercise may be necessary to conserve scarce water reserves.

2.3 Self-weight consolidation

As deposition on the delta or sedimentation within the decant pond continue, the tailings particles come into close proximity and begin to assume the characteristics of soils. Excess pore pressures develop and self-weight consolidation takes place. Critical parameters are the permeability of the tailings sediment (which diminishes as consolidation progresses), and the drainage path lengths. Compared with the process of sedimentation, self-weight consolidation involves relatively small deformations of the tailings at a relatively slow rate.

2.4 Crusting

As ponded water is removed from the surface of the tailings by pumping, breaching of the containment dam or by evaporation, the exposed tailings begin to desiccate and crust. In South-East Queensland, the rate of settlement on crusting may be up to 15 mm per month. Over time, this rate may be halved by the inhibiting effects of rainfall. Desiccation is very effective in reducing the moisture content of the tailings and leads to a surface crust which is much stronger than the underlying tailings due to the high pore water suctions which develop. However, the process occurs to limited depth (depending on the precipitation and drainage conditions) and is substantially reversible on wetting up of the tailings by the deposition of fresh tailings slurry or by ponded rainfall. Its effectiveness as a drying or strengthening process is therefore limited. In addition, with desaturation the permeability of the crusted tailings becomes very low. This inhibits any further consolidation of the underlying soft tailings which tend to remain semi-liquid, unable to support much load.

2.5 Loading

In order to rehabilitate a soft tailings deposit and prevent runoff of contaminated surface water, it is necessary to eventually cover the tailings surface with a layer of fill. The bearing capacity and settlement characteristics of the tailings are therefore important, and will dictate the end use to which the rehabilitated area may be put. Factors influencing the bearing capacity of a typical crusted tailings deposit include its age, the ratio of the average crust strength to that of the underlying uncrusted tailings, and the thickness of the crust. A study of a number of cases has revealed that the best indicator of bearing capacity is the bearing capacity factor N_c applied to the strength of the underlying uncrusted tailings. The onset of bearing capacity failure is marked by a value of N_c in the range 6.0 to 7.1. The N_c value tends to increase as the thickness of the crust increases.

A tailings deposit may also be subjected to loading as a result of raising the embankment by the upstream method on top of deposited tailings. In order to support a reasonable height of fill, it is necessary to take advantage of the strength gain with time resulting from stage construction. Fortunately, the permeability of typical coal tailings allows maximum strength gain to be achieved rapidly (within about 1 month after construction of a given stage). However, there is also the potential for some subsequent softening of the supporting tailings as fresh tailings are deposited against the raised embankment. This effect can be reduced by initially, and periodically thereafter, discharging tailings from the embankment, to deposit coarser particles against the embankment and prevent the excessive ponding of decant water and rainfall runoff against the embankment.

3   RELEVANT ENGINEERING PROPERTIES

3.1 Classification

The specific gravity of coal tailings is far more variable than that of natural soils. This is due to the large difference in the specific gravities of the two solid phase components of coal tailings, namely coal and shaley or clayey mineral matter. Australian coal mine tailings typically have a specific gravity of about 1.75, much lower than that of ordinary mineral matter (about 2.65). This is due to the presence of as much as 60 % fine grained coal (with a specific gravity of about 1.3) in the tailings. Data showing the variation of specific gravity G_s with mean particle size D_{50} for tailings from a number of coal mines is shown on figure 1. The coal mines for which data is included on figure 1 are New Hope and Aberdare Collieries in the West Moreton Coalfield in South-East Queensland, and coal mines in the Hunter Valley Coalfields (de Ambrosis and Seddon, 1986) of New South Wales, Australia. For New Hope Colliery tailings the range of specific gravity values obtained was 1.67 to 1.85, with a mean value of 1.76. For tailings recovered from the main tailings dam and pits at Aberdare Colliery, the range of specific gravity values obtained was 1.56 to 2.48 (with a mean value of about 1.95). The wide range of specific gravity values for the Aberdare tailings is attributable to hydraulic sorting.

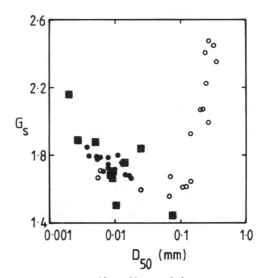

● New Hope data
○ Aberdare data
■ Hunter Valley data

Fig. 1.   Variation of specific gravity G_s with mean particle size D_{50} for different coal mine tailings.

Limited testing of tailings from the Bowen Basin Coalfields in Central Queensland, Australia, revealed specific gravity values in the range 1.57 to 1.72 for Peak Downs Mine and typical values of 1.77, 1.79 and 1.81 for Goonyella, Moura and Riverside Coal Mines, respectively.

Australian coal mine tailings typically have a liquid limit in the range 30 to 45 % and a plasticity index in the range 10 to 25 %, but values for these parameters can vary widely, and some tailings are non-plastic. Atterberg limit data for tailings from New Hope and Aberdare Collieries are shown on figure 2. For New Hope Colliery tailings, measured liquid limits and plasticity index values ranged from 35 to 51 % and 12 to 30 %, respectively. Based on these measured values, the tailings from New Hope Colliery classify (according to the Unified Soil Classification) in the range low plasticity silts ML to high plasticity clays CH, although most samples can be classified as low plasticity clays CL. For Aberdare Colliery tailings, measured liquid limits and plasticity index values ranged from 39 to 43 % and 12 to 18 %, respectively, although some of the samples recovered for testing were found to be non-plastic. The Aberdare Colliery tailings plot close to the Casagrande A-line as low plasticity clays CL or low plasticity silts ML. The different results for the two collieries, located adjacent to each other in the same coal field, can be attributed to the differences in the geological conditions in the seams mined.

Atterberg limit data for Hunter Valley coal mine tailings from de Ambrosis and Seddon (1986) classify this material mainly as low plasticity silt ML, with some CH classification material. For Peak Downs tailings, measured liquid limits ranged from 32 to 38 % (with one sample being non-plastic), and measured plasticity index values ranged from 19 to 23 %. Testing of Moura tailings samples indicated a range from non-plastic material to a material classifying as a high plasticity clay (liquid limit of 54 % and plasticity index of 28 %).

The measured linear shrinkage of New Hope Colliery tailings ranged in value from 5 to 9 % with a mean of 6.6 %. This is consistent with the characterisation of New Hope Colliery tailings as a low plasticity clay.

The particle size distribution of coal tailings depends not only on the seam geology and the mining and beneficiation techniques used, but also on the transport process between the

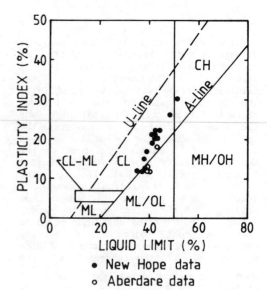

Fig. 2. Atterberg limit data for New Hope and Aberdare coal mine tailings.

washery and the point of final deposition. According to de Ambrosis and Seddon (1986), the coarser parts of tailings particle size distribution curves reflect the design and operation of the specific washery process, while the finer parts of the curves are indicative of inherent material characteristics.

Particle size distribution curves for tailings from a number of Australian coal mines are shown on figure 3. For New Hope Colliery tailings, the clay size fraction (finer than 2 μm) varies from 23 to 40 %, and the sand size fraction (coarser than 2 mm) varies from 6 to 31 %. The remainder comprises silt size particles (2 to 60 μm in size). The range of particle size distributions for Hunter Valley coal mine tailings (de Ambrosis and Seddon, 1986) is similar to that of New Hope Colliery tailings. Aberdare Colliery tailings are finer grained than the other two.

In general, older tailings or tailings produced by old washeries are coarser grained than tailings produced by recently installed or up-graded washeries. Old washeries may not have been capable of separating even fine gravel size coal and this is reflected in the coarse grain size of the waste.

Depending on the sophistication of the coal washery, the maximum particle

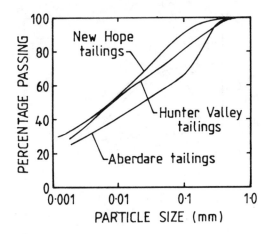

Fig. 3.   Typical   particle   size
distribution   curves   for   New
Hope,   Aberdare   and   Hunter
Valley coal mine tailings.

size of the tailings at different mines
can vary from about 2 mm to as low as
0.06 mm.   Most coal tailings in Australia
have a predominance of silt and clay size
particles   (more than 70 % finer than
0.06 mm).   The finer grained the tailings
and the higher the proportion of clay
particles   they   contain,   the   more
problematic   they   are.   However,   the
intrinsic particle size distribution of
tailings is clouded by the addition of
flocculants   to   the   material   to   aid
sedimentation after deposition.

3.2   Strength properties

The data given in this and the following
section were largely obtained by testing
tailings from New Hope Colliery.   These
tailings closely approximate a purely
frictional   material   with   typical
effective angles of internal friction ϕ'
of   28 °   and   35 °,   for   triaxial
compression and direct shear testing,
respectively.
    A tailings deposit which has been
formed entirely under water, and which
has been allowed to desiccate only when
the storage has reached full capacity,
has a strength profile with depth quite
different to that of a deposit which has
been subject to desiccation during its
formation.   The former case is typical of
deposition into a deep open pit, and is
characterised by a single distinct crust
at   the   surface,   with   near   normally
consolidated   soft   tailings   below   the

water-table.   The latter case is typical
of deposition into a large, relatively
shallow, man-made storage with a gently
sloping base.   At any given time as the
storage   is   being   filled,   a   large
proportion of the surface of the tailings
deposit will be exposed, affording the
opportunity for desiccation prior to
inundation by further fresh tailings
slurry.   As a result a series of remnant
crusts   forms   over   the   depth   of   the
tailings deposit.   Vane shear testing is
appropriate for determining the shear
strength of coal mine tailings in situ,
particularly in the tailings underlying
the crust which are too soft to sample
undisturbed.   Typical vane shear strength
profiles with depth for the two cases
described are given on figure 4.   The
example of the former case is from New
Hope Colliery and the example of the
latter case is from Peak Downs Mine.

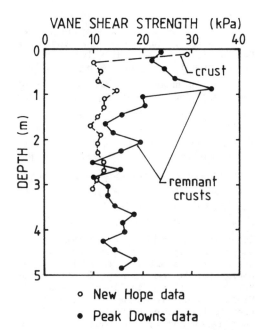

o New Hope data

• Peak Downs data

Fig. 4.   Typical   vane   shear   strength
profiles with depth, without
(New Hope Colliery) and with
(Peak   Downs   Mine)   remnant
crusts.

    In direct shear tests, the friction
angle for reconstituted tailings has been
found   to   decrease   with   increasing
specific gravity of the tailings.   Also,

53

the value of ϕ' on vertical in situ planes in undisturbed tailings has been found to be significantly higher than that on horizontal in situ planes. This has been attributed to a generally horizontal orientation of the long axes of the tailings particles following hydraulic deposition.

3.3 Consolidation properties

In general neither chemical cementation nor weathering effects are significant in coal tailings. Consequently, tailings behaviour during consolidation and desiccation can be interpreted using conventional soil mechanics. Laboratory consolidation testing of coal tailings should be carried out on specimens from undisturbed block samples, to minimise the effects of sampling disturbance.

Typical plots of void ratio e versus the natural logarithm of vertical effective stress $\ln(\sigma_v')$, from laboratory oedometer tests on block samples of New Hope Colliery tailings, are shown on figure 5. A typical value for the compression index C_c of New Hope Colliery tailings is about 0.2, with a value of about 0.02 for the swelling index C_s. An increased proportion of coal in the tailings reduces the C_c. The laboratory determined coefficient of vertical consolidation c_v (rate parameter) for New Hope Colliery tailings is typically in the range 15 to 60 $m^2.yr^{-1}$, and the coefficient of permeability for vertical flow k_v is typically about 3.0E-9 $m.s^{-1}$.

Based on the results of oedometer tests on both vertically and horizontally orientated samples, the ratio of horizontal to vertical permeability for New Hope Colliery tailings is about 5. Field values for the coefficient of premeability were determined by falling head tests and by the back-analysis of trial embankments on tailings deposits. The laboratory determined permeabilities were lower than the field determined values by a factor of about 30. The relationship between the coefficient of permeability k and the void ratio e of coal tailings can be represented by a power function of the form

$$k = a.e^{-b} \qquad (1)$$

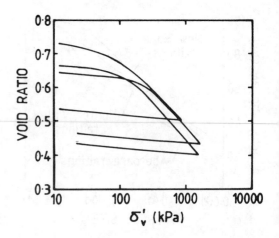

Fig. 5. Typical laboratory oedometer test results for New Hope Colliery tailings.

where a and b are empirical constants determined from the results of oedometer tests on particular tailings, adjusted to represent field conditions.

Long term creep of New Hope Colliery tailings is negligible. This is due to kaolinite being the dominant clay mineral present in these tailings, as well as the creep-resistant aromatic ring structures present in coal.

3.4 Pore water suctions

Matrix and total suctions in desiccated New Hope Colliery tailings were measured using the filter paper method. Osmotic suctions were found to be negligible. This was attributed to the leaching of salts by rainfall infiltration and is not necessarily typical of coal tailings in general. The measured suctions were similar to those occurring in ordinary silty clays, rising to as high as 10 000 kPa for extremely desiccated tailings. At high suctions the tailings are significantly desaturated, so that suction and effective stress cannot be equated.

4 ALTERNATIVE DISPOSAL AND REHABILITATION METHODS

The engineering properties and behaviour of coal mine tailings outlined above allow discussion of alternative disposal and rehabilitation methods.

4.1 Disposal in spoil piles and waste dumps

Traditionally, the ease of handling of coarse waste and the difficulties in handling fine waste has led to the two materials being separated and being disposed of separately. However, if the two materials can be mixed to create a loose packing of coarse waste in a matrix of tailings, a mixture with reasonable engineering properties may result. This could be achieved by placing tailings slurry in the valleys between spoil piles or waste dumps, and then pushing coarse waste into the tailings before significant crusting had occurred, so that effective mixing could take place. By this means, the costs associated with tailings disposal in conventional dams and the eventual rehabilitation of the tailings deposit could be largely avoided. The process of pushing coarse waste into the tailings would constitute the first stage of the necessary re-shaping of the spoil piles and waste dumps. Potential problems with this approach are the loss of return water from the tailings slurry and the possibility that introducing tailings slurry to spoil piles and waste dumps may significantly impair their stability.

4.2 Underground disposal

Old underground workings are potential storages for tailings and other waste materials. Where underground disposal of tailings has been used, little attempt has been made to gain benefits other than the containment of unwanted fine waste. In some cases, the water introduced with the tailings has created serious problems for adjacent active underground workings. However, if old underground workings could be filled by a material with some intrinsic strength, several benefits could be gained. Surface subsidence could be reduced and the waste could provide some lateral support to underground pillars.

A paste comprising tailings slurry, a cementing agent and possibly coarse waste could be suitable. Cole and Figg (1987) specify that the initial strength of the paste must be less than 3 kPa for it to be pumpable and for it to spread sufficiently underground. The paste will undergo about 20 % self-weight consolidation in place. Without cementitious additives the strength of the paste will increase to about 5 to 8 kPa after several years. Ideally, cementitious additives should allow a 30-day delay before ultimately raising the strength of the paste to greater than 20 kPa within 1 year. At 3 kPa strength, a 5 m high deposit of paste would spread about 35 m on level ground, and an 8 m high deposit would spread about 80 m. On ground sloping at greater than 4 °, 3 kPa strength paste would spread until it met an obstacle.

4.3 Potential for re-processing

The potentially extractable product contained within tailings or their potential use as a low grade fuel should be taken into account when disposal methods are considered. The requirements of recovery for re-processing clearly exclude some of the alternative disposal strategies discussed above.

4.4 Optimising crusting

The maximum benefit due to sun-drying can be achieved only if the tailings slurry is deposited in relatively thin layers (0.5 to 1.0 m thick). It must then be left for a period of time to crust. The time rquired depends on the seasonal evaporation/rainfall balance, and may range from 2 to 9 months under Australian conditions. To minimize wetting up of the tailings by rainfall during crusting, deposition should be on a sloping bed to ensure rapid runoff. After crusting, the tailings should be harvested, or the surface rendered impermeable to prevent re-wetting of the crust when fresh tailings are placed. It may be possible to produce an impermeable surface by compaction and/or by adding a cementing agent. In view of the limitations imposed on layer thickness and turn-around time, sun-drying requires vast areas of land and this makes it uneconomic in many cases.

4.5 Optimising Consolidation

The bearing capacity of a tailings deposit is primarily a function of the strength of the soft layer underlying the thin surface crust. Since the strength of the underlying layer is dependent on self-weight consolidation, there is merit in optimising this process. This may be possible by enhancing drainage, and in particular by decreasing the drainage path lengths. However, in field trials conducted in the Hunter Valley Coalfields (de Ambrosis and Seddon,1986), this did not prove very effective. The apparent reason for this is that a filter cake quickly forms in the tailings immediately adjacent to a drainage layer. The permeability of the filter cake rapidly diminishes and this limits further drainage.

5   CONCLUSIONS

The origin of coal mine tailings as a waste product of the processing of run-of-mine coal has been outlined. The physical processes involved in the disposal of tailings slurry and the rehabilitation of tailings deposits have been discussed. Typical values for some of the engineering properties of Australian coal mine tailings relevant to their disposal and rehabilitation have been presented. The most important of these are the strength, consolidation, and permeability characteristics of the material. Other more fundamental properties, including the in situ moisture content and pore water suction, Atterberg limits, particle size distribution, and specific gravity of the tailings, which are indicative of the engineering behaviour of the material, have also been discussed. An improved understanding of the nature of coal mine tailings has allowed rational discussion of alternative methods of disposal and rehabilitation, with the aim of optimizing the use of available disposal areas and facilitating future land use.

6   ACKNOWLEDGEMENTS

Assistance provided by Messrs Peter McMillan and Bruce Cooper of The University of Queensland in performing the laboratory and field testing reported in this paper is gratefully acknowledged. The assistance of the management of New Hope and Aberdare Collieries in providing access to tailings deposits for the purposes of sampling and field testing, is also acknowledged.

7   REFERENCES

Blight, G.E., Thomson, R.R. and Vorster, K. 1985. Profiles of hydraulic-fill tailings beaches, and seepage through hydraulically sorted tailings. Jnl Sth African IMM, May 1985, pp 157-161.

Cole, K.W. and Figg, J. 1987. Improved rock paste. A slow hardening bulk fill based on colliery spoil, pulverized fuel ash and lime. Proc 2nd Int Conf on Reclamation, Treatment and Utilization of Coal Mining Wastes, Nottingham, England, September 1987, pp 415-430. Ed A K M Rainbow. Elsevier.

de Ambrosis, L. and Seddon, K. 1986 Disposal of coal washery waste. Mine Tailings Disposal Workshop Notes, Brisbane, August 1986, pp 4.1-4.80. Brisbane: The University of Queensland.

Williams, D.J. and Morris, P.H. 1990. Comparison of two models for sub-aerial deposition of mine tailings slurry. Trans Inst of Mining and Metallurgy A: Mining Industry, Vol 98, A73-A77.

Reclamation, Treatment and Utilization of Coal Mining Wastes, Rainbow (ed.) © 1990 Balkema, Rotterdam. ISBN 90 6191 154 0

Prediction of desiccation rates of mine tailings

Gareth Swarbrick & Robin Fell

School of Civil Engineering, University of New South Wales, Kensington, N.S.W., Australia

ABSTRACT: Using laboratory and field drying experiments on tailings from five mines, a semi empirical method has been devised to predict the rate of desiccation of mine tailings by the sun. It is shown that after settling, there is a period of constant rate of drying at the evaporation potential (equal to about 85-90% of pan evaporation), followed by a period when cumulative evaporation varies according to the $\sqrt{time}$ ie. the basic sorptivity equation is followed. The sorptivity coefficient is shown to be dependent on the type of tailings, the water content at the end of the first stage of drying, and to a lesser extent on the potential evaporation rate.

1 INTRODUCTION

The properties of mine tailings, particularly those with a large content of clay size particles, can be significantly improved by desiccation, provided the cycle of deposition and drying is designed to allow sufficient time for drying. This in turn is related to the thickness of each layer of tailings, the properties of the tailings, and the climate ie. evaporation and rainfall.

A research project carried out in the School of Civil Engineering at UNSW, has investigated methods of predicting the rate of desiccation, so that the cycle of deposition and drying can be modelled. This will allow prediction of placed density and water content, and through these, compressibility and shear strength.

This paper describes laboratory drying experiments, the method developed to predict the rate of desiccation from the laboratory experiments, and monitoring of field drying behaviour.

2 THE TAILINGS USED IN THE STUDY

The tailings used in the study were obtained from the five mines sponsoring the research.
The tailings all have a high clay size (passing 2 micron) content. The "as-discharged" hydrometer particle size densifications (with dispersant) are shown in Figure 1.

MINE	TYPE OF TAILINGS	COMPANY
Weipa	Bauxite washery	Comalco
Riverside	Coal washery	TDM Riverside
Mt Newman	Iron ore beneficiation	Mt Newman Mining
Tom Price	Iron ore beneficiation	Hamersley Iron
Wambo	Coal washery	Wambo Mining

The tailings soil mechanics properties and mineralogy are given in Tables 1 and 2.

All laboratory testing was carried out on the as supplied tailings. These are considered representative of the bulk of the tailings in the field storages, since the fines predominate.

3 LABORATORY AND FIELD DRYING EXPERIMENTS

3.1 Laboratory equipment

The approach is to use lysimeters (soil containers exposed to evaporation that are periodically weighted to determine loss of water due to evaporation) under a simulated sunlight environment. For the purposes of this paper the lysimeters are referred to as drying boxes.

The equipment developed for the modelling process includes:

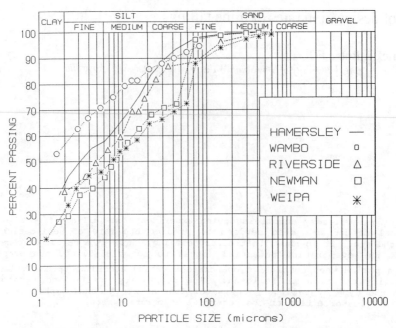

Figure 1. Particle size distribution of tailings

Table 1. Soil mechanics classification of tailings as discharged

Tailings	Water Content %	% Solids	Soil Particle Density t/m³	Atterberg Limits - %		
				Liquid Limit	Plastic Limit	Plasticity Index
Weipa	362	22	2.85	43	26	17
Wambo	411	20	1.86	74	28	46
Riverside	250	29	1.74	44	28	16
Newman	192	34	3.70	33	22	11
Hamersley	169	37	3.50	30	21	9

Table 2. Mineralogy of Tailings

Weipa	Gibbsite(45%), Boehmite(18%), Kaolin(12%), Quartz(17%), Hematite(5%)
Wambo	Na and Ca Montmorillonite, Kaolin, Coal, Quartz
Riverside	Illite, Ca Montmorillonite, Kaolin , Quartz
Mt Newman	Hematite(40%), Kaolinitic shale(50%), Quartz(6%)
Hamersley	Hematite(34%), Kaolinitic shales(52%), Limonite(8%), Goethite(19%)

i)   Drying box. This is 500mm square and designed to adapt to the variation of the tailings depth during drying. The sides of the box are able to be lowered so that a level profile can be maintained and a known constant water evaporation potential can be imposed upon the system. A maximum initial depth of tailings of 500mm can be accommodated. The adjustable sides of the apparatus meant that a suitable method of waterproofing had to be devised to hold the initially saturated material. The box also accommodates a rigid frame to support measurement equipment such as thermometers, tensiometers and settlement gauges.

ii) Support frame. To enable suspension of drying lights, and shielding of tensiometers from intense light rays, a lightweight steel frame was erected to enclose the boxes and provide support for anciliary equipment.

iii) Support columns are used to keep the box raised above the ground to enable access to the underside of the box for weighing. A large balance scale is then able to be wheeled under the box to facilitate the measurement of the mass of the box and its contents.

iv) Light frame. The set of simulation lamps are suspended from the surrounding frame using a chain and adjustable hooks. This allows the frame to be suspended above the box at any desired height.

Figure 2 shows the layout.

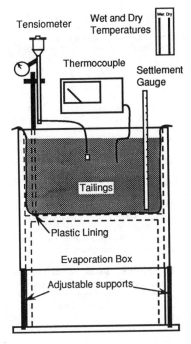

Figure 2   Drying box and anciliary equipment

A series of high intensity 24-volt, 200 watt Quartz Halogen projector lamps in two configurations in combination with two fluorescent ultra violet tubes have been used to simulate the radiation from the sun intercepted by the earth. These sets of lights are used in conjunction with timing mechanisms to enable control of their duration and to record time of filament failure.

Suction or negative pore pressures are measured down the depth of the sample during the experiment. The non-destructive mode of measurement, in theory, means continuous readings. During the experiment, however, the tensiometers are only read once a day as they require recharge of their water which must be followed by one day for equilibration. The tensiometers also tend to exceed measurement range after about one and a half weeks of continuous drying.

Psychrometers were occasionally used to test their performance in tailings and as a check for the tensiometers.

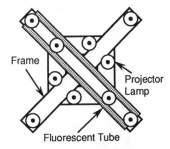

Figure 3   Simulation lights

Filter papers are used to determine the suction profile at depth of extruded core samples which were taken using split 50mm dia thin wall tubes.

3.2 Results

Figure 4 shows plot of cumulative evaporation vs time for tailings from Hamersley and Weipa. These were typical of the experimental results. The plots show cumulative evaporation after the settling phase has been completed, and surface water drained and evaporated from the surface. Drying is either continuous (solid line) or continuous with shutdown periods due to weekends or light failure.

It will be seen that there is a period of constant evaporation rate followed by a decreasing rate. This decreasing rate has been shown to satisfy the sorptivity equation ie.

$$E_{cum} = b \sqrt{t}$$

where E_{cum} = cumulative evaporation after the linear stage – mm

Hamersley : Cumulative Evaporation

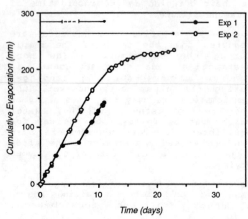

Weipa : Cumulative Evaporation

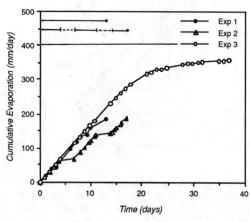

Figure 4 Results of laboratory drying experiments

t = time after the linear stage - days
b = sorptivity coefficient mm day$^{-1/2}$

The laboratory experiments showed that:

• the stage 1 evaporation rate was usually 85-90% of the rate from a free water surface. Rates up to 100% were recorded (for Riverside)
• stage 1 was of the order of 7-15 days, except for Wambo which showed a rather different two phase stage and which has yet to be explained
• sorptivity coefficients for the most reliable tests were
    Hamersley — 21 mm day$^{-1/2}$
    Newman — 31 mm day$^{-1/2}$
    Riverside — 24 mm day$^{-1/2}$
    Wambo — 38 mm day$^{-1/2}$
    Weipa — 25 mm day$^{-1/2}$

However the values are dependent on selection of the time of the end of stage 1. Analysis of the results is complicated by intermittent drying, and the short duration of early tests, but there are indications that the sorptivity coefficient is not a material constant, but is related to the average water content at the end of stage 1 (which varies with thickness of settled tailings at the end of stage 1), and to a lesser extent the potential evaporation rate. The form of equation relating these factors is

$$b = k_2 \, e_p{}^{\alpha_2} \, w_1{}^{\beta_2}$$

where e_p = potential evaporation rate
w = water content at the end of stage 1
α_2 = coefficient ≈ 0.25
β_2 = coefficient $\approx 0.5\text{-}1.25$

Insufficient tests have been carried out to closely define the coefficients at this time and further tests are under way. Tests are also being carried out to assess the effect of intermittent drying on a daily cycle ie. to simulate night and daytime drying conditions. These latter tests have been used to determine α_2.

3.3 Field drying experiments

Field drying experiments have been carried out in 2.5m x 2.5m ponds and at some locations in 50m x 2.5m ponds to assess field behaviour. There is insufficient space to detail the tests and results here, but they do indicate that

— the two stage drying observed in the laboratory applies in the field
— the stage 1 rate of evaporation is similar to the laboratory value (as a proportion of pan evaporation
— the sorptivity approach is applicable to the stage 2 behaviour.

The field tests were complicated by rainfall and leakage from liners which caused differing drainage conditions, and intermittent observations by mine personnel. A more carefully controlled trial has recently been completed at Weipa which will give more detailed evidence of field trends compared to laboratory tests.

4 COLUMN SETTLING EXPERIMENTS

An extensive series of settling tests have been carried out on the as supplied tailings, and using flocculants and dispersants. These are described in Fell (1988). These give settling curves for 300mm high columns shown in Figure 5.

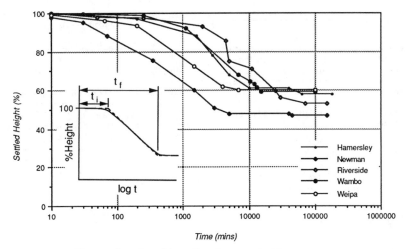

Figure 5. Results of column settling tests

5 METHOD FOR PREDICTION OF DESICCATION

The method is based upon the laboratory and field data and takes into account most aspects of the drying cycle including settlement and desiccation. The effects of daily cyclic evaporation potential has been observed for both laboratory and field situations and is allowed for in the model. Drainage and rainfall have not been modelled directly but are discussed below.

The method may be summarised in the basic steps shown below.

a) Settling phase.

The initial settlement of a given layer of tailings follows a log-linear pattern. The rate of logarithmic decrease in height is assumed to be constant and tending towards some finite layer height H_f at some finite end time t_f. The end layer height and end time have both been shown to be functions of the initial thickness or height of tailings H_i. There exists a point where the rate of evaporation exceeds the rate at which water is released to the surface due to settlement. This point is determined by differentiating the log settlement curve and equating to the available evaporation potential, and gives the settlement time t_s at which a settled height H_s and settled water content w_s may be determined.

b) Stage 1 drying phase

The tailings now undergo drying at a constant rate — stage 1 drying. The water content, w_1, at which stage 1 constant evaporation is deemed to cease can be predicted from the settled water

content and evaporation rate. Three empirical parameters are used to determine w_1, all of which are material dependent. From w_1, the time taken, t_1, to reach this state is determined.

c) Stage 2 drying phase

After the water content falls below w_1, a sorptivity approach is adopted. Here the cumulative evaporation is assumed to be proportional to the square root of time. The constant of proportionality, b, has been shown to be a function of the evaporation rate and the water content at end of stage 1 drying, w_1. Given a time point beyond it, the water content may be predicted using the cumulative evaporation determined by b and the time since stage 2 be~ .

A summary of the method is given below:

Step 1 Calculate settling parameters from 300mm column tests

From the plot of column height vs log time (as in Figure 5) determine

t_i = initial point in time from which settlement nominally begins (days)

t_f = time at which settlement on a linear height vs log time plot ceases (days

k_{t300} = time constant for 300mm column
= $t_f H_i$
= $t_f/300$ for 300mm column

Step 2 Adjust these parameters to take account of actual initial layer thickness using

61

$$k_t = k_{t300} + \frac{(H_i - 300)\ 0.54k_{t300}}{1900 - 300}$$

(This relationship has been calculated from 300mm and 1900mm columns, and could be calculated more generally by doing column tests on the tailings in question at convenient initial heights).

Step 3 Calculate settlement coefficients from column tests using

$$\frac{H_f - H_i}{H_i} = k_h\ H_i^\phi$$

where H_f = settled height of tailings at time t_f

k_h and ϕ settlement coefficients determined from column tests at two different heights (eg. 300mm and 1900mm).

Step 4 Calculate time for settlement - t_s

$$t_s = \frac{k_h\ H_i^{1+\phi}}{e_p\ \log\left(\frac{k_t H_i}{t_i}\right)\ \ln 10}$$

and height after settlement — H_s from:

$$H_s = H_i\left[1 - \frac{k_h H_i^\phi}{\left(\log\frac{k_t H_i}{t_i}\right)}\log\frac{t_s}{t_i}\right]$$

and mass of soil per unit area — M_s using:

$$M_s = H_i\left(\frac{G_s}{1 + G_s w_i}\right)$$

and water content after settlement — w_s:

$$w_s = w_i - \frac{H_i - H_s}{M_s}$$

Tables A1 and A2 give values for the tailings used in the study.

Step 5 Calculate the water content at the end of stage 1 — w_1

$$w_1 = \frac{e_p^{\alpha 1}\ w_s^{\beta 1}}{k_1}$$

using α_1. β_1 and k_1 from Table A3.

In practice these parameters have to be determined from drying tests at different settled heights (and hence different w_s) and evaporation rates.

Step 6 Calculate time to complete stage 1 — t_1

$$t_1 = t_s + \frac{M_s}{e_p}(w_s - w_1)$$

Step 7 Calculate the sorptivity coefficient — b

$$b = k_2\ e_p^{\alpha 2}\ w_1^{\beta 2}$$

using α_2. β_2 and k_2 from Table A4.

Again in practice these parameters have to be determined from drying tests.

Step 8 Determine the air dried maximum dry density ρ_{dmax} by drying tailings in the laboratory, or from dried samples in the field, and the equivalent saturated water content w_f from:

$$w_f = \frac{(G_s\rho_w - \rho_{dmax})}{G_s(\rho_{dmax} - \rho_w) + \rho_{dmax}}$$

Values for the tailings studied and given in Table A5.

Step 9 Depending upon the required end water content condition — w_e, calculate the required parameters eg. time — t_e, final height — H_e, saturation — S, and dry density — ρ_d.

a) To calculate t_e

If $w_s \leq w_e > w_1$ then $t_e = t_s + \frac{M_s}{e_p}(w_s - w_e)$

If $w_e \leq w_1$ then $t_e = t_1 + \left[\frac{M_s}{b}(w_1 - w_e)\right]^2$

b) To calculate S

If $w_f \leq w_e$ then $S = 1$

if $w_e < w_f$ then $S = \dfrac{\left(1 + \dfrac{1}{w_f}\right)}{\left(1 + \dfrac{1}{w_e}\right)}$

This assumes a linear relationship between degree of saturation and volumetric water content, which has been shown to be reasonable from the laboratory test data.

c) Calculate dry density of tailings and final height from

$$\rho_d = \frac{SG_s\rho_w}{(S + G_s w_e)}$$

$$H_e = \frac{M_s}{\rho_d}$$

It should be noted that

• the method assumes constant potential evaporation in any stage. In practice calculations will have to be done on a

Table A1. Settlement Parameters

Mine	Hamersley		Newman		Riverside		Wambo		Weipa	
H_i (mm)	300	1900	300	1900	300	1900	300	1900	300	1900
t_i (days)	0.12	0.52	0.01	0.08	0.19	1.05	0.35	1.63	0.07	0.16
H_f (mm)	186	1026	141	836	171	988	183	988	180	874
t_f (days)	8	80	3.7	35	19	180	16	180	6	69
$\dfrac{\Delta H}{H_i}$	0.38	0.46	0.53	0.56	0.43	0.48	0.39	0.48	0.40	0.54

where $\Delta H = H_i - H_f$.

Table A2. Relational Constants

Mine	Hamersley	Newman	Riverside	Wambo	Weipa
t_i (days)	0.3	0.04	0.6	1.0	0.1
k_{t300}	0.027	0.012	0.063	0.053	0.02
k_h	0.211	0.45	0.31	0.21	0.16
ϕ	0.10	0.03	0.06	0.11	0.16

Table A3. Parameters for estimating w_1

Mine	Hamersley	Newman	Riverside	Wambo	Weipa
α_1	0.75	0.25	0.75	0	0.75
β_1	1.25	0.5	0.5	0.5	0.75
k_1	24	9	22	1	27

Table A4. Parameters for estimating b

Mine	Hamersley	Newman	Riverside	Wambo	Weipa
α_2	0	0.25	0.25	0.25	0.25
β_2	0.75	0.5	1.25	0.5	1
k_2	29	13	18	12	27

Table A5

Mine	Hamersley	Newman	Riverside	Wambo	Weipa
G_s	3.76	3.84	1.74	1.86	2.77
ρ_{dmax}	1.86	1.97	1.00	1.21	1.49
w_f	0.373	0.328	0.74	0.406	0.450

daily basis (or with lesser accuracy) on a weekly, or monthly basis) to allow for seasonal variations

• the coefficients k_h, ϕ and k_{t300} are based upon a particular initial water content w_i. For different initial water contents w_{new} it is suggested that new column tests be undertaken to determine the new coefficients k_h, ϕ and k_{t300}. Alternatively, an equivalent initial height H_{eq} be calculated from

$$H_{eq} = H_i \left(\frac{1 + G_s w_i}{1 + G_s w_{new}} \right)$$

The method then continues using the original initial water content w_i and the new equivalent height h_{eq}. This second method, however, will produce a small error when calculating t_s.

• the method as described does not include the effect of rainfall. Since re-wetting of tailings will not involve significant swell (re-wetting leads to a reduction in the suction pressure and is comparable to rebound on unloading in a voids ratio — log p consolidation test plot) it is adequate to assume that the dry density is not affected, and that drying after rainfall will initially require drying to the water content before the rain, before further desiccation takes place. This can be readily incorporated into a daily model

• in practice tailings are not placed instantaneously as assumed by steps 1 to 4. Provided the deposition cycle time is similar or longer than t_s it is reasonable to assume that settlement occurs concurrent with deposition. In practice for the tailings in the study, it appears as if settlement does finish within a day or two of deposition being finished.

• as shown in Figure 6, the model gives good modelling of the laboratory experiments for the Hamersley and Weipa tailings. The method was similarly able to model behaviour of Newman and Riverside tailings, but was unable to predict the behaviour of Wambo tailings. Further tests are under way to improve the parameter definition for Wambo tailings. Detailed comparison with field tests has yet to be completed.

6 SOME USES OF THE MODEL

The method has been used to estimate the time for desiccation of different layer thicknesses, to either a required strength (previously related to water content by laboratory and/or field tests) or to the plastic limit of the tailings. The results are given in Table 3.

It is of interest to note that

• the time to desiccate four layers of 500mm initial thickness is similar to that for a single 2000mm layer
• the times for desiccation are mostly dependent on the settled water content and the potential evaporation rate

The model can also be used to predict the area required to achieve desiccation to say the plastic limit given the tailings properties, climate and tailings production rate.

Predicted vs Observed : Hamersley Expt 3 — Predicted vs Observed : Weipa Expt 3

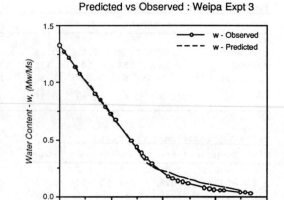

Figure 6   Predicted vs observed behaviour

Table 3. Time taken to dry for a given layer thickness.

Mine	e_p* mm/day	H_i* mm	w_i* (w_s) %	Strength – 15 kPa w %	Strength – 15 kPa t days	Strength – 15 kPa ρ_d g/mm³	Plastic Limit w %	Plastic Limit t days	Plastic Limit ρ_d g/mm³
Hamersley	8.0	2000	194	32	117	1.60	21	132	1.68
G_s = 3.76		1000	(95)		61	1.60		68	1.68
		500			32	1.60		35	1.68
Newman	11.4	2000	305	33	74	1.70	22	80	1.78
G_s = 3.84		1000	(75)		37	1.70		40	1.78
		500			19	1.70		20	1.78
Riverside	5.4	2000	244	36	165	0.87	28	175	0.89
G_s = 1.74		1000	(120)		86	0.87		91	0.89
		500			45	0.87		48	0.89
Wambo	3.9	2000	322	45	715‡	1.05	28	955‡	1.10
G_s = 1.86		1000	(225)		215‡	1.05		276‡	1.10
		500			73‡	1.05		89‡	1.10
Weipa	5.3	2000	410	35	165	1.28	26	173	1.33
G_s = 2.77		1000	(200)		90	1.28		94	1.33
		500			49	1.28		51	1.33

*e_p, H_i and w_i denote the average yearly evaporation rate and initial height and water content respectively. t and ρ_d denote the time taken to reach the specified water content, w, and the dry density at that water content.
‡These values reflect the deviation of Wambo tailings from the model.

7 CONCLUSIONS

The laboratory experiments show that settlement and desiccation of tailings can be predicted from column settling tests and by a two stage drying model. The model is in broad agreement with field drying behaviour but more work is under way to confirm its applicability in the field situation.

In addition to this semi empirical approach, a more detailed analysis using finite difference modelling is being undertaken.

8 ACKNOWLEDGEMENTS

The considerable assistance of the mining companies listed in Table 1 in supporting this research project, and the assistance of the Australian Mineral Industries Research Association in coordinating funding is acknowledged.

Staff from the School of Civil Engineering including Lindsay O'Keeffe, Tony Macken, Frank Mizzi and Russ Hogg have played a major part by undertaking the laboratory testing and assisting in the field program.

REFERENCES

Fell, R. (1988). Mine tailings, dispersants and flocculants. Proc. ASCE Geotechnical Specialty Conf., Fill Structures, Fort Collins, 711-719.

Reclamation, Treatment and Utilization of Coal Mining Wastes, Rainbow (ed.) © 1990 Balkema, Rotterdam. ISBN 90 6191 154 0

Environmental effects of the utilization of coal mining wastes

W. Sleeman
Minestone Services, British Coal Corporation, Hebburn, UK

Coal mining wastes in the U.K. pose a major environmental problem to the industry and the community. Whilst considerable efforts have been made to reduce the problem through reclamation schemes sponsored by Local and Central Government large volumes of material remain derelict on the surface. In addition to these in-situ deposits annual production of wastes from current mining operations is at the rate of 50-60 million tonnes. Whilst the reduction in both current production and material from the historical stockpiles is of benefit there is concern over the possible adverse environmental effects of utilizing these materials. Disturbance to the community from the operations and traffic generated must be considered together with the technical aspects of the suitability of the material for its proposed application. This paper examines both the positive and negative environmental effects of the utilization of these materials as applied by British Coal.

1 INTRODUCTION

Coal mining in the United Kingdom is an old established industry which because of past attitudes has left an inheritance of dereliction throughout the coalfield areas. In the past little regard, by present standards, was given to environmental considerations. An example of the problems which were created by the achievements of the industrial revolution in the nineteenth century and left unsolved is that of colliery spoil heaps. Coal mining being an extractive industry, each mine obviously has a finite life as the coal reserves become exhausted or uneconomic to remove. As collieries close the amount of dereliction increases as there is often no requirement to carry out restoration or environmental improvement to the sites. The major dereliction is the unsightly spoil heaps left, usually adjacent to the mine. Although much has been done by government sponsored activities, many areas remain to be improved. The technical means now at our command enable us to eliminate the problem or at any rate prevent it from spreading. Over the years much experience has been gained in the building of tips and exacting technical specifications are now observed from both safety and environmental viewpoints. The coalfields in the United Kingdom, as identified on Fig. 1, are in locations of relatively high urban develop-

ment with communities having formed around the mine and supplemented by workers from other heavy industries which established themselves to utilize the coal from the mines. With the recent acceleration in the rate of colliery closures the amount of derelict land to be released from operational use by British Coal is likely to increase substantially. In the past twenty years the vast majority of colliery sites and spoil heaps reclaimed have been for low level agriculture or leisure use. Although this use improves the general environment more beneficial use can be gained by using the sites for development. In addition the utilization of colliery spoil from the tips in new developments, or to improve the land form and development of other derelict sites, can have a dual benefit. By using the spoil in developments it not only improves the environment, it also reduces the demand on natural aggregates and thereby the creation of further potential environmental disruption.

1.1 Nature of the material and its origin

Colliery spoil is a by-product of mining coal and of its preparation for markets. It is derived from the rocks - mainly siltstones and mudstones with seat-earths and sometimes sandstones, limestones and other rock types - lying above below and

Fig.1. United Kingdom Coalfields

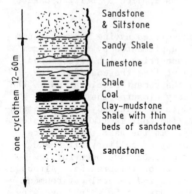

Fig.2. Coal Measures Rocks

sometimes within, the coal seams in the coal measures Fig. 2. During mining operations quantities of these rocks unavoidably extracted with the coal, or in driving the tunnels (roads) which give access to the coal faces, are brought to the surface with the coal. It is then generally necessary to remove some or all of this material to yeild a coal product of the required quality. This separation is carried out with other operations such as crushing, sizing and de-watering, in the coal preparation plants, generally by making use of the higher density of the colliery spoil, roughly between 1½ and 2

times that of the coal. There is usually no immediate use for the separated non-coal material, so it is tipped onto a spoil heap.

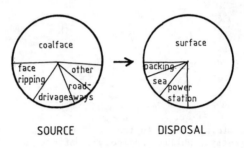

SOURCE   DISPOSAL

Fig.3. Source and Disposal of Colliery Wastes

Until about twenty years ago, and more particularly before the post-World War II advances in mining technology, nearly all colliery spoil heaps were constructed by tipping the spoil loose from transporting containers - aerial flights, skips, wagons or trams. The resulting spoil heaps were nearly always loose and open textured so that air could readily penetrate. If for any reason the residual coal in parts of the spoil heaps became ignited, air could readily reach the burning section and sustain and extend the heating. Before World War II, nearly all collieries were steam powered and had to dispose of hot boiler ashes. There was then no well established market for this clinker as lightweight aggregate, so it was often tipped on the spoil heaps, which were ignited and eventually burned through. The residues of calcined rocks left in these spoil heaps are burnt colliery spoils. In 1955, 261 heaps were found to be burning in England and Scotland. The vast majority of these heaps have since been restored, burnt out or operated for mineral recovery. There are currently very few tips burning, and as British Coal is required to prevent any new tip from being ignited the number should shortly be zero. Colliery discards are largely disposed of in tips, with fine discard (tailings and slurry), being mainly discharged into lagoons. Colliery spoil is currently being produced at an annual rate of 50-60 million tonnes. In addition there is approximately 3000 million tonnes of colliery spoil on the surface, in operational tips, disused and closed tips or at restored sites.

2 UTILIZATION OF COLLIERY SPOIL

Colliery spoil suitable for use can be classified into three basic categories:

1. Unburnt colliery spoil or MINESTONE
2. Burnt colliery spoil or BURNT MINESTONE
3. Modified/Processed colliery spoil

In addition to materials from within these categories, the fines discard produced from coal preparation plants and normally in the form of slurry or tailings are a major disposal problem. The majority of these fine discards are disposed of at high cost and inconvenience to the collieries. In general they are classified as materials unsuitable for any use and are disregarded when assessing sources of supply for markets, however research into potential applications is continuing. Applications and potential uses of materials from each of the above categories are discussed below.

2.1 Uses of burnt minestone

Burnt minestone is acceptable material for use as common fill. However, it is now seen to be somewhat extravagant to use large quantities for this purpose. Since unburnt minestone is at least equally acceptable for use as fill in most applications and is plentiful, widespread and more economic, in large applications. Stocks of burnt material are conserved for other uses in which they are technically more suitable than minestone or are less costly than alternative materials. These uses include: 'special fill material' - for example for capping layer use immediately below formation level in road works aggregate for temporary haul roads, car parks hardstandings and; as alternative to hardcore.

2.2 Uses of minestone

The major use of minestone, because very large quantities are often required has been as imported fill for which minestone has largely taken over from burnt minestone. While the characteristics of available minestone vary from source to source, and sometimes between parts of a spoil heap, laboratory study and field experience have shown that the majority can be readily compacted into stable fills of high dry density.

Recently the sites of many closed collieries have been converted for developments of many kinds. Minestone has been redeveloped under strict control to form terraces on which warehouses industrial premises and supermarkets have been erected.

During the same period there has also been increasing use of minestone imported to other sites for similar purposes. These include: the elimination of surface irregularities on building sites; construction of temporary haul roads, access roads and roadworks over low bearing ground; replacement of silts, peats, soft clays, water-logged and other unsuitable materials, to allow site development to proceed; backfilling of disused quarries and gravel and clay pits to provide building or recreational land; raising of ground levels on low lying sites; blinding and covering of municipal tips; filling of disused canals and docks. Minestone from disused spoil heaps, properly handled - usually by spreading and compacting layers as for highway earthworks - provide good stable ground, strong enough to support many types of structure on suitable foundations, easily trenched for services etc. Minestone will continue to be increasingly used in these ways and particularly reclamation of derelict sites using minestone in bringing derelict low grade land into better use. There are in and around coalfields large tracts of low lying land, some water-logged, some liable to flooding and consequently minimally used. Much of this could be brought into beneficial use by using minestone as fill to raise the ground surface level. Similarly 'new' land may be produced where urgently needed by using minestone as fill on foreshores at existing complexes or at new locations.

2.3 Modified/processed colliery spoil

The major uses of burnt minestone and minestone are basically as fill materials however, processing can enable these materials to be used in more technically demanding applications. These uses include: cement bound minestone where mixing minestone with cement can provide a structural material suitable for use in the stuctural layers of road pavements, brick and cement making, reinforced earth where minestone can perform satisfactorily as the fill material, and the production of lightweight aggregate by sintering minestone.

3 BENEFITS OF UTILIZATION

Problems resulting from past and current surface disposal practices

Well-built modern spoil heaps conforming to thin layers, and properly compacted and sealed should not create air or water pollution problems, however the adverse effects of past disposal practices are glaring and include air pollution, water pollution, safety hazards, and sociological and pyschological impacts.

There is widespread agreement that the need to dispose of colliery spoil in an environmentally acceptable manner is one of the most important, if not the most important issue facing the active coal-fields.

Modern mining techniques have given considerable economic benefits to the industry, but have resulted in vastly increased quantities of spoil being produced.

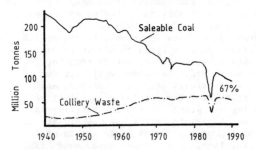

Fig.4. Saleable Coal Production compared to spoil output

In 1947, when coal production was largely unmechanised, the ratio between coal and spoil production was 10:1. By 1981 a highly mechanised coal mining industry had a coal:spoil ratio of 1.7:1, or for every tonne of coal now being produced up to 0.7 tonnes of spoil is brought to the surface.

Most spoil produced is dry solid matter, but increasingly over recent years a proportion of spoil has been in the form of slurry. This form of waste material is the by-product of coal washing in coal preparation plants. Up to 20% by volume of spoil in some areas is in this form.

The most common form of spoil disposal is by surface tipping, close to the pit head. This method of spoil disposal has been practiced from the earliest days of coal mining.

Environmental problems include visual intrusion, loss of land, water pollution by leaching out of soluble material, and the creation of noise and dust by vehicle movements during tipping.

Environmental standards of tipping practice have undoubtably improved in many areas by using improved landscaping techniques for example. However it may be questioned whether this method of spoil disposal can continue to be used with the same frequency as it is used now.

Two basic factors contribute to this view:

a. The volume of spoil produced is now much greater than in the past, and hence the volume of spoil to be disposed of is that much greater, and
b. most suitable tipping sites in the vicinity of pit heads have already been used for tipping in the past.

A further argument in support of this view is that environmentally it is better to dispose of spoil away from the pit head in heavily populated areas.

Other than local tipping adjacent to the mine three alternative methods of spoil disposal can be considered:

a. Long distance or remote spoil disposal.
b. Back-stowage.
c. The use of spoil for construction work etc.

a. There is general agreement that remote disposal is technically feasible; and, subject to suitable means of transport being found, environmentally advantageous in many circumstances. The extra costs involved would prove totally uneconomical to the mining operation.

b. Back-stowage is a system whereby the spoil is returned underground for stowing in the void created after coal (and associated spoil) has been extracted.
The feasibility of stowing spoil underground has been investigated in a number of studies.
Storage of waste in the void created behind the coal face was common in the United Kingdom when less mechanised coal production was used. With modern mining methods the feasibility of back-stowing is problematic.

c. Spoil can, and is, used in a variety of useful ways. It is used for example in construction of embankments for roads, and for reclaiming areas of dereliction. However, the existing market is not secure, in the sense that the demand from the construction industry is variable. There is greater potential for new uses, particularly in place of aggregates. By increased utilization of spoil the problem of disposal can be reduced together with the further opportunity to improve the historical dereliction of old tips by removing them for beneficial use elsewhere thereby enhancing the environment in the coalfield areas. The sections below expand on the problems associated with spoil disposal and alleviation of these aspects can only be an improvement.

3.1 Water pollution problems from colliery spoil heaps

Two general sources of water pollution can result from spoil heaps, physical pollution such as siltation and chemical pollution such as acid drainage. In the former, fine material is carried from the spoil heap in to streams by rainwater run-off causing increased siltation downstream. Most localities require that water containing suspended solids from spoil heaps be collected and treated in the same manner as the processing water produced by coal preparation plants.

A far more serious and complex problem is chemical water pollution. In high-sulphur coal and spoil there is significant potential that the sulphur will react chemically with air and water to produce sulphuric acid unless proper disposal techniques are used.

Other ions including iron, aluminium and manganese are produced as part of the reaction creating acid water. The effect of this on streams has been to lower the pH below tolerance level of many desirable aquatic life species. The heavy metal constituents are toxic singly and may act synergistically, eliminating less resistant plant and animal species. Ferric hydroxide, a precipate resulting from acid drainage, smothers lifeforms and coats stream bottoms, thus reducing percolation and oxygenation of the water, and limiting the available breeding areas for aquatic species.

Conscientious compaction of spoil, construction of water diversion ditches spoil heaps can essentially eliminate water pollution after the use of the tip is terminated.

3.2 Air pollution from spoil heaps

Burning tips were once considered one of the inevitable consequences of coal mining. The putrid smell, the smoke, the noxious gases and dust typical of burning tips unfortunately were once not uncommon. The adverse impact of burning tips on land values and economic development has long been recognised, but the associated physicological or pyschological effects of these noxious, poisonous fumes upon human, animal, or plant communities have never been completely analysed. It is known that emmissions from burning and smouldering waste include carbon monoxide, carbon dioxide, hydrogen sulphide, sulphur dioxide, and ammonia. Gases from burning tips defoliate trees, cause crop damage, and discolour paint miles downwind.

Most serious of all, these emmissions can excerbate problems for people with unhealthy respiratory systems such as those with cronic bronchitis, asthma, or pneumoconiosis. The extent of such conditions, the damage that has been caused or the number of lives shortened or lost has not been calculated.

Spoil heaps may be ignited through a variety of circumstances including spontaneous combustion or careless burning of rubbish on or near the heap. Spontaneous combustion requires voids within the tip for moisture, circulation of air, oxidation of pyrite sufficient to create internal temperatures of at least 450°F, and finally the oxidation of coal.

External sources of ignition can be minimised through a regular inspection programme.

Proper disposal procedures significantly reduces the probability of ignition of a spoil heap, and British Coal have experienced no instances of burning occuring in modern heaps.

3.3 Safety of the surface environment

Historically, little attention was given to spoil disposal. Little or no predisposal site preparation or impoundment design was considered necessary. The result was the creation of many, large, undesigned and poorly constructed spoil heaps and impound-ments. Many of these constituted a safety hazard.

The magnitude of these hazards was brought home to the public after two major disasters: Flowslide of 140,000 cubic yards of waste from a 200-foot tip at Aberfan, South Wales in 1966, and a failure of a waste impoundment at Middle Fork, Buffalo Creek, West Virginia in 1972 when 650,000 cubic yards of water and 220,000 cubic yards of material were transported down-stream. The cause of both slides was improper disposal of coal mine waste. In each case the physical cause was saturation of the waste by water which reduced its stability to such an extent that the material flowed as a liquid. The geological cause of the Aberfan disaster was the disposal of waste over a spring of water. The climatic cause of the Buffalo Creek disaster was a rainstorm which deposited 3.7 inches of rain in 72 hours (a 2-3 year frequency storm). Both incidents were avoidable and inexcusable. A description of the events taken from the official reports follows.

Aberfan, South Wales, October 21, 1966:

At about 9.15 a.m. on Friday, October 21st, 1966, many thoudands of tons of colliery rubbish swept swiftly and with a jet-like

69

roar down the side of the Merthyr Mountain which forms the western flank of the coal-mining village of Aberfan. This massive breakaway from a vast tip overwhelmed in its course the two Hafod-Tanglwys-Uchaf farm cottages on the mountainside and killed three occupants. It crossed the disused canal and sermounted the railway embankment. It engulfed and destroyed a school and eighteen houses and damaged another school and other dwellings in the village before its onward flow substantially ceased.... despite desperate and heroically sustained efforts of the many people of all ages and occupations who rushed to Aberfan from far and wide, after 11 a.m. on that fateful day nobody buried by the slide was rescued alive. In the disaster no less than 144 men, women, and children were killed. Most of them were between the ages of 7 and 10, 109 of them perishing inside the Pantglas Junior School. Of the 28 adults who died, 5 were teachers in the school. In addition, 29 children and 6 adults were injured, some of them seriously. 16 houses were damaged by sludge, 60 houses had to be evacuated others were unavoidably damaged in the initial fall. According to Professor Bishop, in the final slip some 140,000 cubic yards of rubbish were deposited on the lower slopes of the mountainside and in the village of Aberfan. (Report of the Tribunal appointed to inquire into the Disaster at Aberfan, 1967)

Buffalo Creek, West Virginia, February 26, 1972 (see figures 2.8 and 2.9):

Approximately 21 million cubic feet of water was released from the coal-refuse dams on Middle Fork (Saunders, Logan County, West Virginia) beginning at about 8.00 a.m. on February 26.... The previously impounded water then began its wild 17-mile plunge down Buffalo Creek falling more than 700 feet in its race from Saunders to Man.... All homes and structures at Saunders were totally destroyed..... The flood wave travelled from Sauders to Paradee in about 10 minutes at an average velocity of 19 feet per second.... The flood waters arrived at Lorado at about 8.15 a.m. The flood flow was 6 to 8 feet deep on the flood plain and almost completely destroyed the town. A few well-constructed buildings survived, but nearly all homes of wooden construction erected on a slab foundation were demolished.... Flood damage downstream from Amherstadale, although still serious, was far less extensive.

The flooding resulted in the confirmed deaths of 166 persons as of the date of this report (March 12, 1972), total destruction of 502 permanent home

structures and 44 mobile homes, major damage to 268 additional homes along Buffalo Creek from Saunders to Man, West Virginia, a distance of about 17 miles. It was estimated that about 4,000 persons were left homeless. Numerous homes in the Buffalo Creek area that were located above flood plain escaped damage.

The flooding also destroyed about 1,000 automobiles and trucks, several highway and railway bridges, sections of railroad tracks and highway, public utility power cables and poles, telephone lines and poles, and other installations. Mine refuse silt, and debris were scattered for miles along Buffalo Creek. About 30 persons who resided in the Buffalo Creek remain on the missing list (U.S. Department of the Interior, 1972).

Other horror stories have been associated with spoil heaps. Children have been periodically burned, and minor slides have claimed lives.

These unnecessary hazards to a community's security have been brought about by indifferent and irresponsible enforcement of poor legislation. Current good practices of waste-disposal will not result in a dangerous environment for those who live or work near the tips.

3.4 Psychological and sociological effects of past disposal practices

Most travellers in the coal regions of the United States and Europe realise at once that they are in a mining area from the sight of the ever present grey-black tips. Although local residents are said to grow accustomed to such tips and generations have been raised with tips as a backdrop for their homes, the presence of these tips is generally not considered aesthetically acceptable and the tolerance of the local population for them to appear to diminish as the economic dependence of local citizens on the coal industry diminishes. The tips act as serious hinderances for the chamber of commerce and other industrial development groups trying to lure new industry to coal-depleted areas. Proper surface disposal methods can disguise the waste tips so they blend into the surrounding area.

In many areas the coal preparation plants and the spoil heaps are located in valley bottoms adjacent to the transportation network of roads and railroads. At these locations the tips occupy land suitable for development since flat sites near transportation routes are a scares commodity. Foresight in selecting a disposal method could avoid a permanent allocation of this valueable flat bottom land to coal waste disposal and permit its use for development.

Much has been written about the social costs of mining coal. It has been argued that past disposal practices have resulted in rural blights, which, like their urban counterparts, have had a negative impact on the lives of the people who inhabit the region. The overall effect is impossible to measure. Environmentally acceptable surface disposal methods need not mar the beauty of the landscape, pollute the air and water, or endanger public safety. Good disposal practice of mine wastes reduces these problems. However, the alternatives for current or future disposal do not rectify the problems inherited from the past in the form of inactive and abandoned disposal areas.

3.5 Reduction in requirement for new
      quarries

Currently the demand for aggregates in the U.K. is approximately 300,000 million tonnes per annum. This is considerably above the Department of the Environment forecasted demand and predictions by BACMI see the requirement exceeding 350,000 million tonnes per annum by the year 2000. Fig. 5.

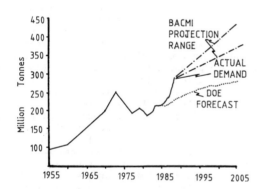

Fig.5. Aggregate Demand - Actual and
        Forecast

To satisfy this demand increased production from existing and new quarries will inevitably result in an increase in the number and size of operations. The environmental effects of this increase are obviously of major concern to the public.

Various committees have commented and reports have been published most notably the Verney Report, on the potential for increased use of secondary aggregates. Little has been done to encourage the application of secondary aggregates and indeed as recently as February this year

BACMI produced a briefing document essentially dismissing the potential for increased use of secondary aggregates.

Some aspects of the BACMI report are valid, particularly regarding technical specifications however it is totally illogical and environmentally damaging to use high quality materials for many applications where secondary aggregates can be satisfactorily employed.

4 POTENTIAL ENVIRONMENTAL PROBLEMS

Utilization involves the same environmental problems as off site and remote disposal of spoil and these must be considered when proposing use of the material on schemes.

Many derelict spoil tips contain material that could be used for construction work.

There are two conflicting views on whether these tips should be worked for the recovery of their economically useful material.

On the one hand the reworking of many old tips creates environmental problems eg. dust; water pollution, as a result of the 'washing' process to recover coal; noise from vehicle movements; and additional road traffic with the possibility of spillage. These problems are sometimes difficult to control.

There are three distinct areas of potential contention:-

a. Operational activity at the Colliery
   Spoil Heap.
b. Transport of the Material.
c. Suitability of the material for its
   intended use.

Generally the transport aspect tends to be the most controversial and is of most interest to the public as it can have the greatest effect on the community at large.

4.1 Operational activity at source

In general, surface tipping of solid wastes and lagooning of liquid wastes give rise to adverse environmental impacts and effects. Of particular importance is landscape and visual intrusion although the extent of this will vary with the surrounding topography.

Other effects may include noise, air and water pollution; disruption or loss of ecological, cultural and heritage amenities; and loss of agricultural productivity and land. Some of these effects may occur during tipping operations and are, therefore, in social terms temporary although tipping may last several decades. Other effects can be permanent.

The timescale of environmental disruption can be increased by the reworking of an old tip for its coal content or the movement of minestone for utilization. The prospect of such operations may delay the transfer of the tip to local authorities for reclamation under DLG powers. In other cases such operations could involve reopening tips or using land where some form of restoration has already been completed, although this unlikely.

Since 1985 in the United Kingdom, planning permission is necessary to excavate and remove minestone from spoil heaps and mineral Planning Authorities can impose requirements and conditions on the proposed working to ensure a satisfactory environmental standard of working on an operation.

4.2 Transport

Transport is generally the major factor considered when minestone is proposed for use both from the environmental and the economic viewpoint. Road transport of minestone can be both costly and environmentally disruptive, particularly to communities near to transportation routes. On the other hand transport by rail or canal may less problematic environmentally, but at higher overall cost.

The considerable quantities of minestone involved require that only bulk handling systems can be considered as appropriate for the movement of material, these being:-

    a. Waterborne Transport
    b. Rail Transport
    c. Road Transport

4.2.1 Waterborne transport

Canals offer an environmentally attractive alternative for the transport of minestone although few collieries or disposal sites are located directly on the waterway system and transhipment facilities are expensive to provide and operate. The advantages of this mode of transport are due to high productivity and low energy costs, claimed to be 25% more effecient than diesel powered rail. Overall costs have been claimed to be 50% cheaper than rail for short distances. Loading facilities may be simple to provide and a quick turnround is not essential due to low demurrage. Unloading is less easy, usually necessitating a grab but sometimes a tippler. Whilst the actual movement of the material can be carried out over long distances with little problems, the facilities and operations at both loading and unloading terminals can be extremely disruptive to the locallity in each case.

4.2.2 Road transport

Without doubt this is the most flexible means of transport, but is also the most environmentally intrusive. It is also the most widely used method of moving minestone.

The particular facet of transport-induced environmental nuisance that is associated with heavy goods vehicles is one which it is impossible to generalise about for different countries, because attitudes are influenced so much by particular circumstances. At one end of an attitudinal scale comes a country such as Canada where vehicle weights and dimensions are among the most liberal in the world, the upper weight limit for an 8-axle combination in Ontario being 140 000 lbs. This, however, meets with little environmental critism in a country with a relatively small population spread over a vast geographical area.

Near the other end of the scale comes the U.K. with a weight limit of 38 tons and massive opposition to heavy vehicles even at this limit, let alone of increased weight; in this case a much smaller country with a dense population and centres of that population spread along most of the major roads on which vehicles travel. It is also relevant that the environmental lobby has been aided in the U.K. by the increased attention in recent years to public involvement into planning procedures, which has enabled public opinion in the U.K. to have more effect than in many other countries.

This increase in the exposure of the public to heavy lorries has occured at a time when there has been increasing concern for the environment in general - and hence it is not surprising that there has been considerable interest in the environmental problems caused by heavy lorries, since all social surveys confirm that these contribute more to disturbance than any other type of road vehicle. Unfortunately, although there are many examples of studies of vehicle noise that deal with lorries, either specifically or indirectly, there are few researchers who have dealt with attitudes to other factors. In the National Environment Survey 1972 lorries were by far most frequently cited as the worst cause of traffic noise nuisance by those who said they were bothered by traffic noise.

Surveys which concentrate on relatively high flow sites, have found that lorries are rarely reported as being heard without niusance being also experienced. This was in sharp contrast to other vehicles, which are more often heard without causing nuisance.

Lorries are in general more noisy than other vehicles. In addition, the relative noise emitted by cars and lorries changes according to traffic conditions. It is clear that in urban conditions, where traffic is less likely to be flowing freely, the noise emitted by commercial vehicles over 3 tons (3050 kg) ULW is likely to be more noticeable in comparison with that of traffic generally.

Heavy diesel lorries have sound frequency spectra which in comparison with those of other vehicles tend to be biased towards the low frequency end. While this difference may not affect the perceived loudness of the noise, it certainly gives lorry noise a different subjective quality from that of other traffic noise, and hence makes it even more noticeable.

Approximately 10 per cent of respondents in the 1972 National Environment Survey claimed to experience traffic vibration often, and a further 27 per cent to experience it occasionally. Of those experiencing vibration often, more than half have reported that it bothered them quite a lot or very much. Although it is experienced much less often than traffic noise, it seems to be more likely to cause nuisance when it is experienced. In fact, the number of people bothered 'very much' or 'quite a lot' was almost as great for vibration as for noise.

The mass of smoke and particulate matter emitted by diesel engines per gallon of fuel burned is approxiamtely 10 times that emitted by petrol engines, and the difference may be more than this under high load with a badly maintained engine. This material is deposited as dust on property adjacent to the road. The exhaust emissions of diesel engines are also much more affected by gradient than are those of petrol engines and have a much stronger odour. Although diesel smoke has not been identified as medically harmful, exhaust particles can sometimes form the centre for condensation of materials which may be harmful, and thus form an avenue for the deposition of these materials in the body. However, the concentration of carbon monoxide and hydrocarbons in diesel exhausts is very much lower than that for petrol exhausts, and only petrol engines use leaded fuel and emit particulate lead in the exhaust.

Many other detrimental effects have been attributed to lorries, some of which are clearly environmental but unquantifiable and others of which would require a complete re-definition of the word 'environment' to be considered as such. In the latter category would come fear,

safety problems and damage to roads, verges and underground mains and services; in the former would come visual intrusion, dust, spray, and severance. It is clear that lorry nuisance involves all the environmental factors associated with other vehicles, mostly in enhanced form, and others besides. Underlying public attitudes to lorries is a widespread and instinctive but unmeasurable feeling that human beings should not be expected to share their living environment with machines that move so close to them with a mass and momentum out of all human scale - perhaps a tenous link with an early ancestor coming face to face with a dinosaur.

4.2.3 Rail transport

Rail freight is the most environmentally acceptable mode of transport and many collieries are rail linked. The system is generally underused, although this may not be the case for certain lines. Since the cost of loading and unloading are very significant factors in the total haulage charge, efficient operations at both terminals are important.

A problem which will arise, is that a number of the colliery surface layouts are already congested and it is difficult to find extra space for further loading facilities to handle the minestone.

Rail transport of spoil might generally be preferred against road transport over long distances on both environmental and energy saving grounds, however the cost of rail transport to the customer may often be higher.

As in the case of waterborne transport, loading and unloading operations for rail haulage can cause environmental problems at both terminal locations due to the concentration of activities involved.

4.2.4 Alternative transport systems

Minestone required for utilization will usually be transported away from the source site by one of the above three methods. Should material be required adjacent or near to the source site the following methods of moving the material can be considered.

a. Aerial ropeway
b. Civil engineering earthmoving equipment
c. Conveyors
d. Pipelines

All these four systems can provide an efficient means of moving material. The environmental effects of their application will generally result in the disruption caused by the colliery activities detailed earlier, being spread over a wider area.

4.3 Suitability of minestone for use

Colliery spoil utilized in developments must be suitable for its proposed use without causing environmental problems due to its inherant characteristics. It is obviously of little if not no benefit reducing or removing one form of environmental blight if the result is the creation of another problem in a different location.

It has been found that by careful pre-selection and refinement, some colliery spoils can meet specifications and many minestones have been used as alternatives for more conventional materials in civil engineering projects.

There has been the fear that if unburnt colliery spoil (MINESTONE) is used for Civil Engineering applications then there could be a risk of burning occuring.

Spontaneous combustion is not considered to be a property of well-compacted spoil fills, since neither oxidation, spontaneous heating nor burning will occur in the absence of air, and it is unlikely that a minor event such as a grass fire would ignite a well-compacted spoil fill.

The presence of sulphate occuring naturally in rocks and soils is well known and particularly high concentrations have been reported in colliery shales.

Minestone is usually composed of 60% clay (minerals such as illite, kaolinite and montmorillonite), 20% quartz and 20% other materials such as iron pyrites (FeS_2), siderite, ankarite and calcite. When freshly dug and brought to the surface it has a pH of 6.5 to 7.

If allowed to weather unchecked some minestones, especially those which have a high content of iron pyrites, will become acidic. Pyrites is ubiquitous in Coal Measures mudstones and can vary from 0.5% to 7% in content. The materials from modern collieries is usually of fine nature with 75% less than 50mm in size and 20% less than 0.5 mm. Its particle size distribution is often very similar to many soils and apart from organic matter and some essential chemicals is often capable of sustaining vegetation as can be seen by the many pit heaps that have a natural vegetation cover.

If waste is to be used for soil making or if existing pit heaps are to be restored then in order to give growth potential the top layer of the minestone must be unconsolidated to allow the access of water,

nutrients and air to the plant roots. This allows for oxidation of any pyrites present and the soil may well become acidic.

In considering sulphate attack on concrete structures it is usually sufficient to determine the sulphate content and pH of the soil and groundwater. If the sulphate content is found to be in excess of the permitted amounts, in terms of percentage SO_3, then steps have to be taken to eliminate the adverse effects on adjacent concrete structures.

Sodium chloride and calcium chloride salts can produce reductions in concrete alkalinity. It has been suggested that the magnitude of the pH reduction is sufficient to enhance the influence of chlorides, one of the primary agents causing corrosion of concrete reinforcement. The determination of chloride content is therefore of importance in assessing the corrosiveness of minestone to embedded metallic reinforcement especially in poorly drained fill.

In the United Kingdom suitable colliery spoils are selected for use from British Coal Corporation sources by Minestone Services and these selected materials are referred to as MINESTONE to differentiate them from other material within the overall category of colliery spoils.

5 CONCLUSIONS

The historical image of bleak conical tips is the manifestation of the coal mining industry's past impact on the environment. The impressions of untreated spoil heaps have resulted in a generalised opinion of mining communities being despoiled which has undoubtedly deterred investment in the areas concerned. Communities dependent almost totally on coal mining have accepted the environmental problems as an inevitable way of life and a factor in remaining in employment. Recently society has become more aware of the environment in general and more particularly when factors impinge upon individual localities interest is polarised.

These aspects are reflected in the formation of Environmental Pressure Groups to promote particular interests. The impact of spoil disposal is obviously of great interest to many of these parties and their views must be seriously considered. The utilization of minestone is clearly much more environmentally acceptable than pure disposal. Minestone used as a landfill material can be beneficial and such use does result in an environmental gain. The use of minestone in place of natural materials has the triple benefit of conserving natural resources disposing of

Fig. 6 Typical Colliery Spoil Heap

Fig. 7 Typical road lorry for Transporting Minestone

Fig. 8 Minestone utilized to reclaim a redundant chemical works site for industrial
      development

'waste' material which is often the cause of unsightly dereliction and finally clearing valuable land for other uses. It is accepted that the use of these materials may have environmental disadvantages as well as advantages. The disturbance caused by haulage being the prime factor against the use of the material. The benefits and disbenefits in any given situation need to be carefully analysed to enable a decision to be made on the best course of action to be adopted. The analysis is not easy or straightforward as there are no generally accepted methods of quantifying some of the factors involved.

In the event a subjective judgement must be made on the environmental benefit or disbenefit resulting from the use of a particular material on a project. At this time Minestone Services carry out an assessment of the benefits of various alternative sources of minestone when identifying materials for projects. The potentially most economic option may often be rejected if environmental problems are evident, usually associated with transport.

This paper has attempted to identify the major environmental aspects involved in examining the utilization of minestone. Whilst problems can result from the use of the material it is considered that the overall long term benefit of removing or preventing dereliction justifies the generally short term environmental disruption that may be involved.

REFERENCES

Rainbow, A.K.M., Composition and Characteristics of Waste from Coal Mining and Preparation in the United Kingdom, United Nations E.S.C.E. Symposium on the Utilization of Waste from Coal Mining and Preparation, Tatabanya, Hungary, 1983.

Proceedings of the Symposium on the Reclamation, Treatment and Utilization of Coal Mining Wastes, Durham, September 1984.

Gibson, J., An appraisal of the utilization potential of colliery waste, Mining and Minerals Engineering, November 1970.

Commission on energy and the environment, Coal and the Environment, HMSO, London, 1981.

Taylor, R.K., Composition and engineering properties of British Colliery Discard, National coal Board, London, 1984.

Rainbow, A.K.M., A review of the State of the Art in Respect to the Risk of Spontaneous Combustion when using Minestone in Construction Works in the United Kingdom, United Nations E.S.C.E. Symposium on the Utilization of Waste

from Coal Mining and Preparation, Tatabanya, Hungary, 1983.

Proceedings of the Second International Symposium on the Reclamation, Treatment and utilization of Coal Mining Wastes, Nottingham, 1987.

BS 6543:1985, Guide to Use of Industrial By-Products and Waste Materials in Building and Civil Engineering, British Standards Institution, London.

Blelloch, J.D., Waste Disposal and the Environment, Colliery Guardian, August, 1983.

Minestone Services, Information Sheets, British Coal, London, 1986.

Department of Transport, Report of the Interdepartmental Committee on the Use of Waste Material for Road Fill, Department of Transport, London, 1986.

Department of the Environment, The Town and Country Planning General Development (Amendment) (No. 2) Order 1985, Department of the Environment, London, 1985.

The Institute of Mining Engineers, Symposium on Mineral Extraction, Utilisation and the Surface Environment, Minescape 88, Harrogate, 1988.

Watkins, L.H., Environmental impact of Roads and Traffic, Applied Science Publishers, London, 1981.

Sherwood, P.T., Roe, P.G., Transport and Road Research Laboratory, The Effect on the Landscape of Borrow-Pits used in major Roadworks, TRRL Supplementary Report 122 UC, Department of the Environment, London, 1974.

BACMI Briefing, Secondary Aggregates, British Aggregate Construction Materials Industries, London, February 1990.

ACKNOWLEDGEMENTS

The Author is grateful to British Coal for the support given in the preparation of this paper and for permission granted for its publication. The views expressed are those of the Author and not necessarily those of British Coal.

Reclamation, Treatment and Utilization of Coal Mining Wastes, Rainbow (ed.) © 1990 Balkema, Rotterdam. ISBN 90 6191 154 0

Land reclamation engineering in mine areas

Bian Zhengfu & Guo Dazhi
China University of Mining & Technology, Jiangsu, People's Republic of China

ABSTRACT: On the basis of research achievements in land reclamation of recent years, combined with the concrete practice experience of land reclamation in China's mine areas, this paper discusses the concept, the objects of study, the research contents and the research ways of land reclamation engineering in a systematic way. In the last part of the paper, some new technology which should be used in the field of land reclamation are introduced.

1 INTRODUCTION

In the end of 1950s, some China's factories and mines have taken the problems of reclamation into account. For example, Xiaoguan ore, which belongs to Zhengzhou aluminium factory, considered the problems of reclamation in the process of design in 1958 and reclaimed thousands mu (1 mu=667 m^2) of abandoned land with waste stone. Bantan (Guangdong) tin ore backfilled with waste soil stripped while mining in 1964 and set a good precedent, which compensated the ore for the loss of land being taken for use at the same year. However, when the reclamation work has been carried out vigorously is 1980s,especially, after The Stipulations of Land Reclamation (State Council of the P.R.C) are promulgated in Nov. 1988. China's coal mine areas have got many encouraging achievements in the field of land reclamation of recent years. With the work of land reclamation developing widly, the practice must have been guided by the theory urgently. No one has expounded the theory of reclamation systematically before, so the authors want to try through the paper.

2 THE CONCEPT OF LAND RECLAMATION ENGINEERING IN MINE AREAS

2.1 The concept of land reclamation engineering in mine areas

Land reclamation engineering in mine areas is defined as a work of renewing and comprehensive treatment on the land destroyed by mining. It includes two procedures of bioengineering reclamation and mining engineering reclamation. Its ultimate aim is to resume productivity of the land. When we discuss the problems of reclamation, the theory of mine ecology, environmental science, pecology and regional planning must be obeyed and thecharastic of mining engineering must be combined.

Because land reclamation is a comprehensive treatment work, its objects of research are not only the destruction surface,but also existing hillock and waste to be exhausted, farm water conservancy facilities,residential planning, land use planning, mine exploitation planning, mining technique etc. To protect the land resource fruitfully, we must prevent the soil of the subsidence areas and adjacent areas from being eroded and prevent the land from being salinized. For underground mining, what it tacklees is surface problems of mining engineering.

2.2 The tasks of mining engineering reclamation and bioengineering reclamation

The jobs of bio-engineering reclamation is to decide to take some appropriate bio-engineering measures to keep ecology balance of mine areas according to the direction of the land use. Its primary technological measures includes: fertilizing the soil;recovering the land with rich soil;putting up subsidiary facilities of agriculture and forestry; selecting farming methods and techniques;seeking optimum crops and saplings etc.

The jobs of mining engineering reclamation are to recover the earth's surface for plant to growth favourably. Its primary technological measures have: heaping the soil of the areas disturbed by mining; filling the pit resulting from subsiding; improving the soil by means of physics and chemical methods; building irrigation works; constructing ways in reclamation district; doing some earlier works in readiness for building;protecting the reclamation district from being eroded etc.

After mining engineering reclamation and bio-engineering reclamation, follow problems should be solved:

1. Reinforcing topography affected by mining; linking the environment of subsidence area with the one of contiguous zone; satisfying the demand for land to change peasant's dwelling place,thus dweller not being far away from the land they live on.

2. Reinstating even improving the productivity of land, thereby the land having better economical value.

3. If establishing overburden and heaping waste reasonably, we may obstruct soil erosion to come into being. As so, acidity stream of water will not happen and ground water will not be polluted in the subsidence area.

4. Digging the ground drainage system equitablly, then the surface will not be waterholl everywhere. Doing as so, it not only protects the ground waterbody,but also keeps the irrigation water in good management and prevents the soil from being salinized as well.

5. Afforesting in subsidence district. this method may not only prevent atmosphere from being polluted, but also prevent the non-stable surface from eroded.

2.3 The relationship between mining engineering and land reclamation engineering

As viewed from system engineering, land reclamation in mine areas is an essential work of mining engineering. Only is land reclamation engineering synchronized with mining engineering, the mine exploitation will be more scientific, reasonable and perfect. Only doing as so, can we put an end once and for all to the problem of land destruction. The relationship between mining engineering and land reclamation engineering is showned as Fig. 1.

As land reclamation engineering in mine areas does not only relate mine department, but also involves many other departments such as urban construction,farming and forestry, water conservancy, enviromental protection, land management etc, it is a frontier science across several disciplines and departments.

3 THE RESEARCH CONTENTS AND WAYS OF LAND RECLAMATION ENGINEERING IN MINE AREAS

3.1 The rules of land destruction and their forcasting methods

When carrying out the researches into this aspect, we rely on the theory of mining subsidence primarily. Besides the parameters such as subsidence factor, subsidence, tilt, curvature, horizontal displacement, lateral deformation,the chief parameters to describe the land destruction also include: the subsidence ratio per ten thousand tons (SR), the waterlogged ratio (WR), the ratio of subsidence area to mining area (AR), the ratio of the subsidence volume to mining volume (VR), extra slope (S), etc.

The parameters stated above relate the geology and mining conditions of a mine. Various mine area shows its parametwrs are different. Under normal conditions, SR of China's mine area in the east coastal plain can be calculated as follow formula:

$$SR = \frac{15.0 \times n \times \cos a}{KMr} \qquad (1)$$

where, n--- influence factor;
    a--- the dip angle of seam;
    K--- working section recovery coefficient;
    M--- working thickness;
    r--- the density of coal.

WR relates water table of mine areas. It is around 30% in China's east coastal plain district, 20% to 30% in China's north.

Statistical data shows that AR is around 1.2, VR around 0.6 to 0.7, S around 1° to 3°The method of influence functions is popularly employed to predict the surface subsidence, its formula is as follows:

$$w(x,y) = W_{max} \iint_F f(x-s,y-t)d_s d_t \qquad (2)$$

where,$w(x,y)$:the subsidence of surface point which coordinate is (x,y);
    w_{max} :the maxium subsidence on extraction of the whole critical area;
    $f(x-s,y-t)$: the influence function of surface subsidence;
    F: domain of integration
    x,y: the coordinate of surface point;
    s,t: the coordinate of underground mining unit within F.

The mathematical models of influence functions usually used are three kinds of distributions,i.e. normal distribution,Weber distribution and hyperbola function.

Other movement and deformation can be drawn from Eq.(2).

The parameters mentioned above represent the rule of land destruction in some extent, but for reclamation engineering only pocessing these parameters is far from use. Dynamic calculation of land area destroyed by mining, the research of determining the level of land destruction is of great significance as well. Preventive measures , such as pre-stripping topsoil etc., can be taken according to the results of dynamic calculation. Various harnessing pattern suitable to various subsidence area can be adopted in the light of the conclusions of classifying research. As so, the direction of subsidence land utilization will be reasonable.

3.2 The planning methods of land reclamation

The land reclamation planning includes two procedures,i.e overall plan and small-area plan.

The crux of overall plan is decision analysis on these problems such as the direction of land reclaimed, engineering measures adopted by reclamation engineering before small-area plan worked out.

The methods of working out overall plan may be qualitative analysis, also quantitative analysis. In accordance with the special characteristics of reclamation, we believe that by means of the method of AHP (Analytic hierarchy process) the decision work will become flexible and convenient. The steps of this method are as follows:

1. Making systimatic analysis and creating the model of hierarchy-structure;
2. Constructing decision matrix;
3. Sorting of single layer;
4. Sorting of all factors;
5. Examining consistency.

The small-area plan is the basis for reclamation engineering to come into being, making the overall plan is prerequisite to it.

The methods of small-area plan have three kinds,i.e economical effectiveness plan, eco-engineering plan and unfied plan of economical and ecological effectiveness . They are stated

3.2.1 Economical effectiveness plan

Principle: Enable the economic effectiveness of the land reclamation to be optimized while satisfying the demands of a great many of constraints.

Contents: Regulate the structure of land use so as to achieve optimum overall effectiveness. Herein the structure of land use refers to the area proportion of various kinds of land used for cultivation, orchard and fish pond etc.

Model: Given: Z- the overall effect; c_j- the economical effectiveness per unit area of each kind of lands; x_j- the area of each kind of lands; then we have:

objective function Max $Z=\sum_j c_j x_j$

constraints $\begin{cases} \sum_j a_{ij} x_j \gtrless m_j \\ x_j \geqslant 0 \end{cases}$

where, a_{ij}, m_j is coefficient.

Constraints here have several kinds as follows: fund constraints,resource constraints, labour power constraints, demand constraints,policy constraints.

In above mentioned models, the crux of the matter consists is how to formulate the constraint equation and how to seek out the parameters (c_j,a_{ij},m_j) of the model.

3.2.2 Ecological engineering planning

Object: When working out the ecological planning for a mine area which is located in an agricultural area, the ecological agriculture must be taken as the object of planning, whereas the landscape of a mine area, which is located in a suburb, must be restored to the level accepted by the resident there.

Principle: In the light of the principle of food chain and the principle of trophic circulation, a structural model which is suited to the features of agricultural production of the mine area, will be designed, then an optimum design must be conducted on the structural model of the system with the method of system analysis.

Contents: The ecological engineering planning consists of planning of system structure and the technological planning. The structure planning consists of trophic structure planning, vertical structure planning and time structure planning etc. Among them the trophic structure planning is the basic one, of which the optimum design must be conducted firstly.

Procedure:

1. In accordance with the feature of destroyed land, the rule of ecological change of the mine area and the experience extracted from the running of ecological agriculture of his own area or other area, one can find out the model of trophic structure suitable for production of his own area.
2. The optimization of the model of trophic structure.
3. Make an assessment on the improved

structure model.

4. In accordance with the improved structure model, design the plane structure, the vertical structure and the time structure.

5. Make a general assessment on the entire structure design.

6. Make technique planning.

Ecological reclamation patterns which have been used successfully in China's mine areas have space exploitation pattern, self-purified pattern of system and the pattern of absorbing nutrient circularly.

3.2.3 Integrated planning of ecological and economical issues

Adding the ecological constraints into the model of method of economic effectiveness planning, one can obtain the method of integrated planning of integrated one. The ecological constraints consist of the one conformed with the requirements of general ecological standard and the one conformed with the requirements of ecological engineering planning.

3.3 The techniques of land reclamation

Mine development falls into underground mining and opencasting, so the reclamation technique falls into the one for underground mining and the other for opencasting. Since reclamation engineering had not been considered on schedule in the process of mine exploitation, spoil reclaiming becomes an important work of reclamation engineering. Spoil reclaiming is evident different from reclamation of subsidence area and the pit resulting from opencasting, consequently, when studied, the technique of spoil reclaiming is one of important research contents. These techniques are stated separatly as follows:

3.3.1 The reclamation technique for underground mining

Main engineering measures for underground mining reclamation have four kinds as follows:

1. taking special mining technique measures;

2, lowering ground water level;

3. improving methods of cultivation and irrigation;

4. raising the elevation of surface.

Special mining technique measures include using shrinkage mining and mining with filling sand to reduce surface subsidence etc.

How left the waste underground for filling is being under research.

Raising the elevation of surface has two methods: one is filling subsidence pit with waste, fly ash etc., the other comprehensive treatment by raising the shallow land with digging the deep.

Thereforce, the techniques of underground mining reclamation include special mining techniques, lowering ground water table, filling subsidence pit and raising the shallow while digging the deep etc.

3.3.1.1 The method of filling subsidence pit

The waste includes spoil, fly ash, rubbish etc. The spoil includes those to be removed from underground and those having become hillock on surface. The backfill method varies from the direction of land use after filled.

1. Land reclaimed used for architecture It's technique process of filling with waste is as follows: (1) filling in slices; This method must roll and press foundation in slices. (2) filling in overall thickness and self-unloaded; It's handling method in force is dynamic consolidation. The weight of tamper used is ten tons. The temper is lift up ten meters high. There is automatic dropped arrangement in crane.

2. Land reclaimed used for agriculture and forestry. besides the work of filling with waste and levelling, the work of covering with rich soil must be done by means of machine or by hand. Rich soil is that pre-stripped and heaped aside. Hydraulic dredger can be used to draw mud and cover the land reclaimed with it.

Reclamation engineering with fly ash must strip surface earth to construct dam of ash store firstly. After filling work has been finished, rich soil of the dam can be used for covering and levelling work can be done. Reclamation technique with other wastes is similar to those of with spoil and fly ash.

It is of a common step for the land reclaimed used for farming that after covering work have been done as steps stated above, the work of fertilizing soil must be done step by step also.

3.3.1.2 Raising the shallow land with digging the deep

This method suited to shallower subsidence. Part of subsiding land is transfomed into meticulous breeding fish pond, the other raised to plant riceoor lotus root or fodder crops or used for building fowls' and animal's sheds. The engineering can be finished by two main ways, i.e by hand or by hydraulic dredger. The later way is popularily adopted now. It is of two merits,i.e. cheap cost and good effectiveness of levelling. Not only has the reclamation cost relation to the method of raising and digging, but also the depth. Quantai colliery (Xu zhou mine area) reclaimed 1095 mu land, digged and raised 1.5 to 2.0 meters high, its cost is 1507 Yuan per mu. By adoping this method, the scale harnessed varied from difference mine area. The experience of Xuzhou mine area is that 3 mu waterlogged ground is harnessed into 2 mu fish pond and 1 mu farmland. Since waterlogged ratio of Kai nuan mine area is smaller than Xuzhou, It's quantity of farmland harnessed is more than Xuzhou also.

3.3.1.3 Lowering ground water table

It's main measure is digging irrigation canels and ditches to dredge surface water logged and control ground water table or cut off external source of water. The parameters of canals and ditches have been discussed in relevant writings. In addition to the work stated above, water conservancy facilities,such as sluice, must have been constructed across these canals and ditches. Deeper subsidenceareas only can be transformed into fish ponds.by using this method. Lastly the mine ecosystem becomes into aquatic and terrestrial coexist ecosystem.

Besides these, afforesting in mine areas may achieves the goal of lowering water level.

3.3.1.4 special mining technique

Some special mining techniques are ripe, when used, only is economic assessment made. If it is more economical of money by taking special mining techniques than by adopting surface reclamation measures, special mining techniques are adopted.

The method of leaving waste underground is of bright prospect.

3.3.2 The reclamation technique for open-casting

The reclamation technique for opencasting falls into five parts:
1. Dividing working district reasonably;
2. Stripping and heaping topsoil;
3. Heaping of soil and backfilling and levelling of mided out space;
4. Covering the land reclaimed with rich soil;
5. Comprehensive utilization and re-plan-

ting of the land reclaimed.

According to the difference of mining method,transportation method and geological condition, the reclamation method for open-casting falls into track hauback reclaimable methods;trucks hauback reclaimable methods;reclaimable methods by using scraper; no-wheels hauback reclaimable methods;the block reclaimable method of contour stripping;roll-over reclaimable method.

3.3.3 The reclamation technique for hillock

There is two ways for reclamation of hillock. One is eliminating hillock by laying a double-track to backfill the subsidence with waste or by using waste as building matierials, this way is usually suitable for noxious waste. The other way is levelling hillock slightly to plant trees to achieve the goal of beautifying the environment.

3.4 The method of land reclamation assessment

3.4.1 Requirements of assessment

1. The crux of assessment is determining the productive force of land reclaimed.
2. Evaluate the degree of land destruction and the effectiveness of reclamation with systematic viewpoints to achieve the unity of economical benefit, environmental benefit and social benefit.
3. The goal of reclamation is guiding the work of reclamation.

3.4.2 Target system

It includes three kindes, i.e. economical target; ecological and environmental target and social target.

3.4.3 Objective of assessment

1. Determine the effectiveness of different programme to be good or bad.
2. Determine the productive force of land before and after reclaimed.

3.4.4 Contents of assessment

1. Land quality assessment reflecting the natural characteristic of land.
2. Land economic assessment reflecting the social nature of land.

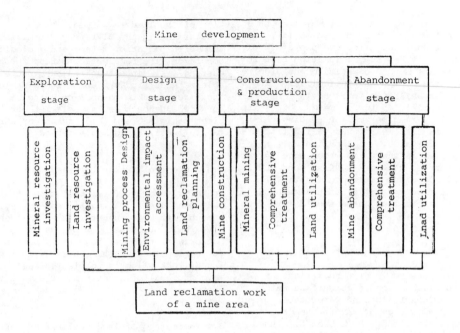

Fig. 1 The relationship between mining engineering
and land reclamation engineering

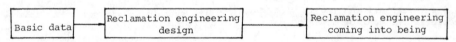

Fig. 2 Chart of reclamation programme

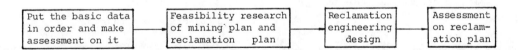

Fig. 3 The procedure of reclamation
engineering

3.5 Land reclamation policies and land reclamation countermeasures

The main foundations in law are The Stipulations of Land Reclamation; Land Management Law of P.R.C; Minerals Resource Law of P.R.c; Environmental Protection Law of P.R.C.

But since we are lack of fund and have not special organization in charge of reclamation presently, the work of land reclamation is difficult to be carried out. Therefore, the work of land reclamation policies and land reclamation countermeasures must have been studied.

4 THE PROGRAMME OF LAND RECLAMATION IN MINE AREAS

As already discussed, the work of reclamation is one of important work of mining engineering, but it is different from mining engineering also. It has a intact programme of itself as Fig. 2.

The factors what follow in the next must be considered during initial stage of reclamation.

1. geographical features of mine area;
2. environmental factors;
3. the phisical and chemical nature of the earth's surface and stratum of mine area;
4. mining method.

The procedure of reclamation engineering design is as Fig. 3.

To make reclamation plan come into being includes following work:

1.coordinating the relationship between mining department and local government;
2. raising fund;
3. organizing labour power and matierial power;
4. reclaiming according to plan until the work finished.

5 APPLICATION OF NEW TECHNOLOGY TO RECLAMATION ENGINEERING

5.1 Remote sensing

By means of remote sensing, investigation of land resource and the work of gathering mine information can be finished quickly; exactly and intactly; sufficiently.

5.2 Application of computer

Computer supplies reclamation planning design with good aid means. It may be used for calculating, evaluating, drawing etc.

5.3 Biological technology

How to apply biological technology is being nder research. There are also practical examples to fertilize soil and to reclaime hillock by means of biological technology.

6 CONCLUSIONS

1. The concept of land reclamation engineering in mine areas is presented in this paper, those problems which can be solved by reclamation engineering and the relationship between mining engineering and reclamation engineering are discussed also.

2. The reseach contents of reclamation engineering are discussed in detail in this paper and some concrete methods of research are presented.

3. The programme of land reclamation engineering is presented in the paper.

4. Application of new technology to the field of land reclamation engineering is prospected in the paper.

REFERENCES

R.J. Sweigard and R.V. Ramani 1986. Reginal comparison of postmining land use practice. Mining Engineering 9.

L.B. Phelps and L.Holland 1986. Siol compaction in top soil replacement during reclamation. For Presentation at the SME Annual Meeting.

Bian zhengfu 1989. Mine exploitation and land reclamation. For Presentation at the Annual Meeting of Mining. (in Chinese) .

Bian zhengfu 1989. Reclamation pattern research of mining area with high phreatic water level. The thesis For master degree. (in Chinese) .

Lin jiacong 1988. The work for mine surveyors during reclamation. Mine Surveying. (in Chinese).

Lin zhengji and Yan xing 1986. The principle and method of land management. China people university publisher. (in Chinese).

Reclamation, Treatment and Utilization of Coal Mining Wastes, Rainbow (ed.) © 1990 Balkema, Rotterdam. ISBN 90 6191 154 0

System approach to the management of coal mining wastes in the Ostrava-Karviná coal field of Czechoslovakia

M.Vavro, P.Ganguly & V.Dirner
Technical University of Mining, Ostrava, Czechoslovakia

ABSTRACT:Ostrava-Karviná coal field produces annually nearly 18 million mt of wastes from different mining processes which makes a waste to clean coal ratio of 0.77. Major utilization of wastes has been found in land reclamation, stowing operation and others. Future projections show that waste availability will fall far short of the total reclamation need of the region in the coming years. No particular technological problem in waste management is foreseen. However, success in waste management will mainly depend on availability of financial resources to meet the technological needs to fulfill the socio-economical and ecological requirements of the region.

1 INTRODUCTION

Ostrava-Karviná coal field, a part of the renowned Upper Silesian coal field, is the prime producer of hard coal in Czechoslovakia.In 1988, the region produced over 22 million mt of clean coal against the na - tional production of a little over 26 million mt of hard coal.

Mining activity started in the 18th century and intensive deep mining over the years has led to the production of nearly one milliard mt of coal and nearly the same amount of mining wastes. These two aspects together have created large subsidence troughs and waste dumps which significantly contribute to the ecological problems in the region. Serious thoughts were first given in the early 70's to tackle these problems and works are pre - sently continuing. The paper highlights the present approach and shows the future directions in the area of waste management.

2 PROBLEM STRUCTURE IN THE HANDLING OF COAL MINING WASTES

Understanding the problem structure requires a close look at the pro - duction pattern of clean coal and the wastes. Table 1 reveals the trend in clean coal and waste production over the years.

Table 1. Trend in clean coal and waste production (in million mt).

Year	1970	1975	1980	1985	1987
Cl.coal prod.(A)	23.9	24.4	24.7	22.9	22.8
Waste prod.(B)	14.0	15.4	16.2	16.9	17.5
Ratio B:A	0.59	0.63	0.66	0.74	0.77

It is obvious from the table that the rate of increase in waste pro- duction is higher than the rate of increase in clean coal production. Reasons can be attributed to the gradually increasing depth of mi - ning (average depth 670 m, maximum depth 1360 m), predominance of thin seams (43.2 % reserves in seams thinner than 1.2 m, minimum thickness worked 0.4 m), complicated geological structures and other fac - tors. It is worth mentioning that in recent years, the cross-sectional area of stone headings has gone up by 8 % and the presence of stone bands in the mined out seams has risen from 9.4 % to 10.7 %. These factors coupled with the application of progressive technology in the area of exploitation, development and transport have conside - rably aggrevated the problem of waste management in this region.

Leaving aside the geological factors, reluctancy in stowing operation has significantly contributed to the problem of waste management. In 1987, only 13.9 % of the produced wastes was used for stowing. Future projections do not show any upward trend in this area. Reasons for gradual reduction of stowing are (Dedek 1989):
- high cost of pneumatic stowing,
- slow return on invested capital,
- low productivity.

As a result, great subsidence troughs have been created in the Karviná region which produces 70 % of total production of the region from relatively thicker seams. The region produces only 40 % of the total wastes. The Ostrava region generates 60 % of wastes and covers only 30 % of the total coal pro - duction.This creates a paradoxial situation which requires large scale movement of wastes from Os - trava region to Karviná region.

Major portion of the waste is transported to the subsidence troughs or to the waste dumps. Modes of transport are: aerial ropeway, belt conveyor, trucks and railways. Aerial ropeways are fast becoming out of date. Belt con - veyors are used for large-scale handling over short distance and for overcoming natural obstacles. Bulk transport, 12 million mt in 1988, is by trucks. Rail transport, 4.5 million mt in 1988, is going to increase considerably in the coming years and truck transport of wastes will accordingly be reduced.

2.1 Utilization of coal mining wastes

Coal-bearing formations in the region are characterized by the pre- dominance of sandstone (40-60 %). Siltstones and claystones are present in considerable quantities, whereas presence of conglomerate is fairly low. Distribution pattern of these rock types lead to the particular characteristics of wastes generated from different mining processes in relation to SiO_2 and Al_2O_3 content of the wastes. Basically mining wastes are divided in two broad groups: wastes gene - rated by underground mining pro - cesses and wastes generated by coal processing plants. In 1987, the region produced nearly 18 million mt of wastes from these two sources. It is necessary to point out that although a considerable part of the mining wastes is produced by coal processing plants, only 25 % of such wastes is used for stowing (Ganguly, Tejbus 1989).
It needs to be stressed that stowing as an operation, is used in this region for technological needs, safety of workings and only in very limited cases for surface protection. If ecological problems, and surface protection are to obtain priority, then there is no other way than to increase the amount of stowing. Presently only 10 % of total production comes from stowing faces. If the wastes from the processing plants are used for stowing, it is possible to mine 30 % of coal by stowing. In that case cost of operation will increase and productivity may somewhat decrease. However, if it is not done then the short-lived gains obtained by the reduction of stowing may in the long run lead to a much higher cost to be paid for the removal of the damages done.
The maximum use of mining waste is in reclamation of mined out areas. This eventually requires large-scale movement of the wastes and the top soil. The need of re - clamation is a top priority fon this region and experience indicates that the problem can be technically managed provided proper fund is allocated for this purpose.

Long-range plan reveals that it will be necessary to reclaim 127 million m^2 of area in next 60 years. With the present rate of waste generation, the availability of wastes has been estimated to 530 million m^3. These figures point to the insufficiency of waste material for the projected reclamation work, as with this quantum of waste it will be possible to achieve only 4.2 m high layer of wastes over

this vast area, whereas the average depth of subsidence trough in
this region varies from 2 - 20 m.
Therefore, it will not be possible
to fully reclaim the land and
there is a necessity for selective
reclamation. Long-range plan for
reclamation, therefore, plans to
reclaim only 61.26 million m2
area in the next 60 years.

It is stated that about 28 % of
the wastes is dumped (Ganguly,Tej-
bus 1989). With regard to the dumping area, attention is paid to
dump the wastes in the existing
subsidence troughs or in places,
where such troughs will be formed
in the near future. This is done
in agreement with the total con -
cept of land reclamation in re -
lation to the future landscaping
of the region. Whereever in future
waste dumps are to be formed, attention will have to be paid to
the following basic factors: legal
requirements and construction regulations related to the protection of agricultural and forest
lands, factors bearing unfavou -
rable influence to the surroundings during or after the construction of the dump, bearing characteristics of the flow of the dump,
thermoactive characteristics of
the wastes, resistance of the
dumped material to air leakage,
drainage characteristics of the
area, precaution against sponta -
neous combustion etc. (Sciulli et
al. 1986).

The rest of the mining wastes
generated are used for engineering
constructions and manufacturing of
different construction materials.

In connection with the complex
utilization of wastes in this coalfield process "Haldex" is worth
mentioning. The process (originally Hungarian) is basically used
to recover coal from the wastes generated by the processing plants
and to use the secondary processed
wastes in the construction industry. In this coal field, last year
the Haldex process was used to process more than 0.5 million mt of
wastes from the processing plants.
The process generated 73 117 t of
coal and the secondary wastes were
used for construction works.

2.2 Future directions in the mana-
    gement of coal mining wastes

Present social, political and economical changes in Czechoslovakia
will eventually call for vital
changes in the industrial stru -
ctures of the country and will definitely effect the coal mining
industry. The extent of changes is
difficult to precast today, but it
is evident that coal and wastes
will be produced in the coming
years (may be in reduced quantity).
Strategies are to be formed to effectively utilize the wastes. Following major steps need to be the
essential parts of the future waste
management policy in the region.
- Waste management policy should
  aim on the most effective use of
  mining wastes, that is, for reclamation of mined out land,stow-
  ing of workings and use in con -
  struction works.
- Reclamation of land should re -
  spect the total concept of re -
  gional landscaping and recla -
  mation, appropriately divide the
  region according to the degree
  of subsidence (final) and fix
  priority for reclamation depending on the socio-economical and
  technological needs of the individual locality.
- Considerations should be given
  to the increased use of waste materials for stowing and necessary infrastructures need to be
  built for this purpose.
- Increased utilization of wastes
  (heat value less than 13 MJ/kg)
  for the manufacture of construction materials.
- Restrict the creation of new
  dumps and utilize the present
  dumps in relation to the concept
  of total regional planning and
  landscaping. Dumps which do not
  fit to the future landscape of
  the region should be liquidated.
- Attention is to be paid to prevent spontaneous combustion in
  the dumps.
- Increased use of Haldex for recovery of coal and to look for
  other suitable processes to re -
  cover maximum possible coal material from the wastes.
The modification of the present

waste management policy or the formulation of a new one will require not only new concepts but will need to respect numerous constants in the socio-economical and technical area. It will be necessary to solve many problems like evolution of cheaper technology of stowing, greater application of ecologically acceptable modes of transport,finding out suitable dump site(for temporary storage), reform of legislations regarding land acquisition, etc. However, under the present circumstances, it seams that the greatest problem will be to find out adequate financial resources to tackle the problems of waste management in an effective way.Hopefully, some new legislations will pave the way to ensure the funds needed for such work.

3 CONCLUSIONS

From the environmental standpoint, the effective utilization of coal mining wastes can play a great role in the Ostrava-Karviná coal field. The utilization of such wastes must have two broad purposes: firstly as materials to be used in connection with the preventive measures against the mining damages in the form of stowing material and fi - nally as the material for the removal of damages already done in the form of material for recla - mation work. It is necessary to prevent the creation of new dumps and, where it is impossible to handle or utilize the wastes in any other way, it is necessary to take adequate precautions to prevent further ecological damage and to fit the dump in the future landscaping of the region.

Keeping in view the extent of damages in this region, the present production of coal and wastes in this region and their projection for the future, it is possible to conclude that the wastes generated in the field can be adequately and properly used within the geogra - phical framework of the coal field. However, the task of total recla - mation will require a lot of money and a huge quantity of material will have to be imported from outside.

REFERENCES

Dedek, M. 1989. Production of coal and utilization of wastes in the concern OKD. Proceedings of the Conference on Management of Coal Mining Wastes in the Ostrava-Karviná Coal Field, Ostrava.

Ganguly, P.K. and Tejbus, V. 1989. Possibilities to use coal mining wastes for environmental pro - tection in the Ostrava-Karviná coal field. First Nation Conference on Ecology and Mining, Ostrava.

Long-range plan of the Ostrava-Karviná coal field 1983. Mine Planning Institute, Ostrava.

Sciulli, A.G., Ballock, G.P. and Wu, K.K. 1986. Environmental approach to coal refuse disposal. Min.Engng. 38, 3.

Reclamation, Treatment and Utilization of Coal Mining Wastes, Rainbow (ed.) © 1990 Balkema, Rotterdam. ISBN 90 6191 154 0

A planning and reclamation policy for the retorted oil shale of West Lothian for the 21 century

A.S.Couper
Lothian Regional Council, Edinburgh, UK

ABSTRACT: In England and Wales, colliery waste is referred to as shale. In Scotland, the word refers specifically to oil shale. It was mined from 1860 to 1962. The large pinky red tips, highly visible from the Edinburgh to Glasgow motorway, the M8, are the only reminders of an oil industry that once competed on the world markets.

Despite a strategic planning policy that encourages the use of oil shale waste for construction, and fourteen years of intense activity rehabilitating dereliction in Lothian, an estimated 100 million tonnes remain within a very small area. With the motorway building programme in the east of Scotland substantially complete, there is a need to evolve an integrated strategic planning policy that controls extraction, but is driven by the rehabilitation programme. Such a policy demands fresh approaches.

1 HISTORY OF OIL SHALE MINING

Oil shale was mined in Lothian for just over a 100 years from 1860 to 1962. It was mined specifically to produce oil by a retorting process. Oil production alone did not sustain the industry; it was the "downstream" products like paraffin wax, sulphate of ammonia etc, which helped to make it profitable.

1.1 Oil shale

In the Lothian, Lanark and Fife areas of Scotland, oil shale is found near the base of the carboniferous system. Due to techtonic movements, the strata, thought to have been formed in lagoons, was folded and faulted and occurs as domes and basins, some of which appear as outcrops on the surface. (Fig. 1)

Both opencast and mining methods were used to exploit it. The mining operations, in some instances, were carried out at a depth of more than 330 metres and a mine could start off from the surface on a decline, or vertical shafts could be sunk to strike a seam or seams. These were from 1.2. to 3.6 m in thickness, and the accepted method of mining was by "stoop and room". High extraction rates were achieved by removing the pillars and adopting a retreat mining method. Long wall working was used but it was not common. Apart from boring for shot firing, there was little mechanised working because of the strong "leathery" nature of the oil shale.

Interesting, because the Scottish oil shale retorts could not handle fines, coal cutting machinery was not used as it produced too much fines. The ideal size of lump was 75-100 mm. Attempts to utilise the fines did not work well in the older retorts, but later versions could. Before the industry closed, heating the oil shale insitu was tried, but the poor yield made the process uneconomical.

The reserves remaining unmined are estimated to be 120 million tonnes, of which, it would be practical to get 65 million tonnes. This could yield 5 million tonnes of oil, which is in the order of 5% of the present day annual UK consumption of crude oil.

1.2 James Young's influence

The Scottish Oil Industry owes its origin to a coal oil industry which was started in 1851 by James Young, at Bathgate in West Lothian. From a particular coal seam the cannel coal or torbanite seam in Bathgate, Young obtained a yield of 120 galls/ton of crude oil by heating the coal in a retort at a low temperature. On futher distillation, this produced a range of products including naphtha, burning oil and

lubricants. This success led Young to
apply for a patent for his process, Patent
No 13292 of 17 October 1850. Realising
that the through-put by 1860 at his
Bathgate oil works was over 10,000 tons of
coal, and the amount of cannel coal was not
limitless, Young began thinking about
producing oil from bitumenous shales or oil
shales, which existed in large quantities
in Lothian Region. These were identified
by the first geological survey of the area
(Bald 1847) and used by local people as a
source of fuel before that date. Young
prospected in the areas of West Calder and
Midcalder, identifying outcrops of oil
shale on the surface, where he
systematically began buying tracts of land.

1.3 The birth of the shale oil industry

In 1864 Young's patent expired in the UK.
There was also a flood of cheap oil
products from the USA. Undaunted, Young
built another oil works at Addiewell, near
West Calder in 1865. The foundation stone
was laid by Young's friend, David
Livingstone, who was home from Africa at
the time. It is alleged to be the largest
in the world with a maximum through-put of
8 million gallons of oil per annum. The
two works were formed into a company called
Young's Paraffin Light and Mineral Oil Co.
Ltd. At both works oil shale was heated
batchwise in horizontal retorts, coal being
used to fire them. The growing demand for
paraffin led him to design and produce
lamps. As retort design improved the
surplus gas was used to fire the retorts,
gradually phasing out the use of coal.
Coal continued to be used by the oil
companies, but was mostly used for
electricity generation.
    In 1870 there were at least 30 oil shale
companies operating in Scotland, because
Young's patent had expired in 1864,
allowing anyone to set up an oil plant. By
1871, it is recorded that there were 31 oil

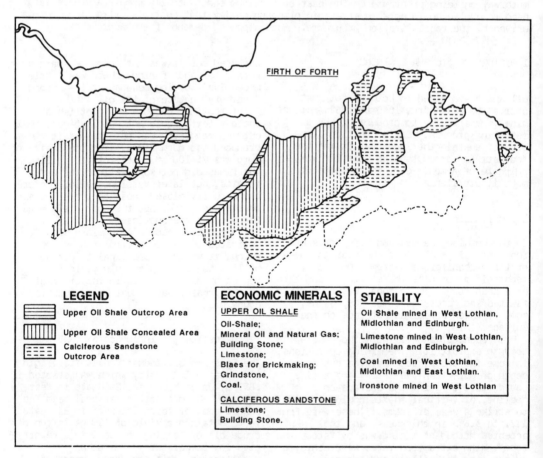

Fig. 1 Upper oil shale strata in Lothian (lower oil shale omitted for clarity)

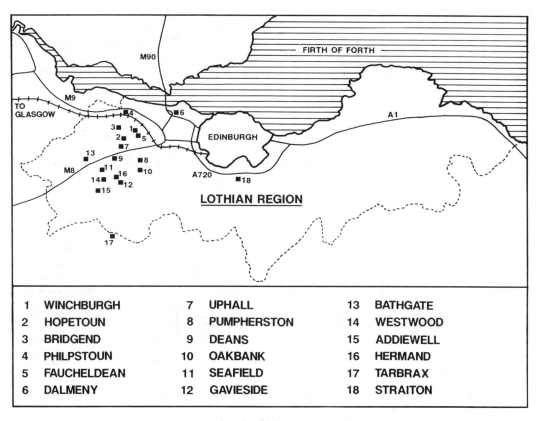

1	WINCHBURGH	7	UPHALL	13	BATHGATE
2	HOPETOUN	8	PUMPHERSTON	14	WESTWOOD
3	BRIDGEND	9	DEANS	15	ADDIEWELL
4	PHILPSTOUN	10	OAKBANK	16	HERMAND
5	FAUCHELDEAN	11	SEAFIELD	17	TARBRAX
6	DALMENY	12	GAVIESIDE	18	STRAITON

Fig. 2 Oil shale refineries operating in 1894

shale companies in operation, with a
through-put of 800,000 tons of oil shale
per annum. By 1894, the number had
dwindled to 18 (Fig. 2), but with improved
mining and retorting operations, the
through-put of oil shale rose to 2,000,000
tons per annum. The through-put of oil
shale continued to rise until it reached a
peak of over 3,000,000 tons per annum,
prior to and during the period of the first
World War. However, the yield of oil per
ton of shale diminished as the richer seams
of shale which lay near the surface, became
worked out.

Oil shale was generally classed as
paraffinic. Although the refining process
was complicated because of nitrogen,
phenols and other unstable compounds, the
process in the later years produced a wide
range of products. (Fig. 3)

1.4 Decline and closure

In 1879, the average yield of oil had
dropped to 34 galls/ton. By 1910, this had
dropped to 27 galls/ton and the output of

shale oil in that year being just over
7,371,000 gallons (273,000 tons). This was
because the industry was working the poorer
seams of oil shale. Early retorts were
10-13 metres high built in rows. The
design of the retorts evolved until the
ultimate design built in 1942, which used
steam and air injection to improve the
overall heat distribution throughout the
oil shale, on its downward passage through
the retort. These retorts processed up to
1,200 tons of oil shale a day, were 11
metres high and operated at a maximum
temperature of 785 c. They were built in
parallel rows with as many as 50 in a row.
(Fig. 4)

During the first World War, the average
production of shale oil was 250,000 tons
per annum. The end of the War, however,
meant that the shale oil companies had to
face financial reality, which was that they
were no longer able separately to compete
with imported oil. They formed themselves
into the Scottish Oils Agency. This became
Scottish Oils Ltd, and while Churchill,
when Chancellor of the Exchequer in the
government in 1928, exempted indigenous oil

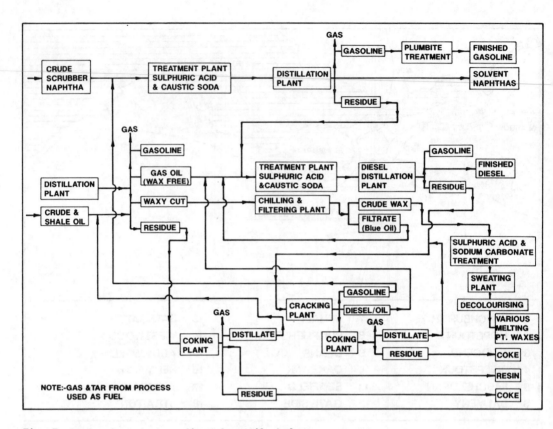

Fig. 3 Range of products refined from oil shale

producers from paying duty on motor gasoline, did slow up the decline of the industry, it did not halt it.

By 1950 there were only 12 operating mines left, supplying 4 works producing crude oil, which supplied the crude oil to a central refinery at Pumpherston.

The oil shale industry in Scotland finally shut down in 1962. At its peak it had employed 10,000 and perhaps as many as 40,000, indirectly dependent on the industry for their livelihood.

2 LEGACY OF DERELICTION

The only reminders today of this once prolific industry, are the numerous highly visible tips of spent oil shale, that one sees from the M8 and M9 motorways and the Edinburgh to Glasgow railway lines. They are highly visible to business people and tourists alike, travelling on the main communications corridors. Their extraordinary pinky red colour and sheer

bulk is what catches the eye. (Fig. 5)

There are no refineries or retorts left, only brick rows of miner's houses in some of oil industry's towns and villages. One unique feature of the tips are the "clinkers", which do not occur in colliery waste tips. Clinkers are due to fusing of hot spent oil shale and are evident as large boulders, which can be more than $2m^3$.

2.1 Scale of the oil shale problem

Today there are 17 major oil shale tips left covering 395 hectares of land. This represents an estimated 100 million tonnes of spent oil shale within an area of 55 square kilometres. (Fig. 6) All are derelict. It represents 19% of Lothian's total derelict problem.

Large areas have also been rendered unstable by mine workings. These areas of underground and surface dereliction present a serious impediment to surface development.

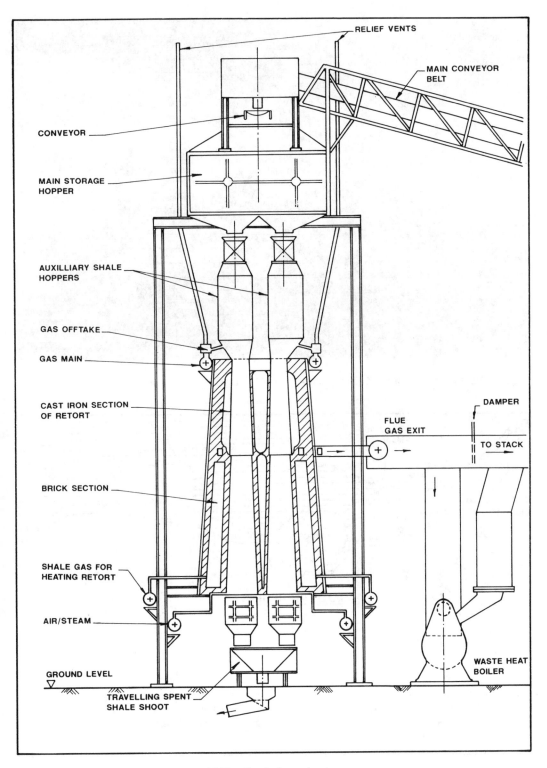

RELIEF VENTS

MAIN CONVEYOR BELT

CONVEYOR

MAIN STORAGE HOPPER

AUXILLIARY SHALE HOPPERS

GAS OFFTAKE

GAS MAIN

CAST IRON SECTION OF RETORT

DAMPER

FLUE GAS EXIT

TO STACK

BRICK SECTION

SHALE GAS FOR HEATING RETORT

AIR/STEAM

WASTE HEAT BOILER

GROUND LEVEL

TRAVELLING SPENT SHALE SHOOT

Fig. 4 Cross section through a 1942 oil shale retort

Fig. 5 Westwood tip, West Calder, known locally as the Five Sisters

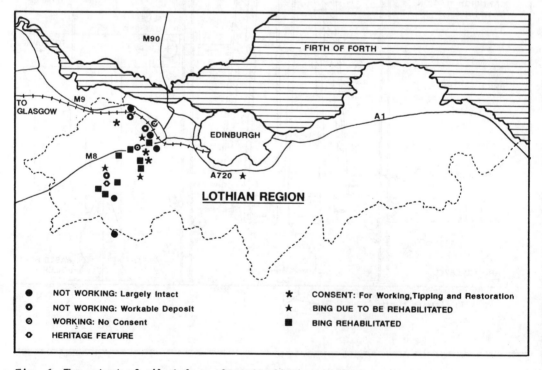

Fig. 6 The extent of oil shale surface dereliction in 1990

2.2 Unstable condition

What is also significant about all of them, is that they are predominantly bare, unvegetated, despite lying undisturbed for at least 30 years, if not between 50 and 80. Clearly the overhead ropeway style of tipping produced side slopes at the maximum angle of repose, and the subsequent effects of weather prevent a stable condition being developed and therefore, a natural greening of their side slopes. The exception to this is the side the ropeway was on itself. The vertical height of these continuous bare side slopes, varies from 40-60 metres. The scale is major.

3. NEED FOR A STRATEGIC POLICY

In Scotland the Regional Councils are responsible for strategic planning and the District Councils for local planning matters. Strategic planning guidance takes the form of a Structure Plan, which is the social and economic development framework for the regional area, for a five year period. The current Lothian Region Structure Plan 1985 is in the process of being reviewed, and a new plan about to be prepared. Lothian's plan covers the district areas of West Lothian, Midlothian, East Lothian and the City of Edinburgh. The 1985 Structure Plan contained two policies to deal with dereliction, one for reclamation and one for extraction.

3.1 Reclamation Policy

Lothian Region's derelict land problem is significant in Scottish terms. Recognising the scale of the problem and the value of a strategic approach, a policy framework was evolved and remains today, a vital component of the Regional Council's commitment to social and economic regeneration. The 1985 Structure Plan stated, in paragraph 10.26, the policies for rehabilitation and the reasons.

"In 1974 it was estimated that Lothian contained 15% of Scotland's derelict land. Despite the considerable efforts of the Regional and District Councils and the Scottish Development Agency to rehabilitate such derelict land, there is still a programme of more than 1,050 hectares of land needing treatment. More of this dereliction is now within settlements and therefore more complex and costly to treat, and consequently greater resources are needed to tackle this changing problem.

EP 22 The Regional Council will support,

encourage and, where appropriate, pursue the continued rehabilitation and restoration of derelict land within Lothian.
EP 23 The Regional Council's programme of derelict land renewal will concentrate upon priority areas, where the programme can support other social, economic and environmental objectives:
Bathgate area
Edinbrugh Green Belt area
Esk Valley area"
The priority areas reflected the greater degree of coalfield dereliction, but Bathgate had both colliery waste and oil shale dereliction.

3.2 An extraction policy

What the stategic plan also recognised was that oil shale waste, was a valuable resource and should be used to offset the need for quarrying natural resources. The 1985 Structure Plan stated in paragraphs 10.27 and 10.28, the policies on mineral workings and the reasons:-

"There are mineral deposits workable by opencast methods which should be protected from sterilisation by permanent development. Areas where these minerals occur have been delineated and should be included in local plans.

EP 25 The Regional Council will encourage:
(i) the removal of minerals to avoid sterilisation in areas where development is proposed;
(ii) the working of more than one mineral from an operation;
(iii) the use of waste materials as an alternative to those occurring naturally.
The District Councils are encouraged to use, to the full, their powers under the Town and Country Planning (Minerals) Act 1981, to achieve environmental improvements to existing planning consents, and to bring mineral operations under planning control.

EP 26 District Councils' should ensure that mineral workings and site restoration to a planned afteruse, are carried out to the highest standards".
The Town and Country Planning (Minerals) Act 1981 was not implemented in Scotland until 1988. This was to have a significant effect on extraction of oil shale waste. Previously a tip was regarded by some as a "chattel" in Scottish Law. However the "chattel" status was challenged on a umber of occassions as this implies a placing with a view to future use. When deposited

the retorted oil shale waste was not viewed as a usable waste.

Extraction of retorted oil shale waste had been allowed to procede because a view was held that it was a mineral operation over land, something which, in planning law is impossible. Had extraction been more properly viewed as an engineering operation, control within the planning system would have been possible. Unfortunately the mineral working overland stance was never seriously challenged by the planning authorities.

When the Central Scotland motorway programme was under construction, large numbers of tips were operating, because no planning consent was required.

4. NEED FOR CHANGE

Since the 1985 Structure Plan came into operation a number of changes have taken place, some of which are a result of Government initiative. The need for a rehabilitation policy however still remains vital today.

4.1 The pace of rehabilitation

The pace of rehabilitation of derelict land continues steadily despite, government funding fluctuations, but the pace of extraction has seen a sharp decline, not only with oil shale, but colliery waste.

Whilst the 1988 Scottish Vacant Land Survey indicated a downturn in the total amount derelict, the scale of the problem still placed Lothian second in Scotland,

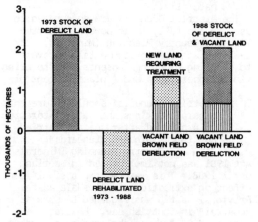

Fig. 7 Survey results for Lothian, 1973 - 1988

with 2041 hectares, only some 417 hectares less than the survey of 1973. (Fig. 7) Analysis of the figures show that, if one takes off the new dereliction due to closure of industry in that period, little has been done to rehabilitate oil shale dereliction.

The total amount rehabilitated by the Council with Scottish Development Agency funding, between surveys, was 688 hectares of land, at a total cost of £14,2 million. This represents 65% of the total rehabilitation effort within Lothian. (Fig 8) Of the 688 hectares rehabilitated, only 144 hectares were oil shale tips. The vast majority activity has been within coalfield areas. Of the 144 hectares, 119 hectares were rehabilitated by the Council through the rehabilitation programme, and 25 hectares by the private sector, by extraction of the waste.

4.2 The pace of extraction

The contribution from the private sector over the same period, through extraction was small, only two tips reaching exhaustion, amounting to 25 hectares. From the three oil shale bings in the Council's ownership the pattern of a downturn in the rate of removal of oil shale can be seen. Over the same period 5.6 million tonnes was extracted but only one is operational today, at a rate of less than 100,000 tonnes a year.

The marketing strategy is interesting. One major operator, still active, sees the oil shale waste as a way of keeping his lorry fleet running, so more or less giving it away. Another saw it as complimentary to a range of construction materials they sold. Others try to make money from it. Nevertheless one could not fail to observe that there were too many sources chasing a declining market, especially as the major roads programme came to a close with the completion of the Edinburgh City Bypass February 1990.

There were more disadvantages than advantages.

1. Until 1988, tips were opened and abandoned as necessary, with little consideration being given to the appearance and safety of the operations. The removal of material from a tip may reduce its physical bulk, but, if badly operated, would not reduce its adverse impact upon its surroundings. On many sites, the net result was an exacerbation of dereliction.

2. The removal of material from tips responded to market forces to provide a low cost fill material.

3. The timescale for exhaustion was often

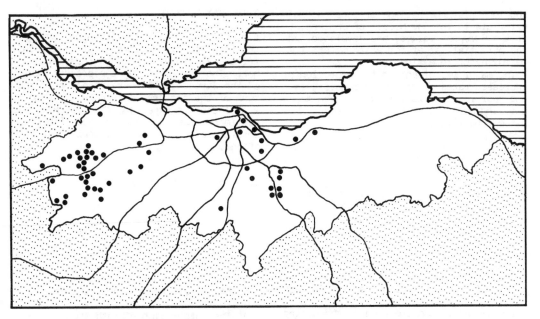

Fig. 8 Derelict land rehabilitated from 1976 to 1990 by Lothian Region

extemely long.

On the other hand, the principle of extraction is an example of recycling a waste material. It results in the saving of natural resources, which could be used for the same purpose. With increasing importance being placed on recycling, it is possible that the government could legislate, to require greater volumes of recycled material to be utilised by the construction industry. The removal of oil shale waste is a small part of the recycling debate.

4.3 Issues arising

In May 1991, the role of the Scottish Development Agency, for whom the Council has operated as agent since 1976 on the rehabilitation programme, will be assumed by Scottish Enterprise and its local company, Lothian and Edinburgh Enterprise Limited. Legislation currently going through Parliament provides the same remit to Scottish Enterprise to rehabilitate derelict land. What influence this will have on the Council's rehabilitation programme and its stategy remains to be seen.

The 1988 minerals legislation, on the other side of the strategy, permits control over oil shale extraction operations in tips. It offers two approaches.

The first is the, "allow operations in all tips" scenario, the second is the, "restrict the number of operational tips" scenario.

There is concern that the first scenario would reduce the bulk of oil shale waste, but would not necessarily reduce the impact of the tips extracted. Given the anticipated demand, it is likely that a competitive market would allow little margin for careful extraction, and result in sites being left in a partially worked condition between contracts.

The second scenario would reduce the intensity of competition, and should ensure a steady market for those consented operations. The concentration of operations should assist to advance the exhaustion, and subsequent restoration date

It is generally accepted, that an undisturbed tip is usually less intrusive than one which has been operated. However, their physical bulk and unnatural appearance, detracts from the local landscape and reduces the potential to attract inward investment and new residents. The control over the release of tips, should be coupled with a programme designed to reduce the impact of the tips, yet retain their status as a resource.

The value of oil shale as a resource is recognised in the Central Belt of Scotland, but this is a restricted market area. The potential of wider markets, has not been considered since the Scottish Development Department study of the 1960's. Since then the long distance haulage of construction materials to the South East of England, and to the European market, has developed and the potential market for waste material may be worthy of re-evaluation.

The major impediment to the promotion of increased rates of oil shale removal, is the poor road and rail links between the tips and both inland destinations and harbour facilities.

4.4 Influence of other new government initiatives

Recent government proposals to investigate improved links from Edinburgh to the south, may create an increase in demand for oil shale for embankment construction, but only for a period of time. No timescale for construction of the "Fast link" from the M8 to M74 motorway, or the upgrading of the A1 and A7 Trunk Roads to dual carriageway, is known. It could be a timescale of 26 years or more.

The second is the creation of the Central Scotland Woodland Company, whose express remit is to aforest derelict land. What influence this will have on implementation of strategic policies is not known, but it does offer a potential solution to green the bare, pinky red faces of the oil shale tips, in a way that could protect the resource for the future.

5 THE WAY FORWARD

A planning and rehabilitation policy for the retorted oil shale tips of West Lothian for the 21 century could be formulated now. The issues facing the present review of the Structure Plan are revelant to the turn of the century, particularly as no new legislation is expected, the 1981 Minerals Act implemented in 1988 and Scottish Enterprise legislation have only just taken effect.

5.1 Planning objectives

The issues explored can be chrystalled into three objectives:-
1. Rehabilitation remains vital, with 73% of the oil shale dereliction remaining untreated. To continue investment in rehabilitation must remain a priority.
2. The promotion of recycling and the possible resultant growth of the use of oil shale waste.
3. The integration of both the rehabilitation and recycling goals into a controlled and coordinated programme, is the most effective way forward.

5.2 A rolling programme

A submission to fund a five year rolling

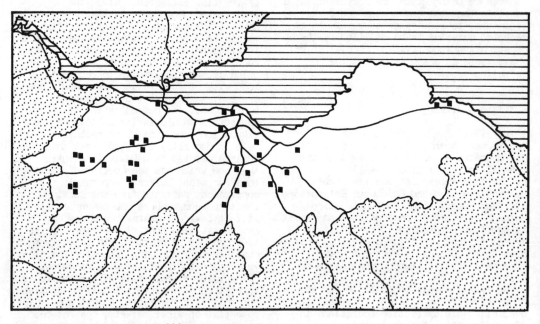

Fig. 9 Proposed 5 year rolling programme

programme to reduce Lothian's dereliction, has already been submitted to Lothian and Edinburgh Enterprise Limited. (Fig. 9) A new, sustained programme, is fundamental to the future of the local economy and of benefit to the region as a whole. West Lothian, because of its oil shale dereliction, must be a priority. Investment in enhancing the image of West Lothian will do much to encourage investment in new development opportunities, arising from a programme of action.

5.3 Integrated promotion

The increase in interest in recycling suggests, that a positive stance should be taken with regard to the use of waste materials, but that the stance be qualified in relation to the removal of oil shale from tips. The qualification should make it clear, that the release of tips should be controlled, to ensure the early exhaustion of those nominated, and in controlling it, assist in minimising the impact of lorry traffic generated by tip

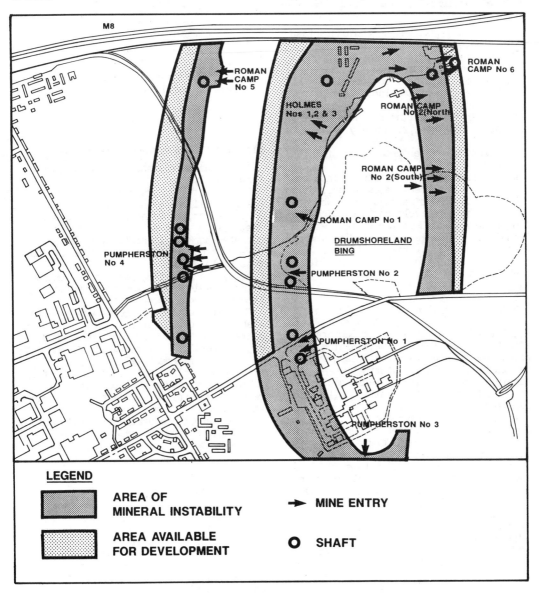

Fig. 10 Extent of oil shale working at Pumpherston

99

removal.

The adoption of a promotional stance would require the additional qualification, that expenditure should be made on improved road and/or rail links, to assist the movement of materials, without any increased impact upon settlements.

5.4 Coordination

A controlled release of oil shale tips must be in concert with the rehabilitation programme priorities. This would have many benefits:

1. Continuous extraction and gauranteed exhaustion.
2. Encourage a more planned approach to extraction and restoration.
3. Control over lorry movements.
4. Earlier use of the land after restoration.
5. Preservation of the resource represented by those relatively undisturbed tips.
6. Create a greener appearance of those preserved.

With a resource that stretches into the 22 century, at current rates of extraction, the establishment of aforested or wooded sides of the tips would not prejudice the long term potential of the waste. With the techniques available today, an investment in greening would, within 10-15 years, significantly soften the image and bulk of these pinky red tips.

5.5 An alternative practical solution

Significant areas of West Lothian are affected by shallow oil shale mining. The acceptance of need to treat underground dereliction, could be a further means of acheiving the overall strategic goal of rehabilitating the West Lothian oil shale dereliction.

In Dudley in the West Midlands, the Department of Environment is funding work in progress, treating underground dereliction, by injecting a paste of minestone into the old limestone workings. Such was the extent of oil shale working, at Pumpherston, east of Livingston, that up to five seams of oil shale were worked amounting to 10m in depth, within an overall depth of 30m. (Fig 10)

Having oil shale to hand at Drumshoreland, a consented source for the next forty years, there is an opportunity to explore whether a similar exercise to that in Dudley could be undertaken.

The Dudley experience may not be

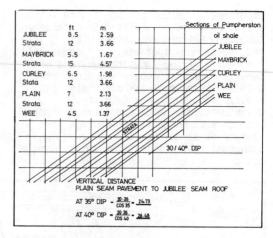

Fig. 11 Cross section through the Pumpherston workings

practical for a number of reasons:-

1. The waste oil shale is too angular to flow readily.
2. The workings are often flooded so a paste of oil shale and water would not run.
3. The nature of the workings is such that cavity volumes are likely to be small isolated pockets in contrast to the cavernous nature of the limestone of the West Midlands.
4. By the retorting process oil shale bulks-up which means the total volume exceeds the cavity space left.

The value of this exercise would be to stabilise the old workings in the five seams which because of their depth and angle, sterilise a large area of land with potential for industrial development. (Fig. 11) A practical solution like this would take pressure off greenfield land and release land for development in accordance with Structure Plan policy.

Seen against the practical problems more investigation and research would be necessary before one could apply the Dudley concept to oil shale workings.

ACKNOWLEDGEMENTS

This paper represents the views of the author which are not necessarily those of Lothian Regional Council. The author wishes to thank the Director of Planning for allowing publication of this paper and to the staff in the Landscape Development Unit for their helpful assistance and comments.

REFERENCES

Cameron, I.B. and McAdam, A.D. 1978. The
    oil shales of the Lothians, Scotland:
    present resource and former workings,
    London: HMSO.
Cook, F.M. 1980. The Scottish Shale Oil
    Industry, Edinburgh.
Scottish Office 1974. Derelict Land Survey
    1973. Edinburgh: HMSO.
Lothian Regional Council Department of
    Planning 1986. Lothian Region Structure
    Plan 1985. Edinburgh: Lothian Regional
    Council, 52-53.
Scottish Office 1975. The Scottish
    Development Agency Act 1975. Edinburgh:
    HMSO.
Lothian Regional Council Department of
    Planning 1989. Structure Plan
    Alteration No 1 Opencast Coal.
    Edinburgh: Lothian Regional Council.
Lothian Regional Council Department of
    Planning 1989. Strategic Policies;
    Progress Report: Edinburgh: Lothian
    Regional Council, 7-8, 67-81
Lothian Regional Council Landscape
    Development Unit Planning 1989.
    Scottish Vacant Land Survey 1988.
    Edinburgh: Lothian Regional Council.
Industry Department for Scotland 1989.
    Towards Scottish Enterprise. Edinburgh:
    HMSO.
Scottish Office Roads Directorate 1990.
    Routes South of Edinburgh; Report on
    Public Consultation and the Government's
    Decisions. Edinburgh: Scottish
    Office.
Scottish Office 1988. Central Scotland
    Woodlands. Edinburgh: Scottish Office.
Lothian Regional Council Landscape
    Development Unit 1990. A five year
    programme to reduce Lothian's derelict
    land. Edinburgh: Lothian Regional Council.
Dudley Metropolitan Borough 1984. Abandoned
    limestone workings, a strategy for
    action, the Council's policy and
    programme. London: HMSO.
Lothian Regional Council Department of
    Planning 1984. Drumshoreland industrial
    area study. Pumpherston, West Lothian.
    Edinburgh: Lothian Regional Council.

Reclamation, Treatment and Utilization of Coal Mining Wastes, Rainbow (ed.) © 1990 Balkema, Rotterdam. ISBN 90 6191 154 0

The application of system engineering to land reclamation planning

Lin Jiacong & Bian Zhengfu
China University of Mining & Technology, Jiangsu, People's Republic of China

ABSTRACT: As viewed from system engineering, the authors deem that the reclamation of land destroyed by mining is an important work of mining engineering, and it must be done in every stage of mine development. This paper pelaces the emphasis on the methods of land reclamation planning. The land reclamation is a complex system engineering and it is relevent to several factors as social, economical, ecological and environmental. Consequently, the viewpoints of system engineering must be considered in process of planning. Here presented are three kinds of methods of land reclamation planning, i.e. the method of economical effect planning, the method of ecological engineering planning and the method of integrated planning formed in consideration of both of the ecological and economical issues. The paper also presents some examples of application of above mentioned methods in Yian Colliery of Xuzhou area in Jiangsu China.

1. INTRODUCTION

In China's coal mining, underground operation is the prevailing mode of production. As a result of surface subsiding and waste heaping, the land in mine area is destroyed rather seriously. Especially in China's east costal plain area, one of the main area of agriculture production, where the terrain is smooth and the water table is high, a series of social, economicil, ecological and environmental problems is arisen from that the

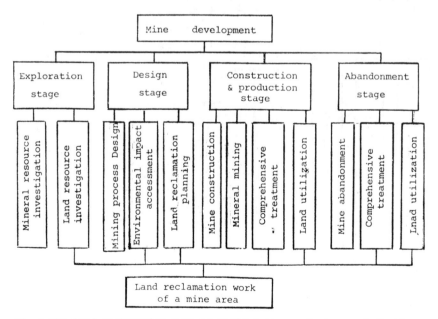

Fig.1 The land reclamation work at various stage of mine exploitation

surface structure, such as residence, industrial buildings, roads and public institutions etc. are destroyed by land destruction, and from that the terrestrial ecosystem is changed with the variation of topography and geomorphy, and from that the acquatic ecosystem is varied by water body destruction. Above all things, here the land destruction is the crux of the matter. The approach to solve this problem is how to reclaim the land properly and reasonably. The land reclamation of a mine area should be tackled in a comprehensive way in the light of the mine area ecology, land economics, environmental science and the theory of district planning etc.

From the viewpoint of system engineering, the land reclamation is an indispensable important work in the whole process of mining. Only synchronized land reclamation engineering with mining engineering, will the process of mine exploitation be more reasonable and can the problem of land destruction be resolved entirely and thoroughly. The relationship between land reclamation and mining engineering is depicted in figure 1.

Since the land reclamation is a complex problem of engineering and technology, like the problem of other engineering, it must be resolved in accodence with the flow chart as shown in figure 2.

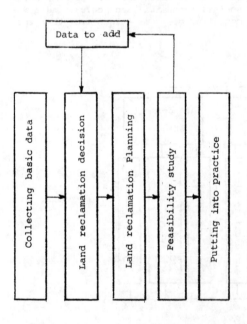

Fig.2 The flow chart of land reclamation of a mine area

What follows in the paper are: a discussion of design procedures of optimum engineering in the process of land reclamation planning; the criteria and basic data necessary for land reclamation of mine area with high water table; a description of three methods of planning with some examples of application.

2. THE THEORETICAL BASIS TO BE RELIED ON FOR WORKING OUT A LAND RECLAMATION PLANNING

2.1 Theory of mining subsidence

Based on the theory of mining subsidence, we can locate the range of destroyed land, draw up the isogram of land subsidence and deformation, calculate the cubic metre of earth necessary for land reclamation etc. According to the forecast we can also take preventive measures to protect the soil of mine area from being eroded and salinized.

2.2 Environmental science and ecology of mine area

The mining subsidence will inevitably lead to a change of ecosystem of the mine area and even lead to the loss of its ecological balance. The mine area with high water table is liable to accumulate water, when subsided, and is prone to change from terrestrial ecosystem into aquatic and terrestrial one. And the vegetation of the earth's surface is damaged and the environment is polluted as well. Consequently, the substance, the energy and the information flow of the ecosystem will lose its balance. So it is necessary to exploit the subsided land by means of the knowledge of environmental science and the exology of mine area.

2.3 System science and theory of district planning

The land reclamation relates to a great many of factors. In order to enable the entire process of land reclamation to be conducted scientifically, it is essintial to apply the theory of system science to ensure the substance, energy and information flow of the system to have a maximum output while the resources are limited. When working out a district plan, the following principles such as the principle of integrity, the function structure principle and coordination principle etc. must be adhered to and so it does in working out the land reclamation plan.

2.4 Land economics

The productive capacity of land varies with the category of the soil and the geographical position where the mine is located. Moreover, being restricted by the factors of the population and the social economy of the mine area, the urgency for the people to reclaim the land is also different.

3. REQUIREMENTS OF LAND RECLAMATION PUT FORWARD IN LINE WITH CHINA'S AGRICULTURE DEVELOPMENT STRATEGY

Most of coal mine areas of high water table is located in China's agricultural area of high yield. The mining of the coal in these areas will cause innegligible influence on its economy. Hence, the land reclamation must be conformed with the requirements of the agriculture development strategy.

3.1 Practising ecological agriculture in the light of the specific conditions of the mine area

China's agriculture is now developing along the direction of ecological agriculture. Though the modes of ecologicol agriculture of coal mine area are different from each other for the sake of their respective conditions, there is a simplified common mode as shown in figure 3.

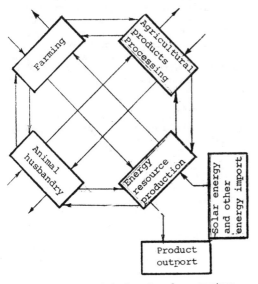

Fig. 3 The simplified mode of ecosystem of coal mine area

To develope ecological agriculture in mine area is to achieve the growth of production and the ecological balance, and to achieve the utilization and regeneration of energy and the economic effect as well. For realization of this purpose, there are two approaches to be adopted. Firstly, to develope and utilize as many subsided lands as possible, not to leave them out of cultivation. Secondly, to strive to have a rational production distribution as to utilize the energy in a multi-level way.

3.2 Making use of resource superiority of local district to develope economy of the mine area

The agriculture of the mine area should first serve itself. The feature of its agricultural structure should be as follows: to farm vegetable, melon and fruit, and industrial crops; to run animal husbandry with meat, egg, poultry, fish, milk etc. as the main products; to run the processing trade of agricultural products and side-line products, and the service trades of transportation, preservation and storage, packing and selling etc. By relying only on such a structure can we bring the resoure superiority of the local district into full play to develope the economy of the mine area.

3.3 Bring soil erosion under control and preventing land from salinization

The area disturbed by mining falls into sloping field and water-logged ground. If they are not to be tackled or not to be cultivated year in year out, the soil on the slope is liable to be eroded and the water-logged ground is easy to be salinized. Furthermore, when the drainage system is out of order or the water table of the entire mine area is changed due to the underground mining, the soil over this area is liable to be salinized as well. The erosion and salinization is very harmful to agriculture production. The best measure to prevent the land from salinizing and to bring the soil erosion under control is to practise ecological agriculture.

4. METHODS OF LAND RECLAMATION PLANNING

Here presented are three kinds of methods suitble for land reclamation planning of mine area. The examples of its application will be introduced later on.

4.1 The method of economic effect planning

4.1.1 Principle
Make the economic effect of the land reclamation to be optimized while satisfying the demands of a great many of constraints.

4.1.2 Contents
Regulate the structure of land utilization so as tp achieve optimum overall effect. Herein the structure of land utilization referes to the area proportion of various kinds of land used for cultivation, orchard and fish pond etc.

4.1.3 Model
Given: $Z(t)$ -- the overall effect in t years; $Cj(t)$ -- the economic effects per unit area of each kind of lands used in t years; $Xj(t)$ -- the area of each kind of land used in t years, then we have:

Objective function

$$\text{Max } Z(t) = \sum_j C_j(t) \cdot X_j(t)$$

Constraints

$$\sum_j a_{ij}(t) \cdot x_j(t) \gtreqless m_j(t)$$
$$x_j(t) \geqq 0$$

If the coefficient of t years is a constant, the above mentioned model can be reduced to:

$$\max Z = \sum c_j x_j$$
$$\begin{cases} \sum a_{ij} x_j \gtreqless m_j \\ x_j \geqq 0 \end{cases}$$

It is a simple model of linear programming and it can be readily solved. The crux of the matter consists in how to formulate the constraint equation and how to seek out the parameters (Cj, a_{ij} m_j) of the model. therefore, it is necessary to make some investigations on them. Hereafter presented are some kinds of common constraints which were met frequently during the land reclamation.

(i) Fund constraints   A large number of funds must be expended for land excavating and for soil fertilizing in the process of land reclamation, but the investment in doing so is limited. In order to achieve the optimum economic effect with such an investment, it will be constrained by funds. For example, to reclaim 1 mu* of subsided land requires an investment of a_{11} Yuan and to dig 1 mu of fish pond requires an investment of a_{12} Yuan and so on and so forth, but the general investment os m_1 Yuan, then we have

$$a_{11}x_1 + a_{12}x_2 + \cdots \cdots \leqq m_1$$

(ii) Resource constraints   No matter what wastes (fly ash from power plant, refuse or other solid scraps) are used for backfilling the subsided area, the total amount of such wastes is limited. In addition, when running agriculture, the local water resouce is also limited. For example, it is known from experience or calculation that to restore 1 mu of land requires a_{21} cubic metres of earth, but to dig 1 mu of pond, we can obtaine a cubic metre of earth, Now the filling materials of all kinds come to m_2 cubic metres, then we have

$$a_{21} - a_{22} \leqq m_2$$

(iii) Labour force constraints   Given: m_3 -- workdays to be put into land reclamation per year; a_{31} and a_{32} -- workdays to be put into restoring land and digging fish pond per mu respectively, then we have

$$a_{31}x_1 + a_{32}x_2 + \cdots \cdots \leqq m_3$$

(iv) Demand constraints   The surface subsidence of mine area leads to a decrease of crops output and increase of nonagriculture population, whereas their demand on grain and oil etc. is growing increasingly. Thus the area of land reclamation will be costrained by them. For example, the mine area requires n kg of grain, the yield per unit area is P kg, the area of undestroyed land is S mu, of which the yield per unit area is q kg, then the restored land X_1 must satisfy:

$$x_1 \geqq \frac{n - sxq}{P}$$

(v) Policy constraints   In accordance with the policy of the state and the local regulations, there are also a great many of constraints to be adhered to. For example, the proportion of afforestation in mine area must amount to η, then we have

$$X \text{ forest} \geqq \eta \cdot \text{area of mine}$$

In above mentioned models, if there is a need to show the economic effect, one might assume Z to be the ratio of output and input.

4.2 The method of ecological engineering planning

4.2.1 The object of planning
When working out the ecological planning for a mine area which is located in an agricultural area, the ecological agriculture must be taken as the object of planning, whereas the landscape of a mine area, which is located in a suburb, must be restored to the level accepted by the resident there.

----

*mu -- Chinese area unit, 1 mu=666 m²

4.2.2 Principle   In the light of the principle of food chain and the principle of trophic circulation, a structural model which is suited to the features of agricultural production of the mine area, will be designed. Then an optimum design must be conducted on the structural model of the system with the method of system analysis.

4.2.3 Contents   The ecological engineering planning consists of planning of system structure and the technological planning. The structure planning consists of trophic structure planning, plane structure planning, vertical structure planning and time structure planning etc. Among them the trophic structure planning is the basic planning, of which the optimum design must be conducted firstly.

4.2.4 Method and procedure

(i) In accordance with the feature of destroyed land, the rule of ecological change of the mine area and the experience extracted from the running of ecological agriculture of his own area or other area, one can find out the model of trophic structure suitable for production of his own area.
(ii) The optimization of the model of trophic structure
Transforming the model of trophic structure into the form of vector diagram, then with reference to it one can write out the adjacent matrix of the vector diagram $A=[a_{ij}]$
Where

$$a_{ij}=\begin{cases} 1 \text{ as the line segment goes from } A_i \text{ to } A_j \\ \quad (A_i \text{ has an effect on } A_j) \\ 0 \text{ as } i=j \text{ or } A_i \text{ has no effect on } A_j \end{cases}$$

According to A to find out reachability matrix R
Assuming C = A + I, where I means unit matrix
Perform Boolean operation on C until $C^r = C^{r-1}$ $(r \leq n-1)$, then we have R = $[C^r]^T$
· From R write out the reachability set $R(A_i)$ and prefictor set $G(A_i)$. $R(A_i)$ consists of the cell corresponding to the column, whose entire elements is equal to 1, of the raws, corresponding to the A_i, of reachability matrix, while $G(A_i)$ consists of the cell corresponding to the raws, whose entire elements is equal to 1, of the columns, corresponding to the A_i, of the reachability matrix.
· Find out the cell of the highest stage. if $R(A_i) \cap G(A_i) = R(A_i)$, then A_i will be the cell of the highest level.
· Cross out the cell of the highest level

which has already been found out and then seek the cell of the highest stage of the remaining subset. Deducing by analogy, we can find out the cells contained in every depth.
· From the reachability matrix R, form the corresponding structure model.
· Form the structure model form the improved trophic structure model.
    iii. Make an assessment on the improved structure model.
    iv. In accordance with the improved structure model, design the plane structure, the vertical structure and the time structure.
    v. Make a general assessment on the entire structure design.
    vi. Make technological planning.

4.3 The method of integrated planning of ecological and economical issues

Adding the ecological constraints into the model of method of economic effect planning, one can obtain the method of integrated planning of ecological and economical issues. The ecological constraints consist of the one conformed with the requirements of general ecologicol standard and the one conformed with the requirements of ecological engineering planning.

4.3.1 The constraint conformed with the requirements of ecological standard

When working out the integrated planning, it is necessary to take the inner coordination in to account on the one hand, and it is also necessary to have a coordination, called the external one, with the development planning of the neighbouring district and the locality on the other hand. The inner coordination means that there are both land and water, and there can be planted trees and flowers as well. Thus we can attain the category of land utilization essential for the planned area. The reason why we must take the external coordination into acoount is that if we regard the planned area as a subsystem, the planned area with the neighbouring disricts will form a system, and if we want to achieve the ecological balence of the planned area, there is a need to take the external coordnation into account. For example, the periphery of a planned area is a factory district with serious pollution, it is not suitable for the planned area to raise poultry and livestock there.

107

4.3.2 Constraiats of absolute area

In conformity with the requirements of agricultural production or with the data of historical climate of a district, a certain amount of water area and planted area is needed to make the climate environment there be mantained stable and the demand of developing production be satisfied. This belongs to the category of such a constraint. For example, there is an area where the drought happens frequently, the subsided land there must be transformed into a reservoir ready for agricultural irrigation in the time of dry season.

4.3.3 Propotional constraint

In order to ensure the rational circulation and the full use of substance and energy, require a certain ratio between the various lands. For example, to raise a certain number of oxen and sheep needs X_1 mu grass to be grown. In addition to X_1 mu grass land, X_2 mu forage crop land needs to grow as well. Hence, the proportional constraint of X_1 and X_2 is obtained.
The ecological constraints can be divided into quantitative and non-quantitative ones. Among them the quantitative constraint can be expressed in the form of constraint equation, while the non-quantitative one can be expressed in the process of structural planning and technological planning.

5. THE APPLICATION EXAMPLE OF THE PLANNING METHODS

We have made a plan for about 1000 mu subsided land of Yian Colliery in Xuzhou mine area.

5.1 The general conditions of the planned area

There is an isogram of surface subsidence of Yian Colliery. A land reclamation plan has been worked out for the subsided land (about 860 mu) in the range of the mine field. The population of the planned area is over 700. Formerly the terrain of this area is smooth, And the land within this area was all good farmland suitable for planting sereal crops such as maize, soybean, cotton, wheat, rice and peanut, and industrial crops. The height mark of the average water table is 34.5 m and the highest one is up to 35.0 m. During the pluvial period, the waterlogged rate is over 30%. The schematic isogram of surface subsidence is shown in figure 5.

The said planned area is not far from Xuzhou city. This is a favourable condition. On the basis of an investigation, the agricultural products, especially the vegetable and aquatic products are in great demand. The social and economical condition of the districts next to the planned area is as follows: There are only 0.1 mu cultivated area per capita in part of the area, due to the cultivated land being subsided on a larg scale by coal extraction; The local peasant has already excavated a number of fish ponds to raise; The local peasant has no certainty of success in a new system of cultivation, so they did not pay much attention to it.

5.2 The economical effect planning

In accordance with the demand of production and the resource now available, the constraints can be classifid into four main kinds, i.e. the total quantity amount constraint, the allocation quantity constraint, the absolute quantity constraint and the non-negative constraint.
The data for the total quantity constraint are listed in table 1.
The allocation quantity constraint includes: To raise 1 mu fish needs to allocate 0.5 mu forage land, namely

$$X_6 \geq 0.5 \ X_3$$

The absolute quantity constraint includes: Being restricted by the level of technology and by the source of forage etc., the piggery and the chicken farm have the following constraints:

$$X_1 \leq 20$$

$$X_2 \leq 40$$

The planned area needs 200,000 kg cereals. Given that, the per mu yield is 600 kg, then the constraint of planted area is as follows:

$$X_4 \geq 333$$

The non-negative constraint includes: X_1, X_2, X_3, X_4, X_5, X_6, $X_7 \geq 0$

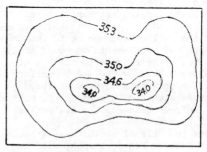

Fig.5 Schematic isogram of surface subsidence

Table 1 The total quantity constraint

Direction of utilization	Net income yuan/y·mu	Reclamation investment yuan/mu (1)	Labour input man/y·mu (2)	Service land mu/mu (3)	Amount of land area (4)	Cubic metre of earth per mu (5)
Piggery X_1	14000*	34500**	3*	0.03	1	-200
Chicken farm X_2	10000*	21180**	4*	0.03	1	-200
Fishery X_3	800	1500	0.1	0.01	1	700
Cereal crops X_4	600	2000	0.1	0.01	1	-400
Orchard X_5	800	1200	0.02	0.02	1	-200
Forage land X_6	167	1200	0.05	0.005	1	-400
Service land X_7	0	700	0	-1	1	0
$\sum c_i x_i$		≤172	≤300	≤0	=860	≥0

*To raised two pigs needs 14 square metres of land. The income per year is 150 Yuan.
One can raise 40 pigs a year.
   To raise 10 chikens needs 1 square metre of land. The income of a chicken is 1.5 Yuan.
One can raise 1500 chickens a year.
**It includes the investment for building the piggery and chicken farm.
   The piggery counts 50 Yuan per 1 square metre, and the chicken farm counts 30 Yuan
per 1 square metre.

The solution of the following two kinds of cases has been conducted by using LP program.
   i) Taken into account are only the directions of utilization of X_3, X_4, X_5, X_6, X_7;
   ii) Seven kinds of directions of utilization as X_1 to X_7 have been taken into account.
The result of calculation shows:
   1) If take the fishery, the crops planting and the orchard into account, we have

$X_{fishery}$=296.1 mu, X_{cereal}=333 mu,
$X_{orchard}$=74.3 mu, X_{forage}=148.1 mu,
$X_{service}$=8.5 mu.

There needs an investment of 1229.9 thousand Yuan. The number of employment is 72 and the cubic metre of earth is 207 thousand. The cut and fill keep balanced. The maximum net income is up to 520.9 thousand Yuan. Calculating in terms of the compound interest rate i = 15%, the time period for recovery of investment is 3.1 year and the income per capita is 744 Yuan.
The advantages of this plan consist in that the subsided land can be fully developed for cereal crops planting, fruit and woods growing, and for aquatic products raising; The volume of earthwork keep balanced; the funs and resources availble to the planned area can be fully utilized. But there exist some problems i.e. the number of employment is not greater; and the income per capita is lower. Even in this case, the economic benifits are pretty good in comparison with the countryside not located in the subsided area.
   ii) In addition to the above mentioned directions of utilization, take poultry and livestock raise into account, we have

X_{pig}=20 mu , $X_{chicken}$=40 mu,
X_{fish}=211.5 mu, X_{fruit}=468.9 mu,
X_{forage}=105.8 mu, $X_{service}$=13.8 mu.

There needs an investment of 2133 thousand Yuan. The number of employment is up to 251 and the volume of earthwork is 143,000 cubic metres. The cut and fill keep balanced. The maximum net income is up to 1242 thousand Yuan. Calculating in terms of the compound interest rate i=15%, the time period for recovery of investment is 2.2 year and the income per capita is up to 1774.27 Yuan. The advantages of this plan is that the

snbsided land can be fully developed for
fruit and woods growing, aquatic product
raising and poultry and livestock raising;
the volume of earthwork is smaller, whill
the mumber of employment and the income per
capita are much greater. But there exist
some problems i.e. the fundsand resources
offered only by the plannde area is insufficient. For one thing, it is needed to
raise funds. The initial investment is 2183
thousand Yuan. There are only 1720 thousand
Yuan in table 1. It still needs 463 thousand Yuan; For another thing, from among
the constraint conditions, we have taken
into account only the resource of forage
anailable to raise fish, while to raise the
pig and chicken also needs a great quantity
of forage purchased from other district;
Furthermore, the grain ration of the local
resident remains to be solved. If we can
find a way to solve these problems, the
plan may be rated as a good one. Otherwise,
the fist plan still must be chosen as a
good one.

   ii) The approach to the problems of the
second plan is to conduct a coordination
through overall arrangement, i.e. if there
are cultivable lands in a still larger area,
the lands can be built into cereal and
forage bases. The development of the planned area must be carried out in accordance
with the second plan.

5.3 The method of ecological engineering
planning

The planned area is one with high water
table, of which the major work to be developed is to tackle the waterlogged land and
the sloping land in the subsided area. The
reclaimed land, when used for planting, is
always with a problem of poor fertility.
So, there needs to take measures to enhance
the fertility of the soil. A trophic stru-

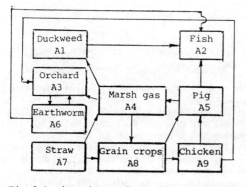

Fig.6 Design chart of trophic structure

cture, as shown in figure 6, is designed
in accordanc with the experiences now available to some mine areas.
   From figure 6, make a vector diagram as
shown in fig 7.

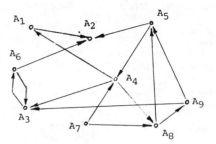

Fig.7 Vector diagram

From figure 7, write out the adjacent matrix
A of the vector diagram and perform the
following calculations C=A+I, R=(C^r)$^{\overline{T}}$, then
we have

$$
A = \begin{bmatrix}
0 & 1 & 0 & 0 & 0 & 0 & 0 & 0 & 0 \\
0 & 0 & 0 & 0 & 0 & 0 & 0 & 0 & 0 \\
0 & 0 & 0 & 0 & 0 & 1 & 0 & 0 & 0 \\
1 & 0 & 1 & 0 & 0 & 0 & 0 & 1 & 0 \\
0 & 1 & 0 & 1 & 0 & 0 & 0 & 0 & 0 \\
0 & 1 & 1 & 0 & 0 & 0 & 0 & 0 & 0 \\
0 & 0 & 0 & 1 & 0 & 0 & 0 & 1 & 0 \\
0 & 0 & 0 & 0 & 1 & 0 & 0 & 0 & 1 \\
0 & 0 & 1 & 0 & 1 & 0 & 0 & 0 & 0
\end{bmatrix}
$$

$$
R = \begin{bmatrix}
1 & 0 & 0 & 1 & 1 & 0 & 1 & 1 & 1 \\
1 & 1 & 1 & 1 & 1 & 1 & 1 & 1 & 1 \\
0 & 0 & 1 & 1 & 1 & 0 & 1 & 1 & 1 \\
0 & 0 & 0 & 1 & 1 & 0 & 1 & 1 & 1 \\
0 & 0 & 0 & 1 & 1 & 0 & 1 & 1 & 1 \\
0 & 0 & 1 & 1 & 1 & 1 & 1 & 1 & 1 \\
0 & 0 & 0 & 0 & 0 & 0 & 1 & 0 & 0 \\
0 & 0 & 0 & 1 & 1 & 0 & 1 & 1 & 1 \\
0 & 0 & 0 & 1 & 1 & 0 & 1 & 1 & 1
\end{bmatrix}
$$

From R, conduct stratification analysis on
the model of trophic structure, as shown
in table 2. After the first operation,
obtain the cell A$_7$ in the first level,
namely in the highest level.
Cross out the elements of the row and column
where the A$_7$ lies. Then perform calculation
according to above mentioned method and
obtain each cell in second level as A$_4$, A$_5$,
A$_8$ and A$_9$. We can also abtain each cell in
third stage as A$_1$, A$_3$ and A$_6$, and each
cell ﾠ in fourth level as A$_2$ (see table 2).
From the reachability matrix, draw out the
model of optimized structure as shown in
figure 8.

Table 2 The stratification analysis of the model of trophic structure

Cell	R(A_i)	G(A_i)	R∩G
1	1, 4, 5, 7, 8, 9	1, 2	1
2	1, 2, 3, 4, 5, 6, 7, 8, 9	2	2
3	3, 4, 5, 6, 7, 8, 9	2, 3, 6	3, 6
4	4, 5, 7, 8, 9	1, 2, 3, 4, 5, 6, 8, 9	4, 5, 8, 9
5	4, 5, 7, 8, 9	1, 2, 3, 4, 5, 6, 8, 9	4, 5, 8, 9
6	3, 4, 5, 6, 7, 8, 9	2, 3, 6	3, 6
7	7	1, 2, 3, 4, 5, 6, 7, 8, 9	7
8	4, 5, 7, 8, 9	1, 2, 3, 4, 5, 6, 8, 9	4, 5, 8, 9
9	4, 5, 7, 8, 9	1, 2, 3, 4, 5, 6, 8, 9	4, 5, 8, 9
1	1, 4, 5, 8, 9	1, 2	1
2	1, 2, 3, 4, 5, 6, 8, 9	2	2
3	3, 4, 5, 6, 8, 9	2, 3, 6	3, 6
4	4, 5, 8, 9	1, 2, 3, 4, 5, 6, 8, 9	4, 5, 8, 9
5	4, 5, 8, 9	1, 2, 3, 4, 5, 6, 8, 9	4, 5, 8, 9
6	3, 4, 5, 6, 8, 9	2, 3, 6	3, 6
8	4, 5, 8, 9	1, 2, 3, 4, 5, 6, 8, 9	4, 5, 8, 9
9	4, 5, 8, 9	1, 2, 3, 4, 5, 6, 8, 9	4, 5, 8, 9
1	1	1, 2	1
2	1, 2, 3, 6	2	2
3	3, 6	3, 6	3, 6
6	3, 6	3, 6	3, 6

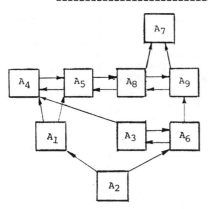

Fig.8 model of optimized structure

From the model of structure, draw out the model of optimized trophic structure as shown in figure 9

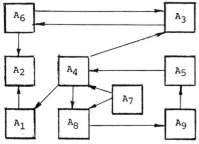

Fig.9 Model of optimezed trophic structure

Having had the model of optimized trophic structure, we can conduct the plane design, the space design, the time design and the technological planning. Some interpretations on the model of optimized trophic structure are given below:

i) The optimized model still more conforms to the theory of multilevel utilization of energy and to the theory of food chain and leads to a full utilization of energy in multilevel. The four cells in second level has just formed a circulation -- the core of circulation utilization of energy of the system, of which the land area occupied by each link must be arranged in a right proportion and the land structure must be also adjudted in accordance with the method of economical effect planning without consideration of this constraint.

ii) To put the straw into circulation is the need to fertilize the soil. There are two ways to do so. One is to put it back directly into the fields, the other is to put it into the methane-generating pit to ferment. It is assumed that the latter can satisfy both sides. On the one hand, the methane can be obtained, on the other hand, the methane sediment can be used for fertilizing the soil and for raising the fish.

iii) In comparation with the method of the economical effect planning, the said method adds a link of methane producing and enable the waste of various link to be fully utilized.

111

5.4 The method of integrated planning of ecological and economical issues

It is clear that both the above methods can solve a number of problems encountered in land reclamation, but there still exist some problems. Therefore, based on the method of integrated planning the following plan has been worked out on the planned area:

On the basis of the economical effect planning, the following constraints have been taken into account:

$$X_2/X_1 \geqq 0.8$$

to ensure sufficient chicken excrements for pig raising

$$\begin{cases} X_4+X_6 \geqq 30X_2 \\ X_4 \quad \geqq 333 \end{cases}$$

to ensure grain ration of the resident and chicken forage

$$X_5 \geqq 100$$

to ensure forested rate

The land used for building the methane-generating pit may be included into the service land.

The results of calculation are:

$X_{pig} = 18.02$ mu, $X_{chicken} = 14.42$ mu

$X_{fish} = 284.96$ mu, $X_{cereal} = 432.46$ mu

$X_{fruit} = 100$ mu, $X_{service} = 10.15$ mu

The total amount of investment is 2346.6 thousand except the investment of methane-generating pit. The number of employment is 185. The volume of earth work is 199,000 cubic metres. The cut and fill keep balanced. The maximum net income is up to 963.9 thousand Yuan, except the econumic benifit of the waste utilization. Calculating in terms of the compound interest rate i=15%, the time period for recovery of invesment is 3.2 year and the income per capita is up to 1377 Yuan.

It is evident that the results of this method of planning is more reasonable for the planned area. It not only solves the problem of the grain ration of the resident and the source of forage, but also takes account of the constraint of ecological criteria and attains a full utilization of the waste of varions links. The problem here is that the initial investment is greater.

It is common to the land reclamation. No matter what method of planning is adopted, there exists the resouce of funds to be solved. Consequently, it is suggested that the plan abtained in accordance with this method of planning must be put into effect while woking out a plan of land reclamation for the mine area.

To sum up, the method of economical effectiveness planning is suitable to the mine area where the economical effectiveness is emphasized, while the ecological effectiveess is not required highly in terms of the overal planning. The method of ecological engineering planning is suitble for the mine area where the ecological effect outweighs the economical effect. Wheresa, the method of integrated planning of ecological and economical issues takes account of the advantages both of the above methods. It is rather rational to adopt this method on condition that the entire reclamation work of the mine area has no overall planning.

REFERENCES

Stanislaw C. Mularz. 1979. Reclamation Problams in the Upper Silesia Mining District, Poland. Ecology & Coal Resource Dev.

M.M. Singh and Bhattacharya. 1987. Proposed criteria for assessing subsidence damage to renewable resource lands, Mining Engineering, Mar.

R.J. Sweigard and R.V. Ramani. 1986. Site planning proess: Application to land use potential evaluation for mined land, Mining Engineering, June.

左军，层次分析法中判断矩阵的间接给出法，系统工程，1988年第6期

许若宁等，层次分析法中 Fuzzy 判断矩阵的建立及其排序，系统工程，1988年第9期

林家聪等，采矿企业造地复田时的矿山测量工作，矿业译丛，1987年第1～4期

林增述和严量，土地管理原理与方法，中国人民大学出版社，1986年6月

Application perspectives of ecological function of reclamations in the conditions of economy restructuring in Czechoslovakia

H.Čížková
Institute of Industrial Landscape Ecology of CSAV, Ostrava, Czechoslovakia

ABSTRACT: On an example of the Ostrava-Karviná area with underground bituminous coal mining the paper explains the importance of ecological role of reclamation measures taken to eliminate damages in landscape caused by mining and related activities, or to create a new, post-mining landscape. Reasons for an existing and unsatisfactory situation in the implementation of ecological function by reclamation are briefly outlined. Based on this analysis and with regard to new perspectives resulting from recent social changes in Czechoslovakia the main solution criteria are formulated, especially in the organizational and administration as well as economic spheres.

1 INTRODUCTION

Underground mining and bituminous coal preparation always represent a significant interference with the structure and function of ecosystems in the landscape, which more or less endangers the ecological stability of large territorial units of mining areas and causes a mediated secondary disturbance of social and economic processes occurring in these areas. For instance, in the Ostrava-Karviná Coalfield with an area of about 325 km^2 presently due to a long-term activity 26 km^2 are strongly devastated and more than 50 km^2 are seriously deteriorated (out of which 5.4 km^2 are covered by mine waste and flotation waste spoil banks, 8 km^2 are covered by lagoons and waste water settling ponds). The scope, complexity and mutual relation of mining effects necessitate adequate measures to be taken in form of comprehensive repairing and reconstruction and recovery activities (usually done by reclamation) which would meet not only economic targets but also equally important ecological and social targets. Each measure taken should therefore fulfill (defined with a certain simplification)
a) economic function, representing an increased economic potential of

the given territory, either directly (by recovery of deteriorated and devastated areas for agricultural, foresting or water management utilization), or indirectly (creation of conditions for an increased efficiency and quality of other economic activities in a mining area);
b) social function, oriented at improving of conditions for social utilization of mining area (creation of more favourable conditions for living, for people's work, housing and recreation);
c) ecological function, consisting in promoting the ecostabilizing links in a mining area both by a reasonable improvement of essential features of ecostabilizing landscape structure elements (especially non-agricultural landscape verdure) and by improvement of conditions for conservation and development of ecologically significant interactions in the territory concerned.

2 PRESENT LEVEL OF IMPLEMENTATION OF RECLAMATION ECOLOGICAL FUNCTION IN THE OSTRAVA-KARVINÁ COALFIELD

Requirements for a greater balance of all functions of reclamation activities and for a comprehensiveness of repairing and regenerating measures within the landscape rise in dependence on the extent and quality of prevention measures applied, minimizing (or at least reducing) the scope and importance of damages caused by coal mining effects:

a) if an efficient prevention of environmental damages is implemented, then the reclamation activities are aimed at correction of such local disturbances which could not be avoided (reclamation should be a means facilitating natural recovery processes of ecologically important elements and links in the mining area and be a tool for optimizing the functional structure of landscape utilization - including rational use of primary and secondary regional resources); the core of ecological function lies in this case in the "microlevel" of landscape ecology (i.e. for example in recovery of water regime, in intensification of pedogenetic processes and succession of plants on reclaimed localities, etc.);

b) if for any reasons the prevention effect is insufficient (or if no prevention measures are taken), the reclamation is focused on a general transformation of large territorial units characterized by a varying degree of devastation and deterioration of ecological conditions and by a considerable (often even antagonistic) differenciation of interests referring to a future utilization of reclaimed territory (reclamation becomes then a potential tool of landscape creation as a whole which should meet efficiently and qualitatively its economic, social and ecological role); the ecological function is implemented both in the "microlevel" (see above) and in the "macrolevel" of landscape ecology (recovery and optimization of large area ecological structures and their functions). With regard to a minimum prevention practised in the Ostrava-Karviná Coalfield (which is explained mainly by apparently high financial requirements of stowing, or by an exceptionally complex application of controlled production advances), reclamation here should secure a full reconstruction and recovery of large deteriorated territory units. Reclamation activity (ensured since 1962 by the Concern enterprise OKD-Rekultivace Havířov as contractor and mostly also projector and financed by a specific Concern fund

of damages and compensations, administered by the Concern management of the Ostrava-Karviná Collieries) is oriented at fulfilling the economic functions: it is focused especially on the recovery of economic damages on agricultural and forest land, or eventually on the property of ofganizations and persons (see Table 1).

Table 1. Extent of rehabilitation improving of landscape sanitary conditions, removing of houses destroyed by coal mining etc.) and reclamation activities in the Ostrava-Karviná Coalfield in the period of 1962 - 1989

Period	Reclamation (ha)	
	agricultural	forestral
1962-1965	313,4	217,6
1965-1970	76,2	178,1
1971-1975	306,2	100,2
1976-1980	344,8	184,8
1981-1985	381,2	73,7
1986-1989	568,3	59,5
1962-1989	1990,5	813,9

Period	Reclamation (ha)	
	fruit farming	water management
1962-1965	78,9	-
1965-1970	32,7	-
1971-1975	-	-
1976-1980	-	5,8
1981-1985	-	11,3
1986-1989	-	3,35
1962-1989	111,6	20,45

Period	rehabilitation (ha)	costs (mill. crowns)
1962-1965	269,1	55,013
1965-1970	217,3	86,243
1971-1975	21,7	161,069

Period	rehabilitation (ha)	costs (mill. crowns)
1976-1980	17,6	351,275
1981-1985	6,0	548,081
1986-1989	39,1	479,091
1962-1989	570,8	1680,772

The fulfillment of social and ecological function is "automatically" taken for granted by actually doing the reclamation - this proclaimed formal precondition is however not well-founded by an objective evaluation and continuous checking. On the contrary, the undervaluation of the significance of the reclamation ecological function brings about also in the Ostrava-Karviná Coalfield an occurrence of other environmental damages caused by a selection of ecologically inadequate reclamation target, or by endagering, degradation and devastation of relatively non--disturbed ecosystems in unsuitable reclamation performed or in the so called "substitutional reclamation" (Čížková and Smolík 1989). In case the ecological function of reclamation is neglected already in the planning and projection stage, then the result are economic losses which are difficult to prove - subsequent superficial reclamation of landscape that does not suit the natural conditions of the given territory represents both a threat to own economic objective of the reclamation (the yields achieved are lower than envisaged, etc.) and low efficiency of the energy, work, material and financial means used.

3 REASONS OF AN UNSATISFACTORY LEVEL OF APPLYING THE RECLAMATION ECOLOGICAL FUNCTION IN THE OSTRAVA--KARVINÁ COALFIELD

The reasons for an unsatisfactory level of respecting and implementing the ecological function of reclamation in the Ostrava-Karviná Coalfield cannot be seen only in the activity of the reclamation enterprise OKD--Rekultivace. The complex of the basic and essential causes is much more serious - closely related to the system of the living environment protection in Czechoslovakia and with its projection into direct and indirect tools of economic and social development management.

Although since November 1989 a whole series of hopeful social changes have been taking place, most of the existing, from the environmental point of view unsuitable regulations, rules and conditions affecting the management of reclamations resulting into the application with undesirab-le consequences, are still existing. Among the main causes are:

a) Insufficient fulfillment of the goverment's ecological role in the past 40 years, which resulted into a neglect and undervaluation of the significance of ecological values and relations and in the subsequent substantial preference for short exploitation economic interests to ecological interests. The consequence was a development of economic structure that does not correspond our natural conditions and possibilities, an inadequate institutional securing of the living environment protection including the absence of efficient checking, or information links and the existence of incomplete, disharmonious legal protection of the living environment and its components subject to the given principles (many of the still functioning legal standards regulating reclamation activities - for instance the law of agricultural land protection - represent in their consequences a serious obstacle to implement the ecological role of reclamation).

b) Low environmental efficiency of direct and indirect economy management tools, enabling a ruthless utilization of natural resources and ecologically unfavourable behaviour of production organizations and other "landscape users". Directive central management, represented by inadequately coordinated system of national economic plan, territorial, area and branch planning, accepted only passive protection of mostly quantitative parameters of individual components of the living environment (for example, the area of agricultural land, etc.). Equally indirect economic management tools (taxes, credits and prices, fees and payments, indemnification and deliveries, subsidies), designed on the basis of the marxist theory of value (which is not able to calculate the price of natural resources and processes), have been practically inefficient in the sphere of the living environment control. The consequences were a general economic inefficiency of active protection and creation of living environment (i.e. also inconvenience of "non-productive" ecological reclamation), neglecting the ecological prevention and stimulating the environmentally harmful utilization of natural resources, including landscape area.

c) Unsatisfactory level of availability of objective, high quality and speedy information on the status and development of the living environment and its components (or on the extent and importance of negative anthropogenic effect on the ecostabilizing significant elements and links in the landscape), resulting from a long-term low level of social demand (the economic sphere did not need these data essentially, public interest has been artificially suppressed). The available information (for instance from an integrated information system about a territory, from a register of air pollution sources, from an information network of hygiene service and inspection bodies, etc.) is incomplete, incompatible and insufficient for responsible management of the economic and social development of regions in accordance with their natural and ecological prerequisites (in the sphere of reclamation for instance for the determination of an optimum reclamation target, meeting the needs of a balanced development of mining landscape).

When considering in addition to these social phenomena also the consequences of aspects specific for the Ostrava-Karviná Coalfield (particularly a generally low labour culture in mining organizations - see Musil, 1989; inappropriate organizational integration of the reclamation enterprise into the mining Concern - see Čížková, 1989; strong informal influences of the so called strong industries on the national administration bodies, deforming the region development according to the industry demands - see Vavroušek, 1989), it is obvious that an ecological implemantation (and thus also a comprehensive landscape shaping) of the reclamation role was almost impossible. If in spite of this situation certain partial results have been achieved, it was thanks to a high environmental responsibility, extraordinarily initiative activity and personal self-devotion of some employees from the establishments concerned.

4 WAY OUT TO IMPROVEMENT OF THE SITUATION

The consequences of the last November events in Czechoslovakia opened possibilities for an essential, principal solution of the existing conditions on the general social, regional and industry level. If the "prerevolutionary" efforts for improvement were only inconsistent partial attempts, limited by barriers of totalitarian power, then the present time offers a unique, hardly repeatable chance to achieve a comprehensive program of the ecological awareness of our society. The term of ecological awareness is conceived as an adaptation of social culture (including ways of world perception and of the resultant ideas, rules of social behaviour and procedures securing interaction and solutions of conflicts between society elements, material tools for achieving social targets) to anthropogenic changes of natural environment. It is a system of mutually linked modifications of national social important values and criteria, relevant for the determination of social development targets (or in selection or creation of ways and means for their achievement), resulting into a conscious acceptance, respecting and application of ecological principles and laws in all spheres of social life, i.e.

a) in the political sphere implementation of the government ecological role aimed at achieving ecologically favourable and economically acceptable coexistence of the society and natural environment;

b) in the legislation and jurisdiction sphere the creation of generally valid and specifically applied comprehensive act of the protection and shaping of living environment and ecologically oriented amendment of acts rugulating activities of individual and organizations in relation to the living environment and its components (with regard to the conditions of market economy and the existence of more forms of production means ownership);

c) in the sphere of economy management and development of economic theory a strong orientation at a general reduction of power consum-

ption (mainly accomplished by structural changes in the national economy), at increased power production economy and at a gradual environmentally more favourable transformation of energy base; a special attention is given to changes in economy management tools
- creation of dynamic price system (projecting for instance into coal prices also the ecological costs related to coal mining and preparation), creation of an environmentally efficient, locally, chronologically and materially differentiated tax and delivery system (including chargex for utilization of natural resources) and of other stimulating and sanction means;
d) in the information and monitoring sphere establishment of information system about the statut and condition and development of living environment, of independent inspection and application of public checking, elegoration of a standard system determining a permitted anthropogenic pollution level of territorial units, etc.;
e) in the social sphere preparation and training of changes in the value orientation of population towards ecologically more considerate way of securing their basic material and existence needs, creation of attitude of responsibility on part of individuals and social groups to living environment.
Specifically for strenghtening the ecological role of reclamation in the Ostrava-Karviná Coalfield it is necessary to perform (in addition to the above mentioned general social measures) modification of organization of reclamation activities (for example by establishment of reclamation enterprise as an independent economic subject, providing paid reclamation services beyond the industrial departments scope) and an amendment of economic stimulation measures (elaboration of an evaluation method of comprehensive reclamation efficiency including the ecological efficiency, economic preference of reclaiming activities increasing the efficiency of ecological stability elements and links of a given territorial system in mining landscape, etc.). Technical measures-shall be also indispensable, preferring the utilization of ecologically more favourable biological reclamational methods based on

intensification of natural pedogenetic and succession processed and minimizing the consumption of power and materials by maximizing the input of information on landscape and principles of its ecological stabilization.

5 CONSLUSION

The objective of the paper is not to present an exhausting review of the given problems (which was not possible due to a limited extent) but to point out one significant, often unjustly neglected and underestimeted aspect of reclamation activity in mining landscape. At the same time the paper in its attempt at a comprehensive, system concept tries to demostrate the complexity, linkage and mutual relationship of an apparently highly specific problem as the application of the reclamation ecological function is together with current social, regional and industry changes presently undertaken by Czechoslovakia.

REFERENCES

Čížková, H. 1989. Ekologizace řídicích a plánovacích nástrojů hospodaření v devastované krajině. Ostrava: Výzkumná zpráva Ústavu ekologie průmyslové krajiny ČSAV.
Čížková, H. a Musil, L. 1989. Ekologizace přímých a nepřímých nástrojů řízení rekultivační činnosti v OKR. Sborník z celostátní konference Doly a životní prostředí. Havířov XVI/1-11.
Čížková, H. a Smolík, D. 1989. Ecological Reconstruction of Sites Disturbed by Underground Mining of Coal. Mineral Planning, 41, December 11-13.
Komárek, V. a kol. 1988. Souhrnná prognóza ČSSR do r. 2010. Praha: Prognostický ústav ČSAV.
Koncepce rekultivace krajiny narušené těžbou uhlí v OKR 1987. Praha-Ostrava: VIDEOPRESS MON.
Musil, L. 1989. Vliv sociální organizace pracovního společenství černouhelného dolu na vztah pracovníků k ekologicky relevantním prvkům těžebního procesu. Ostrava: Výzkumná zpráva Ústavu ekologie průmyslové krajiny ČSAV.

Vavroušek, J. 1989. Vztah systému
řízení společenského reprodukč-
ního procesu k životnímu prostředí. Ekonomický časopis 37,
No. 2: 141-156.

Reclamation, Treatment and Utilization of Coal Mining Wastes, Rainbow (ed.) © 1990 Balkema, Rotterdam. ISBN 90 6191 154 0

Coal mine enterprise – An artificial component in natural ecosystems

N. Davcheva-Ilcheva
Institute of Ecology, Bulgarian Academy of Sciences, Sofia, Bulgaria

ABSTRACT: The autor's scientific point of view on the problems of ecology and negative contribution of coal mining is presented in the present paper. A brief analysis of technogen ous and natural ecosystems is made. The distroying and contaminating effect of coal mining, as an artificial component in the ecosystems, is considered. The main trends for the introduction of ecology in the coal mining technogenic system are given. The future prospects of coal mining and its wastes, from an ecological point of view, are pointed out.

1 ECOLOGICAL PROBLEM

The causes for the emergence of the ecological problem, in a social aspect, are considered to be the demographic boom, urbanization, rapid progress of industry, power industry, transport, industrialization and chemical application in agriculture, etc.

From a scientific point of view this problem is a result of man's gross and incompetent interference with Nature in production and other life activities. In his whole creative activity man constructs his own systems, which he transplants into the natural ecosystems as new components, in this way changing their structural composition and turning them into *ecoanthropogenic systems*.

The anthropogenic systems, which are technogenic by nature, radically differ from the natural ecosystems. Their inclusion into the natural ecosystems as an artifial component leads to the formation of a new, in respect of structure and functioning, *ecotechnogenic* system. It is one of the varieties of the ecoanthropogenic systems.

The Technogenic systems created by man completely answer his requirements and capabilities. They are subordinate to social, economic, technical and other principles and laws. In contrast to the natural ecosystems, they are not complied with the natural principles and laws such as, for example, metabolism, conservation of energy and matter, etc. They do not possess *self-regulatinc abilities* (self-purifycation, addaptation, etc.). That is why, in their co-existence with ecosystems, they greatly contradict the latter and lead to different negative ecological consequences.

Deep scientific studies and analysis of technogenic systems and natural ecosystems spow that basic differences exist not only among these systems but also in themselves. Since the natural ecosystems have been studied, systematized and described in details, I am going to briefly review our ecological analysis on technogenic and ecotechnogenic systems.

2 TECHNOGENIC AND ECOTECHNOGENIC SYSTEMS

The great variety of technogenic systems is based on their designation to meet with certain needs of society, on the raw materials they process, the object of their activity, their structural and functional differences, etc.

From ecological point of view, depending on their effect on natural and other anthropogenic ecosystems, they may be grouped conditionally into several categories: technogenic systems of mainly *polluting effect* - where priority belongs to hazardous (harmful) wastes and emissions such as thermo-electric power-stations, road transport, water transport, etc.; technogenic systems of mainly destructive effect - during their operation the ecosystems are extensively destroyed - most typically represented by open-cast mines, large building sites, etc.; technogenic systems of destruction effect - these systems by their functions destroy for a long period of time having permanent consequences on the ecosystems of a given geographic region of the world - to this kind belong the military systems such as the nuclear and hydrogen-bombs, and other kinds of military equipment and technologies.

Quite often the effect of the technogenic systems is *combined*. When the degree of independence or

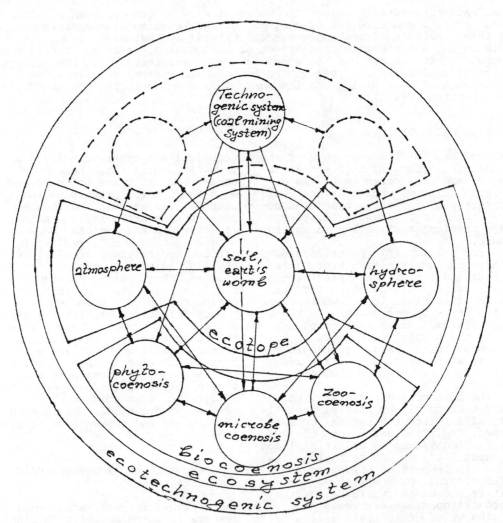

Fig.1 A model structural scheme of an ecotechnogenic mining system

combined negative effect is so high that it discontinues the self-regulation of the ecosystems, they start ruination and in the long run are also destroyed and disappear.

Another important criterion for the ecological classification of technogenic systems is their ability to use metabolism. It is called *recycling* in the production activity of society. While metabolism is valid for all natural ecosystems, recycling is possible only for certain technogenic systems. These are part of the mining industry such as metallurgy, some chemical and textile productions, etc. They are of a great socio-ecological importance, because they play the role of "health-officers" of the public production.

2.1 The coal mining enterprise - an artificial component of the ecotechnogenic system

From a modern ecological point of view the coal mining enterprise is an artificial component in the structure is an artificial component in the structure of natural ecosystems with which it forms *eco-technogenic coal mining system.*
It is one of the varieties of the e ecotechnogenic systems.

A model structural scheme of such a kind of ecotechnogenic system is shown in Fig.1.

Since the coal mining technogenic system is subordinate to social, economic, technical and other laws of maximum satisfaction of society's requirements for effective and conditioning raw materials, and not to the natural laws and principles governing the ecosystems, their inclusion into them grossly interferes their functioning.

The coal mining technogenic system has great many specific features in common with the other systems for production of minerals. The territories of the mining technogenic systems and their relation to the natural ecosystems are naturally predetermined since the occurrances of their raw materials are determined by Nature. In this respect they are similar to the natural ecosystems. This is a substantial shortcoming. The choice of an ecologically consistent construction site for a

large part of the technogenic systems offers the possibility of alternative ecological decisions. These productions are multi-waste and one of the biggest contaminators and destroyers or the natural environment. Their basic activity is related to the production of unrestorable natural resources thus is destroying the earth crust. Another characteristic feature is that their recycling is inapplicable.

Open-cast mines have a particularly destructive effect on the ecosystems. Their production destroys the flora and fauna, contaminates the soil, air and water in the mining regions. Of great significance here is noise abatement caused by the use of technical exuipment (machines, blast works, etc.). Underground mines get deep into the earth's crust, violating its natural conditions, the existing relations, composition and operation of ground waters, etc.

Besides the general features in sommon with other mining systems, coal mining has its own specific features. The most substantial are worth noting. Its end product (mined coal) in the process of subsequent treatment and consumption undergos permanent physical (aggregate) changes and transformations (in to heat, electric energy, etc.), as a result of which it cannot be recycled. That is why, coal resources may be characterized not only as non-recoverable but also irrevocable (non-recycling).

Gas emissions, which share a considerable part in the total coal mining pollution, also have their specific features. The sources for their formation are the coal gas in the underground mines and readily inflammable coal waste heaps in the regions of mining enterprises. The inflammable heaps are not only a Source CO_2 but also of sulphur compounds from the pyrites and large heat amounts. Thermal contamination is also increased by the mining kettlebottoms the temperature of which is higher than that of the environment. So that coal mining enterprises have their contribution to the greenhouse effect.

The role, which waste heaps play in ecology, is also increased by the fact that thay occupy vast areas and in the presence of soluble components (metals, etc.) they contaminate the water in the mining regions.

The wastes from the mining enterprises are included into the trophic chains of the ecosystems and have an adverse affect.

A model functional scheme of non-ecolized coal mining technogenic systems is illustrated in Fig.2.

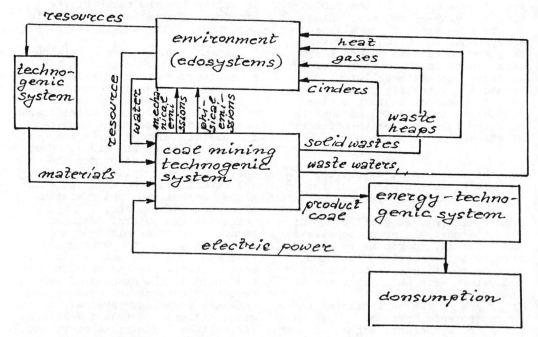

Fig.2 A model functional scheme of a non-ecolized coal mining technogenic system

The self-defending and self-regulating abilities of the ecosystems are greatly disturbed and decreased under the effect of construction and functioning of coal mining systems. The rapid changes caused by them favour adversely the realization of the adaptive abilities of the ecosystems and often lead to their extinction. All stated above to a certain extent characterizes the great ecological role played by the coal mining enterprises.

3 ECOLOGY OF COAL MINING TECHNOGENIC SYSTEM

The strong destroying and polluting effect of coal mining technogenic systems necessitate urgent ecology measures. This process should be in accordance with their specific features and should be carried out in several basic trends. One of these trends is the improvement of the existing technologies and technical exuipments aimed at the restricting their negative effect on the environment and at decreasing hazardous wastes and emissions. Utilization of solid waste products and reclaiming of coal waste heaps and demaged terrains are widely applied already. The selfregulating and self-protection abilities of the recepient ecosystems, into which the coal mining technogenic systems are transplanted, should be improved. This can be done by preliminary ecological preventive measures which would guarrantee

122

the proper functioning of self-regulation and self-protection of the ecosystems during the construction and operation of the coal mining enterprise.

Favourable ecological opportunities are offered by desert mining kettlebottoms. They are suitable sources of domestic, building, industrial, mining and other kinds of waste materials. In using them for such purposes, however, particular attention should be paid to the water draining from the zone in orter to avoid the contamination of the waters with the wastes. Another important problem, in making

such decisions is the due and proper land reclamation of these depots.

However, the main trend of this ecology campaign is the creation, in principle, of new wasteless, pure and acology-complied mining technologies (e.g. underground bacterial gasification, energetics, etc.). Also urgent ecological decisions should be made in order to restore the damaged mining terrains. An example of a functional scheme of an ecology-complied coal mining system is given in Fig.3.

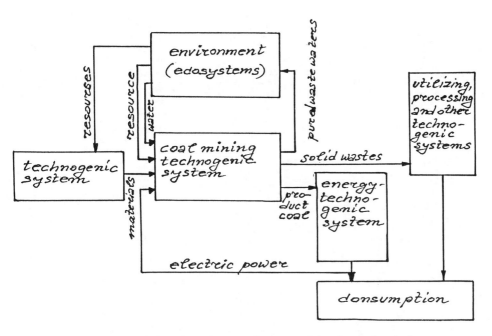

Fig.3   A model functional scheme of an ecolized coal mining technogenic system

4 FUTURE RESPECTIVES

Non-recoverability of coal resources the larg negative ecological concequences during coal mining, power processing and non-recyclability of coal, absolutely necessitate their substitution with restorable and inexhaustible natural and energy resources. Human scientific research in this sphere shows that the prospects belong to wind, water, geo-

thermal, biogenic, and most of all to solar energy. The Sun, which is the main source of energy in nature, is also the future ecological source of energy for human society. The rapid utilization of solar energy requires the foundation of a unified world institute of energy for integrating the research efforts of scientists all over the world. This indicates that the utilization of coal wastes

as raw materials in civil engineering and other activities is of a
temporary nature and that "suitable
substitutes of equal importance
should be duly looked for.

Reclamation, Treatment and Utilization of Coal Mining Wastes, Rainbow (ed.) © 1990 Balkema, Rotterdam. ISBN 90 6191 154 0

The reclamation state of dumping grounds after hard coal exploitation in Poland

Z. Harabin
Ministry of Environmental Protection, Natural Resources and Forestry, Poland

W. Krzaklewski & M. Trafas
Academy of Mining and Metallurgy, Cracow, Poland

ABSTRACT: A need for reclamation in hard coal mining as well as methods and pace of this activity in the years after 1980 have been presented in the paper. Factors that determine a degree of difficulties in reclamation work have been discussed and the elements that call for a change been pointed to in order to improve the situation.

1 INTRODUCTION

Hard coal mining in Poland has long traditions reaching the 19th century. Exploitation developed in a dynamic way particularly well after the 2nd World War (Fig. 1).

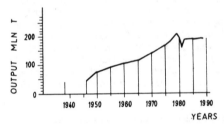

Fig. 1 Coal output in Poland from 1946

It was connected with the development of both heavy and power industry basing on the combustion of hard and brown coal. Exploitation carried out by the deep working method causes a number of changes in the environment, i.c. origin of mine and dressing dumps and hydrological changes of soils (depression of the ground water level or origin of overflow lands that are the result of postexploitation deformations of the surface). Moreover, changes can include water reservoirs and courses as the result of the thrust of saline or polluted in a different way, waters.

At present, about 11 000 ha situated mainly in Upper Silesia belong, to the enterprises of hard coal mining. Dumping grounds covering over 3100 ha in total are a big part of this surface. 0,4 t waste occurs on each mined tonne of coal. That

amounts to 72 – 75 mln t annualy at the existing level of output. They are not utilized to suitable degree (Table 1) and, therefore, the surface of dumps is still increasing (Krzaklewski, Harabin 1988).

Table 1. Utilization of mining waste in mln t

Method of utilization	Years		
	1987	1995	2000
Filling of underground mine workings	9,3	17,7	19,0
Processing by Haldex Co.	4,3	4,6	4,3
In engineering works and as building materials	3,2	4,4	4,4
Transfer to central and local dumping grounds	58,9	72,9	73,8
Others	2,6	5,6	5,3

The number of dumps is considerable – about 270. Most of them are small, located in the vicinity of individual mining enterprises. About 580 ha are occupied by the so-called central dumping grounds where waste coming from different mines is cumulated there. In some areas, especially, in the Rybnik Coal Basin, the terrains strongly subsiding as a result of exploitation (surface subsidences go down to 20 m and, after the year 2000, to

30 - 35 m) are filled with waste.The dumping grounds originated in this way and spread accidentally on the surface are a threat to the surrounding terrain. It takes long (till the completion time of subsidences) to build them and, due to it, their reclamation is delayed.

Prediction says that in the years 1986-2000 hard coal mining will include about 3000 ha soils, among them, about 1600 ha arable soils and about 600 ha forests. They will be given mainly to the waste storage, e.g. the surface of central dumping grounds will increase up to 2400 ha. Perhaps due to political and economic changes occuring in Poland, surfaces taken by hard coal mining will be limited.

2 CONDITIONS AND PRINCIPLES OF DUMPING GROUND RECLAMATION

Since 1966 legal regulations establishing a duty of waste ground reclamation as a result of industrial activities have been obligatory. A year ago reclamation was carried out only as an experimental activity. In the period before 1980 hard coal mining reclaimed about 4000 ha. An accurate establishment of the surface area faces difficulties due to the missing documentation or its scatter.

Dumps of gangue prevail among the reclaimed terrains.From the so far experiments it follows that, above all, thermal activity, shape of relief as well as physical and chemical properties of cumulated waste are the factors that decide about a degree of the reclamation difficulty.

2.1 Thermal activity

Over 25% of the total dump number is thermally active due to the failing technology of coal benefication and storage of gangue with fine coal, sludge and coal dust. Consequently, the content of combustible substances in dumps can even exceed 30%. A self-ignition can easily occur when easily weathered pyrites and marcasites are present.Technology of dumping ground storage is the only applied method that has been used so far and which prevented thermal processes. It makes an access of air to the interior of the dump complicated. The technology depends on the layer, mechanical density of waste. The concept of limiting the threat by the thermal activity due to the storage of the Carboniferous waste combined with smoke-box dusts elaborated in the Institute of Environmental Engineering in Zabrze has not been put into practice so far.

Some dumps can be used for coal recovery, dressing into a building material and others. Such work is done by the Polish - Hungarian joint venture Haldex. It is expected that other enterprises of this kind will come into being. When such work is done, there is a danger of an increase in air access into the interior of the dump and catching a fire (thermal activation). There are the following two examples: the dumps of the mine "Czerwona Gwardia" in Czeladź and that of the mine "Bolesław Smiały" in Łaziska.

Thermal activity makes the reclamation of the dumps impossible till the time of their burning and cooling. On the other hand, the dumps that are not threatened by thermal activity and burned can be reclaimed biologically.

2.2 Shape of relief

Dumping grounds are shaped as overground level forms (dumps) or underground level forms (fills). The fills do not create difficulties in their reclamation, while dumps, especially those of heights over 20 m and considerable inclinations of slopes, sometimes more than 35°,are terrains requiring much work and extremely thorough activities white introducing vegetation. An incorrect shape of relief is frequently observed in small and older dumps. Central dumping grounds are generally inclined and height of slopes, top and shelf relief are correctly shaped.

2.3 Properties of cumulated gangue

On dumps mainly waste of Carboniferous rocks is cumulated, i.e. mudstones, sandstones and coal shales. From the reclamation point of view, a participation of big fragments of $\emptyset > 20$ cm is essential. A big Quantily of this fraction influences the reclamation possibilities in a negative way. Susceptibility to weathering e.g. some shales weather very quickly providing a clayish-dustish residual soil of great potential fertility.

Generally, waste is characteristic for little content of Ca, P and N and contains relatively much Na,Mg,K and S, the latter occuring usually in the sulphide form (Harabin,Strzyszcz 1979, Strzyszcz 1988) The presence of Na compounds can lead to an increase of the reaction even over 9,0 pH. This phenomenon takes place in the first phase after the location of waste on the dumping ground. In turn, a

possibility of acidification to the reaction below 3,0 pH is connected with the
content of sulphides. Both toxic waste due
to its increased content of sulphides or
acidified toxic waste ought to be located
inside the dumping grounds. On the other
hand, the formations of the most profitable charakteristics for management ought
to be cumulated in the surface layers.
This observation is often, however, not
taken into account during the formation
of dumping grounds.

2.4 Degree of reclamation difficulties

Taking into consideration the properties
discussed above, the constructions different from one another as far as a degree
of reclamation difficulties is concerned
are distinguished (See Table 2).(Krzaklewski,Harabin 1988).

Table 2. A degree of reclamation difficulties

	Reclamation		
	very difficult	difficult	easy
D u m p s			
Height of dumps	> 20 m	10-20 m	<10 m
Inclination of dump slopes	> 35°	18-35°	< 18°
Content of the fraction over 20 cm	> 60%	30-60%	< 30%
Density Index	< 08	08-09	> 09
Content of combustible fragments	> 18%	7-18%	< 7%
F i l l s			
Content of the fraction over 20 cm	> 60%	30-60%	< 30%
Density Index	< 08	08-09	> 09
Content of combustible fragments	> 18%	7-18%	< 7%

3 TRENDS AND METHODS OF RECLAMATION

In the so far reclamation activity,affo-
restation, stand density and special park
management are realized, above all.
One ought to understand afforestation as
complex planting of forest species exclusively. They can be planted on dumps of
easy or, medium difficult at most, reclamation conditions not designed for liquidation (dismantling) situated outside
urban and estate agglomeration and on dumps next to forest complexes. According
to the elaborated concept of the intensive
surface reclamation (in the Institute of
Environmental Engineering in Zabrze),the
work can be carried out while taking two
assumptions into consideration :

1. Surface of dumping grounds will not
be covered with soil formation or humus
(cuttings will be planted directly in
the waste cumulated on dumps,
2. The so-called preliminary vegetation
is given up while the so-called final vegetation is introduced,
Taking into account the forest character
of plantings, properties of Carboniferous
rocks as well as site requirements of individual species in the composition of
planned cultivations one ought to consider the quantity 60-70% of the following:
Larix europea, Quercus rubra, Quercus
robur, Acer pseudoplatanus, Acer platano-
ides, the quantity 20-30% of Tilia corda-
ta, Alnus glutinosa, Fraxinus excelsior,
Fraxinus americana, Robinia pseudoaccacia,
Pinus nigra, Pinus strobus. Moreover,
Prunus serotina, Rhamnus frangula, Sorbus
aucuparia, Viburnum lantana, Corylus ave-
llana, Crataegus oxyacantha, Crataegus
monogyna, Evonymus europaeus ought to be
introduced in the quantity 15-20%,
On the other hand, in the composition
of stand density non-forest species can
occur. These are:

1. Goal stand density of the landscape,
recreation or sanitary-protective character planted on dumps not designed for
liquidation,
2. Temporary stand densities planted on
dumps planned for liquidation (dismantling) during 15-20 years.
In the case of temporary stand densities
one ought to apply vegetative cuttings,
mainly paplars and willows. The best results were obtained while introducing the
following variations : Populus x.'H-194;
Populus x.'H-275; Populus x. canadensis
'Grandis: It is **essential** to fertilize the
soil correctly, mainly, with N and P and
it ought to be intensively applied in the
first year of cutting growth (Harabin 1978,
Harabin Strzyszcz 1979).
Fast biological fixing of slopes can take place as sod formation. It depends on
sowing of suitably selected mixtures of

grasses and papilionaceous plants when combined with the correct mineral fertilization, mainly, nitrogenic. It ensures the formation of compact sod already in the time of 1-2 years (Patrzałek 1984). Practically, it is the only possible way to accept the method of the aesthetics correction and limiting of noxiousness of dumps that will be liquidated, e.g. in the course of 5-10 years.

However, it can be emphasized here, that Poland is, perhaps, the only country in Europe that has given isolation (covering) of dumps with fertile soil. The results of this activity are also applied in West Germany.

4   EVALUATION OF RECLAMATION STATE IN HARD COAL INDUSTRY

The first experimental work on the reclamation of dumps of hard coal mining were undertaken in Poland in the fifties (Skawina 1958). Some reclaimed then, surfaces underwent destruction, e.g. the mentioned dump of the mine "Czerwona Gwardia" and some other still exist being compact stand densities.

Dynamic reclamation of dumping grounds and management in hard coal industry is presented in Table 3.

Table 3. Reclaimed areas and areas for reclamation in ha

| to 1980 | Years | | |
	1981-85	1986-90	1991-2010
~ 4000	865	1000	1850

It follows that the pace of the work is more or less the same and predictions for the years to come point to its decrease. At this pace a big surface of dumping grounds remains nonreclaimed and due to a constant increase in surface (the subject discussed in Chapter 1) arrears will also increase. Destruction of already reclaimed terrains by undertaking such activities as partial dismantling or uplifting of the existing dumping grounds must be disturbing.

The species composition of vegetation introduced during reclamation is generally correct in the case of stand densities arranged on dumps adjoining the mines. On the other hand, a selection of species is often accidental in the case of afforestation.

One of the reasons is the lack of the satisfactory quantity and quality of cuttings of suitable species of trees and shrubs.

Also the structure of established cultivations and the form are frequently incorrect. In a number of cases the reclamation work is done carelessly, e.g. cuttings are planted too shallow in the holes filled with thick fragments causing thus withering of young trees.

There is too little individualization of programmes since it calls for carrying out of specialist investigations in the properties of the dump. The conventional approach to the problem becomes the the source of failures for reclamation. It is also the reason of non-utilization of the newly elaborated fast and cheap reclamation methods.

The general problem is the lack of crop cultivation on the reclaimed surfaces what brings unfavourable effects.

5   CONCLUSIONS

As it was stated earlier, the effects of reclamation in hard coal mining in Poland are not satisfactory. It can be partially justified by the fact that considerable environmental changes are observed in mining regions. Repair of damaged constructions, regulation of the flow of courses and drainage of the surface engage a big human and equipment potential to remove mining damages what places reclamation activities at some later instance. Possibilities for reclamation to be applied are not satisfactory although specialist enterprises were established for that purpose.

The following can be considered elements indispensable for an increase in effectiveness of reclamation in hard coal mining:

1. Increase in a degree of waste utilization what will influence a decrease of the threat by thermal activity and limiting of the surface occupied by new dumping grounds.

2. Origin of a greater number of enterprises specializing in the biological reclamation of dumping grounds.

3. Occupation of the greater surface by the temporary reclamation where vegetation, whose zole is mainly sanitary and protective, is introduced for a short time (to 20 years). This problem concerns, among others, big surfaces of central dumping grounds that formed gradually,

sometimes for several years, are not reclaimed to the moment of building completion.

4. Utilization of achievements of scientific centres dealing with reclamation
problems on a greater scale. In particular, it concerns the practical application
of fast, effective and cheap methods.

It ought to be expected that reclamation
in hard coal mining will become essential
in the changing conditions in Poland at
the time, when the society is very sensitive to the problems of environmental
protection and new organization plans for
this activity will be made.

REFERENCES

Harabin, Z. 1978. Zastosowanie zrzezów
topolowych do przejściowego zagospoda-
rowania zwałowisk towarzyszących górni-
ctwu węgla kamiennego. (Application of
polar cuttings to transitional menage-
ment of dumps connected with pit-coal
mining). Zabrze: PAN Instytut Podstaw
Inżynierii Środowiska.
Harabin, Z. and Z. Strzyszcz 1979. Dyna-
mika przyrostu wysokości wybranych od-
mian topoli w latach 1976-77 w warun-
kach centralnego zwałowiska odpadów górnictwa węgla kamiennego "Smolnica".
(Dynamics of the height increase of chosen sub-species of poplar trees in the
years 1976-77 in conditions of the central dumping site of coal-mining residues "Smolnica"). Archiwum Ochrony Środowiska 2 : 79-93.
Krzaklewski, W. and Z. Harabin 1988. Oce-
na stanu rekultywacji w przemyśle wydo-
bywczym. Maszynopis wykonany na zlece-
nie Instytutu Geologicznego w Warszawie
w ramach CPBP 04. 10. 04. (Evaluation
of the reclamation state in exploitation industry. Typescript made at the
order of the Geological Institute in
Warsaw within the CPBP 04. 10. 04.).
Patrzałek, A. 1984. Zdolność darniotwór-
cza mieszanek traw i motylkowatych wy-
siewanych na zwałowisku odpadów węgla
kamiennego oraz ich wpływ na proces
wietrzenia gruntu. Archiwum Ochrony Środowiska 3-4 : 157-170.
Skawina, T. 1958. Przebieg rozwoju proce-
sów glebotwórczych na zwałach kopalni-
ctwa węglowego.(A course of the development of soil-creating processes on coal mining dumps). Roczniki Gleboznaw-
cze 7 : 149-162.
Strzyszcz, Z. 1988. Przyrodnicze podstawy
rekultywacji hałd po kopalnictwie głę-
binowym węgla kamiennego. (Natural basis of dump reclamation in underground
hard coal mining). Zeszyty Naukowe AGH
1222 : 159-173.

Reclamation, Treatment and Utilization of Coal Mining Wastes, Rainbow (ed.) © 1990 Balkema, Rotterdam. ISBN 90 6191 154 0

Solution of renewal of forest cover on Zlate Hory tailing pond

Miroslav F. Podhajsky
Ore Research Institute, Mnisek pod Brdy, Czechoslovakia

ABSTRACT

Intensive industrial activity leads with us to environmental damage and devastation. Environmental interrelations become frequently a limiting factor in mining and processing of ores and industrial minerals due to the volume of produced tailings till full extraction of all reserves. Reclamation is one of the efficient solutions how to liquidate these new formations of the landscape. But we must be fully aware of that chemical composition of dump and tailing pond materials varies very much and is not mostly favourable to plants to take roots. Since each region is isolated, it represents an isolated case and requires individual solution. The submitted contribution documents this experience on a tailing pond belonging to mine extracting polymetallic ore. The main extracted minerals were chalcopyrite, galena, sphalerite with elevated contents of Fe, Mn, Cd and constant trace elements such as Ag, Co, Ga, Hg, Ni, exceptionally In and Sn. There are considerable contents of sometimes gold-bearing pyrite. The tailing pond put out of operation contains approx. 2.5 mil. tons of flotation treatment tailings. High content of sulphide ores caused also their accelerated oxidation in free contact with atmosphere not only on beach surface but also on downstream side of dam. Significant intoxications of the covering topsoil material appeared to be unsurmountable barrier especially from the economic viewpoint.

Many examinations, sample analyses and investigations of many variants of vegetative pot tests have been performed. Because marked toxicity of the substrate did not enable efficient coming up of vegetation by means of direct or over-top-soil coverage, conclusions have been drawn on the necessity to use a suitable, efficient and cheap interlayer. Gangue from mine development works was used. The principle of the solution consisted in simple re-covering of the beach surface and tailing pond slopes with a layer of mine gangue and, in this way, in transformation of the unsoluble problem of biological reclamation of a toxic tailing pond to biological reclamation of sterile gangue dump.

After taking this measure and heaping subsequent layer of fertilizable soil with partial admixture of topsoil, conditions of successful coming up of grasses after hydrosowing and taking roots of pioneer wood species additionally sown on an area of min. 6 ha have been formed. Today's favourable state of all seedlings confirms that the whole cycle science-research-project-realization has successfully been accomplished on this location.

INTRODUCTION

Overall impairment even devastation of natural environment proceeds today to industrial and mining activities; this is mostly reflected in impairment of purity of atmosphere, damaging of water regime and disporportional occupation of soil funds.

Also in mining and mineral processing of metal ores and non-metallics, environmental relationships often become a limiting factor. When planning mining and corresponding mineral processing of the ore mass, the main parameters are size and shape of the deposit, location of the plant and volume of the produced tailings up to full extraction of reserves. The project of siting tailing dumps and ponds is based on proofs of the possibility to deposit or otherwise use all tailings from known commercial and prospective reserves. Today's aim is the highest concentration in one space where all tailings will be deposited and the deposition will be effectively rehabilitated so that the least environmental impairment occurs.

For these reasons, its foundation and operation should be planned with respect to its gradual suitable embodiment into the picture and future of the landscape. In this direction, those depositions best satisfy which serve for liquidation of undermining effects and ground levelling. Technical and biological reclamation is used for embodiment of dumps, heaps and tailing ponds into the landscape. Technical reclamation includes arrangement of shape and ground /erection of depositions of table or terrace shape/ whereas biological reclamation leads to formation of suitable conditions for development and growth of cultural plants, i.e. to providing vegetation of specified /sometimes also non-specified/ character of pretreated objects.

MATERIALS AND METHODS

Even when does not exist a general method for vegetating ore dumps and tailing ponds, the mostly marked isolation from densely populated areas and intensively cultivated areas limits the development of these areas and spaces for recreational and agricultural purposes. This is predominantly forestal reclamation according to the character and site of occurance. It should be all the time born in mind that chemical analysis of dump and tailing pond materials is different and mostly not favourable for vegetation to get footing. As far as the materials are not positively sterile, then stabilization of these depositions by vegetation is in many cases braked by presence of toxic substances, heavy metal salts and weak acids and bases. Also the contents of zinc, nickel, copper and other metals destroy the vegetation growth. Also unwanted concentration of dissolved salts is of great importance as we have become convinced by investigation.

Namely, each area is unique, constitutes an independent case and requires individual evaluation. For these reasons, it is necessary to make in advance a detailed analysis of the situation, to neutralize some tailings before sowing or even to leach out toxic matters. Water spraying is often intoduced for improving the growth of the sown plants which, simultaneously,

washes toxic matters out of the tailing.

The method of direct vegetating dumps and tailing ponds without their covering with a thick layer of soil is of prime importance from the viewpoint of the lack of topsoil and subsoil. Such reclamation using direct renewal of vegetaion should be aimed at a simple conception - providing a permanent plant cover being kept independently and requiring no attandance which will fulfil some or all of the following aims:
- limitation of surface dust nuisance;
- prevention from erosive effect of wind and water carrying off surface soil particles;
- limitation of water volume infiltrating through tailings contaminated with acid and heavy metals;
- aesthetic improvement of immediate region;
- acceleration of nature coming in these anthopogeneously damaged places.

This aim can be achieved, according to the nature of definite use of tailing deposition, by final shaping within the range of mine or surface construction works provided for mosly by mines at their own cost. Afterwards, a part of biological renewal of the soil fertility proceeds in dependence on the character of the substratum, territory, climatic and other conditions. Generally, formation of artificial soil with subsequent formation of plant cover seems to be the most effective. For supporting and accelerating the soil-forming processes on areas temporarily get rid of matured biocenoses and with practically missing biological productiveness, it is necessary to put on at least minimum layer of fertile soil or fertilizer or subsitute mulch, peat and others with microbial flora. This process after previous pH control or detoxication can be characterized as a primary measure of naturally protective character.

Biological reclamation can proceed in two directions. The first is subsequent population with natural succession where the tree cover usually has unsuitable composition and insufficient involvement. The second direction is plantation of wood species.

In biological reclamation of forestal character, those sorts of wood species, shrubs and herbs should be chosen which not only will grow under aggravated conditions but which will also reinforce the gangue by their root system not only at level but also on slopes and, in this way, prevent effectively from wind and water erosion. For this purpose, those wood species, shrubs and herbs can be used on concrete localities which were dominants of the indivdual phytocenological analysis.

Agricultural recultivation in circumstances of mining of ores and other raw materials is a rare phenomenon. Most frequently, we meet with foundation of pastures and we should not disregard, in this connection, the possibility of saturation of biomass with heavy metals especially in reclamation of tailing ponds which is not suitable for application to food production. The most rational seems to regard the whole crop from the reclaimed areas of tailing ponds as microelemental complementation of fodder base for cattle and other farm animals or as cheap microelemental fertilizer.

OWN OBSERVATIONS AND TESTS

The Ore Research Institute in Mnisek pod Brdy is engaged in the problems of re-habilitation of tailing depositions since 1977 and in reclamation problems within the activity of general management Ore Mines and Magnesite Works since 1979. Two governmental and two branch research themes enabled to acquire not inconsiderable theoretical and practical knowledge which cannot be compensated by no matter how through study of domestic or foreign literature. And this is, unfortunatley, very poor according to the performed searches. Moreover, foreign literature is pronouncedly under the sway of advertisement as well as of fears of potential competitors.

In 1981 to 1985, we have elaborated two comprehensive studies / approx. 550 pages, 230 figures/ of mining spaces of North-Moravian region and the whole Czech Basin within the scope of the finished research theme of development of science and technique "Investigation of the Method for Protection of Some Environmental Qualities of Landscape on the Site of Ore Mining and Mineral Processing".

Analysis of locally damaged landscape for more than 120 localities is given here with a suggestion of necessary measures. The value of the whole work can be evaluated only after study of five corresponding volumes in the long-term aspect not only to 1990 but beyond this year as well. And especially here, it has been found that each locality represents an independent specific problem and should be solved using an independent creative process and this make reclamation after mining and mineral processing difficult.

Let us attempt to document this fact on an example of one solved case of forestal reclamation.

Locality of the mine Zlate Hory of the national enterprise RD Jesenik includes non-ferrous metal deposits bedded in the series of Vrbno of the Desen vault mantle.

The district proper is situated in the northern part of this series in massive "Pricny vrch" and belongs to geomorphological zone of Hruby Jesenik. Metamorphites of lower Devonian are here represented by various types of quartzites /sericitic, chloritic, biotite-chloritic, horny, etc./.

Chalcopyrite, galena, sphalerite with elevated contents of Fe, Mn, and Cd dominate among the main ore minerals; Ag, Co, Ga, Hg, Ni and, exceptionally, In a Sn are premanent trace elements. Quartz predominates in gangue, to a lesser degree ankerite and barite. Pyrite / sometimes goldbearing/, partially also pyrrhotite, can be regarded as pre-hydrothermal.

For this reason, dumps and, especially, tailing ponds of Zlate Hory mine mineralogical characteristics made of sericitic and sericite-chloritic quartzites, crystalline limestones and, sporadically, phyllites. The tailing pond 01 with approximately 2.5 mil. tons of mineral processing tailings is already more than 900,000 tons of tailings /Fig.1/.

After previous failures in application of hydraulic sowing on non-treated substratum, more systematic investigation of analytical data and characteristics including their subsequent evaluation was proceeded to. Groups of samples were taken step by step, subjected to agrochemical analysis and evaluated.

Quality of the effluent discharge under the tailing pond into receiving water was investigated in order to verify its suitability as industrial water for sprinkling the beach of the tailing pond for protecting it from unwanted dust nuisance and as irrigation water for possible cultivating purposes.

During the investigation of pedological character of the deposit substratum, also the changes of some of its components were investigated in dependence on time, depth of deposition, the effect on the covering soil and others.

Variability of pH value of the samples taken on the downstream side of the dam of the tailing pond 01 can be considered as our own discovery. 25 samples which represented average quality of material from the depth of 0-10 cm at each 5 m of length have been taken here already in 1980. Identical samples for evaluation of the time factor effect on possible changes have been taken under analogous conditions in the IVth quarter of 1984. Values summarized in Table I were determined by laboratory analysis.

The oxidation process of the tailing pond substratum was verified down to the depth of general rootage. Capillarity of toxic solutions in vertical as well as horizontal direction and intoxication of the covering soil at the contact with underlaying substratum were investigated.

All these tests have confirmed that we meet in the tailing pond substratum with chemically inert as well as chemically active componenets. The latter are exceptionally significant for the conditions of Zlate Hory mine. In addition to the surrounding rock containing calcium and magnesium carbonates, sulphides of heavy metals primarily create due to their treatment /griding/ with their residual contents conditions for oxidizing and hydrolytic processes. After washing out in the tailing pond, residual contents of sulphides after ore mass flotation oxidize relatively quickly under heat releasing; this accelerates, in turn, further chemical reactions.

Oxidation of sulphides proceeds according to the simplified equation:

$$2 \ S_2^{2-} + 7 \ O_2 + 2 \ H_2O \longrightarrow 2 \ SO_4^{2-} + 2 \ H_2SO_4$$

Herewith, sulphuric acid is released which is a strong reagent so that it causes further decomposition of sulphides or, likewise, carbonates and other chemically active components of the substratum. Arising heavy metal sulphates hydrolyze under the conditions of neutral or weakly acid reaction, further sulphuric acid is released and metals are separated out like hydroxides or hydrated oxides. If such docomposition of all reagents proceeds which were bonded with free acid, strong acidification of the medium, significant drop of pH value and dissolution of separated metal hydroxides up to concentrations which are strongly toxic for organic processes proceeds.

This intoxication affects the covering topsoil as well as waste water. Therefore, concluding the investigation the following evaluation could be formulated.

In consideration of the fact that no genuine soil is involved but a material which is produced as waste in mining and mineral processing, the obtained results cannot be quite well compared with average values of the surrounding forest soils.

PARALLEL ELVALUATIONS

They have been dealt with in the last time in the area of old mining operations at the polymetallic deposit Zlate Hory also by Raclavsky and Raclavska[2] who found out, using X-ray diffraction and infrared spectroscopy, mineralogical composition of soils in relative degrees 1 - 5 for contents

of chlorite, illite, kaolinite, monmorillonrite, feldspar and quartz. Representation of the individual soil types and subtypes as distinguished by Pelisek[3] on sampled profiles and mineralogical composition of the individual types can be evaluated from Table II. Average contents of metallic elements in soils /with standard deviations/ are summarized in Table III. The established results show that concentrations of heavy metals in soils in the areas of geochemical background outside the mineralization zones of the ore district of vegitation by the content of toxic metals from soils proceeds. On the other hand, the soils in the areas of geochemical anomalies on the outcrop of ore bodies of these formations affected by medieval mining operations have high contents of toxic metals which are received by plants.

In literature, there are very few data on allowable concentrations of heavy metals in soils on reclaimed areas and, moreover, these data differ very much. The highest allowable concentrations for agricultural soils /100 - 125 ppm Cu, 50 to 100 ppm Ph or 200 - 300 ppm Zn/ could be a certain clue.

The samples of the tailing pond substratum are, in principle, biologically sterile inorganic material having toxic contents of sulphides and ferrous iron. Larger part of samples has extreme toxic Ph values. The samples practically nearly do not contain main plant nutrients. With respect to these facts, the analyzed material cannot be easily reclaimed by mere neutralization and fertilization with mineral fertilizers. A suitable method is, for instance, covering and mixing upper layers of this substratum with overburden soil, thorough liming with ground limestone or ground slag and fertilizing with organic fertilizers.

RESULTS AND DISCUSSION

The results of analytical investigations and studies enabled to formulate the project of primary measures for rehabilitation and reclamation purposes. As it appeared, toxicity of the substratum does not allow direct and effective growth of vegitation. Therefore, it is not only necessary but also efficient to protect beach and slopes by making up gangue with later covering with fertilizable soil having thickness at least of 0.3 to 0.5 m. After making up and levelling, to sow in groups, for instance in mosiac arrangement, alder/ Alnus incana/, birch /Betula verrucosa/, mountain ash /Sorbus aucuparia/ and larch

/Larix decidua ssp. europea/. It may be presumed that, in turn, birch and alder will be self-grown in these areas and birds will spread the mountain ash. After sufficient growing-up of the undergrowth, the gaps are filled out with grown-up seedlings of Norway Spruce /Picea excelsa/ and sylvan beech /Fagus silvatica/.

SUMMARY AND CONCLUSION

Solution of conditions of successful reclamation and its realization

Covering of the beach and, later on, of slopes of the tailing pond with a layer of mine gangue appeared to be the most effective antierosive measure. Technology of dump erection in the area of bridging over the Blue Creek was recommended so that the quickest connecting communication to the beach would be achieved. The gangue ring of the crest having width of 40 m from the north-east part of the deposition connected by transport with the on-dip adit was made-up as the first. Successive growth of the body stability enabled formulation of further suggestions of technical reclamation methods for the whole object.

10,000 cu.m of soil was obtained in the next year after mucking out the sports pond of the town Zlate Hory as a base of fertilized and fertilizable surface layer. In that time, the project of gradual construction of the new plant Zlate Hory has solved the place for deposition of material arising from capital construction excavations /new plant, relaying of main road, etc./ necessary for finalization of technical reclamation /humus deposition/.

The year 1983 is a turning point in solution of reclamation of the tailing pond 01 of the plant Zlate Hory. Research solution and further experimentations confirmed us in the opinion that the project is realizable at optimum economic cost. Solution is based on a simple principle:

"Cover the surface of tailing pond beach and slopes with a layer of mine gangue at the height of at least 1 to 3 m and, in this way, convert the problem of biolgical reclamation of toxic tailing pond into the problem of biological reclamation of sterile gangue dump".

At the same time, possible future varients of covering with green were realized in a fenced part of the plant using vegetation pilot test. Afterwards, they have served for the project organization as verified input data on biological reclamation of sterile gangue. Subsequently, making-up the redeposited soil /fertile, however mostly only fertilizable/ in the height of 50 cm over the all surface, Figs. 2 and 3.

The year 1984 confirmed further preliminary considerations as to cultivation and successful performance of technical reclamation on an area larger than 8 ha; differenciated removal of soil from the new places of capital construction continued.

In the year 1985, project of biological reclamation brought to the end including complete budget. Closing session of the coplex rationalization team Zlate Hory in December 1985 came to the conclusion that technical reclamation of the tailing pond 01 has been finished and biological part of the reclamation began since spring 1986, Figs. 4 and 5.

The cycle science-research-project-realization has been successfully closed at this locality.

REFERENCES

1.Podhajsky M.F.: "Investigation of the methods for protection of some environmental qualities of the landscape on the sites of ore mining and mineral processing" /in Czech/ Research theme of the development of science and technique No. UVR 243-134, Annual Report 1982, Annual Report 1985, Final Report 1985, Mnisek pod Brdy

2.Raclavsky K., Raclabsks H.: "Soils in the area of old mining operations on polymetallic deposit Zlate Hory"/in Czech/. Proceedings of the Symposium Mining pribram in Science and Technique, Pribram 1987, p. 1-12

3.Pelisek J.: Elevation Zoning of Soils in Middle Europe /in Czech/. Ed.: Academia, Praha, 1966, 366 pp.

TABLE II

Mineralogical composition of soils

Group	Profile sample	Soil type	Chlorite	Illite	Kaolinite	Montmorillonite	Feldspat	Gypsum	pH
1	2	3	4	5	6	7	8	9	10
1. Beige									
1.1	1-30	ochre forest soils	++++	+++++		+++	+++	+	3.3
	1-34	-"-	++++	+++			++	+++	2.9
	1-38	-"-	++++	+++	+		+	+++++	4.3
	1-39	-"-	+++++	+++			++++	+++++	4.1
	2-16	-"-	+++	+++++			++	+++	6.9
1.2	1-7	rusty-ochre forest soils	+	+++		++++	++++	++	4.5
	1-33	-"-	+	++		++++	+++	+++++	4.5
	1-37	-"-	+	++++		+++	+++++	++++	4.3
	2.8	-"-	+	+++		+++	++++	+++++	3.6
1.3	1-2	rusty-forest soils	+	+		+	++	+++++	3.6
	1-3	-"-	+++++	++++			+	+++	3.6
	1-5	-"-	++++++	+++++			+++	+++	3.5
	2-1	-"-	++++	+++			++	++	3.7
2. Brown									
	1-8	brown-rusty forest soils	+++	+++++			++++		3.2
	1-9	-"-	+++++	+++++			+++++	3.	3.4
	1-10	-"-	+++	++++++			++++		3.4
	1-11	-"-	+	++			+++		3.4
	1-13	-"-	+++	+++++			+++		3.3
	1-14	-"-	++	+++++			+++		3.3
	1-17	-"-	++	+++++			+++		3.4
	1-24	-"-	++	++++++			+++		3.2
	1-27	-"-	+++	++++++		+	+++		3.3
	1-28	-"-	++	++++++			+++		3.5
	1-29	-"-	++	+			++++	+++	3.3
	1-31	-"-	++	+++++		++	+++	+++++	3.3
	1-35	-"-	++				++++	++++	3.2
	1-36	-"-	++	+++		+	++++		3.4

TABLE II continued

1	2	3	4	5	6	7	8	9	10
	1-2	brown-rusty forest soils	+++	++		+	+		3.3
2.2	1-25	chocolate-coloured brown forest soils	++	++++	+++	+++	++		3.3
	1-26	mountain soils	++	++++	++	+++	++	+	3.3
3.	Brown-grey								
3.1	1-12	brown-grey forest soils	++	+++			++	++++	3.7
	1.20	podzol	+	+++++			+++		3.3
	1-21	-"-		+++++			++		3.3
	1-22	-"-	+	+++++			+		3.3
	1-23	-"-	+	+++++			+		3.2
	1-31	brown-ochre forest soils		+++			+	+++++	3.2
	2-4	-"-	+++	++++			+		4.0
	2-5	-"-	+	+++++			+		3.6
	2-6	-"-	++	+++++			++		3.4
	2-7	-"-	+	+++++			+	+++	3.5
	2-8	-"-		+++			++	+++++	3.3
	2-9	-"-	+	+++++			+	+++	4.6
	2-11	-"-		+++++			+++	+++	4.1
	2-12	-"-		+++++			++	+	4.5
	2-14	-"-	+	+++++					
3.2	1-4	-"-	+++++	+++++		++	+++	+	3.4
	1-6	-"-	+++++	+++++		++	++	+	4.7
	2-13	-"-	+++++	+++++		+	+	+++++	6.8
	2-15	-"-	+++	+++++				+++++	6.9
3.3	1-16	-"-	+++	+++++			+++	+++	3.4
	1-18	-"-	+++	+++++		++	+++	+++	3.2
	1-19	-"-	+++	+++++			+++	+	3.2
3.4	1-1	grey-brown forest soils	+++	++++		+++	+++		5.3

137

TABLE I

Sample No.	Sample taking 1980 pH/H$_2$O/	pH/KCl/	Sample taking 1984 pH/H$_2$O/	pH/KCl/
1	4.1	4.90	2.30	1.70
2	2.70	2.80	2.20	1.70
3	3.00	3.35	2.30	2.05
4	3.00	3.25	2.25	2.05
5	4.20	5.30	2.25	2.15
6	5.60	6.75	2.30	2.20
7	5.10	6.15	2.45	2.80
8	2.55	2.60	2.50	2.35
9	2.70	2.85	2.40	2.25
10	2.60	2.65	2.30	2.15
11	2.85	3.05	2.45	2.35
12	2.65	2.85	2.35	2.25
13	3.00	3.30	2.60	2.50
14	3.90	4.60	6.00	5.90
15	5.50	6.05	2.75	2.70
16	4.95	5.75	3.00	2.95
17	5.70	6.25	2.40	2.35
18	4.55	5.75	2.40	2.35
19	3.50	4.45	2.30	2.26
20	5.70	6.30	2.25	2.10
21	6.35	6.80		
22	5.15	5.85		
23	6.30	6.95		
24	7.05	7.80		
25	7.10	7.85		

the resting 20 of the slope up to the dsm crest have already been covered with gangue and partly with soil

TABLE III

Content of metallica elements / ppm/ in soils /arithmetic mean and standard deviation/

Element	Soils	
	Background	Anomaly
Cu	33.5 ± 21.4	132 ± 165
Pb	49.8 ± 25.8	364 ± 492
Zn	70.9 ± 29.8	334 ± 323
Mn	335 ± 454	578 ± 754
Fe	9,989 ± 5,595	19,145 ± 10,469
Ni	18.6 ± 10.4	29.8 ± 19.8
Co		
As	14.7 ± 11.6	53.1 ± 28.8
Ba	12.0 ± 9.0	778 ± 498
Sr	44.0 ± 36.4	252 ± 33
Zr	180 ± 47	252 ± 68

Fig. 1 A view of sludge pit 03 / in use / and 01 / out of use /
of the Zlaté Hory Mine of the Jeseník Ore Mine Corporation

Fig. 2 Vegetation pot trial within the area of the mine / plaster pots /

Fig. 3 Vegetation pot trial within the area of the mine / plaster pots /

Fig. 4 A stage of technical and biological reclamation of the 01 sludge
        pit / view towards the mine /

Fig. 5 A stage of the start of biological reclamation on two thirds of
        mud settling pond 01

Reclamation, Treatment and Utilization of Coal Mining Wastes, Rainbow (ed.) © 1990 Balkema, Rotterdam. ISBN 90 6191 154 0

Use of dolomitic marbles of Konstantin deposit Stare Mesto pod Sneznikem mine – National Enterprise Jesenik

Miroslav F. Podhajsky
Ore Research Institute, Mnisek pod Brdy, Czechoslovakia

ABSTRACT

When performing technical and phytocenous examinations of waste heaps after mining and mineral processing in Northern Moravia, a dump of dolomitic marbles mined as overburden of a graphite opencast mine attracted our attention. With respect to extraordinary quality of this waste and to the possibility to produce magnesian -calcareous fertilizer from it which is deficient in Bohemia, its utilizability has been investigated.

The graphite deposit is situated on western side of Velke Vrbno vault of Northern Moravia as a part of continuous band of productive carbonatic and schistose metamorphic rocks skirting practically the whole circumference of the vault. The main overlaying rock within the final geometry of the opencast mine is crystalline dolomite occurring particularly in the central part of the opencast mine. It is grey-white, finely- or middle-grained rock in macroscopy, covered with sporadic rusty iron-oxide components on divisional planes.

In the analytical study, three samples have been taken from the dump of dolomite, namely calcareous dolomite, tremolitic dolomite and graphitic dolomitic limestone. The following values have been determined by analysis:

	MgO % mass	CaO % mass	Fe_2O_3 % mass	$\frac{MgO}{CaO}$
Sample No 1	22.01	31.08	0.48	0.7082
Sample No 2	21.91	29.77	0.25	0.7363
Sample No 3	22.23	29.20	0.22	0.7613

In further examination, further samples were treated and their semiquantitative spectral analysis was performed. For verification of treatability of this material to powdered fertilizer, the attempt at comminution in a hammer mill with KVD rolls was carried out twice. Decisive operations such as charging, repetitive crushing, sample taking and time metering were performed.

Conclusions confirming favourable breakability of the material, practically 50% of fractions up to 1 mm were obtained after first repetition, can be drawn from the results of screen analysis of averaged nine quartered samples; the first three samples represent primary crushing, further three samples repetition and the last three samples represent practically the second repetition.

Laboratory- and pilot-scale tests verifying the effect of this material on the change of pedological soil properties with the impact on green mass yield were performed for supporting the argument for application advantageousness of dolomitic marble in agriculture and forestry. Also vegetative pot tests have shown favourable effect on the tested soil after 1.5 year.

The investigation confirmed application advantageousness of this disregarded waste in agriculture with the following positivities:

- overburden ratio increase by 8.1% is possible due to economic effect at profitable sale of crushed dolomitic marble;
- substantial decrease in further occupation of forest soil;
- favourable effect on yields of agricultural plants grown in industrial scale due to saturation of macroelements Mg and Ca;
- increase in resistance of forest cultures by pH regulation of soil;
- also subsequent effect on water streams in the immediate region is not negligible.

When performing the first technical and phytocenous examinations in North Moravia in 1982, waste heap of dolomitic materials mined as overburden of the graphite deposit Konstantin attracted our attention /Fig. 1/. With respect to extraordinary quality of this waste and the possibility to produce magnesian-calcareous fertilizer which is deficient in the whole Czech Socialist Republic, an agreement has been entered into with the management of the Stare Mesto pod Sneznikem mine belonging to the National Enterprise Rudne Doly Jesenik to start the solution of this problem is to continue it with the co-workers of the mine. Therefore, a Complex Rationalization Team was founded which set out to make thorough examination of the deposited material, to determine its qualitative parameters and to propose its use. Besides, environmental protection of the landscape and the adjacent covers was aimed at with reducing initially required occupation of the forest soil.

Brief geological and petrographic characteristic of the area

The graphite deposit Konstantin is situated on the western flank of the silesicum area of the so-called vault of Velke Vrbno of Proterozoic /Misar, Z./ or Late Paleozoic /Kveton P./ age as a part of continuous band of productive carbonatic and shistose metamorphic rocks skirting practically the whole circumference of the vault.

It is a north-east continuation of the early mined deposit Herman-North and is situated approximately 500 m west of Velke Vrbno. The deposit was examined at length of approximately 750 m down to the depth of 100-130 m. Laboratory, model and industrial scale tests predestined this deposit enabled opencast mining /Fig. 2/ of two thirds of all geological reserves. The graphite seams are bonded here to continuous carbonate horizon and are developed at its footwall.

The overburden rocks of the graphite deposit Konstantin are divided in four groups:
- overburden soil and fertilizable strata such as clay, slope and eluvial strata, colluvia of the underlying rocks;
- overburden of overlaying rocks containing crystalline dolomites, crystalline dolomitic limestones, subsidiary amphiboles, paragneisses;
- overburden of underlying rock containing keratophyric metamorphic tuffs, paragneises, crystalline dolomites, dolomitic marbles;
- interdepositalrock includes graphitic schists, dolomites, keratophyric metamorphic tuffs, amphibolites.

The main overlaying rock within the final geometry of the opencast mine are crystalline dolomites /dolomitic marbles/ which can be found mainly in the central part of the opencast mine. It is a grey-white, fine-grained to middle-grained rock in macroscopy, covered with sporadic rusty iron-oxide components on planes of division. The rock fracture is even until conchoidal, the fracture edges are sharp. The rock has massive texture. Under microscope, carbonate /dolomite/ forming xenomorphically confined grains is 0.09-1.0 mm. Subsidiary occurance of quartz and amphiboles / termolite is also common/. Accessories are calcite, muscovite /sericite/, forsterite. According to differential thermal analysis and gravimetric analysis, this rock can be indentified as dolomitic marble, poor in impurities, contianing 89.4 to 94.3 % of dolomite - $CaMg(CO_3)_2$/Fig. 3 and Fig. 4/.

It can be commonly verified / according to chemical analyses/ that the content of $CaMg(CO_3)_2$ decreases from south to north where dolomite is partly replaced by calcite, quartz and amphiboles. Analogously, also the changes in transversal direction from overlying to underlaying strata can be determined, ie. from west to east, where $MgCO_3$ component decreases and the $CaCO_3$ conponenet increases. For carbonates in close overburden of the graphite deposit in its northern part, content of graphitic substance is characterisic which finely impregnates the whole rock and gives it a grey colour. Also alternation of pale and dark thin beds can be seen which gives the rock a characteristic "lamellar" appearance.

The carbonates in the overlaying strata of the graphite seams form a marked stratigraphic horizon and are only locally significantly contaminated with layers of amphibolites, metamorphic tuffs and small seams of lean graphite and graphitic schists. During mining for industrial use, these contaminating layers are mined selectively. Deep oxidation of all rocks, graphite not excluding, is evident in all explored and exposed areas. The depth of the oxidation zone is approximately 50 m in vertical direction. Dolomitic marbles in the south and in the central part of the mining field are affected more heavily. Their oxidation is characterized by rock breaking in form of sand, by formation of debris cones and overall worse stability of quarry faces. Oxidation of carbonate rocks in the northern part of the deposit is less evident. Small karst phenomena in this part are connected with increase in $CaCO_3$ component and decrease in $MgCO_3$ component.

The whole complex of Konstantin deposit rocks is subjected to heavy tectonics what is characteristic for the whole vault of Velke Vrbno. Direction of bedding of carbonatic overburden of graphite seams is from north to south, inclination ranges between 10 and 80° towards west. Three joint systems diagonal to the direction of beds are more marked than foliations. The joints are mostly free without filling. Rarély, veinlets of pure carbonate, calcite $CaCO_3$, occur. According to measurements of structural and tectonic elements in overlaying rocks, it may be stated that tectonics of dolomitic marbles is in principle larger in the southern part.

Analytical studies of waste dolomites

The first orientation evaluations of the heaped dolomitic marbles date back to 1979. The following results were achieved when analyzing three samples, namely sample No.1 - calcareous dolomitic marble, sample No.2 - tremolitic calcitic marble and No. 3 - graphitic dolomite-calcitic marble:

According to analyses performed, the examined materials are suitable, as to their chemical composition and reactivity for production of ammonium salpetre with limestone and as magnesium-calcareous fertilizer in ground state.

Spectrographic analysis gave the following results: (Table 3)

Analyses of new samples were repeated in 1983 in order to find possible deviations in the examined quality after elapsing three years. Semiquantive evaluation of the decisive elements by means of a traditional method were performed. The following data were obtained: (Table 4)

Table 1

	MgO %mass	CaO %mass	Fe_2O_3 %mass	MgO CaO
sample No1	22.01	31.08	0.48	0.7082
sample No2	21.91	29.77	0.25	0.7363
sample No3	22.23	29.20	0.22	0.7613

Furthermore, the reactivity tests in surplus of 5% nitric acid HNO_3 were performed. The following results were obtained from samples weighing 20 grams each:

Table 2

Decompo- sition time	Sample No1	Sample No2	Sample No3
	% of the non-decomposed residue		
30 minutes	24	11	15
3 hours	11	5	7.5

Table 3

	10^1	10^0	10^{-1}	10^{-2}	10^{-3}	?	traces
Sample No. 1 ground	Ca,Mg	-	Fe,Mn	Si	Sr,Pb Al	Cu,Ag	-
Sample No. 1 lumpy	Ca,Mg	-	Fe,Mn	Si,Cr	Pb,Al	Ag	Cu
Sample No. 2 ground	Ca,Mg	-	Al,Fe Si,Mn	-	Sr,Pb	B,Ag	Cu,Cr
Sample No. 2 lumpy	Ca,Mg	-	Al,Si Fe	Mn	Sr,B	Ag	Pb,Cu Cr

143

Table 4

Sample No.	Contents of elements in % approx. /according to subjective classification/						
	10^2-10^1	10^0-10^0	10^0-10^{-1}	10^{-1}-10^{-2}	10^{-2}-10^{-3}	?	traces
1	Ca,Mg	/Si/	Al,Fe/Mn/	-	Cu,Pb Sn,Cr	-	-
2	Ca,Mg	/Si/	Al,Fe/Mn/	-	Cu,Sr	-	-
3	Ca,Mg	-	Fe,Si	/Al/,Mn	/Cu/	-	-

Qunatitiative evaluation of the same material gave the following results:

Table 5

Sample No.	SiO_2	Mgo	CaO	Fe	Al_2O_3	Mn	Cu	Zn
1	0.85 0.87	20.71 20.46	31.64 31.64	0.55 0.55	0.65	0.068	0.017	0.0034
2	1.14 1.11	20.71 20.71	30.94 30.93	0.55 0.55	0.54	0.071	0.014	0.0034
3	0.98 1.05	20.71 20.71	30.93 30.93	0.55 0.55	0.65	0.072	0.024	0.0012

Concluding chemical, mineralogical and petrographic analysis of the samples of heaped waste, the following facts could be emphasized:
- The examined rocks can be identified as dolomitic marbles with few admixtures. They contain 89.4 to 94.3 % of dolomite according to differential thermal analysis and gravimetric analysis.
- Dolomitic marbles differ from each other in macroscopy and colour. The sample No.1 is nearly white with very low content of the graphitic component whereas the sample No.2 is grey, heavily impregnated with graphite.
- The dominant material is dolomite $CaMg(CO_3)_2$, quartz and amphiboles, common aluminosilicate Ca, Mg, Fe and tremolite $Ca_2Mg_5[(OH,F) | Si_4O_{11}]_2$ may be considered as subsidary minerals/Rosler H.T./. Moreover, calcite, muscovite / sericite/ and forsterite $Mg(SiO_4)$, pyrite, chalcopyrite and goethite are accessories in the sample No.2.

- Spectral analyses confirmed the elements present in the dominant mineral as well as in accompanying minerals.

Investigation of comminution of waste dolomitic marbles

Berfore starting the comminution tests of dolomitic marbles of the graphite deposit Konstantin, the results of laboratory and pilot tests of breakability carried out in a single-stage jaw crusher D 160/3. were studied. Simultaneously with the samples of Stare Mesto, limestone samples of Vitosov serving as reference standard were subjected to the same comminution. Graphical evaluation of the results of both tests including mutual comparison shows striking coinsidence of both size distribution curves /Fig. 5 and Fig. 6/. It may be stated that fraction under 1 mm represents nearly 18 %, fraction 1.0 - 2.5 mm less than 10 %, fraction 2.5 - 5 mm represents 22 %; the remainder, i.e. 50 %, is fraction over 5.0 mm. The identity of breakability coincides exceptionally for both sorts of various material. A marked dissimilarity of both materials is seen when pre-crushing lime and dolomitic lime in the same crusher D/160/3. Whereas the fraction up to 1 mm attains for CaO location Vitosov 20 % of total content, the same fraction for MgO location Stare Mesto attains up to 60 %.

The comminution test of dolomitic marbles from the over-burden of Konstantin deposit was prepared directly in the opencast mine Stare mesto pod Sneznikem. After assembly of a hammer crusher with rolls of KVD series, crushing tests of hard graphite were made. The screen analysis gave the following results:

Table 6

Sample	Fragmentation sizing of graphite				
Ø	0-1 mm	1-2.5 mm	2.5-5 mm	5-10 mm	10-x mm
Industrial graphite	29.8	11.7	21.5	18.5	18.5

Only a part of the afternoon shift could be available to verification of functional ability of the hammer crusher with the KVD rolls when the industrial-scale tests with dolomitic marble were performed. However, the significant operations, i.e. cleaning, fuctioning tests, feeding and repetition crushing including sampling and timing have been realized. After detailed evaluation, the following results were obtained:

Table 7

Evaluation of dolomitic marble for agricultural use			
Sample No. 1 /up to 1 mm/		Sample No. 2 /up to 3 mm/	
active pH	9.25	active pH	9.50
exchange pH	8.25	exchange pH	8.50
K content, mg/kg	11.00	K content, mg/kg	19.00
P content, mg/kg	24.70	P content, mg/kg	16.50
Mg content, mg/kg	540,00	Mg content, mg/kg	610.00

From the results of screen analysis made on averaged nine quartered samples from which the first three represent first crushing, the next three first repetition and the last three samples practically the third passage of the material through the hammer crusher, a conclusion may be drawn confirming favourable breakability of dolomitic marbles. After first crushing, there is still insufficient amount of usable material corresponding to the Czechoslovak Standard CSN 72 1220 as to largest percentage under 1 mm. But already the first re-crushing /1st repetition/ gives nearly 50 % of the fraction under 1 mm in average what is a very favourable result for this comminution method. The third crushing does not substantially improve the qualitative properties of the substrate and, therefore, it does not seem to be economically advatageous in this comminution method.

145

Possibilities of use of the crushed waste in agriculture

Possibilities of use of treated dolomitic marbles in agriculture as magnesian-calcareous fertilizer are not new. The first official considerations may be dated back to 1977; North Bohemian Chemical Works, concern Chemopetrol, works Lovosice belonged to the first interested organisation. Also the Ministry of Agriculture and Food of the Czech Socialistic Republic has shown an interest; it assessed the need of 150,000 tons per year for the period of the 7th five-year plan. In spite of it, attention is drawn already in 1981 to really possible withdrawl of 50,000 tons per year as a maximum if this quantity would be withdrawn by agricultural enterprises. Transport in tank trucks of the agricultural enterprises is a distance of 50 km and to greater distance in railway wagons /LAV/ RAJ being the property of supplier is recommended as the most advantageous. Agrochemical Enterprise Sumperk resident in Bludov belonged to the first interested organisations in North Moravia which confirmed to withdraw 15,000 tons per year in bulk from bins into the tank trucks. Also the Agrochemical Enterprise Zamberk has shown similar interest with average withdrawal of 10,000 tons per year.

An important condition for starting production and commercial expedition was and still is keeping to high-quality comminution in griding fineness according to the Czechoslovak Standard 71 1220 as to the fraction up to 1 mm. Also the dolomite price for this purpose has not been fixed and it would be necessary to fix it according to the decree 137/73 86, & 76. Orientation price has been estimated to 90 to 100 crowns per ton when grinding under 1 mm.

Investigation checking tests

In order to lend support to the argument of the advantage of crushed dolomitic marbles use some laboratory and pilot scale tests were performed investigating the effect of dolomitic marbles and limestones on the change of pedologic properties of soils and on the green mass yield.

The pilot scale test verifying the possibilities of crushed dolomitic marbles having various particle size for fertilizing of meadows /especially fraction 1 to 3 mm/ has beencommenced in 1982 on a selected meadow of uniform agricultural cooperative Palkovice, district Frydek-Mistek.

In early spring of 1982, another test has been commenced investigating decomposition of the dolomitic marble and its effect on soil. Topsoil was placed in vegitation pots divided by perforated foil. Crushed substrate having grain size 0-1 mm was applied on the top of the pots amounting to 50 q per 1 ha. Substrate with grain size 1-3 mm was used in orther pots under the same conditions. Then, the foil separated two layers of topsoil having thickness of 10 cm; the top layer was in direct contact with crushed dolomitic marble.

When commencing the test, the soil sample was taken and its pedologic anaylsis was made as given in Table 9.

146

Table 8

Particle size analysis of dolomitic marble, weight of the sample 1000 kg

Sample	Crushing	over 10mm	over 5mm	over 4mm	over 3mm	over 2mm	over 1mm	%
I	1	477.00	77.10	6.90	17.70	23.05	54.60	
II	2	394.00	86.60	6.50	13.70	23.20	63.20	
	3	370.90	104.10	8.25	17.00	27.15	65.90	
	Ø	417.27	89.27	7.22	16.10	25.10	61.20	
	%	42.0	8.9	1.0	1.6	2.5	6.1	62.1
II	1	166.40	79.20	10.30	19.20	40.90	86.00	
	2	320.90	144.50	10.35	21.00	34.50	65.05	
	3	187.40	138.55	12.25	35.10	38.90	83.50	
	Ø	224.20	120.75	11.00	25.10	38.10	78.20	
	%	22.4	12.1	1.1	2.5	3.8	7.8	49.7
III	1	233.0	144.20	15.10	39.00	39.40	80.75	
	2	176.60	141.10	15.00	22.45	37.80	82.95	
	3	133.85	123.65	13.90	26.20	35.90	89.50	
	Ø	171.10	136.30	14.70	29.20	36.20	84.40	
	%	18.1	13.6	1.5	2.9	3.6	8.4	48.1

Sample	Crushing	over 0.5mm	over 0.35mm	undersize	%
I	1	148.00	92.10	104.55	
	2	176.35	108.75	124.80	
	3	172.80	103.00	121.90	
	Ø	165.70	101.30	116.70	
	%	16.6	10.0	11.7	38.3
II	1	217.60	144.75	237.66	
	2	167.40	107.60	172.60	
	3	208.80	132.35	163.15	
	Ø	197.90	128.20	191.10	
	%	19.8	12.8	19.1	51.7
III	1	197.70	125.00	130.35	
	2	217.15	143.16	163.80	
	3	236.60	158.15	121.85	
	Ø	217.10	142.20	138.70	
	%	21.7	14.2	13.9	49.0

Table 9

	Soil analysis when applying dolomitic marble, fraction 0-0.1 mm		Soil anaylsis when applying dolomitic marble, fraction 1-3mm	
	upper part of the pot	lower part of the pot	upper part of the pot	lower part of the pot
pH/H$_2$O/	9.3	11.0	9.5	9.6
pH/KC1/	8.3	10.8	8.5	8.5
need of limiting	-	-	-	-
K mg/kg	11.0	12.0	19.0	20.0
P mg/kg	24.8	24.8	16.5	16.5
Mg mg/kg	540	540	610	610
S mval/100 g	-	-	-	-
T mval/100 g	-	-	-	-
V %	-	-	-	-
Cox %	0.76	0.81	0.27	0.23
humus %	1.31	1.40	0.47	0.40

The first check has been performed in 1983.
Partial samples were taken from the pots
and subjected to the comparison analysis.
It was aimed at recording the first already
occured changes /Table 10/.

Table 10

	Soil analysis when applying dolomitic marble, fraction 0-0.1 mm		Soil analysis when applying dolomitic marble, fraction 1-3 mm	
	upper part of the pot	lower part of the pot	upper part of the pot	lower part of the pot
pH/H$_2$O/	7.3	7.4	9.5	7.4
pH/KC1/	6.9	6.9	6.9	6.8
need of limiting	-	-	-	-
K mg/kg	424.0	500.0	444.0	500.0
P mg/kg	414.0	429.0	447.0	500.0
Mg mg/kg	-	-	-	-
S mval/100 g	94.8	89.4	91.6	91.0
T mval/100 g	17.5	89.7	94.4	91.0
V %	97.2	99.4	97.0	99.3
Cox %	1.35	1.46	0.70	1.34
humus %	2.33	2.52	2.21	2.31

When comparing the results with values for material at the beginning of the test, favourable pH control /both values/ and increase in average humus content can be observed.

In addition to these laboratory-scale pilot tests, further tests have been performed at the customers. For instance, Agrochemical Enterprise, works Zamberk, performed on the farm Kraliky Cerveny Potok towards the end of 1983 spreading of 70 tons of crushed dolomite. In 1984, the consumption in the Czechoslovak State Farm Kraliky attained already 1,500 tons with very good handling and technological properties. Various types of distributor spreaders were tested as well; distributor spreader type D-037, product of GDR, proved to be the most suitable /Fig. 7/.

Also on a strip of land 2 ha in Horni Lipka, a comparison test with dolomitic marble of Stare Mesto pod Sneznikem and ground limestone from lime works Vitosov was performed. Similarly, an analogous test was prepared on the plot No. 7213 with area of 1 ha in Hadca. The results of the individual tests were observed and proved to be comparable with those attained with magnesian-calcareous fertilizers used in agriculture until now.

Concluding recommendation

Solution of the whole problem was orientated to giving sufficiently comprehensive and reliable data for elaborating a report which could be further discussed in superior authorities up to the level of Federal Ministry of Matallurgy and Heavy Engineering Praha.

From mining viewpoint, our paper would not be complete if the favourable effect of utilisation of waste as a secondary raw material on the main activity of the mine, i.e. graphite mining, would not be evaluated. The Mining Law assigns to carry out mining operations so that they provide for the highest recovery of the deposit substance. The stripping ratio is the limiting factor in our case. The parameter used until now was 4.71 cu.m per 1 ton of graphite. When making simplified calculation of cost items for stripping and mining including profits from sale of dolomitic marble, increase in the stripping ratio of overlaying and underlaying gangue minerals /mostly dolomitic marbles again/ from 4.71 cu.m per 1 ton to 5.09 cu.m per 1 ton could be achieved what represents at least 8.1 %.

The environmental profit from decrease in occupation of forest soil is beyond all doubt. Use of a material from local sources for application in agriculture and pH control of soils, even when in the proximity of the works, would affect positively the agricultural production.

The magnesian-calcareous fertilizers brought from elsewhere could be at disposal in other regions. Another application would be possible for pH control of forest soils in Jesenik region.

Consequently, the environmental profit is without any dispute. More detailed informations which could not be set forth in this paper because of text limitations are given in the Annual Report on solution of the partial problem of the research theme of the Ore Research Institute "Favourable from te environmental viewpoint use of dolomitic marbles of the Stare Mesto pos Sneznikem mine, Ore Mines Jesenik, National Enterprise, 1985".

References

1 Research team of 8 co-workers of the Ore Research Institute, Mnisek pod Brdy, headed by PODHAJSKY, M F: "Favourable from the environmental viewpoint use of dolomitic marbles of the Stare Mesto pod Sneznikem mine, Ore Mines Jesenik, National Enterprise" /In Czech/ Annual Report on solution of the partial problem of the research theme.

2 KVETON P: Stratigraphy of crystalline suites in the neighbourhood of North-Moravian graphite deposits, Sbornik UUG, Vol. 18, p. 277-336, Praha 1951

3 MISAR Z: Stratigraphy, tectonics and metamorphism of crystalline suites in the southern part of Keprnik vault /In Czech/ Rozpravy CSAV, 1-79, Praha 1972

4 MISAR Z: Explanatory notes to the geological map, sheet Olomouc. Desen vault. /In Czech/ Geofond, Praha

5 Rosler H I: Lehrbuch der Mineralogie, VRB Deutscher Verlag fur Grundstoffindustrie, Leizig, 1980

Fig. 1: Waste heap of dolomitic and calcitic
marbles

Fig. 2: Opencast mine of Konstantin deposit
incl. overburden of dolomitic
marbles

150

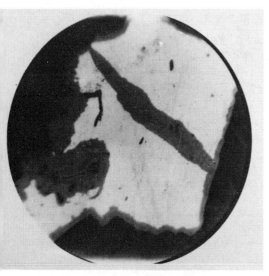

Fig. 3: Goethite hem on pyrite grain

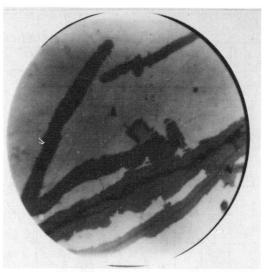

Fig. 4: Pyrite alteration along cracks in goethite, a chalcopyrite grain in the centre

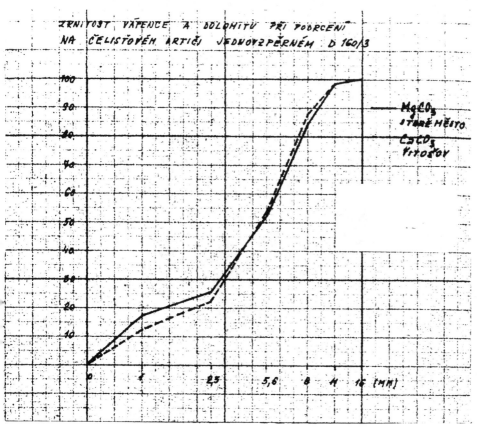

Fig. 5: Size distribution of limestone and dolomitic marble after crushing in single-stage jaw crusher D 160/3

151

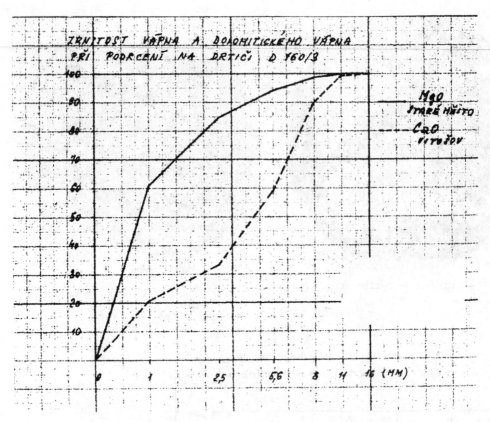

ZRNITOST VÁPNA A DOLOMITICKÉHO VÁPNA PŘI PODRCENÍ NA DRTIČI D 160/3

MgO STARÉ MĚSTO
CaO VITOŠOV

Fig. 6: Size distribution of lime and dolomitic lime after crushing in sigle-stage jaw crusher D 160/3

Fig. 7: Distributor spreader D-037 A 01 on the undercarriage of vehicle T 815

152

Reclamation, Treatment and Utilization of Coal Mining Wastes, Rainbow (ed.) © 1990 Balkema, Rotterdam. ISBN 90 6191 154 0

Prognosis of contaminants leaching from colliery spoils and its effect on the aquatic environment

I.Twardowska
The Polish Academy of Sciences, Institute of Environmental Engineering, Zabrze, Poland

A.Szczepański
Czestochowa Technical University, Poland

J.Tejszerski
The Polish Academy of Sciences, Institute of Chemical Engineering, Gliwice, Poland

ABSTRACT: On the grounds of investigations on the mechanism of generation and leaching of soluble constituents from colliery spoils, a mathematical model has been elaborated for prognosis of the pollution potential from spoil tips or earthworks where minestone is applied. The model is comprised of two major correlative parts: (1) LHI - a model of leaching and transport of contaminants within a spoil tip; (2) SP-2 and FLOW - models of water flow and contaminants migration in ground water in the vicinity of a spoil tip. The results of verification have proved the practical applicability of these models.

1 INTRODUCTION

Coal mine spoil deposition often brings about serious and long-lasting ecological problems in coalfields all over the world, that is evidenced by a rich bibliography. The scale of negative effect on the aquatic environment in the vicinity of spoil tips has been confirmed also by own results of examination of pore solutions in minestone and a survey of water quality in the area of tips site carried out for several years. It has been shown that even a spoil layer 1.5-2.0 m thick can be a constant source of water contamination. Earthworks where minestone is applied should be likewise considered as a potential source of aquatic pollution.

At the stage preceding design and construction of a new tip or earthworks it is necessary, therefore, to elaborate a reliable prognosis of leaching behaviour of soluble species in colliery spoils and to assess the effect of a tip on the aquatic environment for a period of at least 15-20 years under actual hydrogeological and hydrological conditions. Such prognosis has to be a basis for selection of the optimum variant of a tip site and methods of tip construction considering the environment protection requirements.

For this purpose, on the grounds of long-term investigations on the mechanism and dynamics of generation and leaching of soluble constituents from colliery spoils, as well as reviewing several existing mathematical models of pyritic systems (North et al 1972), hydrogeochemical reactions (Parkhurst et al 1981), water flow (Szymanko et al. 1977-1980, 1980) and contaminants migration in ground water (Małoszewski 1977, 1978), a mathematical model for prognosis of the effect of colliery spoil tips on the aquatic environment has been elaborated. This model consists of two major correlative parts:

1. A model of leaching and transport of species within and out of a spoil tip (computer program LHI);

2. Models of water flow and contaminants migration in ground waters in the vicinity of a spoil tip (computer programs SP-2 and FLOW).

In this paper, the basic concepts and the conceptual framework of these models have been presented.

2. MODEL LHI

2.1 General assumptions

A mathematical method and computer

programs in Fortran 1300 code for the leaching and transport model LHI are comprised of five major subroutines. They include:

1. Identification of kinetics parameters of constituents leaching in the unsaturated zone (subroutine LHID, Fig. 1) or the saturated zone of a spoil tip (subroutine HPID) based on the results of column experiments.

2. Simulation of species leaching from a tip - subroutines HSIM 1, HELSIM (Fig. 2) , HSIM 2 (Fig. 3)

The model of leaching and transport of constituents is based on the following simplifying assumptions:

1. In the leaching process participate as reagents or reaction products: in the solid phase - chlorides, iron sulphides, carbonates (dolomite, calcite, siderite),gypsum, goethite and exchangeable ions in clay minerals (Ca^{2+}, Mg^{2+}, Na^+, K^+) ; in the liquid phase - ions of calcium, magnesium, alkalies, iron, hydrogen, chlorides, sulphates and carbonates.

2. Three phases are considered: a solid phase (coal mine spoils) and two liquid phases - a boundary water film (w_m) and leaching water transporting dissolved species from rock material ($w_n - w_m$).

3. A trespass of species from rock to a boundary water film is mastered by the kinetics of iron sulphide oxidation. Other reactions are considered to be very fast ones, of negligible kinetics. Species concentrations are being estimated from phase eqilibria or calculated from mass balance if their loads are lesser than those required for equilibrium.

4. Aqueous species migration from the boundary film to leaching water is a diffusion process; its kinetics is proportional to the concentration gradient of a constituent in the both phases.

5. Diffusion kinetics depends on the extent of a constituent reduction in spoil material due to leaching, and in the unsaturated zone also on the water infiltration rate, assuming that these parameters determine the effective surface area of mass action.

6. Two wariants of water flow are considered: in the unsaturated and saturated zones ('dry' and inundated tips) .

7. Water flow is a one-dimensional vertical infiltration with neither water nor aqueous species exchange between horisontal sections; this simplification has been applied to avoid further complications in the just complicated model.

8. The leaching water flow in the unsaturated zone is of a plunger character; an initial phase of filling of adsorption (w_m) and retention (w_r) capacity of rock material with infiltration water reaching the definite layer of the vertical profile of a tip is taken into consideration.

9. Parameters of each phase are assumed to be unified in the horisontal cross-section of the detached part block of a 'dry' tip and variable along the direction of water flow. A large-area spoil tip of variable composition is simulated as several independent vertical sections blocks . The mass exchange in rock and a boundary water film occurs only in the direction perpendicular to water flow.

10. Parameters of each phase are unified in the entire inundated layer or in its sections described in the point 9.

11. Water flow (ω) through a rock layer is constant; in the case of simulation calculations, dynamic phenomena resulted from its changes are neglected.

2.2 General mathematical concepts

A scheme of species leaching in the unsaturated zone of a tip is shown at the Fig. 4A.

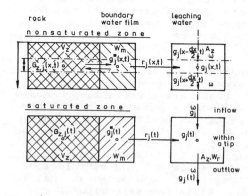

Fig. 4. Schematic illustration of species leaching from spoils in unsaturated (A) and saturated (B) zones.

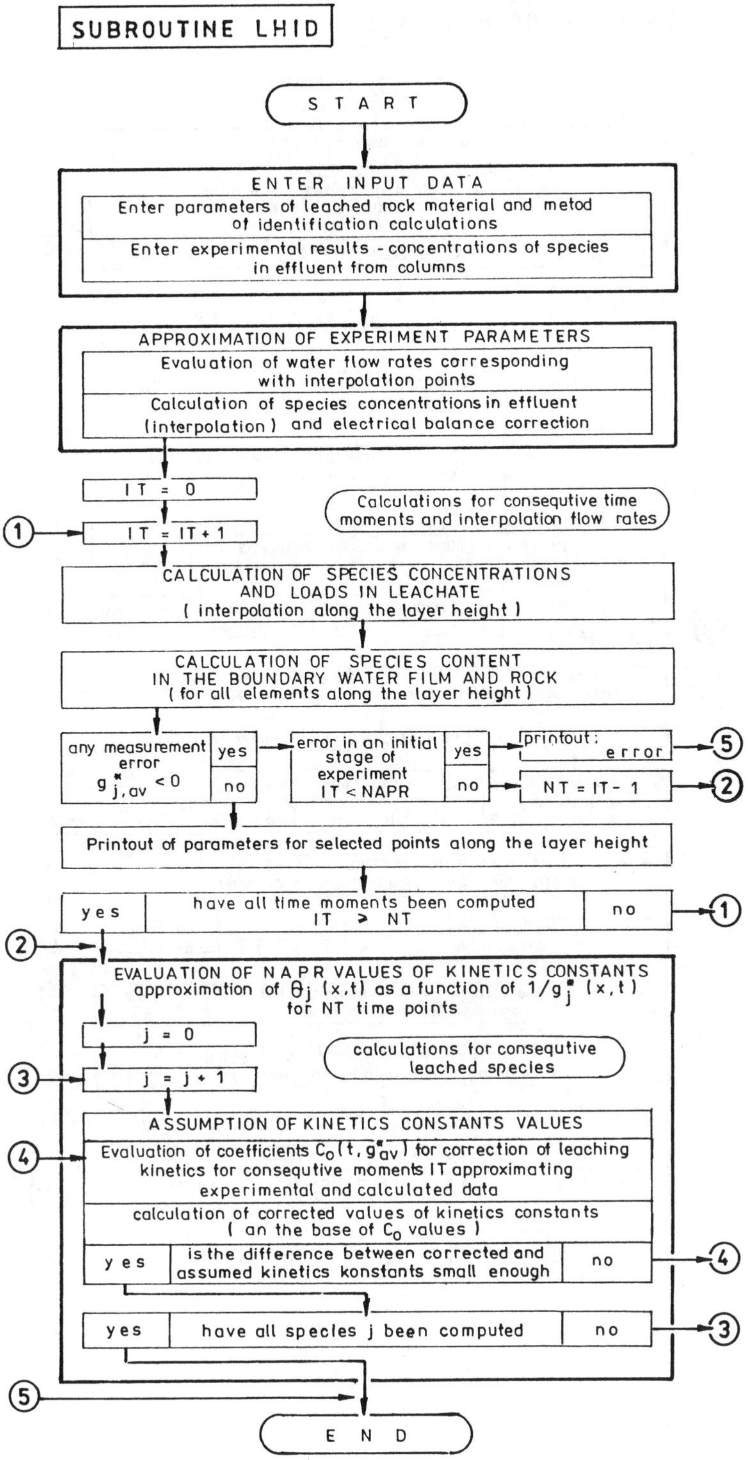

SUBROUTINE LHID

START

ENTER INPUT DATA

Enter parameters of leached rock material and metod
of identification calculations

Enter experimental results - concentrations of species
in effluent from columns

APPROXIMATION OF EXPERIMENT PARAMETERS

Evaluation of water flow rates carresponding
with interpolation points

Calculation of species concentrations in effluent
(interpolation) and electrical balance correction

IT = 0

Calculations for consequtive time
moments and interpolation flow rates

(1) ──→ IT = IT + 1

CALCULATION OF SPECIES CONCENTRATIONS
AND LOADS IN LEACHATE
(interpolation along the layer height)

CALCULATION OF SPECIES CONTENT
IN THE BOUNDARY WATER FILM AND ROCK
(for all elements along the layer height)

| any measurement error $g^*_{j,av} < 0$ | yes | error in an initial stage of experiment $IT < NAPR$ | yes | printout: error | → (5) |
| | no | | no | $NT = IT - 1$ | → (2) |

Printout of parameters for selected points along the layer height

| yes | have all time moments been computed $IT \geqslant NT$ | no | → (1) |

(2) ──→

EVALUATION OF NAPR VALUES OF KINETICS CONSTANTS
approximation of $\theta_j (x,t)$ as a function of $1/g^*_j (x,t)$
for NT time points

j = 0

calculations for consequtive
leached species

(3) ──→ j = j + 1

ASSUMPTION OF KINETICS CONSTANTS VALUES

(4) ──→ Evaluation of coefficients $C_0 (t, g^*_{av})$ for correction of leaching
kinetics for consequtive moments IT approximating
experimental and calculated data

calculation of corrected values of kinetics constants
(an the base of C_0 values)

| yes | is the difference between corrected and assumed kinetics konstants small enough | no | → (4) |

| yes | have all species j been computed | no | → (3) |

(5) ──────────→

END

Fig. 1. Flow chart of the subroutine LHID.

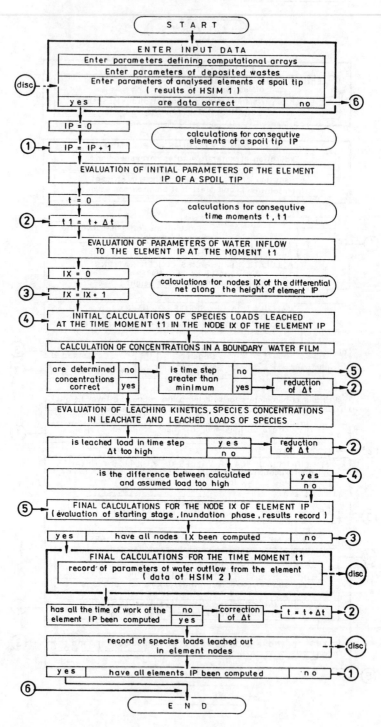

Fig. 2. Flow chart of the subroutine HELSIM.

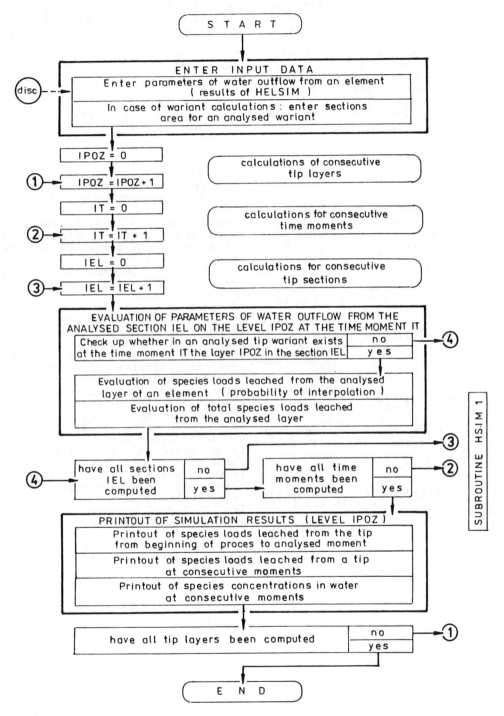

Fig. 3. Flow chart of the subroutine HSIM 2

Processes at the solid-aqueous phase boundary result in the establishment of concentrations $g_j^*(x,t)$ dependent upon either equilibrium conditions or the concentration $G_j(x,t)$ of the species (j) in rock at the time moment (t) on the level (x). To the mean volume ω of water reaching the analysed level (x) of a tip passes an amount g_j of each species j determined by the kinetics of diffusion $r_j(x,t)$. For unit volume of infiltration water, mass balance for the species j is as follows:

$$\frac{\omega \cdot \partial\, g_j(x,t)}{\partial x} = r_j(x,t) \qquad (2.1)$$

and for a unit element of a tip together with boundary film, mass balance can be expressed as:

$$\frac{\partial\, G_j(x,t)}{\partial t} = -r_j(x,t) \qquad (2.2)$$

According to the assumptions 1,2, 4,5, kinetics of diffusion can be described by the equation:

$$r_j(x,t) = [\,g_j^*(x,t) - g_j(x,t)\,] \cdot \theta_j(x,t)$$
$$(2.3)$$

where

$$\theta_j(x,t) = \wp_j\left(\frac{G_{z,j}}{G_j^o}\right) \cdot \psi_j\left(\frac{\omega}{\omega^o}\right)$$

Function $\wp_j$ reflects an effect of the exposed surface of a constituent dependent upon the extent of its reduction due to leaching, and correction coefficient ψ_j shows an effect of deviation of infiltration flow rate with respect to the basic mean value $(\omega)^o$.

Estimation of unknown values and parameters in the equation (2.3) on the grounds of experimental data constitutes an essence of the identification.

Integration of the equations set $(2.1) - (2.3)$ for various conditions constitutes an essence of the simulation of the leaching process at a spoil tip.

Process of species leaching in the saturated zone can be formulated as follows (Fig. 4B):

An inflow of leaching water (ω) to the particular element of the saturated zone mixes with pore water (w_c) in this element. The con-

centrations of species (j) in outflow are resulted from the proportion of an admixture. A mass balance equation for each species (j) concentration in water for each element of the saturated zone of a tip is:

$$w_c \cdot H\, \frac{dg_j(t)}{dt} + \omega\,[g_j(t) - g_j^o(t)] =$$
$$= r_j(t) \cdot H \qquad (2.4)$$

Mass balance and diffusion kinetics equations for the saturated layer are similar to these for the unsaturated zone.

2.3 Identification of kinetics parameters of leaching process (subroutines LHID and HPID)

Model parameters identification is based on experimental data column experiments and comprised of three stages (Fig. 1) :

Stage I - elaboration of experimental results to get a consistent set of concentration data for each constituent in the solid phase and leaching water in all points of an assumed time-and-space net interpolation calculations . The aim of experiments is a receipt of the input data for evaluation of a diffusion kinetics in a process of constituents leaching, according to the equations $(2.1) - (2.4)$. Concentrations of species in leaching water $dg_j(x,t)/dt$ or $dg_j(t)/dt$ and corresponding leached loads $G_j\, x,t$ are to be estimated experimentally. An applied way of identification belongs to integration methods. Process of leaching is modeled simultaneously in several columns of different height filled with spoils leached with unit doses of water corresponding to a ten-year average daily precipitation (Twardowska 1981). Initial mineral composition of spoils, concentrations of soluble constituents in rock and leachate are estimated. Separately, sulphide reactivity r_s is measured. The way of carrying the experiments out causes an unconsistence of results. To obtain the necessary leaching kinetics data (dg/dx) for particular time moments, a time-and-space interpolation of the experimental concentration data has to be applied. Species concentrations in the solid phase in the course of leaching are being asses-

sed on the basis of balance calculations using measured loads of leached species. For identification
calculations, a differential method
is applied for the assumed time-and-
-space net.

For the unsaturated zone, by the
adequate dislocation of interpola-
ted data, an initial stage of a
gradual filling of the retention capacity of rock material ($w_r - w_n$)
by first portions of leaching water
is taken into account (Fig. 5)

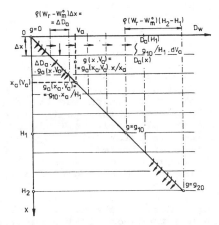

Fig. 5. Schematic illustration of
the initial stage of leaching.

An inflow load of species added to
a calculated element of a tip at
this stage is considered as an initial negative value of a total leached load.

An illustration of the leaching
process at the initial stage (Fig.5)
shows also the way of mutual adjustment of the experimental results
for different columns.

Step II - evaluation of species
concentrations in a boundary water
film in the nodes of the assumed
time-and-space net. According to
the assumption 3, they are calculated either as equilibrium values
or these dependent upon the content
of the particular species in spoil.
A set of equations defining the
composition of solution in the boundary water film in this model includes 14 basic equations for mass
balance, mineral equilibrium, mass
action, electrical neutrality and
conservation of electrons (Twardowska et al. 1988). It enables calculation of species concentrations in
the boundary water film (g_j^*) and in
the solid phase ($G_{z,j}$) on the basis

of a known total content in spoil
(G_j). This value that is equal to
a difference between the initial
content G_j^0 and the leached load
$G_j^1(x,t)$, is computed for each node
(x,t) of the time-and-space net.
The loads of generated dissociated
sulphuric acid $G_{SO_4}^{prod}(t)$ are calculated the same way. Processes at the
phase boundary, equations used in
the model and methods for evaluation of these values are described in
a number of comprehensive publications (Garrels and Christ 1965,
Parkhurst et al. 1981, Twardowska
1981, Twardowska et al. 1988).

Stage III - a proper identification, i.e. evaluation of leaching
kinetics as an exponental function
of the extent of species reduction
in spoils due to leaching $\theta_j(x,t)$
or $\theta_j(t)$ in the equation (2.3).
Identification is derived from species concentrations in leaching water, boundary water film and the
solid phase for each node of the
time-and-space net, assessed in the
steps I and II.

On the grounds of the presented
conceptual framework, computer programs for kinetics parameters identification of species leaching in
the unsaturated (program LHID, Fig.
1) and saturated zones of a tip
(program HPID) have been designed.

2.4 Simulation of species leaching
    from a tip (subroutines HSIM 1,
    HELSIM, HSIM 2)

A set of subroutines HSIM 1, HELSIM
and HSIM 2 simulates processes of
species leaching from tips of different construction by defining
parameters of water and spoil for
different sections blocks of a tip.
As output data, loads of leached
constituents in water outflow from
the particular tip sections as a
function of time, as well as mean
concentrations of species in leachate are calculated.
Subroutine HSIM 1 isolates from a
tip homogenous sections and groups
together sections of the same course of leaching process, assessing
their parameters.
Subroutine HELSIM (Fig. 2) simulates leaching of homogenous sections
of the unit cross-section area 1 m².
Subroutine HSIM 2 (Fig. 3) evaluates leaching parameters for a whole
tip. All three subroutines communica-

te with each other, but can also work independently.

For the subroutine HSIM 1 , information on the construction of a tip is required. A simulated tip is divided into sections of defined area; for each section values of ground, top and water-table levels, time of filling and a kind of material in the defined layers are being given as input data. Sections of the same course of leaching process are selected according to the scheme of a tip construction. Groups of the homogenous sections zones of the same starting point of leaching and the time of covering with upper layers of spoil are isolated (Fig. 6).

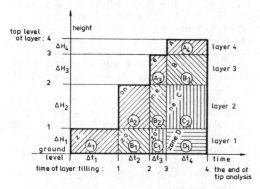

Fig. 6. Selection of homogenous zones in a spoil tip.

The task of the subroutine HELSIM Fig. 2 is an evaluation of the constituents content in spoil, the leached loads and the average concentrations of aqueous species in leachate from a tip as a function of time. Simulation is carried out separately for the blocks of the unit cross-section, representative for the homogenous zones selected by HSIM 1. For each zone of a tip leaching conditions are defined. Inflow of pure or contaminated water from the upper layers of a tip is considered. The process may run from the beginning or from any other definite time moment. Leaching in the unsaturated and saturated zones, a gradual inundation of a tip toe with a constant velocity, an initial phase of filling of the retention capacity as well as a probability of presence in one section of two parts of different kinetics of pyrite oxidation are considered.

A differential method of leaching simulation is based on the general mathematical model of the process. Evaluation of the leached loads of species is carried out consecutively for the definite time intervals, the major part of calculations being an iterative assessment of species concentrations in a boundary water film $g^{*}(x,t)$ – according to the assumptions presented above.

The purpose of the subroutine HSIM 2 (Fig. 3) is an evaluation of the time array of the leached constituents contents in the whole volume of a tip. The loads of leached major species and their mean concentrations in leachate from a tip in the definite time intervals at the stage of construction and for the assumed period after the tip is completed are also being calculated. This task is being solved by the summation of simulation results for the homogenous sections of a tip according to the schedule of a tip construction.

The results of calculations serve as input data to the model of contaminants transport in the aquatic environment in the vicinity of a tip (computer program FLOW).

3 MODELS OF WATER FLOW AND CONTAMINANTS TRANSPORT IN THE AQUATIC ENVIRONMENT (PROGRAMS SP-2 AND FLOW)

3.1 The model of a hydrodynamic field

The prognosis of contaminated water flow consists in time-and-spatial solution of transport equations in a dispersion model (Małoszewski 1977, 1978) . One of the most important tasks of calculations is the definition of the hydrodynamic field as a function of time. The predominant method of assessment of contaminant migration is a separate solution of hydrodynamic problems in first order, and next of migration questions – after the reliable model of the hydrodynamic field is derived. The necessary output data from the hydrodynamic field evaluation for the construction of the contaminants migration model are as follows:

1. The definition of time-and-space variability of external and internal border conditions.

2. The conductivity distribution in a water-bearing layer in the filtration field.

3. The pressure distribution in the field for spatial or space-and-time system - for steady or unsteady water flow.

4. The privileged directions of ground water flow.

5. The water particles velocity in a rock structure.

6. The possibility of derivation of current lines and borders of a ground water stream from the point or area of existing or expected contamination.

For definition of a hydrodynamic field in the foreground of coal mine spoil tips, the program SP-2 has been applied, that constitutes a basis for the prognosis of contaminants migration in a continuous ground water stream calculated by the program FLOW. The mathematical model SP-2 is of the nature of a two-dimensional filtration field in conditions of a steady ground water flow, as the program FLOW that uses its output data, also serves for modeling of two-dimensional contaminants extention. The SP-2 takes into consideration a variability of internal and external border conditions of the filtration area, unstability of filtration parameters k and n_a, as well as inflow to the tip bed of a stream of contaminated waters of different intensity and species concentration with a simultaneous feeding of the filtration area by atmospheric precipitation (Szymanko et al. 1980)

There is also a possibility of application of other computer programs enabling solution of ground water filtration problems under conditions of unsteady flow, variable in time, and taking into account the complexity of geological structure and hydrogeological conditions. For hydrogeologists disposal is a whole set of the computer programs SP and EP from the program library (Szymanko et al. 1977-1980) 1980 for derivation of the hydrodynamic field model in any moment and point in a flat, two-dimensional area (x,y)

3.2 The model FLOW for description of contaminants extention in the vicinity of tips

The computer program FLOW presented

in details elsewhere (Małoszewski 1977) is constucted on the grounds of two-dimensional solution of a dispersion equation for a coördinate scale (x,y) optionally oriented in space. For calculations, Crank-Nicolson's differential scheme is applied. A computation can be accomplished for the hydrodynamic field variable in time and for the simulation of contaminants extention, either migrating together with water stream or undergoing decomposition or adsorption according to a linear adsorption isotherm.

The general assumption is, that aqueous species are instantly and homogenously admixed in the entire layer. The contaminants transport is calculated in a plane (x,y) as a function of time (t). To the program FLOW are entered as input data:

1. The initial concentrations of contaminants leached from a tip - in contact points of a tip with water-bearing layer.

2. The hydrodynamic field calculated by the program SP-2.

3. The spatial distribution of a ground water-table as a function of time.

4. The spatial distribution of a hydraulic depression k/n_a calculated by the program SP-2.

The results of calculations, after the model identification, are:

1. The distribution of contaminants concentrations in time and space.

2. The distribution of contaminants concentration for selected moments in all nodes of the net block centres .

3. The isoline maps of selected contaminant concentrations.

The modified models SP-2 and FLOW have been integrated with the model LHI. For this purpose, to the original program FLOW some modifications have been introduced:

1. As input, the output data of the program LHI, i.e. flow rate of effluent infiltrating into a tip bed from a single section of a tip and concentrations of the analysed ions in leachate are being entered.

2. The program evaluates actual mean concentrations underneath each section block of a tip, where contaminated water infiltrates. This way, a ground water flow rate underneath each block, flow rates and concentrations of contaminants in effluent from each block that enter ground

water, as well as actual contaminants concentrations in ground water underneath each block are considered.

4 VERIFICATION OF MODELS

The validity of the designed and applied models has been verified on the simulation of ground water contamination in the vicinity of Smolnica tip (Fig. 7) for the suveyed period from 1968 to 1982.

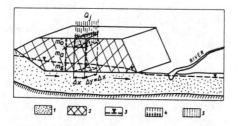

Fig. 7. Detachment of sections in an elementary block of a spoil tip. 1 - water-bearing strata; 2 - spoil tip; 3 - water-table; 4 - infiltration from atmospheric precipitation; 5 - unsaturated zone

Parameters obtained from the field survey, as well as hydrogeological and hydrological data for the tip site were used for the identification. In the computer program SP-2 the natural hydrogeological conditions of the area and the effect of effluents from the tip as an additional source of aquifer feeding an external border condition were taken into account. As a result, the pressure distribution in the filtration area under assumed border conditions was obtained (Fig.8).

These data were used as an input function for evaluation of contaminants migration by the program FLOW. For the evaluation of another input function contaminant injection , the output data calculated by LHI were used. To the program FLOW following input data were entered:
1. The hydrodynamic field in the assumed time intervals.
2. The contaminant concentrations in effluents from the tip.
3. The spatial distribution of coefficients k/n_a .

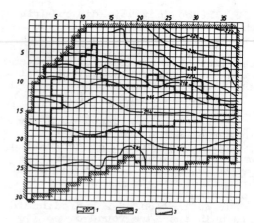

Fig. 8. The hydrodynamic field of Smolnica tip calculated by the program SP-2.
1 - hydroisohipses (m o.s.l.) ; 2 - border of the model SP-2; 3 - border of the tip;

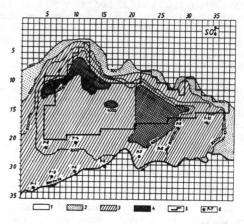

Fig. 9. Distribution of sulphate concentrations in the vicinity of Smolnica tip calculated by the program FLOW, situation on 31.10.1982 . Sulphate concentrations: 1 - below 200 mg/dm^3; 2 - from 200 to 1000 mg/dm^3; 3 - from < 1000 to 4000 mg/dm^3; 4 - above 4000 mg/dm^3; 5 - tip outline; 6 - blocks with the record of sulphate concentration changes in the course of calculations;

As a result of calculations, it was obtained:
1. The pattern of extention of ground waters containing contaminants in concentrations exceeding the maximum standard level (Fig. 9).

162

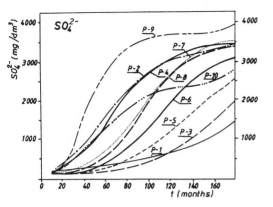

Fig. 10. Sulphate concentrations in the selected blocks P-1 - P-10 according to Fig. 9 of the Smolnica tip foreground, calculated by the program FLOW.

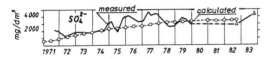

Fig. 11. Comparison of sulphate concentrations measured and calculated by the program FLOW (Smolnica tip, block P-2, see Fig. 9) .

2. Printouts of the distribution of contaminant concentrations in ground water in the vicinity of Smolnica tip in the assumed half-year intervals with an exemplary pattern at the end of calculations (Fig. 9) .

3. Charts of the contaminant concentration changes in the selected blocks of the bed and foreground of Smolnica tip (Fig. 10) .

The results of calculations show a good agreement with the field measurements (Fig. 11) .

Further works have been carried out for the improvement and generalization of the presented models.

ACKNOWLEDGEMENTS

This work has been supported financially by the Research Problem Fund 10.2 and the Central Program of Fundamental Research 03.11 coördinated by the Institute of Environmental Engineering of the Polish Academy of Sciences, Zabrze, Poland.

REFERENCES

Garrels, R.M. and C.L.Christ 1965. Solutions, minerals and equilibria. New York: Harper and Rowley.

Małoszewski, P. 1977. Numerical two-dimensional solution of the equation of contamination transport in ground waters in Polish . Cracow: Reports Inst. Nuclear Physics 943/PM.

Małoszewski, P. 1978. Prognosis of the contaminants migration in ground waters in Polish . Cracow: Reports Inst. Nuclear Physics 1035/PM.

Morth, A.H., E.E.Smith and K.S.Shumate 1972. Pyritic systems: a mathematical model. Washington: Office of Research and Monitoring, U.S. EPA B2-72-802.

Parkhurst,D.L., D.C.Thorstenson and L.N.Plummer 1981. PHREEQE - a computer program for geochemical calculations. Washington: U.S. Geological Survey.

Szymanko, J. et al. 1977-1980. Library of computer programs, No.1-5. Warsaw: Wyd. Geologiczne.

Szymanko,J. et al. 1980. The program block SP-2. In J.Szymanko et al. Mathematical model of the two-dimensional flow in the porous medium. Steady régime, square net. Library of computer programs No.4. Warsaw: Wyd. Geologiczne.

Twardowska, I. 1981. Mechanism and dynamics of coal mine spoil leaching at tips in Polish . Prace i Studia No.25. Wrocław-Warsaw-Cracow-Gdańsk-Łódź: Zakład Narodowy im. Ossolińskich, Wyd. PAN.

Twardowska, I., J.Szczepańska, S. Witczak et al. 1988. The effects of coal mine spoil on the aquatic environment. The estimation of environmental contamination, prognosis and prevention in Polish . Prace i Studia No.35. Wrocław-Warsaw-Cracow-Gdańsk-Łódź: Zakład Narodowy im. Ossolińskich, Wyd. PAN.

Reclamation, Treatment and Utilization of Coal Mining Wastes, Rainbow (ed.) © 1990 Balkema, Rotterdam. ISBN 90 6191 154 0

Static leaching of coal mining wastes

J. Gonzalez Cañibano & J. A. Fernandez Valcarce
HUNOSA, Dirección Técnica, Oviedo, Spain

A. Falcon & J. L. Ibarzabal
Ministerio de Obras Publicas, Demarcación de Carreteras del Estado, Oviedo, Spain

J. M. Rodríguez Ortiz
E.A.T., S.A., Madrid, Spain

J. A. Hinojosa
Ministerio de Obras Publicas, Dirección General de Carreteras, Madrid, Spain

ABSTRACT: The results obtained in static leaching tests with coal mining wastes are detailed. Various parameters were measured: pH, conductivity, solids in suspension, Cl^-, NO_3^-, PO_4^{3-}, $SO_4^=$, Fe, Ga, Se, Cd, Cu, Pb, Zn, As, Cs, Ni, Mn, Al, Cr, Cr^{3+}, Fe, B, $S^=$, NO_2^-, CN^-, NH_4^+ and phenols, to determine changes in their concentratic at different times. It is deduced from these results that the coal wastes tested may be considered non-contaminating materials.

1 INTRODUCTION

Although coal wastes may meet the technical requeriments for uses such as earthworks, a series of other characteristics have to be determined before they can be deemed suitable from the ecological point of view. These include self-combustion potencial, contamination, structure corrosion or attacks on other materials, and so forth.

This paper describes the lixiviation tests carried out and the results obtained, so as to determine the contamination power on the water bathing the coal wastes, an important factor not only in spoil heap but also in certain coal waste uses such as embankments and dams.

2 TESTS PERFORMED AND RESULTS

2.1 Selection of coal wastes

Since the contamination produced by wastes derives from the rainwater falls on the spoil heaps or on the sites such as highway embankments in which they deposited, leaching tests were carried out to determine whether or not contamination occurred, and if so, the degree and importance of the same.

In the first place, a selection of coal wastes was made which included the different types and production areas encountered in Spain, as follows:
. Sample 1/2: Fresh hard coal washery wastes from the North of Spain.
. Sample 3: Minestones from the North of Spain.
. Sample 4: Fresh hard coal washery wastes from the Northern Meseta of Spain.
. Sample 6: Fresh anthracite washery wastes from the Northern Meseta of Spain.
. Sample 7: Spoil heap wastes (proceeding from hard coal washeries) from the South of Spain.

On each of these samples a series of analyses and tests was carried out before and after the lixiviation tests so as to make the relevant comparisons.

2.2 Testing procedure

For lixiviation, cilindrical columns 150 cm high and 15 cm in diameter were used, with a capacity of 26.5 liters.

Once the columns had been loaded with their corresponding sample, distilled water was added until each sample was covered to a level 5 cm above the solid part.

24 hours later, the columns were purged and the liquid extracted. The volumen extracted was replaced

in each case with distilled water. 24 hours later the whole process was repeated, and again at 24 hours intervals for several days.

Samples were taken of the liquid obtained from each column, and of the distilled water used for replacement, for analysis.

2.3 Results of the tests

2.3.1 Percolation

Table I sets out the data obtained for loading, purging and liquid replacement, from which the percolation of each of the samples tested is deduced.

In sample 3 the lixiviation liquid was obtained after 96 hours, with a consequent variation with respect to the others.

2.3.2 Particle size distribution

Particle size analysis was carried out for each of the samples before and after the lixiviation tests were carried out, from which it could be determined whether or not there was fine-particle drag, particle agglomeration, waste degradation, etc.

The following conclusions may be drawn from Graphs 1, 2, 3, 4 and 5, which represent particle size distribution before and after the lixiviation tests for each of the samples:
- degradation exists in some sizes greater than 20-40 mm, as there is an increase in the percentages which pass through the meshes present at this interval. This may appear in the more shaley rocks, such as silstones, as they are submerged in water.
- in sizes less than 20-40 mm, washing due to the water is greater than the generation of particles passing through this mesh interval.
- fine particle washing occurs. However, the percentage passing through 0.074 mm is practically the same before and after the lixiviation test.

2.3.3 Chemical Analysis of lixiviation water.

Contaminating elements

Chemical analysis of contaminating elements was carried out on each of the samples (for heavy metals for example) to determine whether the water added dissolved or dragged with it the same. Table II shows the results of the chemical analyses before the lixiviation tests.

Table III sets out the data obtained from analysis of the water used in the tests and that resulting from lixiviation at different times. From these it can be seen that there is no dissolution of potentially contaminating elements, as the lixiviates show the same contents as the water used to carry out the tests, with the exception of magnesium (Mg) and aluminium (Al) in some samples, though even here values are insignificant, being very small and similar to what would occur with other earths. Moreover, after 3-4 tests these values maintain the same values as those of the original water.

Anions and cations

Furthermore, other components have been identified in the water resulting from sample lixiviation, on the basis of the "tables of the characteristic parameters which are to be considered as minima in the evaluation of waste treatment" as published in the B.O.E. (Official Gazette of Spain) No. 103, dated 30-4-86, as there may be present other contaminating components as well as heavy elements, such as those deriving from nitrogen or sulfur.

Table IV shows the values obtained from the analysis of anions and cations carried out both on the original water used in the lixiviation tests and on that of the lixiviates, from which it can be deduced that:
- in the case of Cl^- there is in some samples and in the first lixiviates slight dissolution, but its value is insignificant, or becomes so on the third or fourth lixiviation.
- there is dissolution of the sulfate ion $SO_4^=$, which diminishes with time, as can be seen in Graph 6.

The greates lixiviation of the $SO_4^=$ ion occurs in wastes of sample 3, followed by sample 4, and after that samples 1/2, 6 and 7, as can be appreciated in Graph 6, from which it will also be seen that after 10 days' lixiviation concentra-

TABLE I. PERCOLATION

Sample	Initial volume of wates added (liters)	Volume of liquid extracted after (liters)	Percola tion
1/2	8,4	5,85	Very good
3	6,0	–	Very bad
4	7,0	4,75	Good
6	10,0	6,70	Fair
7	10,0	6,45	Fair

TABLE II. CHEMICAL ANALYSIS OF POTENTIALY CONTAMINATING ELEMENTS BEFORE THE LEACHING TESTS.

ELEMENTS ppm	S A M P L E				
	1/2	3	4	6	7
Cu	52	45	70	36	48
Pb	24	34	39	58	5
Zn	134	121	160	90	69
As	16	26	43	78	17
Ni	69	55	70	54	60
Co	23	24	30	20	70
Mn	650	765	768	406	1450
Ga	29	27	33	33	30
Cd	< 0,1	< 0,1	< 0,1	< 0,1	< 0,1
Se	1,05	0,97	1,75	1,38	0,67
Hg	<0,05	<0,05	<0,05	<0,05	<0,05

tions have considerably diminished.

Sulfate ion concentrations diminish considerably at the tenth day of lixiviation.
- there is no lixiviation in the other anions and cations, or in phenols, or if there is, it is insignificant amounts, less than one part in a million, which disappears in a short time.

pH

The values of pH –which gives an idea of the nature of the materials of which the coal wastes are made up- are set out in Table V, measured in the water used to carry out the lixiviation tests and in the lixiviates obtained in each of the samples 1/2, 3, 4, 6 and 7.
It is deduced from this Table, and from Graph 7, in which the values obtained in each of the samples are represented, that:
- all the samples show an increase in pH in the lixiviates with respect to the original water used, passing from 5.9 to almost neutral (7.2) or basic (9.2).
- all the samples show the same sequence, that is, the pH value found in the first lixiviation water is maintained or varies by ± 0.4 in the remaining lixiviation waters at different times.
- the lixiviation waters of the whole set of samples remain within the limits pH = 7.2 and pH = 9.2, which shows the basic character of the materials making up the wastes, and is in agreement with their petrology.
- the most basic value is that of the wastes in sample 6, and the least are those of samples 4 and 3, which correspond to the values of anion and cation content, and

specificallly with the values for
S pyritic (0.5 - 0.37 - 0.63 - 0.35
and 0.24 % for samples 1/2, 3, 4,
6 and 7 respectively), as the greater
this content, the more acid (or
less basic) the lixiviate waters
will be.

Conductivity

As with pH, Table V sets out the
conductivity values of the lixiviation
waters of each of the samples mentioned.
It can be seen from this Table,
and from Graph 8, in which the values
obtained are set out, that:
- in all samples there is a progressive
decrease in conductivity, which
is logical since the passage of
water lowers ion concentration (anions
and cations) until it becomes practically constant.
- the lixiviate samples presenting
the highest concentration are those
of sample 3, and the lowest those
of samples 6 and 7, coinciding with
the highest and lowest concentrations
of ions $SO_4^=$, Cl^- , etc) presented
by the lixiviates, as can be seen
in Table IV and Graph 6.

Solid in suspension

Table V shows the content of solids
in suspension of the lixiviation
waters of each of the samples tested.
It can be deduced from this Table,
and from Graph 9 in which the result
obtained are set out that:
- in all the samples, the content
of solids in suspension decreases
with time, as is to be suspected,
since fine particle drag will less
the more water has passed. However,
from the eighth day it begins to
rise slightly as a result of the
breaking-up and degradation of the
coarse particles of siltstone and
other easily degradeable rocks,
as can be seen from the sample particle
size analyses represented in Graphs
1 to 5 inclusive, which show particle
size distribution before and after
lixiviation tests.
- sample 6 is that with the highest
solid contents, due to the greater
degradability of the rocks it comprises.
At the same time, there is an increase
in the content of solids in suspension
with respect to the seventh day.
This coincides with the findings
reflected in the particle size distribu-

tion curves, as while in the 100-40
mm mesh this sample suffers a loss
of 9%, sample 4 loses 4%, sample
3: 4%, sample 7: 5.55%, etc., this
breaking-up of coarse particles
by water action giving a greater
production of fine particles which
are carried away in the water, since,
as can be seen from the particle
size analyses, fine particle percentages
maintain the same values as those
before the lixiviation tests.
The increase in concentration
on the eighth or ninth day indicates
that this is when coarse particle
break-up occurs.

4 CONCLUSIONS

In the light of the above, and comparing
the values obtained with those require
by the Spanish Law of Waters (Law
29/1985) of the 2nd of August 1985
and B.O.E. (Official Gazette of
Spain) No. 103 of the 30th of April
1986, it can be affirmed that coal
wastes are not contaminating materials.

ACKNOWLEDGEMENTS

The authors would like to express
their gratitude to MOPU (Direccion
General de Carreteras), HUNOSA and
OCICARBON for the facilities afforded
them in the elaboration of this
paper and for financing the project
"The use of coal mining wastes in
earthworks", Dña. Josefina Grela
Vàzquez por the typing and Don Ceferino
Fernandez Cuetos for the drawings.
They would also like to thank
the Laboratory and Mineralurgy Department of E.N. Adaro de Investigaciones
Mineras, S.A., where the lixiviation
tests were carried out.

TABLE III. CHEMICAL ANALYSIS OF POTENTIALLY CONTAMINATING AGENTS IN LEACHING LIQUIDS (ppm)

ELEMENT	WATER	24 hours leaching Samples				48 hours leaching Samples				72 hours leaching Samples				240 hours leaching Samples		
		1/2	4	6	7	1/2	4	6	7	1/2	4	6	7	1/2	3	7
Ga	<0.02	<0.02	<0.02	–	<0.02	<0.02	<0.02	<0.02	<0.02	–	–	<0.02	–	–	–	–
Se	<0.03	0.04	<0.03	–	0.04	<0.03	<0.03	<0.03	<0.03	<0.03	<0.03	<0.03	<0.03	–	<0.03	–
Cd	<0.2	<0.2	<0.2	–	<0.2	<0.2	<0.2	<0.2	<0.2	<0.2	<0.2	<0.2	<0.2	–	<0.2	–
Cu	<0.5	<0.5	<0.5	–	<0.5	<0.5	<0.5	<0.5	<0.5	<0.5	<0.5	<0.5	<0.5	–	<0.5	–
Pb	<0.2	<0.2	<0.2	–	<0.2	<0.2	<0.2	<0.2	<0.2	<0.2	<0.2	<0.2	<0.2	–	<0.2	–
Zn	<10	<10	<10	–	<10	<10	<10	<10	<10	<10	<10	<10	<10	–	<10	–
As	<0.5	<0.5	<0.5	–	<0.5	<0.5	<0.5	<0.5	<0.5	<0.5	<0.5	<0.5	<0.5	–	<0.5	–
Co	<3	<0.3	<3	<3	<3	<0.3	<3	<3	<0.3	<0.3	<3	<3	<3	–	<3	–
Ni	<3	<3	<3	<3	<3	<3	<3	<3	<3	<3	<3	<3	<3	–	<3	–
Mn	<3	<3	5.2	<3	<3	<3	3.8	<3	<3	<3	<3	<3	<3	–	<3	–
Mg	<0.1	2.1	210	–	38	1.4	154	3.7	27	1.1	104	2.3	19	–	<80	–
Al	<1	11	<1	–	<1	12	<1	20	<1	11	<1	21	<1	5	<1	4
Cr	<3	<3	<3	–	<3	<3	<3	<3	<3	<3	<3	<3	<3	–	<3	–
Fe	<3	4.9	<3	–	<3	<3	<3	<3	<3	<3	<3	<3	<3	–	<3	–

TABLE IV. CHEMICAL ANALYSIS OF ANIONS AND CATIONS IN LEACHING LIQUIDS   (mg/l)

COMPONENT	WATER	24 Hours Samples				48 Hours Samples				72 Hours Samples				96 Hours Samples					120 Hours Samples				
		1/2	4	6	7	1/2	4	6	7	1/2	4	6	7	1/2	3	4	6	7	1/2	3	4	6	7
Cl^-	<1,0	9,7	7	<0,1	21	14,2	7	<0,1	14	14,2	7	<0,1	10	<3,0	78	<1,0	<0,1	<0,1	<3,0	79	<1,0	<0,1	7,0
$SO_4^=$	<1,0	560	1520	300	240	180	1250	160	170	120	1000	150	1,0	40	2200	540	60	60	40	2300	560	60	40
NO_3^-	<0,5	<0,5	0,8	<0,5	18,2	<0,5	0,7	<0,5	14,4	<0,5	0,6	<0,5	11,8	<0,5	<0,5	0,6	<0,5	8,6	<0,5	<0,5	0,6	<0,5	7,0
$PO_4^{\equiv}$	<0,1	<0,1	<0,1	<0,1	<0,1	<0,1	<0,1	<0,1	<0,1	<0,1	<0,1	<0,1	<0,1	<0,1	<0,1	<0,1	<0,1	<0,1	<0,1	<0,1	<0,1	<0,1	<0,1
F^-	<0,01	2,2	0,2	0,6	0,29	3,6	0,17	0,36	0,24	2,7	0,17	0,56	0,32	2,9	0,56	0,17	0,56	0,22	2,2	0,42	0,11	0,40	0,18
Cr^{6+}	-	<0,1	<0,1	0,14	0,15	<0,1	<0,1	<0,1	<0,1	-	-	<0,1	-	-	<0,1	-	-	-	-	-	-	-	-
$S^=$	-	<0,1	<0,1	<0,1	<0,1	<0,1	<0,1	<0,1	<0,1	<0,1	-	-	-	-	<0,1	-	<0,1	-	-	-	-	-	-
NO_2^-	-	2,1	0,06	0,10	0,24	0,3	0,08	0,09	0,24	0,2	0,07	0,06	0,10	0,15	0,11	0,07	0,06	0,14	0,15	0,05	0,08	0,05	0,10
CN^-	-	<0,01	<0,01	<0,01	<0,01	<0,01	<0,01	<0,01	<0,01	-	-	-	-	-	<0,01	-	-	-	-	<0,01	-	-	-
NO_4^+	-	5,0	2,9	2,9	1,20	3,0	1,5	1,7	0,7	2,4	1,0	1,0	0,5	-	2,3	-	-	-	-	1,5	-	-	-
Phe-nol	-	0,01	<0,01	<0,01	<0,01	0,01	<0,01	<0,01	<0,01	-	-	-	-	-	<0,01	-	-	-	-	<0,01	-	-	-

TABLE IV. CHEMICAL ANALYSIS OF ANIONS AND CATIONS IN LEACHING LIQUIDS (mg/l) (Continuation)

COMPONENT	WATER	168 Hours Samples					192 Hours Samples					216 Hours Samples					240 Hours Samples				
		1/2	3	4	6	7	1/2	3	4	6	7	1/2	3	4	6	7	1/2	3	4	6	7
Cl^-	<1.0	<3.0	106	<1.0	<0.1	7.0	<3.0	43	<0.1	<0.1	<0.1	23.0	50	<0.1	<0.1	<0.1	<3.0	39	<1.0	<0.1	<0.1
$SO_4^=$	<1.0	40	2600	440	75	36	50	1400	420	50	40	26	1400	360	25	25	26	1200	232	25	19
NO_3^-	<0.5	<0.5	<0.5	0.6	<0.5	6.0	<0.5	<0.5	0.6	<0.5	5.2	<0.5	<0.5	<0.5	<0.5	2.6	<0.5	<0.5	<0.5	<0.5	2.3
$PO_4^{\equiv}$	<0.1	<0.1	<0.1	<0.1	<0.1	<0.1	<0.1	<0.1	<0.1	<0.1	<0.1	<0.1	<0.1	<0.1	<0.1	<0.1	<0.1	<0.1	<0.1	<0.1	<0.1
F^-	<0.01	1.8	0.4	0.11	0.4	0.18	1.3	0.35	0.11	0.35	0.18	0.9	0.39	0.11	0.26	0.16	0.7	0.22	0.10	0.17	0.15
Cr^{6+}	—	—	—	—	—	—	—	—	—	—	—	—	—	—	—	—	—	<0.1	—	—	—
$S^=$	—	—	—	—	—	—	—	—	—	—	—	—	—	—	—	—	—	—	—	—	—
NO_2^-	—	0.15	<0.05	0.09	0.09	0.15	0.12	<0.05	0.07	0.06	0.14	—	—	—	—	0.10	—	—	—	—	0.09
CN^-	—	—	—	—	—	—	—	—	—	—	—	—	—	—	—	—	—	—	—	—	—
NO_4^+	—	—	1.0	—	—	—	—	—	—	—	—	—	—	—	—	—	—	—	—	—	—
Phenol	—	—	—	—	—	—	—	—	—	—	—	—	—	—	—	—	—	—	—	—	—

T A B L E V

VALORES FOR pH, CONDUCTIVITY AND SOLIDS IN SUSPENSION OF LEACHING

W A T E R		p H					Conductivity $\mu\,\bar{\Omega}^{-1}\,cm^{-1}$					Solids in suspension (mg/1)				
			4.5					4.5					0.01			
Hours	Samples	1/2	3	4	6	7	1/2	3	4	6	7	1/2	3	4	6	7
24		8.6	–	7.2	9.2	7.9	1550	–	2700	560	520	2716	–	824	20108	1888
48		8.9	–	7.2	9.2	7.9	850	–	2100	350	380	2204	–	668	21852	2076
72		8.9	–	7.2	9.2	8.0	580	–	1550	280	300	2428	–	416	9076	1848
96		9.2	8.1	7.3	9.2	8.2	490	4600	1240	275	230	1420	<0.1	308	3688	956
120		9.0	8.2	7.3	9.2	8.2	460	500	1100	275	220	1304	<0.1	276	3016	412
168		9.0	7.7	7.3	9.2	8.1	500	4500	1150	275	240	492	<0.1	244	2652	336
192		9.0	7.7	7.3	9.2	8.1	400	2800	920	200	190	504	<0.1	1120	4380	348
216		9.0	8.0	7.4	9.2	8.2	375	3100	780	185	180	404	<0.1	872	1276	648
240		9.0	7.9	7.4	9.1	8.2	360	2300	740	175	180	280	416	820	472	580

Leaching liquids

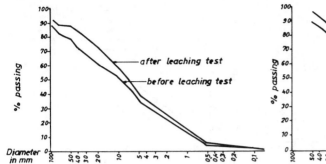

Graph 1. Particle size distribution before and after leaching test. Sample 1/2

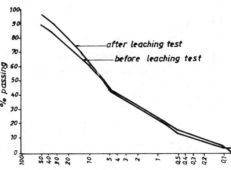

Graph 2. Particle size distribution before and after leaching test. Sample 3

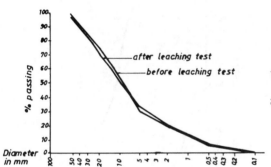

Graph 3. Particle size distribution before and after leaching test. Sample 4

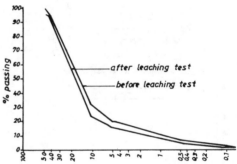

Graph 4. Particle size distribution before and after leaching test. Sample 6

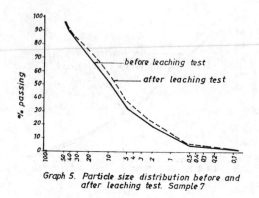

Graph 5. Particle size distribution before and after leaching test. Sample 7

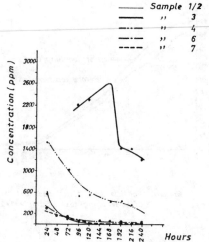

Graph. 6. SO_4 = Concentrations in leaching

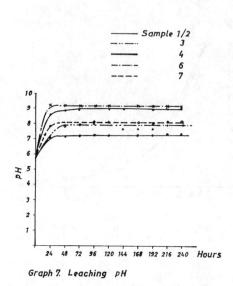

Graph 7. Leaching pH

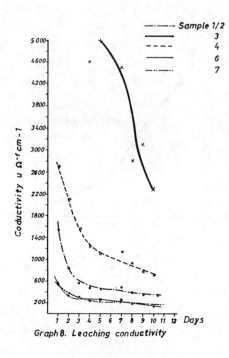

Graph 8. Leaching conductivity

174

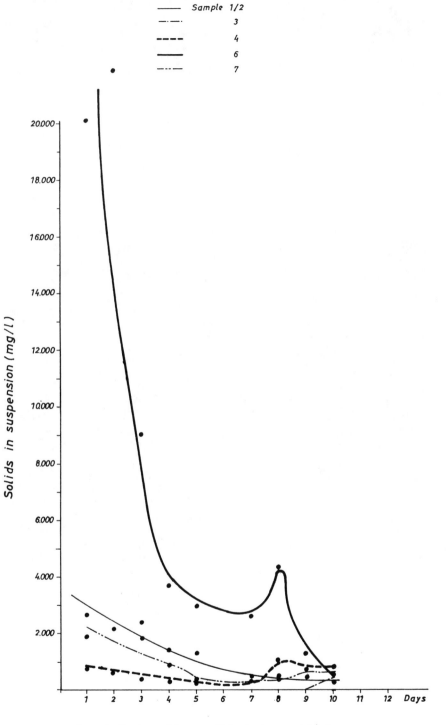

Graph 9. Leaching solids in suspension

175

Reclamation, Treatment and Utilization of Coal Mining Wastes, Rainbow (ed.) © 1990 Balkema, Rotterdam. ISBN 90 6191 154 0

Transformations of chemical composition of pore solutions in coal mining wastes

I.Twardowska
The Polish Academy of Sciences, Institute of Environmental Engineering, Zabrze, Poland

J.Szczepańska
University of Mining and Metallurgy, Institute of Hydrogeology and Engineering Geology, Cracow, Poland

ABSTRACT: The problem of transformation of chemical composition of pore solutions in coal mining wastes has been exemplified on the results of long-term lysimetric leach experiments and field studies on colliery spoils from different stratigraphic series of the Upper Silesian coal basin (USCB). The investigations have shown considerable and differential, qualitative and quantitative changes with time, particularly of chloride salinity, sulphate concentrations, pH and heavy metals content in pore solutions due to weathering. The mechanism of the processes having an effect on these transformations has been explained.

1 INTRODUCTION

Underground coal mining in Poland is the major source of solid wastes that are discharged from collieries and coal preparation plants in the form of run-of-mine spoil, discard, slurry, reject wet fines and tailings. Annual production of colliery spoils in the Upper Silesian coal basin (USCB) ranges from 70 to 80 Tg, some 80 % being deposited at local and central tips or used in civil earthworks as common fill.

Similarly, millions tonnes of colliery spoils are known to be lying in tips throughout other coalfields all over the world. At the same time, applications of spoil in civil earthworks are wide and steadily increasing. Both ways of minestone management, in consequence of exposure to the atmospheric conditions, give rise to weathering processes that usually result in the considerable tranformations of the chemical composition of pore solutions in spoils. A conversance with these transformations is of essential importance for adequate assessment of the pollution potential of the deposited spoils to the aquatic environment, as well as the corrosion impact on concrete and reinforcement in the case of minestone application in civil engineering constructions.

2 SPOIL PROPERTIES

The properties of fresh wrought colliery spoils are largely depend on the regional variability and stratigraphic position of mined coal seams, those of the USCB being in the Westphalian A-D and the Namurian A-C series of the Carboniferous formation. The proportion of run-of-mine and coal preparation discards into the deposited minestone is also of great importance.

Spoils deposited at tips are essentially granular and coarse in nature. Gravel fraction accounts for about 54 wt%. The predominant components of spoils are usually mudstones that amount to more than 80 wt%. The lowest mudstone (from 60 to 75 wt%) and the highest sandstone contents occur in spoils from the youngest Westphalian C-D series.

The permeability to water of the fresh wrought spoils is generally high and varies in the range from greater than 10^{-3} to about 10^{-5} m/s (Skarżyńska et al. 1982). It can be considerably reduced provided that adequate particle size degradation and sealing of voids occurred due to weathering and compaction. It has though been found that minestone from the Namurian series is high resistant to breakdown associated with weathering or mechanical compaction processes. Upwards the pro-

177

file of the USCB Carboniferous sequence, the susceptibility of rocks to breakdown increases and reaches the maximum in the Westphalian C-D series. It determines the permeability of weathered spoils to water that ranges from 10^{-3} to 10^{-7} m/s, being as low as from 10^{-6} to 10^{-7} m/s for the spoils from the Westphalian C series (Szczepańska 1987). Therefore, the weathered spoil material still remains permeable or semi-permeable to water. The natural moisture content of spoils at tips ranges from 2.5 to 14 wt%, that means that voids in the deposited minestone are filled with water to the extent of from 0.3 to 0.5 (Szczepańska 1987).

The fresh wrought spoil discharged in the USCB contains from 0.001 to about 1 wt% of chlorides occurring solely in the pore solution. Chloride content in minestone is a function of chloride concentration in associated underground water and the natural moisture content of rock material. In the USCB it depend upon the regional distribution of underground water salinity and the vertical hydrogeochemical zonality of the Carboniferous sequence (Herzig et al. 1986).

Total sulphur content S_t of spoil ranges from 0.01 to above 10 wt%, mean value being as low as 1 wt%. Sulphide sulphur amounts to some 60 wt% of S_t. Sulphur reactivity expressed in half-life values $t_{1/2}$ ranges from 29 to 10502 days, the mean value accounts for 588 days. It has been found that the highest reactivity show disulphides from the Westphalian C-D series and the lower part of the Namurian A series (Twardowska et al. 1988).

The buffering capacity in colliery spoils at pH 7 is a function of calcium and magnesium carbonate contents and exchangeable ions of cation exchange capacity. In the USCB, no definite regularities in these components occurrence have been observed (Twardowska et al. 1988).

An important property of colliery spoils with respect to the contaminant generation is the permeability to air. The investigation have shown the presence of atmospheric oxygen and occurrence of oxidation conditions within the entire unsa-

turated zone of typical tips in the USCB (Twardowska et al. 1988). These results are in disagreement with assumptions of some authors (Glover 1978).

3 EXPERIMENTAL

3.1 Material and methods

To determine the mode of transformations of the chemical composition of pore solutions in spoils due to weathering, the long-term leach experiments and field studies on the selected spoil material and spoil tips have been carried out. For experiments, two different types of spoils representative for the mined coal seams in the USCB have been applied.

Spoils of the type I have been discharged from the Westphalian C series, that is from the youngest top part of the Carboniferous sequence of the USCB. The rock material is susceptible to weathering and particle size degradation to a silt fraction. Its chloride salinity is very low and ranges from 0.003 to 0.007 wt%. The material is buffered $\xi = 1.50$, though pyritic sulphur concentrations in the freshly produced spoils are relatively high ($S_p = 1.48$ wt%, $S_t = 1.72$ wt%). These spoils come from Siersza mine localized in the eastern part of the USCB, and have been deposited at Siersza-Misiury tip.

Spoils of the type II represent the Namurian C and Westphalian A series. They come from Szczygłowice mine in the western part of the USCB and have been placed on Smolnica tip. The freshly produced material is coarse grained, resistant to breakdown, of the mean chloride content ranging from 0.03 to 0.05 wt%, and low-buffered $\xi = 0.70$. The average pyritic sulphur concentration accounts for 0.79 wt% ($S_t = 0.90$ wt%) and is apparently lower than that in spoils of the type I.

For leach experiments, a lysimetric method has been applied. The initial stage of soluble constituents leaching has been modeled in columns using fresh wrought spoil in layers from 2 to 20 m thick. Flow rate of leaching water corresponded to a ten year average daily precipitation of

4.74 mm.

The pore solution transformations within the 1.5 m thick surface layer of a tip have been studied in situ on three lysimeters 600 mm in diameter. The lysimetric columns were exposed to the natural atmospheric conditions for 22 - 24 months during 1984-1986. Precipitation and seepage water composition were examined.

The lysimeter TS-1 contains 7 years old spoil material of the type I taken from Siersza-Misiury tip. Two lysimeters SM have been filled with spoils of the type II - fresh wrought (SM-1) and 10 years old (SM-2) taken from Smolnica tip.

Field studies carried out at Smolnica tip where spoils of the type II have been deposited, included examination of transformations with time of the buffering capacity of the surface spoil layer and chemical composition of water in the drainage ditches above and below the tip.

3.2 Initial stage of leaching

Mechanism of major species e.g. chlorides and sulphates leaching from spoils and their vertical transport in pore solutions within the unsaturated zone of a spoil tip displays a substantial difference that exteriorizes already at the initial stage of the process. The column experiments have shown that in leaching of soluble constituents

during the vertical flow of infiltration water from atmospheric precipitation through a spoil layer, two stages can be distinquished:

1. Initial stage - in which the retention capacity of spoil is gradually filled.

2. Seepage phase - in which free percolation of water proceeds through the full thickness of a spoil layer.

Chloride anions and associated sodium cations, being fairly mobile and unconstrained by equilibrium conditions in the actual range of concentrations, easily migrate with infiltration water that fills up the retention capacity of subsequent layers of a tip. In consequence, a vertical redistribution of chloride content within a tip profile develops to the extent dependent upon the degree of saturation of the retention capacity of the spoils (Fig. 1) . The high dynamics of chloride migration causes that in a short time after spoil deposition, in upper layers of a tip where thorough exchange of pore solution occurred, the chloride content in pore water is close to nil, whereas the loads of this constituent in the lower, just filled layers substantially exceed the original chloride concentration in the spoils. As a consequence, chloride concentrations in the initial outflow from the spoil layer are much higher than in original pore water, proportionally to layer height and the extent of the retention capacity deficit.

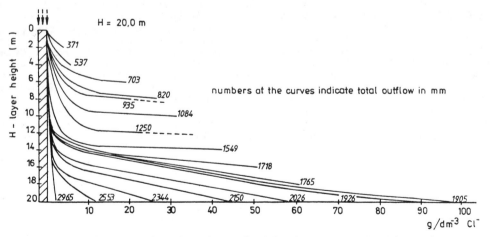

Fig. 1. Vertical redistribution of chloride concentrations in pore solutions in a spoil layer during the filling of retention capacity.

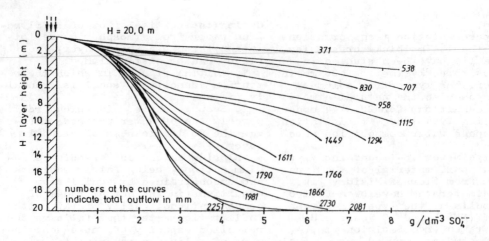

Fig. 2. Vertical redistribution of sulphate concentrations in pore solutions in a spoil layer during the filling of retention capacity (fresh wrought spoil of the type II) .

A pattern of the vertical redistribution of sulphate in pore solution within the unsaturated zone of a tip is a resultant of sulphate generation due to sulphide oxidation and their transport out of the system by infiltration water. In buffered material, it shows also strong limitation by the geochemical constraints imposed by equilibrium with gypsum. In the material of moderate or high Na-Cl salinity, where the bulk of cation exchange capacity of fresh wrought spoil is occupied by alkali ions, in the initial stage of leaching, sulphates in pore solutions are balanced mainly by Na^+ ions and thus their concentrations are unconstrained by gypsum equilibrium and may be much higher. Simultaneously to alkali ions depletion, gradual decrease of sulphate concentrations occurs, up to a relatively constant level determined by the rate of sulphate generation restricted by equilibrium conditions (Fig. 2) . In the initial stage of leaching, the buffering capacity of spoils is generally adequate.

3.3 Seepage phase

The results of lysimetric experiments on the seepage phase under in situ conditions follow the ge-

neral pattern defined by the dynamics of chloride and sulphate leaching, sulphate and acid generation as well as the mode and extent of the buffering capacity of spoils.

In the initial period after seepage occurrence from the 7 years old buffered material of the type I, concentrations of total dissolved solids (TDS) in leachate were considerably lower than those observed in pore solutions at the tip (Szczepańska 1987) due to disturbance of leach conditions during the spoil replacement. In some 4 months the chemical composition of leachate reached the level corresponding to the natural conditions (Fig.3) . The leachates were invariably of SO_4-Mg-Ca type and their pH range was close to neutral (pH 6.8-8.4) . Sulphates occurred in the range from 2000 to 2800 mg/dm^3 SO_4^{2-}, whereas concentrations of chlorides were very low and did not exceed 40 mg/dm Cl^-. The chemical composition of effluents displayed a considerable stability on the level mastered by the carbonate and gypsum equilibria in the system Ca-Mg-C-O-H-S.

Opposite to the leaching behaviour of the buffered spoils of the type I, the chemical composition of leachate from the low-buffered, chloride-bearing spoils of the type II undergo substantial qualitative and quantitative transformations with

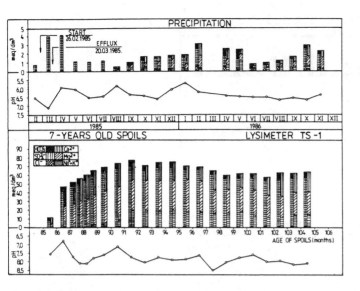

Fig. 3. Transformations of the chemical composition of leachate from the low chloride, buffered, 7 years old spoils of the type I taken from Sier-sza-Misiury tip. Spoil layer 1.5 m thick.

time resulted from the natural wea- thering processes. Washing of solu- ble constituents from fresh wrought spoils by infiltrating precipitati- on water caused fast decline of mo- bile species in leachate, such as chloride and sulphate anions balan- ced mainly by alkali cations (Fig. 4) . In a half year's period adequ- ate to a single water exchange rate for a 1.5 m thick spoil layer, chlo- ride concentrations in leachate de- creased from about 7000 mg/dm^3 to some 700 mg/dm^3 and next fast sta- bilized on the level as low as 50 mg/dm^3. Sulphate concentrations in leachate from fresh wrought spoils declined from nearly 8000 mg/dm^3 to the range from 1000 to 3000 mg/ dm^3 being the resultant of the dy- namics of sulphide oxidation, car- bonate and gypsum equilibria, amo- unt of percolating precipitation water and flushing intervals the greater concentrations were relea- sed with increased intervals bet- ween rainfalls . In the first two years since deposition, the buffe- ring capacity of spoils had been still sufficient to neutralize the released acid loads, and pH values of leachate ranged from 7 to 9 (Fig. 4) .

During 10 years' deposition the chemical composition of leachate from the same spoils changed drama- tically as a consequence of depleti- on of buffering capacity below the critical level, that resulted in acidification of pore solutions (Fig. 5) . The pH values ranged from 4.0 to 5.2, whereas sulphate concen- trations in leachate increased up to 5000 - 6000 mg/dm^3 SO$_4^{2-}$ because of the lack of equilibrium limita- tions. In the next two years of the leach experiment, pH values displa- yed further decrease to the range from 2.73 to 2.96, while the sulphur content also began to show distinct declining tendency in consequence of the gradual sulphide depletion in spoils.

Due to weathering processes, the chemical composition of leachate changed gradually from the initial Na-Ca-Cl-SO$_4$ type into Na-SO$_4$-Cl, and next into Na-SO$_4$ type in the buffered material. Since acidifica- tion of spoils and pore solutions was evolved, also ions Fe^{3+} and H$_3$O$^+$ had appeared in considerable concen- trations and chemical type of water transformed into SO$_4$-Mg-Fe Ca type and next at pH 3, into SO$_4$-Na-H-Mg- Ca type.

181

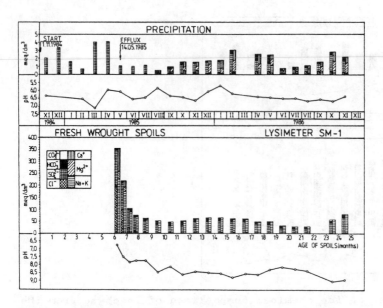

Fig. 4. Transformations of the chemical composition of leachate from the moderate chloride, low buffered, fresh wrought spoils of the type II taken from Smolnica tip. Spoil layer 1.5 m thick.

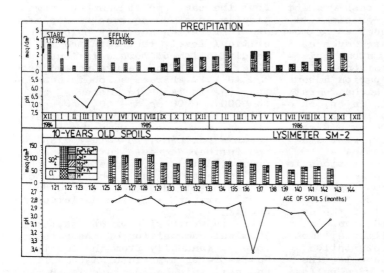

Fig. 5. Transformations of the chemical composition of leachate from the low buffered, 10 years old spoils of the type II taken from Smolnica tip. Spoil layer 1.5 m thick.

At pH values close to 3.0 spoils can be buffered through the displacement of Na^+ and Al^{3+} ions from their position in the lattice of aluminosilicate minerals (albite, kaolinite) by H_3O^+ ions, that causes repeated appearance of sodium ions and occurrence of complex ions $AlSO_4^-$ in the leachate. This process has been reported also by other authors (Van Breeman 1972, Palmer 1978).

Furthermore, acidification of spoils and leachates causes release of particularly toxic and environmentally sensitive species such as heavy metals not only from spoils, but also from any material or soil matrix, which the pore solutions or leachate are in contact with. In examined leachate, exceedingly high concentrations of zinc come from corrosion of the galvanized interior surface of the lysimetric tube (Tab. 1). Other metals that occurred in concentrations many times exceeding the maximum standard levels, were leached from the spoil matrix. In pore solutions of buffered material, heavy metals contents were much lower (Tab. 1).

Table 1. Heavy metal concentrations in leachate from the low-buffered spoils of the type II. Spoil layer 1.5 m thick.

Heavy metals	Concentration range, min/max in mg/dm^3		Max. standard level*
	SM-1	SM-2	
	pH 7.3-8.9;	pH 2.7-3.0	
	Spoil age 2 years	Spoil age 12 years	
Co	0.01 0.02	0.28 0.41	
Cd	0.024 0.039	0.050 0.213	0.05
Cr$_t$	0.00 0.01	0.02 0.46	0.05**
Cu	0.005 0.02	0.12 1.40	0.10
Fe$_t$	0.24 0.48	1.44 5.59	0.50
Mn	0.10 0.19	3.20 4.32	0.10
Ni	0.20 0.22	0.93 1.28	
Pb	0.01 0.05	0.12 0.40	0.10
Zn	0.73 0.96	73.20 111.80	5.0

* For potable water

** For Cr VI

Therefore, the contamination potential of colliery spoils is a function of their buffering properties.

For characterization of susceptibility of spoils to acidification, the term 'buffering rate' $\xi = G_{buf}/G_{ac}$ has been introduced, which is the ratio of the total potential buffering capacity G_{buf} to the total acid generation potential G_{ac} (Twardowska 1981, 1989). The value $\xi \geqslant 2.4$ was found to guarantee the permanent pH $\geqslant 7.0$ of spoils and pore solutions. Spoils with $\xi < 1.5$ were identified as low-buffered material. The critical level of buffering capacity when acidification process started, was estimated at $\xi = 0.60$ (Fig. 6).

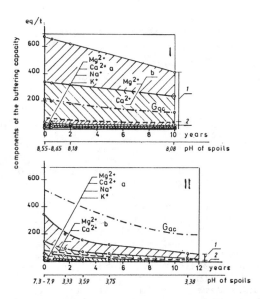

Fig. 6. Changes in time of the buffering capacity of spoils in the top layer of a tip 0 - 20 cm.
I - buffered spoils ($\xi = 3.30$);
II - low-buffered spoils of the type II from Smolnica tip ($\xi = 0.70$).

This way, susceptibility of spoils to acidification and thus the long-term prognosis of the contamination potential of minestone to be deposited at tips or applied in civil engineering, can be accomplished in advance.

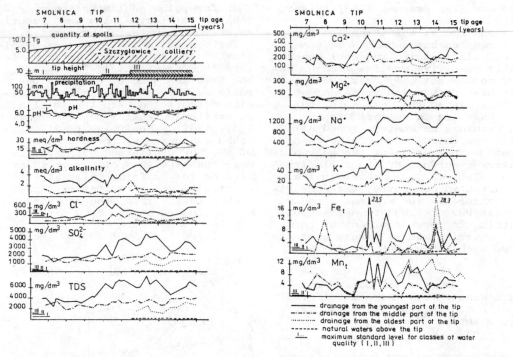

Fig. 7. The transformations of ground water qulity in the vicinity of Smolnica tip.

3.4 The effect of a tip on ground water quality

From the investigations of the intensity of precipitation water infiltration through spoil layers in the unsaturated zone of a tip, it was evaluated that during the examined 10 years' period, pore solution in a layer 1.5 m thick of unbuffered spoils of the type II had been exchanged about 25 times. However, the pollution and corrosion potential of spoils and leachate was still wery high and even reached its maximum. A 10 m thick layer of minestone at the tip would be washed to the same extent in some 65 years. Therefore, spoil tips and other constructions of minestone can be a permanent source of the aquatic environment contamination for many years.

The drainage water survey in the Smolnica tip site confirmed this conclusion (Fig. 7). The highest TDS, chloride, sulphate and alkali concentrations at pH values ranging from 6.0 to 8.0 were observed in the effluents from the youngest

part of Smolnica tip. In time, gradual decrease of drainage mineralization occurred simultaneously with the decrease of pH as an effect of pyrite depletion, but relatively even greater exhaustion of bufferin capacity. The lowest pH values below 5.0 were estimated in the drainage ditch at the oldest part of Smolnica tip , where also the highest concentrations of iron and manganese were obserwed. Though the examined waters were considerably diluted by natural ground water (Szczepańska and Twardowska 1987), the registered range of maximum con centrations of trace heavy metals was still much above the standard level for water of the Ist class of purity $(0.82 - 1.03$ mg Cu/dm^3, $1.89 - 2.70$ mg Zn/dm^3, $0.73 - 1.07$ mg Pb/dm^3, $0.45 - 0.80$ mg Ni/dm^3 and $0.08 - 0.28$ mg $Co/dm^3)$.

4 CONCLUSIONS

1. The results of investigations confirm the high and permanent pollution and corrosion potential of

184

low-buffered weathered spoils deposited even in thin layers, scarcely 1.5 m thick. This negative effect of low-buffered material on the aquatic environment, concrete and reinforcement should be taken into account at tips construction or minestone application in civil earthworks.

2. The short-term pollution and corrosion potential of fresh wrought spoils of high or moderate chloride salinity has to be also considered, though the pollution potential of buffered spoils is generally substantially lower due to equilibrium constrains.

3. Susceptibility of spoils to acidification and thus the long-term pollution and corrosion potential can be assessed in advance by means of buffering rate evaluation.

ACKNOWLEDGEMENTS

This work has been supported financially by the Research Problem Fund 10.2 coördinated by the Institute of Environmental Engineering of the Polish Academy of Sciences, Zabrze, Poland , and by the Central Program of Fundamental Research 04.10.09 coördinated by the Institute of Hydrogeology and Engineering Geology of the University of Mining and Metallurgy, Cracow, Poland.

REFERENCES

van Breeman, N. 1972. Soil forming processes in acid sulphate soils. In Acid Sulphate Soils 66. Wageningen.

Herzig, J., J.Szczepańska, S.Witczak and I.Twardowska 1986. Chlorides in the Carboniferous rocks of the Upper Silesian coal basin. Environmental contamination and prognosis. Fuel 65:1134-1141.

Palmer, M.E. 1978. Acidity and nutrient availability in colliery spoil. In G.T.Goodman and M.J.Chadwick (eds.) ,Environmental Management of Mineral Wastes, p.85-126. Alphen aan den Rijn - The Netherlands: Sijthoff and Noordhoff.

Skarżyńska, K. et al. 1982. Geotechnical examination of wastes in order of their application for raising of bunds for PFA lagoon at Przechlebie. Cracow: University of Agriculture.

Szczepańska, J. 1987. Coal mine spoil tips as a source of the natural water environment pollution (in Polish). Geology, Bull. 35. Cracow: Sci.Bulletins of St.Staszic Academy of Mining and Metallurgy.

Szczepańska, J. and I.Twardowska 1987. Coal mine spoil tips as a large area source of water contamination. In A.K.M.Rainbow (ed.) , Proceedings of the Second Int. Conf. on the Reclamation, Treatment and Utilization of Coal Mining Wastes, p.267-280. Amsterdam-Oxford-New York-Tokyo: Elsevier.

Twardowska, I. 1981. Mechanism and dynamics of coal mine spoil leaching at tips (in Polish) . Prace i Studia No.25. Wrocław-Warsaw-Cracow-Gdańsk-Łódź: Zakład Narodowy im.Ossolińskich, Wyd. PAN.

Twardowska, I. 1989. Buffering capacity of coal mine spoils and fly ash as a factor in the protection of the aquatic environment. Sci. Total Environ. 91:177-189.

Twardowska, I., J.Szczepańska, S. Witczak et al. 1988. The effects of coal mine spoil on the aquatic environment. The estimation of environmental contamination, prognosis and prevention (in Polish). Prace i Studia No.35. Wrocław-Warsaw-Cracow-Gdańsk-Łódź: Zakład Narodowy im. Ossolińskich, Wyd.PAN.

Reclamation, Treatment and Utilization of Coal Mining Wastes, Rainbow (ed.) © 1990 Balkema, Rotterdam. ISBN 90 6191 154 0

The neutralizing effect of joint fly ash and high sulfur rejects disposal

Z.A.Nowak
Central Mining Institute, Katowice, Poland

ABSTRACT : A general description of the coal/rejects production has
been presented.
The mechanism of shale breakdown in water of Polish coal preparation
rejects has been investigated and influenece on technology studied.
A classification of rejects in accordance to shale breakdown capability
is given in connection to the geological structure of the Polish Carbono-
ferious Strata and its geological age. The chemical composition of rejects
is the basic parameter controlling the environmental impact of reject
disposal.It has been found that both sulphur and sal-ts soluble in water
are influcencing shale behaviour changing the acidity of effluent water
and its chemical composition.The effect of fly ash addition on the rate
of leaching of various chemical components had been described based on
a laboratory method simulating the behavior of rejects during its
disposal and an example of this analysis is given.

1 INTRODUCTION

Coal has been for a long time the
main Polish energy source of a
shareof about 80% of the total
primary energy demand.This trend
shall continue for at least the
next two decades with bituminous
coal in the first place. Due to
this energy policy a strong
development of underground bituminous coal production took place,
during the last years -fig 1/1/.
The changing geological conditions
and increased mechanical coal
mining caused a simultaneous raise
of rejects content in the raw coal.
The increasing environmental impact
of coal utilization is one of the
major concerns particulary in
Silesia-the most populated area in
Poland-where almost all the bituminous coal comes from.

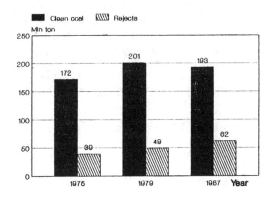

Fig.1.Clean coal/rejects
production

This impact originate mainly from:
-solid rejects disposal-both
from coal mines and power plants;
these rejects being disposed
separately today
- SO_2 emission from power plants
using in Poland exclusively coal
as a fuel.This caused that a multi-

directional research programme has been started concerning :
- evaluation of the maim pollutant sources of the energy sector
- coal desulphurisation and cleaning
- characteristics of solid rejects /shale,ash/
-reject utilization possibilities as secondary raw matirials
-envirommentaly save reject disposal,

Some results has allready been achieved like :
-building material production based on coal shales
-sealing off technology for rejects disposal sides
-application of fly-ash in the cement industry
-usage of coal rejects as road building and construction material and many others.
All this applications unfortunately cover only a very small amount of the total rejects production and the main problem is a save disposal of this kind of rejects.This problem is an increasing one because of the intended extention of small coal cleaning and the production of high sulphur coal rejects due to sulphur reduction in the coal cleaned.Some of the problems are presented in this paper.

2 COAL REJECTS CHARACTERISTICS

2.1 Coal shale behavior in water

The coal seams of the Silesian coal basin are in imbedded in a great variety of rock structures and rock types ranging from very strong water resistant sandstone to very weak and water saturated shales with changing mineralogical composition and physical properties.
It has been found that the behavior of shale in contact with water depends also on the so called "geological age of the rock " in relation to the rank of the imbedded coal / 2, 3 /.

This can greatly influence the coal cleaning process and the methods chosen for washery rejects or tailings disposal.Some of the results have been presented on fig.2.

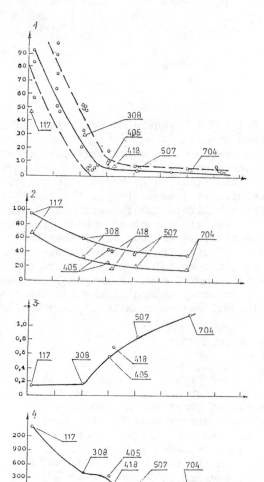

Fig.2. Coal shale behavior in water according to "geological age",1-shale break down in water,% 2-finse yield $<10\,\mu m$, $< 40\,\mu m$, 3-settling velocity of slurry, m/h , 4-filtration resistance 10^{12} m^{-2}, 117, 308 ,418 seam number according to increasing age

188

On the other hand it might help in rejects disposal enabling a better compacting of the reject pile.
To get a quantitative-qualitative discription of the shale be-havior a classification had been created /see fig.3 /and a standard had been introduced for the evaluation of shale in water breakdown /4 , 5 /.

CLASS	BREAKDOWN	PERCENT
I	Very low	below 2 %
II	Low	>2 to 5 %
III	Medium	>5 to 10 %
IV	High	>10 to 20 %
V	Very high	>20 %

Fig.3.Shalebreakdown classification

2.2 Chemical composition of the rejects

One of the characteristic features of the Silesian coal basin is that in some regions and at various depths soluble salts occur in the mine water. This salt water can have a positive influence by inhibiting the shale breakdown /6/ but also negative by contaminating the effluent water from the reject pile.
The carbon content in coal rejects varies in a wide range depending on the kind of shale but also on the coal cleaning process e.g. process efficency,required coal quality and so on. The other conta-minent of great importance for the rejects disposal is the sulphur.
For a great series of reject samples a correlation has been presented on fig.4.bothfor " S " and "C" content /7/. The figures present the result of coarse coal cleaning rejects. The sulphur content is much higher if a special coal desulphurisation process has been applied. It lies then between 5

and 15 % total sulphur of which about 90 % is iron sulfide-pyrite.

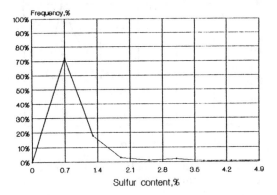

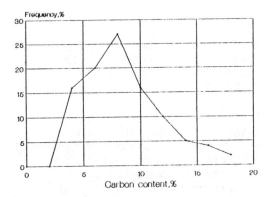

Fig.4 Sulphur and carbon content of coal rejects

This pyrite can be found as free particles or they are impregnating the shale /coal matter of the rejects. The concentration of silica and alumina in the washery rejects are shown on fig.5 /7/

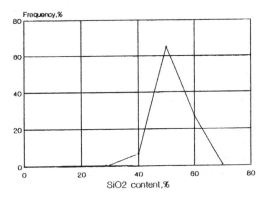

189

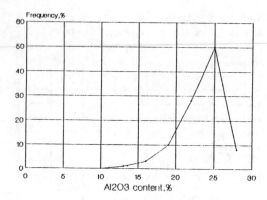

Fig.5 Silica and alumina content
of coal concentration in coal
rejects

3 BEHAVIOR OF COAL REJECTS IN CONTACT WITH WATER AND AIR- A LABORATORY INVESTIGATION OF THE LEACHING MECHANISM

There are many known methods of
rejects disposal. They have to
meet several requirements concerning air and water pollution and
also recently recognized-as
equally important-esthetic shape
fitting into the landscape.
It is very often extremely difficult to meet this requirements particulary for old reject piles.

Serious problems in reject disposal occur especially if the shales
have an increased carbon and
sulphur content. This kind of
rejects must be then subject to
special disposal methods, because
exposed to atmosphere they oxidize
and very often ignite causing
local enviromental problems ;
dangerous air pollution and
subsequently also ground water
nontamination as a result of the
leaching of combustion products.
The oxidizing effects are magni-
fied in the presence of bacteria
/ 8/.
The main parameter influencing
the leaching process of a reject

disposal side exposed to air
flow and precipitation is the
acidity build up of the water
solution in the disposal heap.
The pyritic sulphur content, and
the physical properties of the
reject expressed by the shale
breakdown are of great importance.
A laboratory method has been
applied to simulate the precipitation of a reject pile as a function
of time / 9 /. This method has
been standard-ized and used in
the Central Mining Institute and
the influence of varies parameters
on a great number of samples has
been analysed using this method.
Some of the results have been
presented on fig.6 and 7.

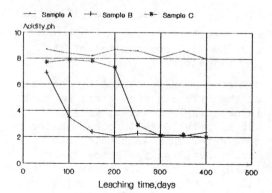

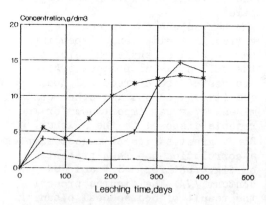

Fig.6 Acidity and leachate
concentration of coal rejects

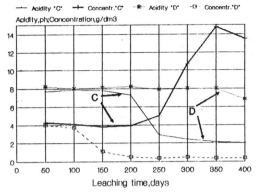

Fig.7 Influence of fly ash addition
on the leanching process

chosen and given on fig.7.It
compares acidity - leaching
functions of two samples :
sample C presented on fig. 6 and
the same sample C prepared as a
mixture with the addition of fly-
ash. The addition of 30 % fly-
-ash has been found as a save
figure for levelling out the
acidity build up in the sample.
This can be seen on fig.7 with
acidity changing from 8 to 2
for coal rejects and almost
constant for the coal reject/fly-
-ash mixture.

The samples chosen for the graphic
presentation on fig 6 and 7 are
of the following characteristic:
sample A is a low sulphur /0,7%/
and low shale breakdown coarse
coal reject ; Sample B represents
the medium sulphur /2 % / and
medium shale breakdown coarse
coal rejects ; sample C is a very
high sulphur /16 % / and relatively
low shale breakdown sample beeing
a product from a selective coal
crushing equipment of a coal fired
power plant. The leaching results
as a function of the simulated
precipitation time has been shown
on fig. 6 for the three samples
mentioned above . The leaching
time simulates the age of a pile
up to 10 years. It can be seen
from fig.6 - where the acidity
build up curves has been shown -
that the decisive factor for the
change of the acidity is the sulphur
content of the sample. In corelation
to this a great leaching process
has been generated with an ever
increasing leachate effluent as can
be seen on fig.6   It should also
be mentioned ,that for the low
sulphur containing rejects a low
leanhing process has been observed.
From the series of simulation tests
aiming at controlling the leaching
process one example has been

4.CONDLUSIONS

1. It has been found that
coal rejects disposal causes series
problems if not managed in a
proper way ,but the enviromental
influence can be minimized if
proper control is applied.
2. One of the main parameters
influencing the behavior of
rejects on a disposal side is its
sulphur content and this sulphur
content can be controlled through
pyrite removal from the rejects
prior to disposal or by the
preventing acidity build-up
of the water in the pile origina-
-ting from precipitation. One of
the solutions applicable for the
acidity control is the addition
of alcaline fly ash to the rejects,
a method not applied yet in Poland.
The advantage of this mixed rejects
disposal method is both the
inhibiting influence on the
leanhing process and a better
compacting of the rejects disposal
side.

3. Shale breakdown in water plays
also an important role due to the
better natural compactment of the
shale - fly ash mixture.

191

REFERENCES

Nowak, Z. 1989.Rational utilization
of coal in Poland Warsaw Polish
Academy of Sciences.
Nowak, Z. 1982. Water economy in
Coal Preparation, Katowice,Śląsk
Nowak, Z. 1988. "Eta geologica",
posizione stratigrafica e proprie-
ta tecnologiche del carbone e del-
lo sterile. La Revista dei Com-
bustibili, Vol.42.
Kozłowski, C. 1967. Shale break-
down behavior and its evaluation.
Katowice CMJ Report.
Polish Standard. 1970. National
Standards Organization Warsaw.
Kurczabiński, L. 1985.The influen-
ce of coal preparation water
quality on solid-liquid separa-
tion.Ph.D. Thesis. Katowice.
Romańczyk, E. 1981-1985.Washery
rejects utilization. Katowice.
Series of Research Reports.
C.M.J.
Le Roux,N, W. 1973.Bacterial oxidation of pyrites. Institution
of Min. Mettalurgy.
Przybyła, I. Szpetman, Z. 1986.
Method for the determination
of leaching behavior of coal
rejects. Katowice. C.M.J.Reports.

Reclamation, Treatment and Utilization of Coal Mining Wastes, Rainbow (ed.) © 1990 Balkema, Rotterdam. ISBN 90 6191 154 0

Granville Colliery land reclamation / coal recovery scheme

A.C.Johnson & E.J.James
Scott Wilson Kirkpatrick, Consulting Engineers, Telford, UK

SYNOPSIS

This paper describes how the former Granville Colliery at Telford New Town, Shropshire was restored to provide land for employment, housing and open space use. In addition, a major site for waste disposal was created and recovery of coal from mine waste was achieved. The project required the close co-operation of British Coal Corporation (BCC), Shropshire County Council (SCC), Wrekin District Council (WDC) and Telford Development Corporation (TDC). It illustrates how a specialist team can resolve conflict and bring forward for implementation a major scheme of derelict land reclamation and mineral recovery while maximising the available opportunities for the long term benefit of the community.

1. INTRODUCTION

The site of the former Granville Colliery lies in the North Eastern part of Telford and until 1986 it formed the single largest tract of derelict land remaining in the New Town Area, covering in all some 162 hectares. The Colliery itself closed in 1979 but the majority of the land remained in the ownership of BCC, it was affected to a substantial extent by various forms of dereliction, ranging from spoil heaps, mineshafts and derelict buildings to areas of ill drained land affected by mining subsidence.

Following closure of the Colliery, BCC formulated a proposal in 1981, in conjunction with SCC, for coal recovery and waste disposal within the site. The scale of the waste disposal facility proposed at the time was, however, considered disproportionate to the needs of the community and was rejected.

Subsequently, in 1984, a feasibility study was commissioned by SCC under the control of a Steering Group of officers from SCC, WDC and TDC. This study was completed in 1985 and considered a wide range of alternatives for the development of the site. The preferred option was subjected to extensive public consultation and examined in more detail to identify the most practical means of implementation.

TDC and the local authorities ultimately agreed a land use strategy for the area compatible with the development of the whole of North East Telford (see Fig 1). A package of proposals was thus compiled comprising: coal recovery from the waste tips; the creation of land for private housing, campus style employment uses, and areas for open space use; formation of part of a strategic road link; and the formation of an above-ground refuse tip site for future operation by SCC. This formed the basis of a planning submission by TDC to the DOE under Section 7(1) of the New Towns Act 1981 for the whole of the 316 hectares of North East Telford; 95 hectares of which comprised the Granville Colliery land reclamation scheme. In parallel with this, BCC prepared a separate Planning Application for the coal recovery operations, which could be implemented should the New Town's submission be rejected.

2. SITE DERELICTION AND RESTRAINTS

The whole of the area under consideration was affected by mining activity which had occurred over the last 150 - 200 years and was incapable of development without major reclamation. Virtually the whole site was covered by mine waste consisting of deposits originating from the Granville

Colliery Mine and other mounds formed in the last century. There were over 150 mineshafts dating back to the 19th century varying in depth from 30 to 140m; 4 hectares of colliery lagoons, waterlogged or flooded land including numerous culverts and watercourses; 6 hectares of suspected shallow mineworkings; derelict buildings, hardstandings, and buried structures at the Granville Pit Head and National Fuel Distributors Coal Depot; 2.5 kilometres of abandoned mineral railway tracks and tip haul roads; and Archaeological remains at the site of the Old Lodge Furnaces (see Fig. 2).

There were also restrictions due to statutory services crossing the site, public and private rights of way and accesses, and existing tenancy/lease arrangements.

3. RECLAMATION PROPOSALS

It was agreed that a comprehensive reclamation scheme was required to achieve the general objectives set out for the area, and that it was necessary for TDC, SCC, BCC and WDC to enter into negotiations to draw up Legal Agreements for land acquisition and implementation of the contract. The formulation and implement- ation of the scheme was to be carried out by TDC, utilising their specialist Land Reclamation Engineering team (who sub- sequently merged with Scott Wilson Kirkpatrick & Partners in August 1987). For this purpose, the whole of BCC's land holding in North East Telford was acquired by WDC using Derelict Land Grant alloc- ation. Land required for development and Waste Disposal purposes was then purchased by TDC and SCC respectively from WDC. BCC had pointed out that four of the mounds within the site contained high proportions of coal (up to 15% by weight) and it was agreed to include coal recovery within the reclamation scheme in a manner which provided economic benefits to all parties.

The main reclamation proposals were as follows:

3.1 Development Land

1. General site clearance, with the exception of mature landscape and other features to be retained.
2. Drainage of waterlogged and flooded land, installation of under drainage, piping of open water courses and replacement of old culverts.
3. Removal of soft material in water-

logged and flooded areas prior to the provision of a stabilising layer of free draining material and/or geotextile membranes, and filling with acceptable material.
4. Excavation of coal bearing spoil and processing for coal recovery. Excavation of inert colliery deposits and re-distrib- ution, together with coarse discard from the coal recovery plant, as fill to produce a suitable land form for future development. Fine discard would be stored in open space areas for use by SCC in the Waste Disposal Site at a later stage.
5. Provide the formation for the Eastern District Road (EDR) corridor.
6. Installation of open ditches and french drains to deal with run-off.
7. Full stabilisation (drilling, grouting and capping) of all mineshafts within future development areas, and landscape standard stabilisation (capping only) of shafts within future open space areas.
8. Stabilisation of shallow mine- workings beneath the EDR corridor by means of drilling and grouting. The invest- igation of other areas of suspect shallow mineworkings was to be delayed until the later development stage, when areas requiring treatment would be stabilised to an appropriate development standard.
9. Soiling, seeding and planting of all open space areas and earthworks batters. Development plateaux would be seeded as an interim measure only.

3.2 Coal Recovery

BCC estimated from analysis of the spoil that some 1.75 million tonnes (T) of waste material would be washed in the processing plant and that this would yield approx. 220,000 T of low grade saleable coal. It was proposed that this would be transported off site and blended with other coals prior to marketing.

The coal recovery operation would take 3 to 4 years to complete and it was envisaged that a minimum of 1,000 T of saleable coal would be produced per week. It was agreed that fines discard settle- ment lagoons would not be acceptable either in planning or environmental terms. As a result, a closed circuit coal recovery plant was proposed comprising:
1. Large barrel washer to separate coal and fines from coarse spoil discard.
2. Screens to separate coarser coal particles (> 0.5mm size) from fines.
3. Cyclones to separate fine coal particles (< 0.5mm size) from fine discard.

194

4. Filter presses to dewater fine discard.

5. Centrifuge to dewater coal particles.

The coal recovery plant was to be capable of producing coarse discard with a moisture content at or near the optimum for the material, so as to facilitate it's use as 'suitable' earthworks material. Fines discard would also need to be dewatered, to a level sufficient for its placement as fill in open space areas.

3.3 Refuse Disposal Site

The Granville Steering Group agreed, that the most suitable location for a future refuse disposal facility was at the site of the former Granville pithead.

SCC had identified an urgent need for waste disposal facilities in the Telford and surrounding areas for the 1990's and beyond, and the site would need to cater for 80-100,000m³ of waste per year over a 20 year period. The proposed site had an overall capacity of 1.8 million m³ and would have a life expectancy of 18 to 23 years, based on the above tipping rates.

The proposals involved the following works to create the required capacity for refuse tipping:

1. Excavation of two existing spoil mounds within the proposed tip area and processing of spoil for coal recovery.

2. Placing of coarse discard returned from the coal recovery plant together with non-coal-bearing spoil and burnt shale to form an oval shaped retaining and screening bund around the tipping area. All material within the bowl would be excavated down to virgin ground level to create the required tipping capacity.

3. Removal of soft clays beneath bund walls to provide a stable foundation.

4. Stockpiling of agreed quantities of fine discard within the bund bowl for future sealing purposes.

5. Diversion of an existing water-course.

6. Soiling, seeding and planting of the outer slopes of the refuse tip bunds.

7. Re-capping of the two Granville pithead mineshafts.

A substantial quantity of coarse and fine discard generated by coal recovery would be stockpiled in landscape areas for future use by SCC as operational materials.

The proposals would not, however, include works necessary to seal the base and sides of the tip prior to waste disposal, nor works required to provide site entrance facilities. These works,

together with hydrological, hydrogeo-logical and ground monitoring assessments would be carried out prior to tipping.

Planning permission for the waste disposal operation would be processed as a separate application by SCC once basic site earthworks had been formed under the Reclamation Contract.

4. CONTRACT WORKS

Having agreed the basic principles and scope of the scheme, detailed discussions were held with all parties to enable the contract to be prepared for competitive tendering. These details formed the basis of Legal Working Agreements and were in addition to the Legal Agreements required for Land Transfer.

The following basic principles were agreed:

1. TDC would, in view of their interest in the development of the area, be responsible for preparing and administering the Contract and supervising the Works.

2. TDC would be responsible for the costs of reclaiming areas of land scheduled for development and associated infrastructure works, including the EDR corridor. This would not, however, include the cost of handling materials required for coal recovery, nor any other costs for which BCC would have been responsible if tip washing operations had been carried out as an independent exercise.

3. BCC would bear all costs associated with coal recovery, including re-distrib-uting and placing coarse discard in areas of controlled fill, and fine discard in open space areas. Recovered coal would be stockpiled for future haulage off site to BCC's own disposal point.

4. BCC would be responsible for safety works to mineshafts found within the coal recovery areas and for the cost of cultivating and seeding an area of completed earthworks equivalent to that disturbed by the coal recovery operations.

5. SCC would be responsible for the cost of redistribution of spoil to form the waste disposal site bund walls and any associated works. SCC would not be responsible for the cost of handling any material required for coal recovery, other than the cost of compaction of any discard used as structural fill.

6. WDC would not be responsible for any costs under the Contract, as any works affecting their land would either be the result of coal recovery or earthworks carried out by TDC for development purposes. WDC would need to be satisfied

concerning the environmental aspects and protection of the archaeological site, in view of their proposals for development of a Country Park in North East Telford.

7. The Contract would be based on the ICE Conditions of Contract and the DTp Specification for Roadworks.

8. Due to the financial responsibilities of the various parties, the Bills of Quantities would be sub-divided into three sections covering BCC, SCC and TDC works costs, and a fourth Bill for site establishment costs which would be apportioned to an agreed formula.

5. CONDITIONS OF CONTRACT

Lengthy discussions were held with BCC concerning Conditions of Contract Clauses normally incorporated within Tip Washing Contracts. BCC had never previously undertaken any Contract where tip washing was part of a Civils Contract and where a substantial area of the site affected by coal recovery was required for immediate development.

A series of Special Conditions of Contract Clauses were added for the Coal Recovery Operation to cover such matters as 'Coal and discard requirements', 'Obligations and responsibilities under the Mines and Quarries Act', 'Quality, quantity and characteristics of coal and content of spoil heaps', 'Method of Working', 'Suspension of Coal Recovery Operations', 'Termination of Coal Recovery Operations', and 'Price Fluctuation' (where the normal Civil Engineering Clauses were modified to apply every 12 months in line with price adjustments for the Coal Industry, and a 'Price Fluctuation' Clause added to cover variation of price for coal production).

6. SPECIFICATION

It was recognised that the strict criteria which would apply for the re-use of coarse discard as fill in development areas, would be crucial to the viability of the coal recovery process. The spoil material also contained a high proportion of clay which gave rise to concern due to the limited amount of open land available for storage of fines discard. This discard normally had a high moisture content, which could present serious handling difficulties.

A quantity of spoil was therefore taken from the site for a trial wash to obtain guidance on the proportion of coarse and fine discard that would arise. This was estimated to be 3:1, and on this basis, it was judged possible to contain fine discard within the open space area and waste tip bowl.

It was concluded that it was practical to achieve a moisture content for the coarse discard in the region of 14% and that the Specification should stipulate the control of moisture by whatever means necessary to comply with the requirements for use as suitable fill. Fine discard would be de-watered using a multi-roll belt filter and it was agreed that a moisture content of 35% or less could be achieved. The Specification required that the material be dewatered to a level where it could be properly handled and placed by mechanical plant to form stable areas of open space.

In view of the variable nature of the materials it was decided that an 'end product' specification would apply. The required state of compaction was defined as being 95% of the maximum dry density determined in accordance with Test 12 of BS 1377, and 'suitable material' was defined as material having a moisture content within $\pm$ 2% of the optimum value determined in accordance with this test.

BCC's requirements for tip washing were included as an additional section, suitably modified to comply with the standards of control required for discard materials and effluent produced. A closed circuit processing plant was specified which included washing, crushing, sizing, centrifuging and other dewatering equipment necessary to prepare the washed products to the required standards, and to maximise recovery of saleable coal (including the <0.5mm fraction). BCC estimated that the three spoil tips were capable of yielding 220,000 T of coal in the >0.5mm size fraction having characteristics of 12.4% moisture, 15.09% Ash, 1.17% Sulphur and 24,320 kJ/kg calorific value. This calorific value would form the basis of the contractual price, with the Contractor not being entitled to payment if the quality fell below 90% of this value.

Additional clauses were included for the treatment of mineshafts and mineworkings. This covered requirements for exploratory operations to locate shafts, drilling and pressure grouting to stabilise shaft infill and shallow working voids, and the construction of reinforced concrete caps. These requirements complied with standards recommended by BCC and CIRIA.

The Contract contained a full landscaping specification.

This allowed for chemical analysis to establish the need for additional nutrients to promote plant growth. Soiling was only proposed for the main batters in the development areas and the outer slopes of the refuse tip bunds. On development areas the raw spoil would be seeded as an interim measure to minimise erosion of the surfaces and to improve the appearance of the site.

Provisions were also made to cover the numerous other restraints, which stipulated requirements for access, public and private rights of way, statutory services diversions, protection of trees, arrangements relating to existing tenancies and leases, treatment and control of run-off and retrieval of archaeological artefacts.

7. BILLS OF QUANTITIES AND TENDERS

Bills of Quantities were prepared in accordance with the DTp Method of Measurement for Roads and Bridgeworks with additional sections for coal recovery, mineshaft treatment and landscape work. As described previously, the work was itemised in four Bills according to the responsibilities of the various parties participating in the Contract.

Six tenders were received; the lowest being £4.8m, submitted by A F Budge Contractors Ltd. The costs for which BCC, SCC and TDC were responsible were submitted to each party for approval and included in the Legal Agreements exchanged for the Contract. DOE approval to the expenditure attributable to the New Town's Development land was obtained and all legal agreements for the Contract and land sale were concluded in April 1986. This enabled the Contract to be let and siteworks to commence on 12th May 1986. TDC's submission under the New Town's Act and BCC's Planning Application for the coal recovery operation were both approved in February 1986.

8. SITEWORKS

8.1 Coal Recovery

The Contractor's initial programme of works allowed for first washing Tips 1 and 2 so as to create space to build the bund walls for the refuse tip site and to maximise the re-use of coarse discard as structural fill in the waste disposal site. Tip 3, which covered an extensive area, would then be washed working west to east, thus allowing surplus spoil to be progressively brought into the area to be compacted as fill, and the route of the

EDR corridor to be formed.

The coal recovery plant was erected during the initial site establishment period and was a Parnaby 8/36 closed circuit washing system supplied and erected by Parnaby Cyclones International.

Coal recovery commenced on programme in August 1986 but it soon became evident that the amount and quality of coal bearing spoil in tips 1 and 2 would not be sufficient to keep the wash plant operating at full capacity. Spoil from Tip 3 was therefore brought in to maintain the plant through-put and coal production. As excavations progressed, substantial quantities of burnt shale and burning spoil were exposed in Tip 2 which would significantly reduce the overall quantity of recovered coal. In addition, other difficulties were encountered, as outlined below:-

1. The moisture content of coarse discard exceeded specified limits for use as structural fill. This was rectified by installing additional water sprays to remove fine clay particles clinging to the coarse discard.

2. Ash and moisture in the coal product were highly variable and often outside specified limits. The coal product was therefore monitored continuously by BCC and the plant adjusted as necessary to maintain accepted standards.

3. Burning spoil materials adversely affected chemical additives used during the coal recovery process and disrupted the 'tuning' of the plant. This caused complete 'break-downs' in output requiring the whole plant be cleaned out and re-adjusted.

4. Coal bearing spoil in Tip 3 was overlying older deposits of spoil. This required the advancing of numerous worked faces at different depths and resulted in a very irregular and difficult site to manage.

5. The moisture content of fine discard proved variable and excessive, with values up to 80% recorded. This was placed in open space areas where it was retained by coarse discard and worked in such a way that moisture was able to drain from it. This, together with the action of flocculants used in the belt press, helped to stabilise the material. Areas of fine discard deposition were capped in the short term using coarse discard, to form areas for open space use, pending re-use as operational material in the refuse tip site.

Up to 2,000 T of coal was produced weekly with the plant capable of processing 200 T of spoil per hour.

Due to the substantial reduction in

amounts of coal bearing spoil, the time allowed for tip washing was shortened considerably. During the latter stages of coal recovery, the Contractor elected to work 24 hours a day and the whole coal recovery operation was completed by August 1988, some 6 months ahead of programme. Recorded data relating to coal recovery is summarised below:

Spoil Processed	800400 m³
Coal produced	175850 tonnes
Discard produced	655000 m³
Ratio of fine/coarse	40%/60%
Fine discard moisture	up to 80%
Coarse discard moisture	Av. 14%
Calorific Value of Coal	
Range	21400 - 27000 kJ/kg
Bulk of Coal	25500 - 26500 kJ/kg
Moisture in Coal	12.6% - 18.5%
Ash in Coal	8.7% - 15.4%

8.2 Refuse Tip Earthworks (Tips 1 and 2)

The major problem faced in this area was the release of sufficient land to commence construction of bund walls. Tips 1 and 2 together with the colliery buildings yards and structures occupied the majority of land designated for the waste disposal site, and the only area of land free of spoil was under lease for agricultural use until April 1987.

Early in the Contract, virtually all coal bearing spoil from Tip 1 was removed to create the initial stockpile of material for the wash plant. The colliery yards and buildings were cleared and these two operations enabled filling of shallower parts of the bund wall to proceed. Progress was complicated by:-

1. The extent of soft clay beneath the bund walls, which had to be removed to depths of up to 3m to reach a firm foundation.

2. The shortage of space available for deposition of additional soft materials. As a result, some of this material was used to form a buttress against the main bund wall.

3. The extent of burnt shale in Tip 2 meant that far less material was removed to the wash plant.

4. The majority of the inert fill was required at the northern end of the bowl in the area occupied by Tip 2, where the height of the bund wall reached a max of 25m and it's base was up to 130m wide.

5. Areas of active burning within shale deposits where special precautions were required to protect operatives and earth-moving plant.

The earthworks operations included numerous field compaction trials on the various materials. Satisfactory results were readily obtained on colliery spoil and shale materials, using a sheepsfoot compactor and towed vibratory roller. Use of discard proved more problematic but the compaction requirements were met by mixing with drier shale deposits, provided the moisture in the discard did not exceed 14%. Burning spoil was placed and compacted in thin layers and inter-layered with inert material.

Most of the material excavated (over 600,000 m³), met the acceptability criteria for use as fill. However, due to additional foundation work and the overall reduction in amounts of discard arising, there was insufficient material available to complete the bunds to their full height. The overall shortfall was approx. 100,000 m³ and it was agreed that this should be completed as a separate exercise utilising surplus material from other contracts.

Drainage works were required to divert the existing watercourse which passed through culverts beneath the tips, and prevent build up of water in the base of the bowl. Several springs and old culverts encountered during the earthworks were also connected to the new outfall.

The existing mineshaft caps at the Granville pit head were set at a level beneath the bund walls and the base of the tip and were not therefore disturbed by the earthworks operations.

The outer slopes of the bund walls were topsoiled and seeded but planting was omitted because the bund walls had not reached their full height. It was agreed that this would be undertaken at a later date when the bunds were completed.

Due to the additional amount of inert spoil (mainly burnt shale) handled, the cost of the refuse tip site earthworks increased significantly, and an additional capital vote was required from SCC.

8.3 Main Reclamation Works

The main area of the site scheduled for development was affected by a number of factors which dictated the programme and progress of the works viz; the number of mineshafts (153) requiring treatment, the substantial area (Tip 3) affected by coal recovery, land under lease, services to be diverted, shallow workings beneath part of the road corridor, and the areas occupied by the coal recovery plant and site offices.

During the first year therefore, only a limited amount of earthworks was carried out to clear areas for road construction, service diversion, and drainage routes. This was followed by work along the main EDR corridor, which involved excavation of a deep cutting (up to 10m), removal of approx. 150,000 m^3 of spoil to provide land for a grade separated junction, and the removal of soft deposits to create a stable platform for road construction.

The material generated was to be placed in the main fill area (Tip 3) but at this stage conflict was occurring with coal recovery operations. It was therefore necessary to rearrange the sequence of coal recovery excavation to enable the cut/fill operations to proceed satisfactorily.

The remaining earthworks operations involved excavation of the large mound on the southern boundary, and the substantial deposits of inert spoil south of Tip 3. This material was placed and compacted in low lying and quarried areas to form the main development plateaux. Cut and fill was also carried out on the mounds in the western area to form the areas designated for housing.

During the excavation work, an existing pond containing a substantial stock of fish was cleared. These were netted and used to stock other pools elsewhere in Telford. A large timber and wrought iron wheel identified as a 'horse gin' was also unearthed and taken to the Ironbridge Gorge Museum for preservation.

Low lying areas, covering approx. 2 ha, were found to be completely waterlogged. Trial holes revealed soft clays and sand and gravel, with a watertable at or near existing ground levels. Removal and replacement of material would have been prohibitively expensive, and the contractor was instructed to lay a Geotextile Membrane over the undisturbed ground, prior to placing an initial stabilising layer of spoil or granular material and subsequent filling. The Geotextile used was Lotrak 35/30.

Two ponds, situated in strategic locations on the site, were retained until the latter stages of the earthworks operations to control run-off and leachate from the site. Outfalls from the site were monitored throughout the Contract period and analysed for suspended solids and chemical content.

Numerous watercourses and old culverts required piping or diversion clear of the main earthworks areas. This drainage work also enabled low lying areas to be utilised for stocking of discard and deposition of unsuitable material. Surface drainage was achieved mainly through open ditches cut along the edges of the de-velopment areas and at the toe of batters. Stoned drains were used alongside perm-anent open space areas where disturbance by future development was unlikely.

Along the EDR Corridor, shallow mineworkings were treated for a distance of approx. 200m. Probe holes were drilled to a maximum depth of 30 metres on a grid pattern and grout injected. Similar treatment was carried out at the position of a proposed underpass.

Mineshaft treatment continued throughout the Contract period. This work was particularly complex due to the large number of shafts involved and the potential conflict between the treatment works and the major earthworks and coal recovery operations. Considerable difficulties were experienced in locating a number of the shafts, and there were isolated cases where shafts were not found despite extensive exploratory work. One such case occurred on the road corridor, where a heavy duty geo-grid reinforcing mesh was used as a safety precaution to protect the road against any possible sudden collapse in the future.

In all, 100 caps were constructed and 90 shafts were drilled and grouted, involving 9,000m of drilling and the injection of 20,000 T of grout. Search operations required over 50,000m of probe drilling. Eight shafts were found open, 33 were disproved and 11 shafts on WDC land were omitted.

A range of exploratory techniques were used including visual inspection, system-atic trenching, probe drilling, geo-physical survey and radar investigations. However, due to the variable nature and excessive depth of spoil deposits, only the more conventional methods (ie excavation or probe drilling) proved to be reliable methods for locating shafts on this site.

All shaft stabilisation and capping work was finally completed in July 1989, but at a substantial increase on estimated costs (£698,000 as against £417,000), resulting from greater than anticipated requirements for exploratory and grouting works.

Grass seeding, including liming, was carried out over the whole site during the Spring and Autumn of 1989. All main batters in development areas were top-soiled beforehand for planting during the 1989/90 Winter season. Relevant statistics for the completed project are summarised in Fig 3.

9. SUMMARY

The Granville Colliery Land Reclamation

Scheme was finally completed in September
1989; 3 months ahead of schedule, despite
the various problems that occurred during
the course of the works. This scheme was
the largest single land reclamation
contract ever carried out within Telford;
reclaiming extensive areas of derelict
land for positive end uses complementary
to the overall development strategy for
Telford. The objectives of all parties
were successfully met.

Item	Estimated	Completed
Coal Produced (T)	220,000	176,000
Earth moved (m^3)	2,750,000	3,000,000
Spoil washed (m^3)	900,000	800,000
Shafts Treated (no)	153	133
Land Reclaimed (ha)	95.1	95.1
- Development (ha)	37.4	37.4
- Open Space (ha)	43.7	43.7
- Waste Site (ha)	14.0	14.0
Expenditure (£m)	4.8	5.1

Fig 3 Scheme Statistics

Development land for residential and
employment uses, together with associated
open space, was created for TDC at a cost
some 40% less than that normally
associated with reclamation schemes in
Telford. A major site for future waste
disposal has been created for the County
Council, which will serve Telford and the
surrounding community for the next 15-20
years, at a time when existing facilities
were virtually exhausted. Coal mineral in
substantial quantities has been recovered
from waste tips for the benefit of BC,
without detriment to the ultimate land
uses. The project would not have been
possible without the close co-operation
fostered between British Coal Corporation,
Shropshire County Council, Wrekin District
Council and Telford Development
Corporation, or without the expertise of
TDC's Specialist Land Reclamation
Engineering team. It demonstrates how
large and complex schemes can be brought
to fruition, given sufficient will to
succeed, and how the overall community can
benefit from such a venture.

ACKNOWLEDGEMENTS

The authors acknowledge the assistance of
Telford Development Corporation, British
Coal Corporation, Shropshire County
Council and Wrekin District Council in
preparing this paper.

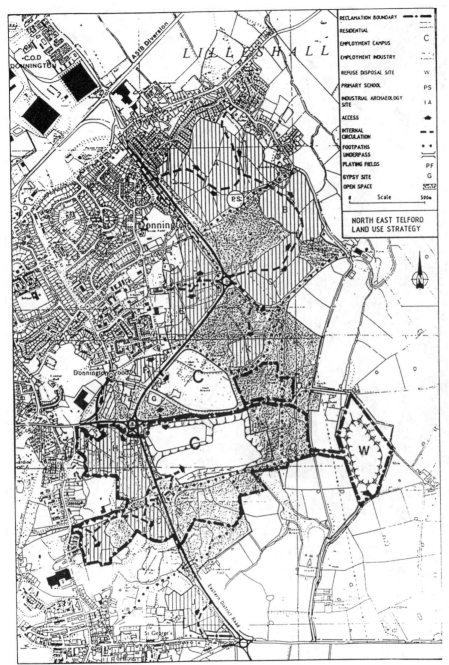

Figure 1   North east Telford land use strategy.

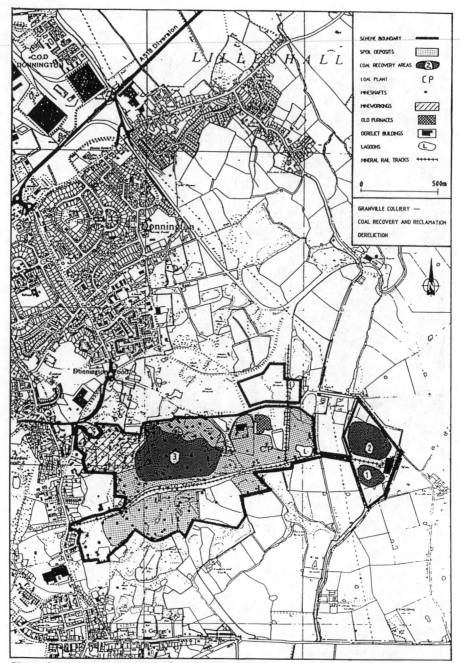

Figure 2 Coal recovery and reclamation dereliction.

Reclamation, Treatment and Utilization of Coal Mining Wastes, Rainbow (ed.) © 1990 Balkema, Rotterdam. ISBN 90 6191 154 0

Influence of waste selection in the dump reclamation at Puentes Mine

A. Gil Bueno & C. Val Caballero
Department of Land Reclamation, ENDESA, Madrid, Spain

F. Macías Vázquez & C. Monterroso Martínez
Department of Soil Science, University of Santiago, Spain

ABSTRACT: The results of a research on the physical properties of materials from excavation headings at the Puentes Mine, Corunna, Spain, are given. The main object of this study is to obtain information on the wastes with a view to achieve a selective arrangement at the final dump surfaces. The application of such materials selection has resulted in final surfaces suitable for vegetation growth. The methodology followed in the reclamation works is also included.

1 INTRODUCTION

The enactment of the first Spanish legislation on the reclamation of natural spaces affected by coal strip mines (1982) and an understanding of the need to reclaim the final surfaces has involved a change in the planning and operation of existing active mines, which had previously perform important materials displacement, before the reclamation activities were started.

More specifically, the reclamation of dump at the Puentes Mine (Corunna, Spain), owned by ENDESA, began in 1980. At that time, when the mine had been operated with the current technology for four years, 11 mt of coal to supply the nearby 1400 MW power plant and 13,7 Mm3 of waste were already excavated.

Such time lag, between the beginning of reclamation work and the start-up of mining has produced a limited availability of top soil, together with the specific problems of this deposit, such as the high acidity of most non-coal bearing layers and the clayish nature of the materials, made evident the need to carry out a preselection of materials to be heaped. This paper presents the results of the study on the quality of the materials extracted from the headings of the Puentes Mine with the main purpose of obtaining knowledge about the spoils leading to their selective arrangement on the final surfaces of dumps.

2 DESCRIPTION OF THE MINE

The mine, located at the Northwest of the As Pontes town (Corunna), is about 6 km long and 2.5 km wide, with a narrower point at the middle dividing the working area in two fields, West and East. Maximum depth is slightly over 400 m.

Total practical reserves were estimated at the start of the works at 280 Mt. In order to achieve such yield, 888 Mm3 of spoils had to be also extracted, including clayish sediments from the Tertiary basin and slates from the deposit edge. The average ratio is 2.8 m3/t.

The entire coal production is used by the 1400 MW power plant installed near the mine, requiring a yearly production of 12.2 Mt to operate 6,500 hours at full steam. This means that, at the indicated ratio, 35 Mm3 of waste are extracted each year.

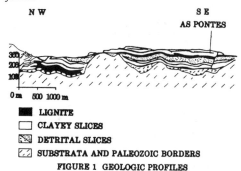

FIGURE 1 GEOLOGIC PROFILES

The deposit is formed by alternating layers, on places folded over, of brown coal and clay material. The high variation of layer thicknesses and qualities goes from pure clay to brown lignite, thus requiring the application of highly strict coal selection criteria. (Fig. 1)

2.1 Extraction method

As it is a multilayer deposit having coal and waste in many different thicknesses, the only practical extraction method is using bucket wheel excavators.

As material transportation system, conveyor belts were chosen. On the bench, one conveyor belt is fitted to each excavator to move the excavated material to a transfer junction where it is classified, coal being conveyed to the power plant and waste to the dump respectively.

Waste is piled in the dump by means of swivel-arm stackers. Slag and ashes from the power plant are also moved to the dump area from the same transfer point where waste is conveyed to be heaped.

At this time, seven bucket wheel excavators and five stackers are available. The average gross extraction capacity of each excavator is 1800 m3/h or 1450 m3/h of selective work is performed.

Two stackers have an average piling capacity of 7500 m3/h (each one) and three stackers have 11500 m3/h (each one).

The overall schematic layout of the operation is shown in figure 2.

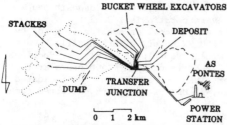

FIGURE 2 DIAGRAM OF THE PUENTES MINE AND POWER STATION COMPLEX

2.2 Dump

The dump of the Puentes Mine is designed to hold 960 Mm3 of waste and will attain a final surface area of 1400 ha. Maximum height, including 12 waste layers, will be in the order of 260 m.

Up to this time, 335 Mm3 of waste have been piled and the existing finished surface is 210 ha, while piling of the following amounts is pending:
- Slates : 198 Mm3
- Clays : 379 Mm3
- Ashes : 49 Mm3

The dump has been designed using the terraces formed by the mine operation system, consisting in subdividing the general slope in a number of partial slopes and ledges allowing better runoff water control. As a result of the slant of the slope, the width of each level varies from 60 to 100 metres. These levels drain either towards the East or the West, according to the general drainage plan. Average inclines of the various surfaces are as follows:
- Berms: 2.5% (cut by short walls and ditches every 100 m of length).
- Slopes between ledges: 20%.

Figure 3 represents the overall dump configuration.

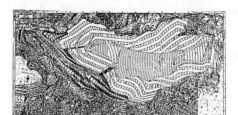

Figure 3.- Puentes Mine Dump

3 DUMP RESTORATION

The reclamation work of the Puentes Mine dump is aimed to achieve the following goals:
. - Creation of a vegetal cover to control erosion and runoff water quality and to promote the formation of productive soil.
- Recovery of mine activity altered landscape.
- Make possible to use the land obtained after reclamation as it was before exploitation.

In order to achieve the above goals, a number of constraints now making significantly difficult the reclamation work, must be overcome. These constraints can be grouped by source as follows:
- Originated by the deposit properties and the extraction, basically affecting the substrate to be restored, characterised by:
. very high acidity of most types of waste: pH = 2;
. inadequate texture: clays;
. lack of nutrients;

. highly heterogeneous materials;
- Caused by the environment: local weather conditions:
. extreme pluviometry, with average annual rainfall of 1700 mm, and peak values of 700 mm, per month; 120 mm per day and 30 mm per hour, which creates:
. extreme erosion;
. substrate saturation;
. need to channel high flow rates;
. high limitation of the year periods where surface preparation and farming activities can take place;
. strong winds with an average yearly speed of 3.5 m/s and 16 m/s peaks.
- From the time lags among start of mining operations, initial design and start of reclamation work, which originates, among other things:
. limited availability of top soil;
. existing final surfaces formed without regard to future reclamation.
On the basis of the situations, objectives to be attained and existing conditions described above, plus the experience gained from the first reclamation stage, a working methodology was adopted, which is sequentialy explained below:
- Infrastructure works: design and construction of tracks and large size drains required to remove high flow rates of water.
- Research and testing: Studies were carried out including the determination of the agro-climatic, edaphological and botanical characteristics of the area surrounding the deposit, for the purpose of both evaluating the environmental impact of the mining operations and allotting land uses.
Additional studies were conducted to evaluate the most adequate soil correction and fertilisation levels, including laboratory, greenhouse and actual field tests.
- Study of wastes obtained from different excavation headings, as described in the following section.
- Analysis of heaped materials: Sampling and testing of materials from all dump final surfaces. The number of samples per ha is 8, sufficient to obtain adequate information for the stated purpose. All are tested for pH, pH oxidation, total S, exchange complex (exchange capacity, H+, Al, K, Ca, Mg, Na) and Al saturation.
Fertility determinations were not carried out as it was verified the waste material contained no nutrients, except for traces of available phosphorus.
As an indication, some typical test results are given:

SAMPLE	pH H2O	I TOTAL S	pH OXIDATION	EXCHANGE CATIONS (mg/100 g)					I Al SATURATION
				Al	Ca	Mg	Na	K	
1	2.80	0.19	2.70	4.68	0.30	0.41	0.32	0.96	70
2	2.50	1.31	2.27	6.27	0.40	0.24	0.52	0.66	78
3	6.58	1.86	3.10	0.10	5.25	1.64	0.47	0.62	3
4	6.87	0.16	6.83	0.14	4.72	1.23	0.15	0.76	2

- Corrections: In view of the impossibility of replacing a layer of top soil on the whole surface of the dump, calcium carbonate and ashes from the plant were applied as a corrective measure.
Overall, the following strategy was established:
pH < 3.5 : ashes added;
5 > pH > 3.5 : calcium carbonate added in 5-15 t/ha proportions;
pH > 5 : calcium carbonate added in < 5 t/ha proportions;
while total sulphur and potential acidity of the sample were also considered.
The ashes proportion may vary between 500-1500 m3/ha, depending on the acidity and the type of operations planned to incorporate ashes to the waste.
- Chemical and Organic Fertilising: Once the acidity was controlled, the substrate is treated with both bottom and surface fertilisers. Initially, as bottom fertiliser previous to seeding, the following fertiliser units (FU) were applied:
- N : 50-60 FU's
- P : 150-160 FU's
- K : 110-120 FU's
Test performed with organic fertilisers have shown excellent results for pasture growing.
- Seeding: Aside from the planned use of the land, all surfaces were seeded with species of grass with the intention of rapidly obtaining a vegetal cover.
Areas assigned to forest or shrub heath were planted with these species: Holcus lanatus, Agrostis tennuis, Festuca arundinacea, Dactylis glomerata, Trifolium repens and Lotus corniculatus.
At the same time as pasture grass, shrub plants are also seeded (Cytisus scopiarius or Ulex europaeus), when tree species are not to be introduced.
Areas assigned to pasture are basically seeded with Lolium perenne and Trifolium repens.
Plantations: In areas destined to forest uses, mainly slopes, the following species are introduced:
. Betula pubescens
. Alnus glutinosa
. Pinus insignis
. Pinus pinaster
. Eucaliptus ssp.
. Castanea sativa
. Quercus robur

205

- Maintenance and results follow-up: In the years following the first farming season, the chemical evolution of restored surfaces was followed up my means of periodical analyses. As a result of such tests, the pertinent corrective materials and fertilisers were applied until the stability of the vegetation could be observed.

4 STUDY AND SELECTION OF MATERIALS FROM EXCAVATION HEADINGS

Materials from the edge and substrate of the deposit are constitued by Precambric and Paleozoic schists, phyllites and quartzites which form a series of bands more or less parallel and with variable thickness. The sedimentary terciary deposits are constituted by a series of lignites, carbonaceous clays, clays, sands and clayish marls with frequent lateral changes of facies and more or less important folds, all covered by quaternary sediments of a more rough texture (see figure 1).

Out of the above materials, both clayish wastes, sandwiched between brown coal layers, and Ordovician outcrops, at the North and West edges of the deposit, are excavated and stored in the dump.

4.1 Ordovician materials: slates

These are constituted by alternating detrital material with great differences in particle size (ludites, phyllites, quartzites) and varied mineral composition. The most noticeable aspect of these materials, from the standpoint of their outdoors behaviour, is the presence of sulphur, generally associated with materials of a finer original texture and higher contents of carbonaceous matter. The differences in sulphur contents relate to the non-homogeneity of the materials and the sedimentation conditions prior to their metamorphism and, with no doubt, there must be an effect associated with the techtonisation stages as materials closest to faults and joints are richer in sulphur.

Up to now, three types of blocks, designated as F1, F2 and CF1, constituting the main excavation volume, have been analysed.

Determinations carried out include pH in water, S% and cationic exchange capacity. Parameters for pH in water and S percentage, are indicative of the current and potential acidity (induced by the oxidating processes of sulphur), and are the most suited and helpful parameters for decision making as to the management of these materials in the dump.

- pH in water: There are clear differences in current acidity in the different slate blocks, in spite of the fact that pH ranges are very broad in all three cases:

Table 1. Average, maximum and minimum values of pH in water in slate blocks under study.

	pH		
Block	average	maximum	minimum
F1	3.51	8.82	2.42
F2	2.76	7.47	2.30
CF1	5.33	9.64	3.95

- Sulphur content: Analytical results shown in figure 4 (frequency distribution graph) evidence the difference in quality among the analysed blocks and the lack of homogeneity within each block.

On the basis of the results obtained in the study of Ordovician materials and in order to perform a subsequent selection and arrangement on the final surfaces of the dump, it was concluded that:

Block CF1 is the most suitable as it generally shows low sulphur contents (<0.2 %) and alkaline reaction.

Block F1 may be adequate provided non-fractured areas are selected as these are the ones showing sulphur contents higher than 0.3%.

Block F2 must be always rejected owing to its high acidic character and high sulphur contents.

4.2 Miocene wastes: Clays and sands

The alternating layers of lignites and clays characterising the Miocene appear in the deposit in varying thicknesses according to their location in the basin, as is clearly seen in figures 1 and 5.

From the lithological standpoint, five types of Miocene materials can be identified: "base conglomerates", of highly variable nature and composition going from mantle-rock or paleosoils to very plastic clayish sediments; marls, rich in atapulgite (Nonn and Medus, 1963) and smectites; "lignites", mostly of the so-called common brown coal type with

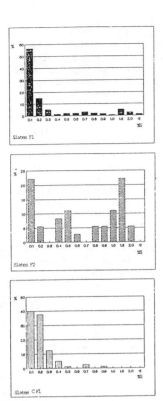

Fig. 4.- Distribution of sulphur contents frequencies in the slate blocks F1, F2 and CF1.

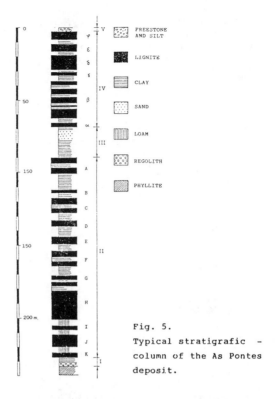

FREESTONE AND SILT	
LIGNITE	
CLAY	
SAND	
LOAM	
REGOLITH	
PHYLLITE	

Fig. 5.
Typical stratigrafic column of the As Pontes deposit.

xiloid, pyropisitic and clay levels; "clays", predominantly of the kaolinite type, that can appear mixed with coal, presenting in such case high levels of biogenic origin pyrite; and "sands", rich in quarz, which tend to be more abundant around the upper sections of the deposit.

Until today, the quaternary materials and materials belonging to the Miocene wastes interlayed between lignite layers φ and B have been characterised as shown in figure 5.

The following test types have been used to characterise the behaviour of these materials; particle size, mineralogic composition, sulphur contents and possible acidification by oxidation (oxidising pH), acid-base behaviour (pH in water) and ionic exchange (effective cationic exchange capability and nature of exchangeable ions).

In view of the mineralogical homogeneity, it was concluded that particle size, pH in water and sulphur contents determinations are the most suitable and operational as selection criteria for this material.

- Particle size: Three texture groups were differentiated:
1) Balanced granules with a sandy tendency (loam to sandy loam): Clay blocks: $\varphi - \varphi$, $\varepsilon - \varphi$, $\delta - \varepsilon$.
2) Fine granules (silty clay loam and clay loam): Clay blocks: $\gamma - \delta$, $\beta - \gamma$, $\beta - \delta$, $\beta - \beta$, $\alpha - A$.
3) Extra fine granules (silty clay): Clay blocks: $\beta - \overline{P}_2$, $\alpha - \beta$, $A - B$.
- pH in water and sulphur contents: The results for these parameters are shown in table 2.

Table 2. Values of pH in water and sulphur contents in waste clayish materials from the Puentes Mine.

BLOCK		AVERAGE PH (INTERVAL)	AVERAGE S % (INTERVAL)
R-α		4.87 (4.54 - 6.66)	0.014 (0.003 - 0.073)
φ-α		7.22 (6.94 - 7.94)	0.022 (0.000 - 0.051)
φ-φ		5.88	0.022
ε-γ		4.27 (3.36 - 8.40)	0.056 (0.001 - 0.142)
γ-δ		4.11 (2.76 - 8.13)	0.041 (0.001 - 0.300)
δ-γ		5.57 (4.51 - 7.83)	0.045 (0.011 - 0.147)
δ-γ		5.68 (5.07 - 7.68)	0.025 (0.009 - 0.060)
β-β		5.82 (5.45 - 6.80)	0.044 (0.002 - 0.221)
β-β		4.26 (3.76 - 6.72)	0.188 (0.047 - 0.404)
α-β		6.13 (4.90 - 8.35)	0.026 (0.001 - 0.126)
α-β		4.98 (5.30 - 8.89)	0.029 (0.001 - 0.239)
α-A		5.46 (4.30 - 7.63)	- (0.269 - 0.978)
α-β	*	4.90 (2.86 - 7.09)	0.030 (0.001 - 0.089)
α-β	*	4.18 (4.61 - 5.85)	- (0.128 - 3.240)
α-A	*	3.30 (2.54 - 7.22)	0.697 (0.554 - 0.822)

* Samples with carbonaceous matter interlayers.

Biogenic pyrites found in clayish sediments are highly reactive so as to minor sulphur contents may cause a sharp pH drop by oxidation, as can be seen in table 3:

Table 3. Sulphur contents, pH in water and oxidation of some samples of waste clay materials.

Block	S %	pH in water	Oxidation pH
β-γ	0.160	5.33	2.72
	0.221	6.31	2.29
β-β	0.175	4.87	2.66
	0.404	5.55	2.26
β-P_2	0.126	4.90	3.07
$\propto$-β	0.269	5.76	2.72
	0.550	4.61	2.10
$\propto$-A	0.128	5.42	2.87
$\propto$-A *	3.240	2.54	1.60
A-B	0.554	7.09	2.90

* Samples taken from faces in contact with lignite.

- Ionic exchange capacity: The properties of ionic exchange are related to the nature of materials and their particle size. As a rule, ionic exchange values of these materials are low in the pH of the samples, but significant higher values can be found when pH increases due to the predominance of variable charges existing in kaolinitic phyllosilicates and carbonous compounds.

Following the obtained results, it is concluded that:

- Terciary clay blocks in the As Pontes deposit are not, by themselves, currently and potentially acidic (S % < 0.1), but their low cationic exchange capacity (<10 meq/100 g) indicates a small tampon capability, thus being highly able to become quickly acidified by any large source of protons.

- Carbonaceous matter, and more specifically the pyrites associated with it, are what acidify these materials and therefore, all contact faces with lignite layers or with unusable coal interlayers must not be used on final surfaces; neither blocks difficult to separate in these areas owing to their scarce thickness shall be used in such final surfaces (block β-β).

- The best blocks for final surfaces are then ε-γ and δ-ε (upper levels in the deposit), not even requiring mixing with coarse fractions owing to their balanced particle size (loam and sandy loam).

- Blocks γ-δ, β-γ, β-δ, $\propto$-A, β-P_2 and $\propto$-β are adequate (particle sizes between silty clay loam and silty clay), but must be mixed to improve their texture which, at their present state, may cause pooling and compaction problems.

- The Quaternary mantle (R), despite its acidity and low exchange capacity, shall be used in final surfaces whenever possible, owing to this adequate texture and to the fact it never contains associated carbonaceous matter.

- The addition of substances to improve the structure and increase the exchange capacity (organic matter) of the stored waste material will be very useful and, in no case, the current pH level shall be allowed to drop until an important amount of organic matter is incorporated either through additions or through biological cycles.

5 CONCLUSION

The application of the selection criteria of waste materials to be dumped at the Puentes Mine, already begun in 1989, has brought about the generation of dump final surfaces capable of being planted more adequately than the ones previously existing and it has practically demonstrated the importance of dumping selected materials as the first step in the reclamation process.

REFERENCES

Bacelar, J. et al. 1988. La cuenca terciaria de As Pontes (Galicia). Su desarrollo asociado a inflexiones contractivas de una falla direccional. Simposio sobre cuencas de régimen transcurrente. SGE 113-121.

Guitian Ojea, F.; Carballas, T. 1976. Técnicas de análisis de suelos. Ed. Pico Sacro. Santiago.

IGME. 1975. Hoja núm. 23 Puentes de García Rodríguez. 1:50.000.

Manera A. et al. 1979. Aspectos geológicos de la cuenca terciaria de Puentes de García Rodríguez (La Coruña). Bol. Geol. Min. XCV, 451-461.

Nonn, H; Medus, J. 1963. Primeros resultados de la cuenca de Puentes de García Rodríguez. Notas y Com. Inst. Geol. Min. Esp. núm. 71, 54-87.

Urrutia Mera, M.M. 1989. Procesos ácido-base en suelos de la provincia de La Coruña. Tesis Doctoral. Facultad de Biología. Universidad de Santiago.

Reclamation, Treatment and Utilization of Coal Mining Wastes, Rainbow (ed.) © 1990 Balkema, Rotterdam. ISBN 90 6191 154 0

Dust monitoring overcomes a burning bing thought too hot to handle, Ramsay Bing, Loanhead

A.S.Couper
Lothian Regional Council & Institute of Occupational Medicine, Edinburgh, UK

ABSTRACT: Colliery tips are called bings in Scotland.
One of them on Lothian Regional Council's reclamation programme called Ramsay Bing, lay in the middle of the town of Loanhead surrounded by houses and industrial buildings, including a high precision nautical engineering company.
Everyone was concerned about this burning tip in the middle of the town but no one had been prepared to take the risk of a major conflagration it if went out of control when rehabilitated. The tip contained ponds of coal slurry which could be sold or maybe on fire.
By a more scientific approach, the confidence of local residents and industrialists was gained and a pilot phase begun with extensive controls and emergency measures built in. The controls were a dust monitoring and supression system. So successful was this, that the whole tip was done. No lost production was caused due to airborne dust and all burning was extinguished without danger to the town.

1 HISTORY OF RAMSAY BING

In the centre of the former mining town of Loanhead lies Ramsay Bing. It was the spoil tip of a coal mine which started in 1890 and eventually closed in 1960. The tip was a dominant and unsightly landmark in the town (Fig. 1). It was on fire causing problems of dust blow and burning nuisance. The dust blow was a particularly serious problem for local industry which the National Coal Board attempted to overcome by damping down but with only temporary success.

A surface temperature survey by infra-red techniques indicated much of the tip was burning at temperatures of over 100 c at the surface. If opened up it would undoubtedly be considerably hotter.

Before the formation of Regional and District Councils in Scotland in 1974, the then Authority, Midlothian County Council, was keen to deal with the threat the tip posed to the community, but shied away from taking the plunge, as it considered the risk of a major problem during rehabilitation, greater than those it presented, as it stood. The County Council considered it too hot to handle.

2 LOTHIAN REGIONAL COUNCIL'S SOLUTION

With the passing of the Scottish Development Agency Act 1975, the government gave the Scottish Development Agency powers to deal with derelict land under Sections 7 and 8 of that Act. In 1976, Lothian Regional Council was appointed Agents of the Scottish Development Agency to undertake a major programme of rehabilitation of derelict land within Lothian. That the Council was able to demonstrate early in the programme, it possessed the skills and experience to successfully tackle burning tips without problems arising during implementation, gave the Agency and the local community confidence, that Ramsay should be dealt with as a priority in the programme. The major difference with previous burning tips was Ramsay's location in the centre of Loanhead (Fig. 2).

2.1 Discussion the key to progress

Public consultation is a part of the Council's procedure for implementing rehabilitation projects, and is undertaken at the planning application stage. In the Ramsay situation, that would be too late and was unlikely to allay local fears.

209

Fig. 1   Aerial view of Ramsay Bing 1986

From the very outset before any survey work was undertaken, it was decided to meet with the industrial concern nearest to the tip, MacTaggart Scott and Company Limited. Their concerns would be the most critical and therefore give the Council and the Agency an indication of the level of concern (Fig. 3). The early meetings were very positive; the Company was keen to see the dust blow nuisance dealt with as this caused serious problems for their high precision manufacturing. They welcomed a permanent solution, but because of the nature of their business, would not be willing to allow the rehabiliation works to proceed, unless they had some measure of control to protect them. What emerged from the early meetings was the need to have an independent monitor to avoid the Company and the Council disagreeing once the rehabilitation works started and the Company then faced with obtaining an interim interdict from court, to resolve matters.

advantages, but it meant reaching agreement beforehand on the mechanisms by which the project could go ahead and when it had to stop. The Company makes naval steering gear components for the Ministry of Defence under contract, and an unenforced stop to production could cause a lay-off of the workforce and if very serious, could cause the closure of the business. Information about the present atmospheric pollution levels within the factory was essential. From their work with asbestos dust, the Institute of Occupational Medicine was

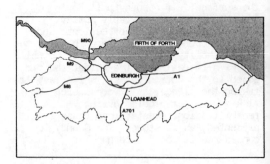

2.2 Independent monitoring

The principle of an independent monitor had

Fig. 2   Location of Loanhead

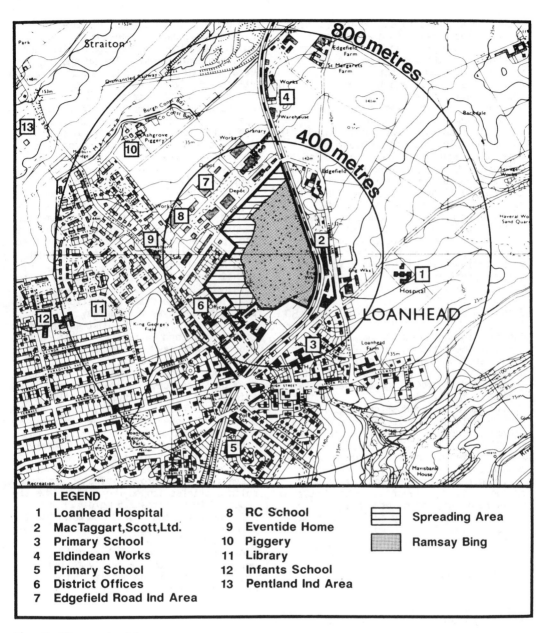

LEGEND

1	Loanhead Hospital	8	RC School	▭	Spreading Area
2	MacTaggart,Scott,Ltd.	9	Eventide Home	▦	Ramsay Bing
3	Primary School	10	Piggery		
4	Eldindean Works	11	Library		
5	Primary School	12	Infants School		
6	District Offices	13	Pentland Ind Area		
7	Edgefield Road Ind Area				

Fig. 3 Places at risk

invited to initially survey the level of
atmospheric pollution within the factory.
This work ultimately led to the appointment
of the Institute as the independent monitor
during the rehabilitation works contract.

2.3 A means of control

While the appointment of an independent

monitor would remove any disagreement
between parties, the Company was concerned
that it had an effective means of control
over the rehabilitation works contract.
Under contract law, the Agency would be the
Employer for the works and the Regional
Council, the Engineer for the works. A
third party under contract law would have
no authority to instruct the Agency's
appointed contractor for the works, unless

a court action was raised by the Company stopping the works. Such an action was what the Company wished to avoid as it wished to see the works done. A formal agreement between all three parties, the Agency, the Council and the Company was the only way all interests could be taken account of, and this was precisely the framework under which the actual rehabilitation works were implemented. It was a formally recorded document and therefore legally binding on all three parties. It also provided the framework for the contract documents themselves. Essentially it protected the Company, by giving powers to the Company and the independent monitor, at the same time allowing the rehabilitation works to proceed in a controlled manner. Special insurance provisions to protect all parties were incorporated in the event that a serious problem arose.

3 INSTITUTE OF OCCUPATIONAL MEDICINE'S ROLE

The Institute of Occupational Medicine was asked to undertake an intitial survey to establish background levels of pollutants around the site of the Ramsay Bing. They recommended a three stage approach as follows:-
1. Measurement of dust deposition rates and of toxic gas levels, eg sulphur dioxide, hydrogen sulphide, carbon monoxide, ammonia and nitrous fume around the perimeter of the tip before removal commenced.
2. Monitoring of dust and fume levels during a first phase of the rehabilitation works ie during the pilot phase
3. Similar monitoring during the final phase of rehabilitation works, if the pilot phase was successful.

3.1 Initial survey

The Institute of Occupational Medicine anticipated that the type of dust would be essentially coarse in nature and the amount in suspension in the atmosphere would depend, not only on the activity on the tip, but also on wind speed and direction, and on humidity and precipitation. The instrument they used for the initial survey around the tip, was the C.E.R.L. Directional Dust Gauge BS 1747 (1972) Part 5. Seven sampling sites were chosen representing the closest businesses or public places. Five of them are shown on the plan (Fig. 4). A compass was used to check the orientation of the sites. Three of these sampling sites were used to

measure nitric oxide, NO_1 nitrogen dioxide NO_2, carbon monoxide CO_1 sulphur dioxide SO_2, hydrogen sulphide H_2S, and ammonia NH_3. The NO and NO_2 were sampled using a Nitrous Fume Sampling Attachment with a Casella personal sampler. For CO, air samples were collected and determined by infra-red spectophotometry. SO_2 and H_2S were meaured by the N.I.O.S.H. Manual of Analytic Methods (1978) and ammonia by Axelrod and Greenberg's (1976) method.

They considered the first results not representative, due to a protracted spell of wet weather. They changed the NO and NO_2 sampling system to incorporate a Mining Research Establishment gravimetric sampler, Type 113A instead of the Casella sampler. Some testing for gases was also done on the tip itself, using a Draeger pump and detector tubes.

Their conclusions from the survey data obtained (Table 1) were as follows:-
1. The average daily deposition of dust per dust gauge was 2.65 mg/day.
2. The majority of the dust was deposited in the North and East ports of the dust gauges, as opposed to the South and West ports in the first exercise.
3. Some of the dust appeared to arise from urban sources. There was also evidence of seasonal influences with the presence of pollen dust etc. A proportion of the dust appeared to be attributable to the tip and, although the proportion was difficult to quantify, it was certainly less than that of the first exercise.
4. Considering the results of both surveys, the background level of airborne solid particles around the site of the tip appeared to be circa 2.5 mg/day.
5. The toxic gas levels found on the tip were below these which might be considered dangerous or harmful to health.
6. The established levels for toxic gases were thought to be due partly to the local urban environment and partly to the presence of the burning tip.

3.2 Monitoring

Although the initial survey found the dust deposition rates around the tip were generally low, the exception was a period immediately before the pilot phase of rehabilitation works started in May 1982, when a remarkably high deposition rate of 1g/day was measured at MacTaggart Scott's, compared with a few mg/day previously. This happened during a windy spell towards the end of an unusually dry period, and was highly visible with clouds of black dust blowing off the tip. Similar dust clouds

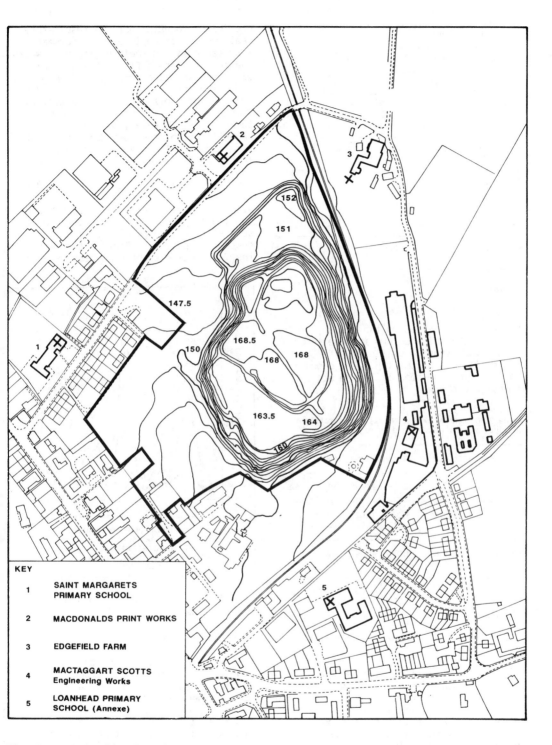

Fig. 4 Dust sampling locations

213

Table 1. Rate of total dust and rainwater collection over 14 day period in May 1980

Port	N	S	E	W	TOTAL
Site 1	+0.80 *2.8	0.50 -	1.28 3.2	0.66 0.7	3.24 6.7
Site 3	0.68 2.8	0.52 0.7	0.94 2.8	0.44 0.7	2.58 7.1
Site 4	0.58 1.8	0.73 1.4	0.61 1.4	0.67 · 0.4	2.59 5.0
Site 5	0.47 2.1	0.49 1.4	0.93 0.4	0.56 -	2.45 3.9
Site 7	0.54 2.6	0.61 1.8	0.52 1.4	0.75 1.8	2.42 7.6
Mean	0.61 2.4	0.57 1.0	0.86 1.8	0.62 0.7	2.65 6.0

+ Dust gauge deposits (mg/day)
* Rainwater (ml/day)

had been experienced by the Company in the past and the nuisance and inconvenience it caused was strong justification for rehabilitation of the tip. The deposition rates during the rehabilitation are summarised below:-

Table 2. Concentrations of nitric oxide, nitrogen dioxide, sulphur dioxide and hydrogen sulphide

Sampling Site	Date	Concentration (ppm)			
		NO	NO_2	SO_2	H_2S
No.3	19.5.80	0.03	0.02	N.D.*	N.D.
	21.5.80	0.01	0.01	N.D.	N.D.
No 5	19.5.80	0.03	0.04	N.D.	N.D.
	21.5.80	0.01	0.00	N.D.	N.D. ·
No 7	19.5.80	0.01	0.03	N.D.	N.D.
	21.5.80	0.02	0.02	N.D.	N.D.

* N.D. = Not detected
Detection limit for SO_2 and H_2S = c. 0.02 ppm

During rehabilitation, measured dust deposition rates were in the range 2-40 mg/day, the highest levels on the east side of the bing at MacTaggart Scotts. Wind direction did have an influence on the deposition results which indicated a predominantly westerly wind. To give MacTaggart Scott assurance that the excess dust pollution would not affect the engineering tolerances on their high

precision products, nor the respiratory health of their employees, measurements were taken within their engineering works. The results ranged from 0.07-0.36 mg/m³, with one exception at 0.54mg/m³ (Fig. 5).

Had the very high levels found in May 1982, occured during rehabilitation, the Company would have had to close down temporarily. However, during rehabilitation, the dust deposition rates remained much lower and the respirable dust levels in the works remained low by normal hygiene standards.

Levels of gaseous pollutants around the site were generally found to be low, although occassional high concentrations of individual gases, NO_2, were observed at some sites but these did not fit a systematic pattern. Because they were low, the Institute discontinued measurements.

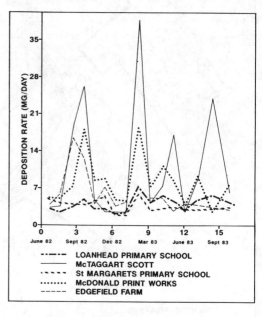

Fig. 5 Dust deposition rates during rehabilitation

4 IMPLEMENTATION

All parties involved, the Scottish Development Agency, the Regional Council and the Company, agreed that a phased approach was the most desirable way forward. A pilot phase to move part of the tip to a spreading area beside Edgefield Road was prepared. Planning permission was applied for and a series of public meetings held to explain the proposals in detail to the community, and to hear what concerns the locals had. The meetings did much to allay concern and subsequently planning consent was granted, without objection. Contract documents for the pilot phase were prepared, encompassing the conditions in the formal agreement. The agreement made it possible to progress this stage.

4.1 Legal agreement

The formal agreement legally binding the Scottish Development Agency, the Regional Council and MacTaggart Scott and Company Limited was signed in April 1982. It contained eight clauses as follows:-

1. The first stated the Company had no objection to rehabilitation of the burning tip.

2. The second described the rehabilitation works as the contract that would be let by the Scottish Development Agency.

3. The third covered the appointment of the Institute of Occupational Medicine as independent monitor, who's findings and conclusions were binding on all parties, in respect of the incidence and levels of dust, smoke, fumes and general pollution. The costs of monitoring were to be borne by the Agency.

4. The fourth clause placed a duty on MacTaggart Scott to have a designated employee as a contact, who had the responsibility of ensuring the incidence and levels of dust would not adversely affect the Company's operations or property. This employee had the authority to request the Engineer, the Regional Council, to instruct an immediate stoppage of the rehabilitation works operation. So that this was effective, there was a duty on the Engineer to be present on site at all times during the works.

5. The fifth clause placed a duty of reporting all incidences to the Institute of Occupational Medicine, who would advise on them, but would not have the power to authorise a stop to the works.

6. The sixth clause extended the powers of MacTaggart Scott, to authorise stoppage in the event of unforeseen circumstances.

7. The seventh clause covered consequential loss. The Scottish Development Agency had to undertake to meet all costs incurred by MacTaggart Scott in providing sufficient insurance to cover consequential loss as a result of the project works being carried out, up to a value to £13,000,000 as well as damage to property, up to a value of £4,000,000. The Engineer, the Regional Council, was also obliged to take out professional indemnity insurance, up to a value of £10,000,000 and the Contractor for the rehabilitation works, insurance under Clause 23(a) of the Conditions of Contract, to cover his negligence, up to a value of £10,000,000.

8. The eight clause bound all the parties to the Sheriff of Lothian and Borders as the final arbiter in the event of a dispute.

4.2 Contract management

The legal agreement placed a strict obligation on the Engineer and the contractor, under clause 4, to be on site at all times during the operations. This was managed on a shift system between the Engineer's Representative and his assistants. A standby arrangement was also adopted so that if an area of the tip went on fire during the periods when the Contractor was not operating, the Engineer's Representative, or his assistant, could be on site within an hour.

For the Contractor there were a number of very specific requirements laid down in the contract documents to ensure the works would be carried under strict control. They were as follows:-

1. Part of the rehabilitation works involved removal of coal slurry simultanueously with the earthworks on the burning tip, and the times for moving that material by road, Edgefield Road, were restricted to outwith peak periods.

2. The Contractor had to take any steps necessary to prevent dust arising from his operations.

3. Before any material was moved the Contractor had to install a 100 mm pipeline from Bilston Glen colliery, 2.4 kms away, along a live railway line, to ensure there was on site a sprinkling system capable of delivering 20,000 l/hour at all times. This was essential to keep the working areas and haul routes damp at all times. In addition a back-up storage tank of 20,000l was required. The Contractor was required to have a minimum of two men available on a 24 hour stand-by basis, to be on site within 2 hours to operate the equipment.

4. If dust was affecting any premises within the vicinity of the tip, whether due to the Contractor's negligence or not, the Engineer had the right to stop the work. For a stoppage of less than 1 hour, or outwith the working day, no payment would be made for the stoppage.

5. The method of working was strictly prescibed, as to what constituted hot burning material and how it had to be treated. Hot material was material over 200°c, measured by thermometer inserted

Fig. 6 Simultaneous removal of burning material and coal slurry

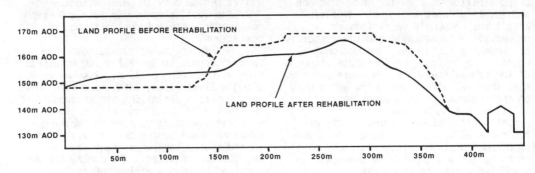

Fig. 7 Longitudinal section through tip

50mm into the ground immediately after a 300 mm layer was removed. The hot material had to be laid in 300 mm layers, rolled by a roller of a static load of 17.5 KN, until it had cooled to below 50 °c. Until it had reached this temperature, no material was to be placed on top.

6. The Contractor also had a requirement to produce a programme with his tender which would show how he proposed to undertake the earthworks to move the burning tip to its deposition place on the field between the tip and Edgefield Road,

simultaneously with extracting coal slurry from the coal slurry lagoons that lay within the top of the tip. The interrelationship of the methods and timing of the two operations was seen as critical to the success of the overall operation (Fig. 6).

7. The Contractor was required to indicate the credit accuring from the coal slurry extraction against his earthworks costs.

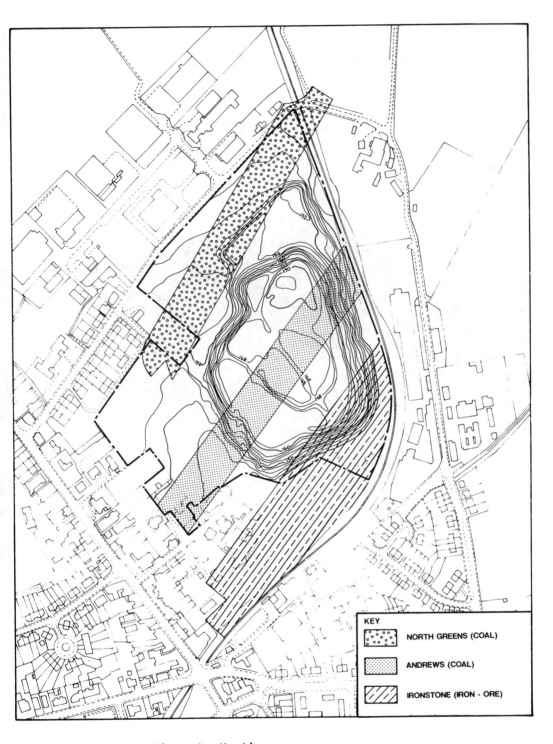

KEY

North Greens (Coal)

Andrews (Coal)

Ironstone (Iron - Ore)

Fig. 8 Shallow coal working under the tip

217

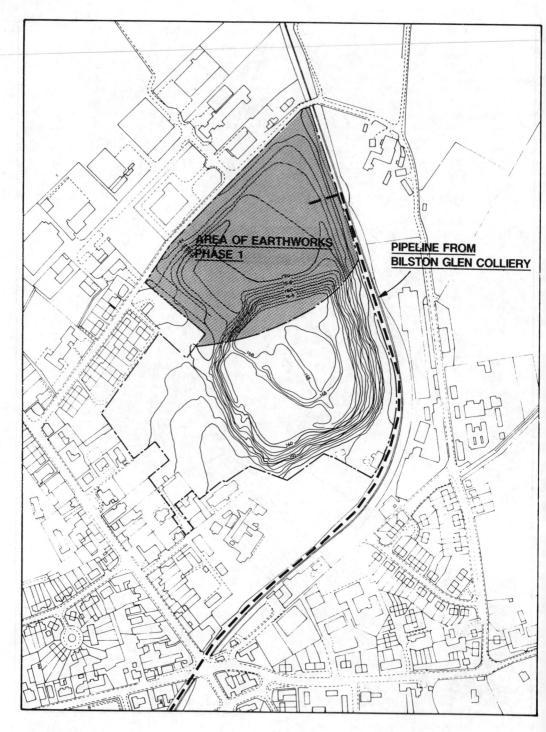

Fig. 9 Pilot phase works

4.3 Design difficulties

Besides the legal and contractual controls, Ramsay presented difficulties of design because it was hemmed in on all sides with very limited scope for reshaping (Fig. 7). Shallow coal workings under the site in the North Greens and Andrews coal seams placed restrictions on any development after-use (Fig. 8). Road access was limited to Edgefield Road that in itself presented difficulties for new development as it was narrow beside the Roman Catholic primary school and reached capacity at peak times due to the industrial estate at the other end.

The alternatives of removal by road to Haveral Wood sand quarry, or the fields east of MacTaggart Scott's factory, were investigated but rejected on the likelihood of increased dust problems due to the removal method, the slow removal rate and the extra costs involved. From a planning point of view, any new land created in either location could only be for agriculture forestry or recreational use, as it lay within the Edinburgh Green Belt.

The design solution adopted to overcome these restriction, was ingenious. Two development sites on raised terraces with a shared access from Edgefield Road were devised, one hidden by building a mound over the North Greens coal working. This allowed the balance of the tip site to be reshaped into a natural hill form as a central feature of a proposed town park.

5 RESULTS

The rehabilitation of the burning tip was carried out in two phases of earthworks, a pilot and a main phase, followed by a planting works phase. The details of these phases are as follows:

1. Pilot phase. Started 19 May 1982, completed 6 June 1984 at a works cost of £119,537. It involved moving 132,000 m³ of the tip (Fig. 9).

2. Main phase. Started 11 March 1985, completed 10 November 1986, at a works cost of £321,529. It involved moving 355,000 m³ of the tip.

3. Planting phase. Started 18 March 1987, completed 10 March 1988 at a cost of £60,349, involving treeplanting, fencing, footpaths.

The total works cost, excluding land acquisition, design and supervision costs and any development costs, was £501,415.

The total site area was 11.9 hectares.

5.1 Achievements

No lost days of production were incurred by MacTaggart Scott during the pilot phase or the follow on main phase. The sophisticated approach adopted of strict control, monitoring and measurement worked. Generally the tip was worked in a south to north or west direction and whilst temperatures of burning were recorded well in excess of 600 c, such was the degree of continuous control over the operation that even when the tip burst into flame, the back-up was at hand to stop work at that point, to damp it down and to continue elsewhere. Stoppage time due to flare-ups was never excessive and from the Contractor's well organised approach to programming the work, was usually compensated by other operations that could be done instead. The Contractor has to be complimented for his very responsible and responsive attitude to the whole operation.

5.2 Costs

The costs of rehabilitating Ramsay in this highly controlled manner were considerably more than double the average cost for rehabilitating derelict land in Lothian. Taking the rehabilitation works cost only, the cost per hectare was £43,225/ha compared to the average of other tip rehabilitation scheme of £19,950/ha, at 1985 prices. Both costs are to the same standard of landscaping work.

Having left the development sites to settle and consolidate by natural means since completion of the earthworks in November 1986, both sites are now in a condition to be disposed off. A recent site investigation proved that the bearing capacity of both sites was suitable for development. The 4.0 hectares of development land created can be marketed for housing development (Fig. 10). This will make a significant contribution to the housing policies of the Lothian Region Structure Plan 1985, in terms of housing requirements within Midlothian District Council's area and in doing so continue the preservation of the Edinburgh Green Belt, at a point when it is at its narrowest.

The value of the development land is estimated at £200,000 at today's prices which would reduce the works costs to a net figure of just over £300,000.

5.3 Conclusion

Rehabilitating the burning tip was a

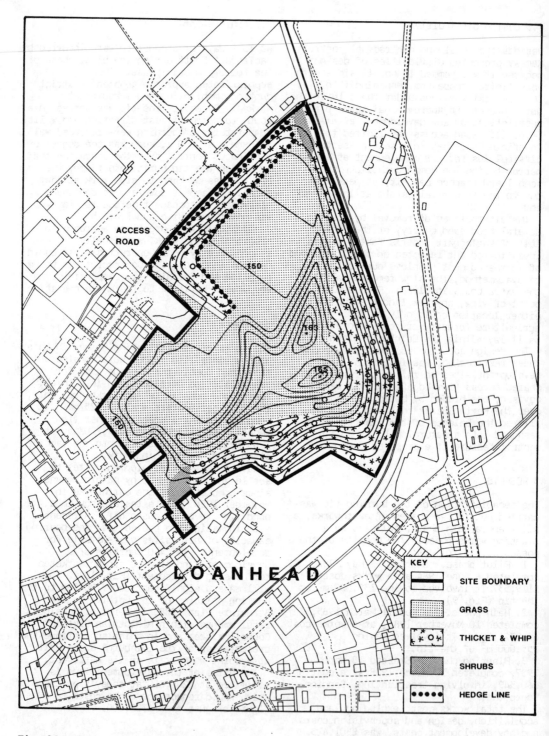

ACCESS
ROAD

150

160

145

150

L O A N H E A D

KEY

SITE BOUNDARY

GRASS

THICKET & WHIP

SHRUBS

HEDGE LINE

Fig. 10 Final shape of the tip showing development areas and planting

success, as it resulted in the permanent removal of dust, smell and fumes that had affected the town of Loanhead for many years. One cannot put a price on the value of the result, other than that the largest employers in the town, MacTaggart Scotland Company Limited, no longer has a threat of stoppage to its business due to dust blow. For this reason alone it was a worthwhile rehabilitation project.

ACKNOWLEDGEMENTS

This paper represents the views of the author which are not necessarily those of Lothian Regional Council or the Scottish Development Agency.

The author wishes to thank the Director of Planning for allowing publication of this paper and to the staff in the Landscape Development Unit for their helpful assistance and comments.

Reclamation, Treatment and Utilization of Coal Mining Wastes, Rainbow (ed.) © 1990 Balkema, Rotterdam. ISBN 90 6191 154 0

Coal discard – Rehabilitation of a burning dump

B.J.Cook
Rand Mines (Mining & Services) Ltd, Johannesburg, South Africa

ABSTRACT: Collieries are traditionally associated with pollution from water, dust and burning dumps. By changing the manner in which the coal discard is stored it is possible to correct this. Air and water pollution legislation, which is becoming more stringent, can also be satisfied and the image of the colliery enhanced.

A case study is presented where tipping on a burning dump was stopped and the dump encapsulated with compacted discard. Methods of placing the discard, compaction standards and control, dump profile, surface water and seepage control, are described.

Results were immediate. 80% of discernible air pollution disappeared rapidly and water pollution was contained and controlled. A year after the project started the colliery was awarded the 1988 National Association of Clean Air trophy. Further, a valuable asset, a low grade source of energy for use at some future date, is being preserved in an aesthetically and environmentally acceptable manner.

1 INTRODUCTION

Welgedacht Colliery is situated in South Africa at Utrecht in northern Natal. The town lies in a picturesque valley within a triangle formed by the three sections of the colliery - Utrecht, Umgala and Zimbutu. Each section has a discard dump. The town is thus very often downwind of one of the dumps. The dump rehabilitation methods described in this paper apply specifically to the Zimbutu section. Similar measures have been adopted at both of the other sections.

Mining at Welgedacht started at the turn of the century at the Utrecht section. The Zimbutu section was commissioned in 1965 and 180 000 tons of coal per month are currently mined. The coal is beneficiated in a preparation plant and a product, with a calorific value (CV) of 6750kcal/kg is produced for both the local and export markets. Approximately 72 000 tons of discard, with a CV of approximately 3800kcal/kg is dumped on the discard dump. This is 40% by mass of the run of mine feed and represents a relative cost disadvantage to the producer as well as a loss of material having a heating value.

Until fairly recently discard was viewed as a waste product and a liability. The dumps commonly ignited spontaneously and, when this did not occur, they were in fact ignited and encouraged to burn to reduce the volume of material. This caused considerable air pollution and frequent complaints from the local residents. In 1984 however, Rand Mines, the mining house controlling Welgedacht colliery, resolved that burning discard dumps were no longer acceptable on environmental or economic grounds. Shortly thereafter a method of shrouding a burning dump with a skin of compacted discard that would smother the existing fires and prevent future spontaneous combustion, was developed.

This paper deals with air and water pollution problems on burning discard dumps and suggests methods to overcome these. The disposal of slurry, the minus 100μm material discharged in a 30% to 50% aqueous suspension, is also discussed and some experiences and results of using these methods are presented. Tipping, spreading and discard compaction techniques are also presented.

All the operations described in the case study are continuing, and will do so until closure of the colliery.

2 LEGAL CONSTRAINTS

2.1 Water Pollution

Most of the water pollution on a colliery is caused by run-off from coal and discard dumps, from the coal preparation plant site and by seepage from these areas.

In South Africa the regulations relating to The Water Pollution Act, and The Mines and Works Act, must be satisfied.

2.2 Air Pollution

The Atmospheric Pollution Prevention Amendment Act and The Mines and Works Act regulations provide the legal framework for the control of air pollution.

The regulations have recently been amended and, as a result, any form of pollution such as dust, sand, smoke or fumes must be prevented from escaping from dumps. This means that the dumping of discard in such a way that it can burn, is now an offence.

3 POLLUTION PREVENTION

3.1 Air pollution

Historically discard was tipped over the edge of the dump. This caused the material to segregate. The fines remained near the top of the slope and the coarse material ran to the bottom. The accumulation of uncompacted coarse material at the base of the dump allowed air to flow into the dump where high internal temperatures soon set up convection currents that promoted combustion of the discard.

All tipping over the side was stopped at Zimbutu and a start was made to encapsulate the entire dump with a compacted skin of discard with a minimum thickness of 10m. The compacted skin, of necessity, had to be sufficiently dense to prevent the ingress of air to the old burning dump and to prevent spontaneous combustion of the newly placed discard. Figure 1 shows a typical cross section through the old discard dump and the compacted skin.

Care was taken to ensure that the skin was founded on a solid foundation for slope stability reasons and to ensure that there was no air passage through permeable material at the base. All soil and soft material was removed from the site with the soil being stockpiled for future use. It is used in a 1m wide layer on the outer surface of the compacted discard to seal the surface and to provide a growing medium for vegetation. Slope stability calculations determined that for Zimbutu, an outer slope inclined at thirty degrees would be stable. The perimeter of the base of the discard was accordingly positioned so that, allowing for a 10m stepback halfway up the dump, the skin would be at least 10m wide at the top to provide adequate working space for construction plant at all times. The stepback was planned to provide access to the dump sides for maintenance of the vegetation. Discard dumping then started in 200mm thick compacted layers with the compaction of each layer being tested and approved before the next layer was placed so as to ensure that compaction was uniform and satisfactory.

Once the compacted discard skin emerged above ground level a strip of topsoil, 1m wide, was included around the outer edge of each discard layer and was compacted with the discard. Vegetation was later established on the soil. An insulating layer , also 1m wide, consisting of burnt-out material or soil was included in the discard layer at the inner face to insulate the newly placed discard from the burning dump.

At all three of the sections at Welgedacht it was found that, once the tipping of discard over the edge of the dump was stopped, discernible air pollution decreased by an estimated 80%. This was mainly because the existing fires were no longer supplied with fresh fuel and also because, as the compacted skin rose up the dump, the remaining burning zones were progressively smothered.

3.2 Water Pollution

Tipping of discard over the side creates a dump profile that sheds rainwater. This erodes the highly erodible discard polluting the environment and also exposing a fresh face to the air which promotes spontaneous combustion. The first measure undertaken in rehabilitating the dump was to control the movement of rainwater. This was achieved by reshaping the top of the old dump by dividing it into horizontal paddocks and by constructing a wall around the perimeter. The paddock and perimeter walls, which were compacted by means of a vibratory roller, were constructed by pushing up in situ material with a bulldozer to form a berm 3m wide and at least 1m high. The paddocks thus formed, with compacted floors, prevent the movement of water on the dump thus eliminating erosion as well as preventing large volumes of water from accumulating at any particular location. The water that collects in the individual paddocks either seeps slowly into the dump through the compacted floor, contributing to a humid inert atmosphere within the dump,

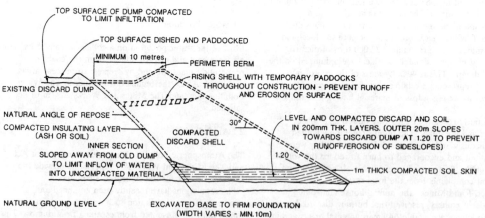

FIGURE 1. TYPICAL CROSS SECTION THROUGH COMPACTED SHELL AROUND OLD BURNING DISCARD DUMP

or evaporates. The perimeter walls prevent water cascading over the side. To meet the requirements of the 1956 Water Act, clean water cut-off drains were constructed at ground level, upslope of the dump and any other dirty areas, to divert all uncontaminated rainwater away from and around polluted areas. Downslope of the dump and polluted areas, seepage interceptor drains, which also served to collect polluted surface water, were constructed. At Zimbutu these drains were excavated down to an impervious sandstone layer at about 1.5m below surface. Thus all polluted water was collected and discharged into pollution control dams where it was either recycled back to the coal preparation plant, evaporated, or treated and discharged. Uncontaminated rainwater was discharged into the storm water drainage system. The dams were sized so that, at closure of the colliery, they will be large enough to evaporate all anticipated polluted water that will accumulate. Figure 2 shows the lay-out of the water pollution prevention measures at Zimbutu.

4 PREVENTION OF SPONTANEOUS COMBUSTION

Coetzee (1985), in discussing the causes of spontaneous combustion of coal, concludes that spontaneous heating can only be averted if the following conditions are satisfied:

1. If the coal is kept in an oxygen deficient atmosphere consisting of nitrogen and/or carbon dioxide.
2. If the coal is kept in a humid atmosphere with excess moisture to dissipate any heat formed and.
3 If the movement of air caused by the daily variation in barometric pressure or changes in temperature is restricted to the outer surface of the stored coal.

These criteria at Zimbutu were satisfied in the following manner:

4.1 Soil

Soil, that was excavated and stockpiled before the placing of discard started, was included in the outer meter of each discard layer and compacted with the layer. It effectively provides a dense outer insulating layer that prevents the flow of air into or out of the discard. It also provides the growing medium for a vegetative cover that will protect the surface from wind and water erosion. It is costly to acquire and is accumulated at every opportunity.

4.2 Placing and spreading discard

The discard was transported from the loading bin to where it was required by means of 16 ton rear or bottom dumpers and spread, so that after compaction a 200mm thick layer was obtained. The material was mixed and

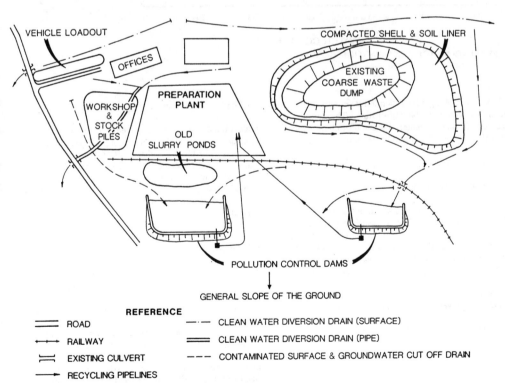

FIGURE 2. LAYOUT OF POLLUTION CONTROL WORKS AT ZIMBUTU

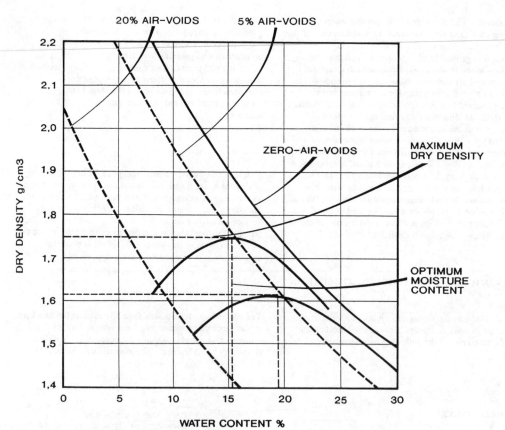

FIGURE 3. LABORATORY COMPACTION CURVES

spread by means of a grader which left a smooth, well mixed surface with no visible segregation. The smooth surface ensured that the material was uniformly compacted and that the dumpers had a good surface on which to operate.

4.3 Profile of the working surface

In order to prevent erosion, the shape of the working surface was carefully controlled. The surface was kept horizontal with the outer 20m around the perimeter and the inner 20m against the old dump rising at a rate of 1 in 20. This prevented any spillage over the edge or the concentration of water against the uncompacted burnt out material of the old dump. If the working site was not absolutely level longitudinally, mountable cross-berms were constructed to prevent any movement of the water. Rainwater that ponded on the working site in this manner has not interfered with operations to date.

4.4 Conditioning the discard

Unless the discard was compacted at near optimum moisture content (OMC) satisfactory compaction was not achieved. The OMC for discard is of the order of 15% as shown in Fig 3 (Coetzee 1985). Discard was delivered from the plant at near OMC and was suitable for immediate spreading and compaction. If compaction was delayed, water was applied by a tanker before compaction started.

4.5 Compaction

The grading of the particle sizes within the material determines the ease with which it can be compacted. Figure 4 shows typical discard sieve analysis results. From these curves one can determine whether the discard is well graded or whether the particles are uniformly sized. Uniformly graded discard has a coefficient of uniformity ($C_u = d_{60}/d_{10}$) of five or less and compacts with difficulty. Well graded discard with a C_u greater than 5 compacts more readily as the voids between the larger particles are filled with the smaller particles and a denser, more impermeable layer results (Forssblad 1981).

226

$$\left(Cu = \frac{d60}{d10}\right)$$

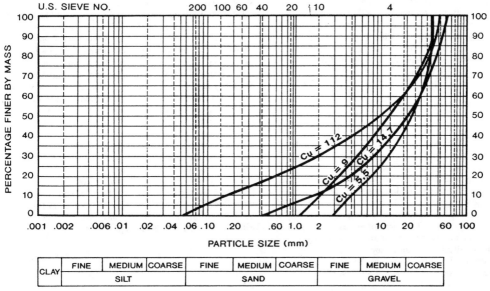

FIGURE 4. TYPICAL DISCARD SIEVE ANALYSIS RESULTS

For practical reasons the largest piece in the layer should not be thicker than half the height of the compacted layer. If the discard contains pieces larger than 100mm, the use of a padfoot drum should be considered for the initial compaction as it breaks up the particles. In many cases this technique can save the cost of a crusher. The final passes of each layer should however be done with a smooth drum so that the surface is well sealed.

At Zimbutu a ten ton self-propelled vibratory compactor was used for compaction. The choice was based on work that had recently been completed by Coetzee and whose unpublished report can be summarised as follows;

4.5.1 Vibratory Compactors

Coetzee tested five, ten and sixteen ton compactors on a 200mm thick layer at OMC. He concluded that a smooth drum 10ton machine operating at a frequency of between 20 and 30 Hertz and with an amplitude of at least 1.8mm is the most suitable machine. He found the 5ton machine to be much too light, the 16ton machine compacted at depth but sheared and loosened the surface ahead of the roller. The sheepsfoot and padfoot drums crushed and loosened the surface while compacting at depth. Lower amplitudes only compacted the upper portion of the layer.

At Zimbutu it was found that a 10ton smooth drum machine was able to achieve satisfactory compaction of a 200mm layer of properly conditioned discard in six passes.

The output of a single machine, with good conditions, was thus about 700ton per hour.

4.5.2 Impact Rollers

Impact rollers are able to compact considerably thicker layers than vibro compactors. Recent further tests by Coetzee at three different collieries have indicated that an impact roller can be used with advantage to compact most areas of the dump. However, because of its lack of manoeuvrability and the speed at which it must operate, a vibro compactor is required in addition to compact the inaccessible areas and also around the edge. Impact rollers are drawn behind a tractor of at least 160kW. The recommended operating speed is between 10 and 12 km per hour. Tests have shown they are capable of compacting a 1m thick layer in 8 passes at 10 km per hour. This gives a theoretical compaction capacity of just on 5000 tons per hour.

4.6 Compaction control

A Dynamic Cone Penetrometer (DCP) was used to control the compaction. The type developed for use on coal and discard stockpiles by the National Institute of Coal Research (NICR) has a 9.75 kg dropweight falling 575mm and has a 30° pointed tip. The total weight of the

227

apparatus is approximately 12 kg. The standards recommended by the NICR are:

After 10 impacts, penetration <200mm.
After 20 impacts, penetration 200 - 300mm.
After 30 impacts, penetration 300 - 380mm.
After 40 impacts, penetration 380 - 460mm.

Compaction tests were done at a frequency of one test approximately every 1 000m².

5 SLURRY DISPOSAL

Coal slurry at Welgedacht was normally discharged into slurry ponds, where it was dried and removed, or deposited underground. Mechanical de-watering was considered but thought to be too expensive. Each of these methods has advantages and disadvantages. To list some disadvantages:
* Ponds are a source of pollution and expensive to construct and to empty. They are not normally at the top of the plant operators' priority list and consequently do not always receive the attention they require.
* After drying, the slurry solids must be stockpiled and compacted in an environmentally acceptable manner.
* Underground disposal can prevent stooping and generally interfere with mining operations.

Slurry is also an energy source and should be stockpiled in such a manner that it can be reclaimed at some future date. Recognising this it was decided to stop underground slurry disposal at Zimbutu. Instead the dump rehabilitation site has been enlarged by some 18 ha and new slurry disposal facilities have been included in the extension. The work has only just started and Figure 5 shows the conceptual design. Two slurry dams, each 1.8ha in area, have been included within the discard dumping area. Placing of discard in the new area will start around the slurry dam site and, once the walls are 1m high, slurry deposition will start. The dams have been sized so that the rate of rise of the slurry will equal the rate of rise of the discard dump. It is anticipated that there will be sufficient capacity for discard and slurry for the next fifteen years. To prevent a rise of the phreatic line within the discard dump the dams have been provided with an underdrainage system around their entire perimeter.

The slurry will be fed into the dams around the outer perimeter and the penstocks have been located close to the inner face. The ponds will thus be kept well away from the outer face of the discard for stability reasons. The slurry will eventually be over 30m deep.

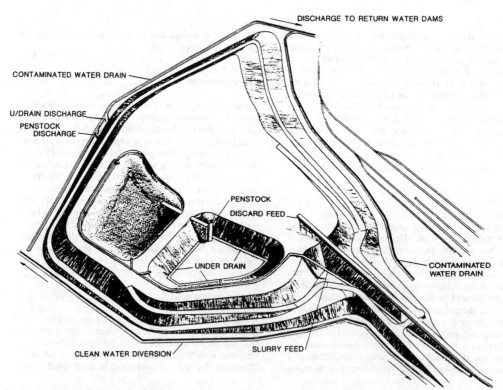

DISCHARGE TO RETURN WATER DAMS

CONTAMINATED WATER DRAIN

U/DRAIN DISCHARGE
PENSTOCK DISCHARGE

PENSTOCK
DISCARD FEED

UNDER DRAIN

CONTAMINATED WATER DRAIN

CLEAN WATER DIVERSION

SLURRY FEED

FIGURE 5. CONCEPTUAL DESIGN FOR DISCARD / SLURRY STORAGE ZIMBUTU COLLIERY

6 VEGETATION

Utrecht is situated in the upland area of northern Natal in the summer rainfall area. The summers are subject to periodic very hot dry spells and the winters are cool with sporadic frosty nights. 80% of the annual rainfall, 750mm/annum, falls between October and March and the annual evaporation, 1680mm/annum, is more than twice the annual rainfall.

6.1 Vegetation establishment

The annual rise of the outer face of the dump was vegetated, under dry land conditions, during the rainy season every year. In addition to providing an aesthetically pleasing covering, vegetation offers the best long term protection against erosion and also improves slope stability. It prevents erosion and thus air and water pollution by:

* binding and retaining soil particles in place.
* trapping suspended solids in the run-off.
* intercepting raindrops and thus reducing their ground impact.
* retarding the velocity of run-off water.
* protecting the soil surface from the force of the wind plus:

In addition storm water run-off from the side of a well vegetated dump is not contaminated and can thus be discharged without treatment. Also any water falling on the top of the dump is retained where it falls by the contour and perimeter walls. No contaminated water escapes from a vegetated stockpile and thus the size of the evaporation dam required after closure can be reduced.

The permanent vegetative cover should ideally consist of plant varieties that:

* have a low nutrient requirement.
* have a low management requirement. They must not require cutting or grazing or burning to prevent them going moribund.
* provide ground cover as opposed to canopy cover.
* recover well after burning.
* are not frost tender.
* will provide food for birds and animals and thus encourage the return of the natural ecosystems.

Before seeding, the soil was ameliorated by a blanket application of agricultural lime and fertilisers designed to achieve a pH of above 6, phosphate levels of 20ppm and potassium levels of 80ppm. During favourable weather conditions a seed mix consisting of various legumes, runner grasses and tufted grasses, to provide some textural relief, was sown. Nitrogen fertiliser was applied three weeks after seeding and again six weeks later during favourable weather conditions at a rate of 30kg of nitrogen per ha per application.

6.2 Annual maintenance

Annually, towards the end of winter, soil samples were taken from the vegetated areas and a full analysis of the macro nutrients was done. The analysis was used to determine the fertiliser programme for the next growing season. The pH and nutrient levels were adjusted if necessary and two applications of nitrogen, each at 30kg nitrogen per ha, were applied. The first application was given at the start of the growing season once the plants had shown signs of root activity. The second application followed six weeks later.

7 COSTS

Over the last three years the spreading, conditioning and compaction of the discard as well as the maintenance of all haul roads, have been contracted out. The costs vary slightly with the amount of discard placed per month but are currently approximately R0.85 per discard ton or 30 cents per run-of-mine ton. The contract is won on public tender and runs for a period of two years. The discard is delivered to the dump by the colliery.

The annual vegetation of the discard dump has also been contracted out to a specialist vegetation company. The current average cost is R5000.00 per ha.

8 CONCLUSION

Colliery management has the responsibility of making the best use of the assets of the company while at the same time preventing the degradation of the environment.

Storing the discard coal in the manner described has preserved a low grade source of energy for use at some future date. The additional cost involved in preserving this asset is less than R1.00 per discard ton. The potential value, in today's terms, is at least thirty times that amount.

In addition the discard storage methods described resulted in a rapid reduction in water pollution and an immediate improvement in air quality. An estimated 80% of discernible air pollution disappeared within weeks. One year after the start of the project Welgedacht was awarded the 1988 National Association of Clean Air trophy.

The new slurry dams will save slurry disposal costs and have stopped interference with underground mining operations. Another source of water pollution has been removed and a large volume of slurry will be preserved for use at some future date in an environmentally acceptable manner.

ACKNOWLEDGEMENTS

I would like to acknowledge the assistance of my friends and colleagues and thank the management of Rand Mines (M&S) Ltd. and Welgedacht Exploration Company Ltd for permission to publish this paper.

REFERENCES

Coetzee, S.D.1985. The reactivity of coal and the prevention of spontaneous heating in coal in stockpiles in South Africa - Second supplement. Proceedings of a symposium "The spontaneous heating of coal". The Coal Processing Society.

Forssblad, L.1981. Vibratory soil and rock fill compaction. Stockholm. Robert Olsson Tryckeri AB.

Reclamation, Treatment and Utilization of Coal Mining Wastes, Rainbow (ed.) © 1990 Balkema, Rotterdam. ISBN 90 6191 154 0

Assessment of appropriateness of reflotation for processing of hard coal slurries from some slurry-ponds in the Ostrava-Karvina coalfield

M. Sikorova
Institute of Industrial Landscape Ecology of the Czechoslovak Academy of Sciences, Ostrava, Czechoslovakia

P. Fecko & H. Raclavska
Mining University, Ostrava, Czechoslovakia

ABSTRACT: Reflotation of coal slurries from Ostrava-Karvina Coal District (Northern Moravia, Czechoslovakia) is evaulated in relationship to maceral composition of coal. Percentage of vitrinite determines flotability of material.

1 INTRODUCTION

Industrial agglomeration of Ostrava is situated in northern part of Moravia (Czechoslovakia). It has area of 1600 km^2 and it is characterized by great concentration of mining, metallurgy, chemical industry and energy industry. In Ostrava-Karvina Coal Basin there is extracted anually 22.7 millions ton of bituminous coal and it is generated 17.5 millions ton of wastes. Slurry ponds in Ostrava-Karvina have at the present time the area of 815 ha.

The main content of this contribution is the discussion of reflotation possibility of coal wastes deposited in slurry ponds of Mine Dukla and Coking Plant Odra (Fig. 1) which belong by their areas and amounts of deposited materials to the largest in Ostrava-Karvina Coal District.

1.1 Characteristics of samples

Waste waters from Coal Dressing Plant of Mine Dukla containing coal slurries from the flotation process are purified in two systems of slurry ponds. Small slurry ponds inside the territory of enterprise are used for cyclic filling and primary sedimentation of waters from the Coal Dressing Plant. Small slurry ponds are formed by system of 7 tanks, from which number 3 serves for sedimentation and 4 for further purification. Samples of weight 10 kg were taken from depth of 0.5 m in nine sampling points of the sedimentation slurry pond with area 4.5 ha and volume of 200 000 m^3.

Slurry pond of Coking Plant Odra has area of 4.5 ha and volume of 1.5 millions m^3.

From this slurry pond 5 samples were taken from different height horizons (samples 1-3 the depth 4 m under the water level, sample 4 from the depth of 8 m and sample 5 from the depth of 0 to 20 cm from dry level of the slurry pond.

Mineralogical, petrographical and chemical analyses were performed on homogenized samples. X-ray diffraction was performed on diffractometer Philips PW 1800. In the samples from the slurry pond of Mine Dukla was determined the presence of clay minerals (kaolinite, illite and in less extent chlorite), further carbonates (ankerite, siderite and calcite), quartz and felspars. In sample from Coking Plant Odra from clay minerals is not present illite and kaolinite prevails over chlorite. The difference is also in percentage of carbonates. Ankerite is missing, mainly is present siderite which prevails over calcite. Further is present quartz and felspars.

Coal petrographical analysis was performed on microscope ORTHOPLAN. Sample from the slurry pond if Mine Dukla contains 46% of vitrinite, 9% of exinite, 35% of inertinites and 10% of admixtures (carbonates and clay minerals). Petrographical structures of sample is on Fig. 2. Sample from the slurry pond of Coking Plant Odra contains 55% of vitrinite, 35% of inertinite and 10% of admixtures (carbonates and clay minerals). Petrographical structure is on Fig. 3.

From chemical analysis it follows that sample from Mine Dukla contains 24.02% A^d and 0.22% CO_2, sample from Coking Plant Odra contains 23.90% A^d and 0.25% CO_2. Chemical analyses of ashes from both localities is presented in Table 1. From mineralogical and chemical analyses it follows that for both samples is typical relatively high content of coal matter and after

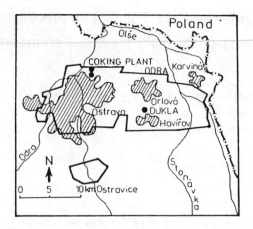

Fig. 1 Map of Osrava-Karvina Coal District with location of slurry ponds Dukla and Odra

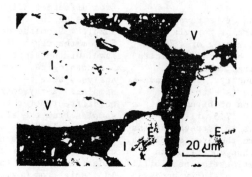

Fig. 2 Microphotography of sample from slurry pond of Mine Dukla (I = inertinite, V = vitrinite, E = exinite)

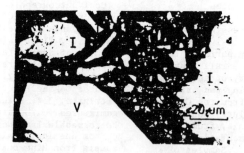

Fig. 3 Microphotography of sample from slurry pond of Coking Plant Odra (I = inertinite, V = vitrinite)

dressing they would represent material suitable as a fuel in thermal power plants.

Table 1. Chemical composition of coal slurries (in weight percent)

	Dukla	Odra
SiO_2	51.09	59.51
Al_2O_3	20.17	16.64
TiO_2	1.24	1.03
Fe_2O_3	14.12	8.11
mnO	0.20	0.11
MgO	2.54	1.50
CaO	5.55	6.89
Na_2O	0.35	0.90
K_2O	2.85	3.33
P_2O_5	0.14	0.64

2 REFLOTATION OF COAL SLURRIES

The reflotation of coal slurries is often very complicated and can be affected by the whole range of factors. One of the main problems is caused by changes of properties of oxidized coal e.g. by presence of humic acids. These problems were studied by many authors. Arnold et al. (1986) and Fijal et al. (1989) have found that humic acids during flotation of coal act as a suppressor because they increase number of hydrophile positions (COOH-, OH-). Increase of hydrophobic effect on the surface of coal particles during flotation can be achieved according to Fijal et al. (1989) by adsorption of aliphatic alcohols which cause decresing of zeta potential. Hydrophobic effect is associated with decreasing of surface of OH groups and it is increasing with length of aliphatic alcohol chain.

During reflotation of coal slurries we have studied influence of slurry density and collector dose that would be optimal for preparation of marketable coal concentrates which can meet requirement of ash content under 11 %.

Flotation tests were performed with flotator VRF-2 manufactured by RD Pribram at following conditions: cell volume 1 dm³, agitation of slurry by collector - 2 minutes, time of flotation 5-7 minutes. As a collector was used FLOTAKOL NX which

Table 2. Laboratory tests of reflotation of sample from slurry pond Mine Dukla

No.of sample	Product	Recovery /%/	Ash content /%/
1	C	39.30	13.27
	T	60.70	35.89
	I	100.00	27.00
2	C	45.80	8.35
	T	54.20	40.77
	I	100.00	25.92
3	C	67.50	12.94
	T	32.50	52.05
	I	100.00	25.65
4	C	49.40	13.45
	T	50.60	37.52
	I	100.00	25.63
5	C	72.81	10.06
	T	27.19	83.44
	I	100.00	30.01
6	C	57.80	21.84
	T	42.20	56.13
	I	100.00	36.31
7	C	66.90	17.79
	T	33.10	64.37
	I	100.00	33.20
8	C	61.10	10.42
	T	38.90	48.24
	I	100.00	25.17
9	C	72.50	21.07
	T	27.50	57.60
	I	100.00	31.30

C - concentrate, T - tailing, I - input

contains also foaming agents as admixtures and therefore foaming agent is not added during flotation. Density of slurry varied in range 50 to 200 g.l-1 with constant collector dose of 1000 g.l-1.

For sample 5 from slurry pond of Mine Dulka best rresults were obtained for density 200 g.l-1, when mass recovery of concentrate was 71.74 % with ash content of 10.58 %. From slurry pond of Goking Plant Odra for flotation tests was selected sample 2. Also in this case, best results were obtained for slurry density 200 g.l-1, when mass recovery was 87.60 % with ash content of 12.89 %.

Influence of collector dosing was tested for both samples with slurry density 200

Table 3. Laboratory tests of reflotation of
sample from slurry pond of Coking
Plant Odra

No. of sample	Product	Recovery /%/	Ash content /%/
1	C	53.90	12.27
	T	43.10	61.80
	I	100.00	33.60
2	C	81.80	8.37
	T	18.20	69.18
	I	100.00	19.44
3	C	86.70	15.70
	T	13.30	91.65
	I	100.00	27.79
4	C	81.90	14.40
	T	18.10	89.56
	I	100.00	27.98
5	C	83.90	8.98
	T	16.10	76.72
	I	100.00	19.89

C - concentrate, T - tailing, I - input

$g.t^{-1}$ in wide range of dosing from 500 to
2000 $g.t^{-1}$. For sample from slurry pond
of Mine Dukla it is possible to regard as
an optimaldose of 750 $g.t^{-1}$ when was
obtained mass recovery of concentrate
72.81% with ash content 10.06%. For
sample from slurry pond of Coking Plant
Odra it is possible to regard as an optimal
dose of 500 $g.t^{-1}$ when was obtained mass
recovery of concentrate 81.80% with ash
content 8.37%.

Becauseresults of flotation tests have
met required parameters for marketable
coal concentrates, same conditions were
applied for samples from all sampling
points (Table 2 and 3). If we compare
table 2 and 3 it is clear that coal from
slurry pond of Coking Plant Odra is much
more suitable for flotation than coal from
slurry pond of Mine Dukla, which corresponds to petrographical analysis. By
this case was proved that quantitative
content of macerals - mostly vitrinite -
influences decisively flotability of
coal (Lynch et al. 1981).
On the basis of obtained results it is
possible to assume that during process of
basic flotation in first 2 minutes flotat-
tion of glossy components of coal takes

place (Klassen 1966) and later flotation
of matt components and dispersed clay
particles takes place. Because by increase
of flotation agents concentratations is
increased content of ash in concentrate
(Mishra 1978), there was further tested
technology of one basic and one refining
flotation for preparation of marketable
concentrates. By introduction of refining
flotation (Table 4 and 5) all samples met
parameters corresponding to quality of
marketable concentrates. This technology
causes pronounced minimazing of matt
components of coal and clay minerals.
Horsley (1951) and Lynch et al. (1981) have
found that matt components of coal are
considerably porous. During longer contact
with water surface pores are filled by
water, surface thus became hydrophilous and
matt components of coal do not flotate.
Clay minerals (kaolinite and illite) are
formed by particles of size under 6 micro-
meters, they are dispersed in water solution
and they are not concentrated in flotation
foam. Arnold et al. (1986) who described
behaviour of clays during flotation of coal,
has found that ability of coal for flotation is not changing at low pH, but in neutral environment layers on the surface of
coal particles which phenomenon causes sup-
pression of coal. As a most suitable they
consider alcaline environment where both
clay minerals and coal particles have negative charge and surface layers are not
formed.

In spite of the fact that results of flotation tests for both samples corresponds to
the requirements for quality of marketable
concentrates, it appears different flotation
ability which is not influenced only by
differences in mineralogical and petrographical composition. With samples from slurry
pond of Coking Plant Odra it was achieved
besides required quality of concentrates
also high recovery of concentrate. Samples
from slurry pond of Mine Dukla have also
required quality, but recovery values
varied widely and they were considerably
lower. it is possible to assume that
changes in concentrate recovery had samples
from the vicinity of mouth of transporting
pipes. These places are characterized by
faster sedimentation speed and thus they
contain higher proporation of coarse-
grained material and higher percentage of
ash. Samples with higher recovery are
mostly from central part of slurry pond,
from places with quiet sedimentation, where
are accumalting particles with lower sedimentation ability.

Table 4. Results of basic and one refining flotation, sample from slurry pond of Mine Dukla

No. of sample	Product	Recovery /%/	Ash content /%/
1	C	19.09	4.70
	BF	72.46	34.44
	RF	8.45	13.40
	I	100.00	26.98
2	C	53.68	6.13
	BF	34.33	56.28
	RF	11.99	16.91
	I	100.00	24.64
3	C	53.08	10.19
	BF	33.52	60.12
	RF	13.40	50.22
	I	100.00	32.29
4	C	61.57	7.97
	BF	26.62	54.10
	RF	11.81	54.09
	I	100.00	25.70
5	C	62.71	5.48
	BF	27.20	78.13
	RF	10.10	52.60
	I	100.00	30.00
6	C	36.21	10.78
	BF	42.85	46.42
	RF	20.94	53.51
	I	100.00	35.00
7	C	17.30	7.45
	BF	68.91	39.80
	RF	13.79	39.02
	I	100.00	34.10
8	C	56.33	6.65
	BF	31.67	50.62
	RF	13.00	34.00
	I	100.00	24.21
9	C	30.05	6.52
	BF	46.68	42.29
	RF	23.27	28.09
	I	100.00	28.24

C - concentrate, BF - basic flotation,
RF - refining flotation, I - Input

3 CONCLUSION

The result of flotation tests from all sampling points are showing that by introduction of one basic and one refining

Table 5. Results of basic and one refining flotation, sample from slurry pond of Coking Plant Odra

No. of sample	Product	Recovery /%/	Ash content /%/
1	C	51.40	10.07
	BF	29.55	73.48
	RF	19.05	35.22
	I	100.00	33.60
2	C	78.75	8.35
	BF	13.10	74.74
	RF	8.15	37.42
	I	100.00	19.42
3	C	79.96	10.70
	BF	14.03	75.98
	RF	6.01	57.03
	I	100.00	22.64
4	C	72.07	11.02
	BF	19.96	85.08
	RF	7.97	27.87
	I	100.00	26.84
5	C	70.15	8.50
	BF	15.15	67.96
	RF	14.70	21.37
	I	100.00	19.40

C - concentrate, BF - basic flotation,
RF - refining flotation, I - input

flotation, it is possible to reach quality required for marketable concentrates of bituminous coal. Refloation of material is advantageous because both dressing plants have flotation line. From high content of coal matter on both slurry ponds it follows that in present time both flotation lines work in great extent not selectively, what is caused by neglecting technological parameters and by outdated technological equipment. This technology would allow to solve at least partly also ecological problems of some from 96 slurry ponds in Ostrava-Karvina Coal District.

REFERENCES

Arnold, B.J. and Aplan, F.F. 1986. The effect of clay slimes on coal flotation. International Journal of Mineral Processing 20: 359-386.

Fijal, T. and Stachurski, J. 1989. Electrokinetic properties of non-oxidized and superficially oxidized coal. Archives of Mining Sciences 34: 356-368.

Fijal, J. and Stachurski, J. 1989.

Influence of aliphatic alcohols on the electrokinetics properties of the coal surface in flotation systems. Archives of Mining Sciences 34: 368-379.

Horsley, R.M. 1951. Principles of coal flotation. fuel: 54-63.

Klassen, V.I. 1963. Coal flotation (in Russian). Moscow: Gosgortechizdat.

Lynch, A.J. and Johnson, N.W. and Manlaping, E.V. 1981. Mineral and coal flotation circuits. Amsterdam: Elsevier.

Mishra, S.K. 1978. The slime problem in Australian coal flotation. Australian I.M.M. Mill Operators Conference, Mt. Isa: 159-168.

Moxon, N.T. and Bensley, C.N. 1987. Insoluble oils in coal flotation. The effects of surface spreading and pore penetration. International Journal of Mineral Processing 21: 261-274.

Sablik, J. 1982. The grade of metamorphism of polish coals and their natural and activated flotability. International Journal of Mineral Processing 9: 245-257.

Shirley, Ch.T. 1982: funamentals of coal benefication and utilization. Amsterdam: Elsevier.

Reclamation, Treatment and Utilization of Coal Mining Wastes, Rainbow (ed.) © 1990 Balkema, Rotterdam. ISBN 90 6191 154 0

Evaluating subterranean fire risks on reclaimed sites

T.Cairney & R.C.Clucas
Liverpool Polytechnic, UK

D. Hobson
W.A. Fairhurst & Partners, Newcastle-upon-Tyne, UK

ABSTRACT: An increasing number of sites which are being reclaimed for re-use have the potential for subterranean smouldering and combustion. Evaluating of the risks such sites poses is important, if the reclamations are to be successful. The traditional use of calorific values for the risk assessment is basically flawed and the authors prefer a non-isothermal combustion potential test, which also pinpoints the critical air flow necessary to support thermal runaway. Use of the combustion test information allows the necessary protection to be included in the basic compaction and preparation of the site in the early stages of the redevelopment work.

1. INTRODUCTION

The re-use in the U.K. of derelict land for modern housing and industrial developments is increasing, particularly since the establishment of urban Development Corporations.

Disused land usually has had a prior industrial history, which often has given rise to a range of contamination problems (Cairney, 1987). Amongst these is the commonplace occurence of zones of high calorific value materials in the near surface layers. Bands of coal and coke intermixed with soil and inert rubble, colliery spoil materials of variable type, and tips of tars, wood and paper wastes are all frequently encountered.

The fear that such materials could spontaneously combust, or be ignited by buried power cables, or the actions of vandals is in fact not unrealistic. The Fire Research Station (Beever, 1989) found that, in the three year period to June 1987, 64 cases of subterranean smouldering had necessitated the attendance of the local fire authorities, and that - of these 64 sites - six had been redeveloped and occupied before the underground heating was noticed.

The difficulties and costs of dealing with subterranean heating, once it has become established, have been described by a number of authors. Drake's 1987 account of the complexity of removing the fire hazard in shallow coal workings under Oakthorpe Village in Leicestershire is especially well-detailed, but could, because of the close relationship to mining activities, be regarded as more of coal mining significance than of importance to the redevelopment of disused land. More worrying was the fire below an industrial estate at Dronfield Village in Derbyshire (Smith, 1989) since there the sub-surface materials were thought to be inert steel slags and foundry sands. Only during the remedial work did it become obvious that bands of coke existed within the predominantly slag foundation materials and that this coke was the cause of underground temperatures that reached 1000°C.

Subterranean heating and smouldering therefore is a matter of practical concern where redevelopments are planned. The effect is not likely to cause any loss of human life, since it tends to be relatively slow and invariably give obvious warnings of its presence (Redpath, 1989), but it almost inevitably will lead to significant material damage. Thus subsidence, structural failures, and the high costs of remedial actions are the likely consequences that have to be avoided.

The need therefore is for a simple and cheap test that allows the assessment of the risk of underground heating and smouldering.

2. CALORIFIC VALUE TESTS

Various workers and control bodies

(e.g. the former Greater London Council) have suggested that a material's combustion susceptibility should be judged on its calorific value. Critical levels such as 7000 kJ/kg have been suggested as indicators of potential combustibility.

This reliance on calorific values, however, seems more to be the acceptance of an existing test method (albeit one that was designed for a quite different purpose) and it is difficult to believe that a measure of carbon content (which basically the calorific value is) is necessarily related to combustibility potential. Even with simple coals the discrepancy is obvious, lower ranking coals tend to have low calorific values but quite high volatile contents and are known to be reactive and easily ignited. Higher ranking coals have lost much of their volatile content and so have higher carbon percentages (and thus higher calorific values) and yet are relatively unreactive and quite difficult to ignite.

Thus despite the ease with which bomb calorimeter tests can be conducted, and the availability of the even simpler proximate analysis (British Standard 1016 part 3, 1973) and Ball's proximate method (1964), which is more suitable for combustible matter set in an inert soil admixture, it is difficult to accept that the measured values (Table 1) in fact indicate any susceptibility to combustion. All the materials in Table 1 in fact failed to burn effectively, even when exposed to a gas burner in free air for 30 minutes duration, yet all have quite high calorific values.

Table 1 - Calorific values by various methods - Hylton Colliery site, Sunderland, U.K.

Sample type	Bomb calorimeter kJ/kg	Proximate analysis kJ/kg	Ball's method kJ/kg
Soil with coaly fragments	9,000	11,500	7,500
Soil with abundant coke fragments	9,000	12,750	8,250
Coaly matter with soil admixture	16,000	14,500	9,000

3. Combustion potential test

An effective test of combustion potential has to mirror the conditions that actually will occur on a site. Whilst scientific opinion is not entirely uniform on the causitive factors for subterranean smouldering, there is agreement that the two critical factors are:

a gradual build up of heat which is retained and not dissipated and,

an adequate supply of oxygen to allow combustion.

The test procedure adopted follows these two factors (fig.1) by:

a) having the test sample in a central tube (32 mm. in diameter) with a governable air flow being passed through it.

b) surrounding the sample to be with a cylinder of an inert reference material (uniform single size sand).

c) heating the two concentric cylinders with a wire wound electric furnace gradually up to temperatures of 600°C.

The apparatus and procedure was adapted from earlier work by Sebastian and Mayers (1973) on coke reactivity.

The resultant temperature-time graph can be examined to establish if ignition and combustion has taken place.

Should the test sample be entirely inert, the results gained from the thermocouple probes, in the sample and in the reference material, will remain parallel (fig.2). The distance between the two curves is simply a measure of the time needed for the heat to pass through the reference sample and affect the test material. Should the test sample be of a combustible material (fig.3) the two curves remain parallel only for as long as it takes for ignition to commence, and then the test sample curve's gradient steepens and usually crosses the reference sample's curve. In detail, and by the use of more sophisticated monitoring equipment, it is possible to record temperate differences between the two materials, and so to identify the onset of oxidation and self heating in the test sample.

The test procedure generally takes less than 2 hours per sample and the furnace itself is easily constructed at low cost.

The value of the test is that combustion susceptibility is directly measured. The lower the tempeature at which the test

238

sample's temperature gradient increases,
the more susceptible the test material
is.

Table 2 - Critical air flow to cause
runaway combustion - Hylton Colliery

Air flow (1/m)	Ignition (°C) (smouldering)	Runaway (°C) (combustion)
0	165	-
1.5	160	-
2.5	145	-
5.0	120	-
10.0	120	-
15.0	120	200
20.0	120	200

A second and practically important
advantage is that the critical air flow
(Table 2), which allows runaway combustion
to commence, can be found by simply testing
samples at different air flow rates. Geotechnical research into soil compaction
has revealed that, at the optimum soil
compaction, (which would be the normal
aim on a reclaimed site and so would be
required of the reclamation contractor)
the air permeability of a soil material
is extremely low. Thus an associated
programme of compaction testing (to British
Standard specificatin BS 1377-1975) reveals
the soil material's dry density value
at its optimum compaction level. This
density value can then be reproduced in
the combustion test material, and the
achievable air flow through the sample
at that density can be measured. In practice, no case has yet been found in which
it has been possible to pass the critical
air flow to allow combustion through a
fully compacted material.

The final advantage of the test method
is that it, in fact, tests materials largely as they exist on the site, and not
after they have been artificially modified
by grinding and sieving, or other ameliora-
tion.

4. Fire Research Station test method

A somewhat similar test method has
been evolved by the Fire Research Station
(Baker 1989).

In this the sample material is ground
to its most reactive particle size (less
than 2 mm. in diameter), placed in a wire
mesh basket and hung in a standard electrical furnace. The furnace is then set
at a known temperature and a thermocouple,
set in the centre of the sample, is monitored. The test is continued until the
sample temperature has reached its peak
level and then fallen back to the oven's

ambient temperature. By testing samples
of the same material at different temperatures, the critical temperature at which
the sample gives out its greatest heat
is determined and so its sensitivity to
combustion is assessed.

Whilst the F.R.S. method offers useful
data on the sensitivity of a material
to combustion, the method fails to mirror
site reality in that:

a) the test material is ground to a much
finer size than is the case in reality

b) the test material is loosely packed
and no assessment of the benefits that
compaction gives are possible and,

c) air is constantly circulated around
the sample, a situation that does not
conform with the actual site conditions.

Thus the F.R.S. method gives a theoretical evaluation of combustion potential,
which may not be relevant to site conditions.

5. Use of the combustion potential test

So far, the test has been utilised
at:

a) a former coal stock yard still underlain
by 1.2 m.thickness of waste coals.
This site has been developed for housing,
since the tests showed that site compaction would prevent the critical
air flow needed to allow combustion
to enter the waste coal. Thus no removal of the waste coal proved necessary.

b) a former coke works stock yard at Gates-
head. Testing revealed the site was
underlain by 2 m. to 4 m. of coals,
ash and shale and everywhere had calorific values in excess of 10,000 kJ/kg.
Combustion testing revealed that even
with quite low air flows, the site
material would ignite at 200°C to 260°C.
The owners then determined to remove
the larger coal fragments by double
washing and to spread the shale/ash
discard over the site. Re-testing
revealed that this discard would not
burn if compacted to normal site
requirements.

c) the tipped area around the former Hylton
Colliery at Sunderland,which is currently being redeveloped for industrial
uses. Despite the fact that subterranean fires have been found on this
site, the testing initially revealed
that no sample could be made to burn,

239

despite heating to 600°C in the presence
of air flows of 2.5 litres/minute. Re-
testing did reveal that some samples could
ignite at relatively low temperatures
and indicated that the site was less uniform than had been believed. A process
of coal washing and site compaction is
now in progress and should provide conditions that will prevent any subterranean
heating.

6. Summary

The risks inherent if subterranean
heating can commence are particularly
serious to the fabric and infrastructure
of a development.
Calorific value testing of materials
from such development sites has not proved
adequate to pinpoint the sensitivity for
combustion.
The combustion potential test described
does allow the identification of the more
sensitive materials and helps in the decision of whether or not these materials
should be removed. A particularly useful
aspect of the test is that the critical
air flow to allow combustion can be identified and site compaction can usually
be so carried out to prevent this critical
value occuring.
Subterranean heating is just one of
the risks that can occur in the redevelopment of a disused site and - like the
other risks that exist - has to be understood well enough so that the developers
and their advisors can judge the level
of work and costs necessary to give a
guaranteed security. What the combustion
potential test cannot do is to give safe-
guards against bad workmanship in the
site compaction or against any later deterioration of the site that permits large
scale air inflows to occur.

References

1. CAIRNEY T. 1987 "Reclaiming contaminated land". Blackie & Sons, Edinburgh.

2. BEEVER P. 1989 "Subterranean fires
   in the U.K. - the problem". B.R.E.
   Information paper IP3/89.

3. DRAKE D. 1987 "Subsurface heating
   at Oakthorpe Village". Inst. Mining
   Eng. S. Staffs Branch.

4. SMITH A.J. 1989 "Investigation and
   treatment of a major subterranean
   heating beneath an industrial estate
   at Dronfield, Derbyshire". B.R.E.
   Research Colloquium, 22 March 1989.

5. REDPATH P.G. 1989. "Containment, spread
   and effect of an industrial site fire".
   B.R.E. Research Colloquium, 22 March
   1989.

6. British Standard BS 1016 pt.3, 1973
   "Method for the analysis and testing
   of coal and coke, part 3. Proximate
   analysis of coal".

7. BALL D.F. 1964 "Loss-on-ignition as
   an estimate of organic matter and organic
   carbon in non-calcareous soils". Journal
   of soil science, Vol 16 No 1.

8. British Standard B.S. 1377, 1975 "Methods
   of test for soils for civil engineering
   purposes".

9. SEBASTIAN J.J.S. and MAYERS M.A. 1937
   "Coke Reactivity - determination by
   a modified ignition point method".
   J. Industrial and Engineering Chemistry,
   October 1937, Vol 29, No 10.

10. BAKER B. "Subterranean fires". B.R.E.
    Research Colloquium, 22 March 1989

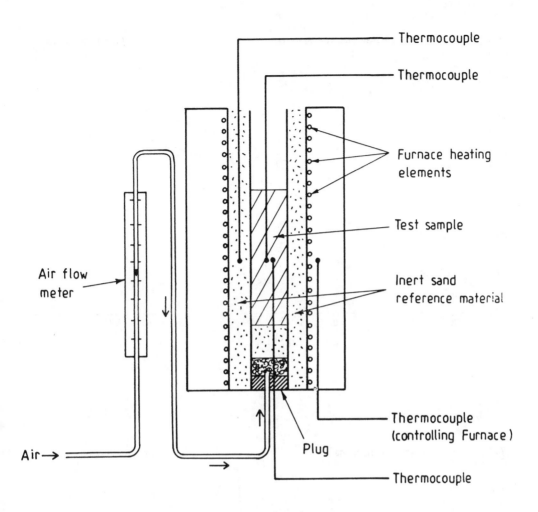

Thermocouple

Thermocouple

Furnace heating
elements

Test sample

Inert sand
reference material

Air flow
meter

Air →

Plug

Thermocouple
(controlling Furnace)

Thermocouple

Figure 1

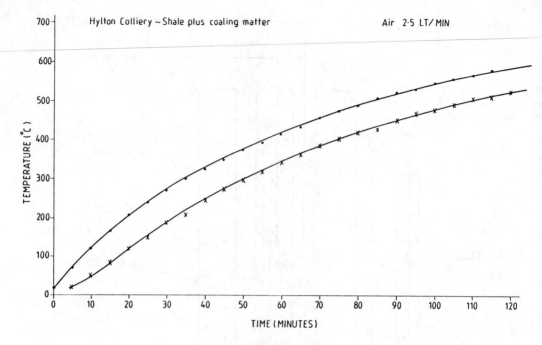

Figure 2

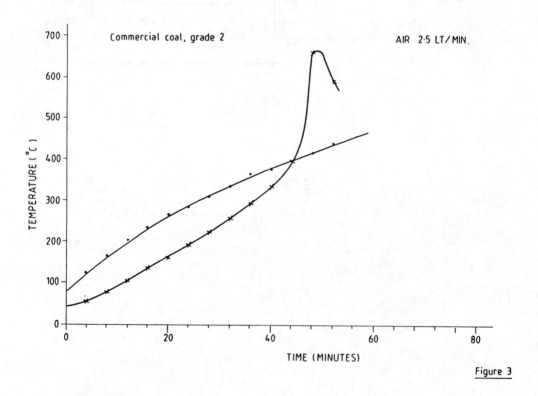

Figure 3

242

Reclamation, Treatment and Utilization of Coal Mining Wastes, Rainbow (ed.) © 1990 Balkema, Rotterdam. ISBN 90 6191 154 0

Tests concerning reclamation, treatment and utilization of coal-mining wastes in waterway engineering

K. D. Hauss & B. K. Mayer
Federal Waterway Engineering and Research Institute (BAW), Karlsruhe, FR Germany

ABSTRACT: Minestone causes many problems in coal-mining. Restoring it in the mine is very expensive, storing it at dumps is not only expensive, too, but also dangerous for the environment. So it was tried to reclamate minestones in embankments etc. The aim of the experiments described in the following paper was to find out particular aspects and measures in utilization of minestone in waterway engineering. The paper reports mineralogical properties established in laboratory experiments as well as the results of field studies.

1 INTRODUCTION

Minestone is a waste material of mining. In dependence on the kind of rock and on the mining technique minestone may be up to 40% of the total workings. In the Federal Republic of Germany those minestones mainly originate from the two coal-mining-areas, "Ruhr" and "Saar". They are named after local rivers.

The main part of minestone is not re-placed in the mine but stored at dumps. So it is a necessity - economically and ecologically - to reuse and recycle the material in mounds, dams, fills and similar buildings. But there are problems occur-ing in the interaction of minestone and water, as pointed out in (1).

To solve these problems concerning the application of minestone in waterway engineering extensive investigations in laboratory and field have been carried out. The examined samples originate from the coal-mining areas "Ruhr" and "Saar" and for this reason are named sample "R" and "S".

2 PETROLOGICAL PROPERTIES

2.1 Mineralogical and chemical composition

At the Mineralogical Institute of the University of Karlsruhe the fresh samples were dried and ground. The quantitative determination of the mineral content was effected by X-ray diffractometry using gauged diagrams with known compositions. The results are shown in Figure 1.

in weightpercent	samples "R"	samples "S"
montmorillonite	10-20 %	20-25 %
illite	12-30 %	20-30 %
kaolinite	15-30 %	25-30 %
gypsum	5-7 %	–
quartz	12-30 %	14-18 %
feldspar	4-10 %	2-6 %
dolomite	3-8 %	10 %
chlorite	5-10 %	–
corrensite	10 %	–
calcite	3 %	–

Fig. 1 Mineralogical content

A portion of clay minerals (montmorillonite illite and caolinite) is - according to the examinations - very high in comparison to other rocks.

The chemical composition determined from two fresh samples, too, is shown in figure 2. Very small concentrations (e.g. Cl^-: 0,1 % or P_2O_5: 0,07 %) have been neglected.

Concerning buildings and ecological aspects mainly those ions are interesting which form sulfates and chlorides. The largest concentration of sulfur was found in fresh material of the "R" – samples as pyrite (FeS_2); while oxidating three times as much sulfate results, that means an average amount of total sulfur of 2,15.

Chlorides are washed out generally

glowing losses	sample "R"	sample "S"
CO_2 H_2O	23 %	13-19 %
K_2O	5,3 %	1,0 %
Fe_2O_3	3,7 %	1,0 %
SiO_2	30 %	54 %
$CaCO_3$	0,4 %	0,16-0,96 %
MgO	0,7 %	1,05 %
SO_3	0,4 %	0,1-0,48 %
S	0,6 %	0,02-0,13 %
CI	< 0,1 %	-
SO_4	0,35 %	0,25 %
total SO_4	1,7-2,15	1,2-1,8
HCI (unsoluble)	69,4-76 %	-

Fig. 2 Glowing losses

during the separation of coal and waste stone and can be traced at the most in the seepage water.

Relatively high glowing losses of sample "R" (23 %) indicate a high amount of fine coal (10 %). With water-contact this part will be solved and will pollute water surface with a cloudy film. Especially at waters with slow run-off this would cause deposition and pollution of the shore-line area.

2.2 Grain size distribution

The grain size distribution is one of the characteristics of filling materials. Many papers (2), (3), (4) (5) give grain size distributions which differ in the amounts of the largest (o 2,0 mm) grain. This has to be traced back partly to different upgrade and coal washing and partly to different ages of the examined samples. Also mechanical stress and grain destruction has to be avoided in deter-mining grain size distribution curves.

Comparisons of grain size distributions will for this reason only be possible (as known from the above-mentioned publications) if the sampling date is known and if there is no mechanical stress afterwards.

Fig. 3 and 4 show curves of character-istical grain size distributions of one year old samples "R" and "S". Sample "S" of fig. 4 is coarser grained (= 15 % coarse gravel, 40 % rocks) than sample "S" of fig. 3, which is an average of several determinations.

2.3 Weathering

The decay of minestone is - according to

older publications - mainly an effect of weathering. It had been assumed that the material decays into clay because of water uptake and aeration.

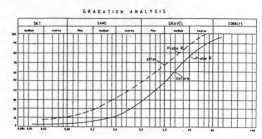

Fig. 3 Grain size distribution before and after weathering, Sample "R"

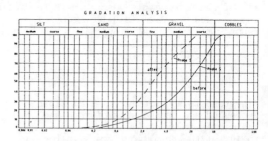

Fig. 4 Grain size distribution before and after weathering, sample "S"

Schmidbauer (2) found an original dump decayed into a loamy material. Annen and Stalmann (3) showed that simultaneously with decay swelling proceeds and that mechanical stress entailed by compaction is an important factor. For this reason Schwab (4) dried and wetted minestone material several times and pointed out that the sand fraction increases generally up to three times with respect to the initial value.

For our own experiments shares of the samples had been exposed for 2 or 3 years in containers to the influence of atmos-phere (temperature variation, frost and sunshine; Fig. 5).

For sample "R" a decay of the coarse grain fraction into more finer grained components was established. The portion of the silt fraction ($\emptyset$ 0.06 mm; Fig. 3) did not increase.

Concerning sample "S" also a decay of coarse grained material (coarse gravel and rocks) into fine grained (fine - medium gravel and coarse sand) was obvious. The portion of the cohesive fraction ($\emptyset$ 0,06 mm) did not increase even under extreme weather conditions.

The above-mentioned experiments only give

Fig. 5 Sample in a container before weathering

qualitative hints on weathering decay because the conditions are not quite the same as in reality.

2.4 Settlement and sinking of minestone

In dam building practically all earth materials settle when first flooded. This settlement is called saturation-settlement or sinking. The load is - in most cases - only the dead weight of the filling material. The extent of settlement is according to (6), (7) and (8) dependent on the grain form, grain roughness, initial water content, the stress state of soil and mainly on the initial compaction. In general they occur during first flooding or wetting and reach with repeated flooding around 1/10 of the initial sinking. The settlement behaviour was examined using a device developed only for this reason by the second author (Fig. 6).

It is - as usual - a cylindrical receptacle with water saturation beginning at the bottom. Below the lid there is an air cushion which works as a flexible element between the rigid chargeable lid and the soil sample. This construction effects that in case of unsymmetrical settlement the surface will be deformed while the vertical stress remains constant.

The test results prove saturation-settlement at flooding rates of 5 - 10 % as shown in figure 7. During practical fit the saturation-settlements could be superimposed

by grain destruction and self-consolidation settlement. Further investigation is needed to judge the deformation behaviour of materials whose grain and grain-edge destruction as well as decay due to water uptake.

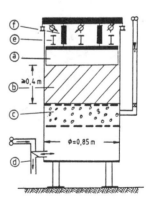

Fig. 6 Device for measurement of settlement behaviour

a air cushion	d water in and outlet
b sample	e measuring device
c filter	f hydraulic presses

In dam building the stress settlement behaviour plays a minor roll, so only two shares of sample "S" (grain size 1 - 12 mm) had been used for settlement tests, whose

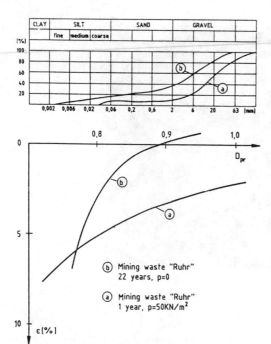

Fig. 7 Settlement due to saturation

(b) Mining waste "Ruhr" 22 years, p=0

(a) Mining waste "Ruhr" 1 year, p=50KN/m²

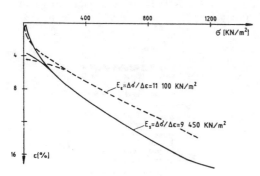

Fig. 8 Settlement test curves, sample "S"

$E_s=\Delta\delta/\Delta\epsilon=11\ 100\ KN/m^2$

$E_s=\Delta\delta/\Delta\epsilon=9\ 450\ KN/m^2$

results are shown in figure 8 as stress-settlement graphs.

2.5 Shear strength

The shear strength determination of sample "R" and "S" was carried out with the large triaxcell of the Federal Waterway Engineering and Research Institute (Fig. 10). The second author had projected the device some years ago. This device (sample diameter D = 30 cm and height H = 70 cm) allows to fill in coarse grained soil samples

	age [years]	dry density qd [t/m3]	internal friction angle θ'[°]	cohesion c'[kN/m2]
samples "R"	1	1,74	37,5	0
	1	1,86	35	65
	1	1,98	29	150
	6	1,60	32	25
	22	1,54	28,5	50
	22	1,74	36	35
	22	1,86	36	50
sample "S"	1	2,06	33,5	50

Fig. 9 Shear strength of coal-mining wastes

Fig. 10 Device for triaxial tests

up to ⌀ D/8 approximate 4 cm if the percentage of largest grains will not exceed 15 % of the total sample. Several samples "R" were used to carry out consolidated-drained shear tests (CD) with different fitting densities and material of different ages (Fig. 9). It compiles received values and results of a test with sample "S" as pointed out in (1).

2.6 Compactability

To examine the compactability thus to gauge fitting densities on the building site several proctor tests were practised while the grains bigger than ⌀31,5 mm had been screened out. Inspecting the results (Fig. 11) the water content of sample "S" (w_{Pr} = 4,75 %) is remarkably low.

3 RESULTS OF FIELD STUDIES

Technical literature states that in minestone with advancing years the clay-stone portion decays increasingly into a clay-stone mixture and finally into

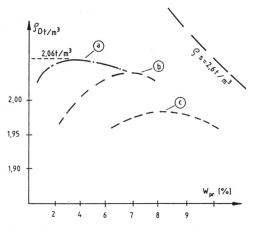

Fig. 11 Proctor compaction curves
a sample "S"
b sample "R"

plastic clay. For this reason existing
dams were studied and samples were taken.
The assumption in technical literature
could not be proved for these dams which
are partly 22 years old. Indeed we
determined a decay of the largest claystones into sand and gravel grain size
but there was established beyond no decay
into plastic clay. This proved laboratory
experiments which had yielded decay only
of the coarse fraction into sand or gravel
grains (see Chapter 2.3).

4 SUMMARY

The tests we carried out have had yielded
that minestone contains partly a large
portion of minerals with ability to swell
which especially due to contact with water
and occurring weathering influences cause
a limited decay into sand - gravel grain
sizes. The sagging and stress-settlement
behaviour in grain rupture susceptible
rocks merely can be estimated on the whole
since swell, decay and grain destruction
influences superimpose each other. Shear
strength experiments as CD-tests did not
yield unequivocal connections between
fitting density and friction angle. Only
for the cohesion of the examined samples a
certain dependence was established. The
compactability was tested with proctor
tests. It cannot be avoided to destroy
grains (and for this reason changing the
grain size distribution, the shear
strength, the sagging behaviour and the
permeability) while consolidating mechanically. Field studies proved the results
of weathering tests. A total decay into
plastic clay could not be found. The

ecological and toxicological effects caused
by pyrite and chloride contents were
examined too. The resulting particular
building measures are treated in other
papers and have to be fixed for each
singular case.

REFERENCES

(1) K-D Hauss & M.H. Heibaum. 1990.
    Minestone in German Waterway Eng-
    ineering. Proc. of 3rd Int. Symp.
    of Coal Mining Wastes, Glasgow.
(2) J. Schmidbauer. 1952. Die Standsich-
    erheit von Bergehalden Bautechnik-
    Archiv, Heft 8.
(3) Annen & Stalmann. 1969. Waschberge
    im Deich- und Dammbau, Gluckauf, 105/
    1969 Nr. 26.
(4) Versuchsanstalt fur Bodenmech. und
    Grundbau, TH Darmstadt. Bericht Nr.
    38/83 uber Untersuchungen von
    Waschbergematerial. R. Schwab.
(5) Bundesanstalt fur Wasserbau, Karls-
    ruhe. Gutachten Nr. 22.4462 uber die
    Verwendbarkeit von Waschbergen im
    Verkehrswasserbau.
(6) Hellweg, V. 1981. Ein Vorschlag zur
    Abschatzung des Setzungs- und
    Sackungsverhaltens nichtbindiger
    Boden bei Durchnassung. Mitt.Inst.
    f.Grundbau, Bodenmech. u. Energie-
    wasserbau, TH Hannover, Heft 17.
(7) Schade, H. 1984. Die Sattigungsset-
    zungen von Steinschuttungen. Diss.
    TH Darmstadt.
(8) Kast, K. & Brauns, J. 1981. Verdich-
    tungs-, Drucksetzungs- und Sattig-
    ungsverhalten von Granitschuttungen.
    Zeitschr. Geotechnik 1981/3.

Reclamation, Treatment and Utilization of Coal Mining Wastes, Rainbow (ed.) © 1990 Balkema, Rotterdam. ISBN 90 6191 154 0

Minestone in German waterway engineering

K.–D. Hauss & M. H. Heibaum
Federal Waterway Engineering and Research Institute (BAW), Karlsruhe, FR Germany

ABSTRACT: In waterway engineering minestone is used frequently as general or additional fill of mounds, earth dams, dikes or embankments. Material properties of minestone from German coal mines are reviewed. Attention is focussed on the interaction of minestone and water: (1) interaction of minestone and canal-, ground- or seepage water and (2) settlement of minestone due to saturation. When saturated the first time (and occasionally further times) earth and rock fill may show a certain amount of settlement caused only by self weight. This phenomenon has to be kept in mind, when designing fills or embankments. Comparison is made of minestone and other fill materials and recommendations are given concerning the application of minestone in waterway engineering.

1 INTRODUCTION

In Germany, coal fields are named after the rivers 'Saar' and 'Ruhr'. The first river denotes the smaller and less important coal mining area. The coal of those mining fields is of lower quality than the coal of the 'Ruhrgebiet', the large mining area extending far beyond the catchment area of the river Ruhr. Coal mining wastes of both areas contain sandstone, sandy schist, shale, claystone or siltstone. Most of the mining wastes consist of shale, a rock material that has to be recognized as a rock of altering strength. Its colour is mostly grey but turns to black in the vicinity of coal seams. It's a moderate hard rock and may swell when wetted.

Mining wastes have always been used as construction material in road- and dike-construction. In 1971 guidelines for road construction using coal mining wastes were published by the governmental authorities (BAST-E9). As early as 1952 Schmidbauer discussed the stability of mining waste embankment slopes and other papers (Carp, 1952 and Annen/Stalmann, 1968) dealt with the construction of dikes.

Using coal mining wastes for the construction of canal dams or fill behind sheet piles walls there are special requirements to be fulfilled. A lot of different boundary conditions have to be considered. On the one hand the long-term stability of the material itself (shear strength, weathering) has to be proved, then the influence to and from the groundwater and the seepage water is of great importance and, last but not least, subsidence and horizontal strain (compression and - being much more detrimental - extension), caused by mining activities, create severe load situations for an embankment. Thus only well known fill materials may be used.

Mining activities cause settlements on the ground surface, but the water table of a canal remains unchanged because of the locks some place not being influenced by mining subsidence (Fig.1). So the dam in a subsidence area may have to be heightened a certain number of times. Since often different materials were used each time the dam had to be heightened (mine wastes containing more or less sandstones, more or less shale), today we find a rather inhomogeneous fill of the canal dams with probably a soft 'core' under better compacted top layers (Fig.2). Improvements in compaction machinery led to increasing compaction with increasing heightening of the dams. Some problems concerning the compaction of coal mining wastes are discussed later.

Heightening embankment dams led to the favourable effect of widening the canal when a trapezoidal cross section was built, since today we need a width of 53 m compared to 33 m in former times.

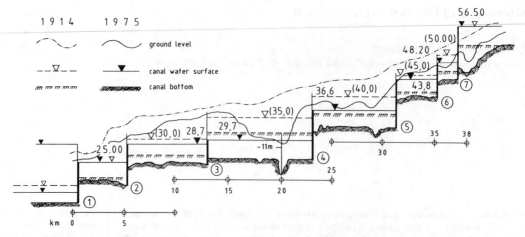

Fig.1: Mining subsidence of Rhein-Herne-Canal (FRG) since 1914 (after Kuhn, 1985)

2 GEOTECHNICAL EXPLORATION

A fundamental study has been carried out
on dam fills of German canal dams (Dort-
mund-Ems-canal and Datteln-Hamm-canal)
containing coal mining wastes. The fills
of these dams were placed 1 up to 22 years
ago.

What was expected to be found was a
considerable amount of shale weathered to
smaller particles, even to plastic clay,
thus reducing the permeability and the
shear strength and increasing the compressibility of the fill - corresponding
observations have been published by
Schöne-Warnefeld (1973). Even if the
weathering should be found to be restricted to the layers near to the surface, it
would be of unpredictable influence on the
direction of seepage flows, since it has
to be taken into consideration that the
earth dam was heightened several times and
there might exist the same number of zones
with material that has undergone weathering (Fig.2).

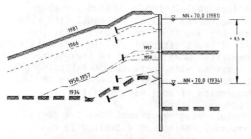

Fig.2: Several times heightened backfill
of a canal dam (8.5 m of mining subsidence
from 1934 to 1981)

As to the studies of the Federal Waterway Engineering and Research Institute
(BAW), the weathering expected was not
found in the relevant explorations. Bigger
pieces of rock were fragmented to pieces
of the gravel fraction, in some cases even
of the sand fraction, but no considerable
amount of real fine grained soil was to be
found (fine grained in the geotechnical
sense, i.e. grain diameter less than
0.06 mm). When loaded mechanically (e.g.
truck wheels), the bigger grains were
covered by a thin soft clay coating.

3 GRAIN SIZE DISTRIBUTION

The original mine waste material used as
earth dam fill shows silt (approx. 10%),
sand (approx. 30%) and gravel and rock
fragments (approx. 60%). The finest material found in a recent test pit contained
3% clay, 12% silt, 27% sand and 58% gravel
and rock fragments.

The weathering effects may be amplified
by truck wheels moving frequently on the
surface of the mine waste fill during
construction (Fig.3). But even in these
limited areas the clay fraction did not
exceed 7% by weight. Fig.4 shows the grain
size distribution of an original fill, a
weathered fill and a fill which was rolled
on frequently with truck wheels during
construction.

4 WEATHERING TESTS

To get more information on the weathering
of coal mining wastes, material from test
pits has been exposed to atmospherical

Fig.3: 'Impermeable' surface due to truck wheel compaction

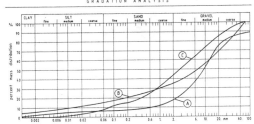

Fig.4: Grain size distribution of mine waste; (A) 1 year old, (B) 20 years old, (C) new fill fractured by truck wheels

influences over about one year. The material has been stored in several small open containers (2 dm³) and was exposed to natural rain and snow, sunlight and temperature changes. The shale was found to be fractured to smaller particles, but always remaining a granular material. Sieving showed now substantial increase of the percentage of fine material (grain diameter less than 0.06 mm).

This testing is considered to give good information on the weathering behaviour, since it represents extreme boundary conditions. In situ, only a top layer of a mining waste fill will be exposed to atmospherical loads and in addition, a fill is usually covered by surface soil and plants, thus being protected against these influences.

5 MINERALOGICAL COMPOSITION

A mineralogical analysis gave quantitatively the following mineral contents of coal mining wastes of the 'Ruhrgebiet':

Clay minerals make up the major part, ranging from 60-75 % by volume. A relatively large amount of it (30-50 %) shows high swelling capacity, which far exceeds the amount of the swelling capacity of

usual fill materials. Therefore, special attention has to be paid to measures avoiding heave of the fill.

15 to 30% of the minerals is quartz and there are some carbonate, pyrit and coal contents, each below 10%.

6 SHEAR STRENGTH

To determine the shear strength of the coal mining wastes triaxial tests were carried out in a big triaxial cell (diameter 30 cm, height 70 cm). Shear strength parameters were determined considering three different void ratios on original and weathered material. Fig.5 shows the results in terms of internal friction angle Φ' and (geotechnical) cohesion c'.

Surprisingly the friction angle decreases with decreasing void ratio, but the shear strength increases as expected, since the cohesion of the material is increasing strongly. It is striking that the cohesion increases linearly with decreasing void ratio as far as it can be said from only three different void ratios.

Since stability is very much dependent on the cohesion, the wide scatter of test results shows the importance of proper compaction and proper tests. Old scarcely compacted mining waste fills showed a specific weight of 14.5 to 15.5 kN/m³, while new material, well compacted, reaches 19 to 20 kN/m³. The unit weight of soil particles remains nearly constant and ranges in between 23.2 and 24.6 kN/m³.

One has to distinguish between the shear strength of new waste material and weathered material, even if the values are similar. The cohesion of new material is due to the strong interlocking of the grains because of the compaction of the material and shows the effects mentioned above. Weathering decreases the interlocking effect but increases 'real' cohesion in the geotechnical sense. Then the geo-

251

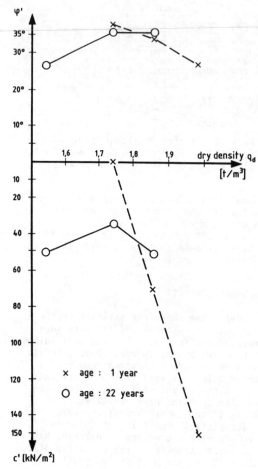

Fig.5: Shear strength of coal mining wastes versus dry density

technical cohesion is more or less constant for a certain material and independent of the degree of compaction (Fig.5).

7 PERMEABILITY

Since the grain size distribution of coal mining wastes varies in a wide range, the permeability of this material is difficult to estimate. Usually, there are coefficients of permeability ranging from $1*10^{-7}$ to $1*10^{-1}$ m/s. But even these extreme values may be exceeded. The permeability is influenced not only by the grain size distribution, but also by factors like compactness (void ratio) or the development of clay coatings, covering the larger granular grains, or of thin layers with high percentage of fine grains on the surface of each layer after compaction or

because of truck traffic.

As to compaction, the age of the material is of great importance, since the older mining wastes (time of deposit, not geological age) behave in a much more brittle manner than fresh material. The brittle fragments are easier to compact, thus reducing the void ratio and lowering significantly the permeability. Following the experience of the BAW, in German canal banks a difference of $1*10^4$ m/s in permeability has to be taken into consideration looking at the highest and the lowest permeability of the fill.

8 SEEPAGE FLOW

The most difficult problem is to estimate the ratio of vertical and horizontal permeability. This is of extreme importance as to the stability calculations (and thus to the safety and the reliability of the whole earth dam). Especially the already mentioned layers of low permeability caused by compaction or construction traffic lead to nearly horizontal seepage surfaces. One example is given by a leaking dam, where the exit point of the seepage line on the downstream side is only 0.4 m below the free water level. The proof of its being only a local aquiclude was given by the fact that piezometric measurements below that layer show no water at all!

Another possibility is a stepwise seepage line, when the water is percolating on a local impermeable layer until a hole in this layer allows it to flow down to the following layer. Since the local boundary conditions are never known exactly, one shouldn't do any other than designing far on the safe side, even though some help may be got by choosing proper grain size distributions of the coal mining wastes (containing no fines, limiting the smallest particle size), thus increasing the permeability in all directions. Weathering after construction of the fill won't be that much of a problem to the inside permeability as has been pointed out earlier.

The usual potential method is not practicable to the problem of seepage flow in mining waste fills. An empirical approach in German waterway engineering is the assumption of the seepage surface at an inclination of 1:10 starting from canal water level, assuming the canal bank or bottom sealing leaking (dashed line in Fig.6). Experience proves this assumption to be sufficient, even if the already mentioned example shows an even less sloping seepage surface.

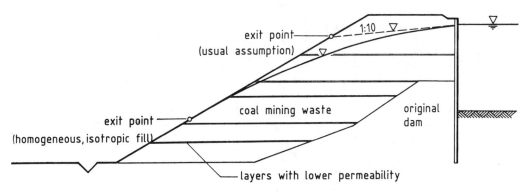

Fig.6: Comparison of the seepage surfaces of an isotropic fill and the usual design assumption for mining waste fill

To avoid pollution, the fill material has to be tested in advance. Additional layers under a fill of coal mining wastes to drain the seepage water to a drainage ditch is not of much use, since the fill may be designed for sufficient permeability (so it may drain sufficiently and a filter layer does not offer a better flow path), and there is no sense in carrying the problem water via the main drain just to the next river. Polluted water of old fills should be tapped and conducted to sewage treatment plants.

9 CHEMICAL ANALYSIS OF WASTES

When using coal mining wastes as fill in geotechnical constructions, discussion starts always on the influence or even danger that is revealed by the chemism of that material. It is well known that mining wastes may contain problem agents, which can be transported to the ground-water or the neighbouring areas by seepage water or rain. Ground- and canal water may be polluted this way and physical properties of the fill material could be changed. In literature, sulfatic and chloridic contents are considered the greatest problem. While chlorides lead to an increase of the content of salt of the water, sulfatic materials may cause corrosion or even damage to steel and concrete structures.

Material taken from test pits showed significant loss of sulfate in the material of the upper layers compared to materials of lower points of the fill.

The relatively small amount of sulfatic agents is due to the fact that mining wastes are firstly stored in heaps and are there exposed to atmospheric influences before being used as fill material. Thus a significant amount of sulfate is washed out already when being placed as fill material. Chlorides are to be found in the mining wastes only to a negligible extent. They are washed out, too, and are to be found only in the seepage water. Chlorides are washed out definitely faster than sulfates.

10 CHEMICAL ANALYSIS OF THE SEEPAGE WATER

Even in fills existing 10-20 or more years, the content of chlorides in the seepage water reached values up to about 100 mg/l. The content of sulfate varied between 180-425 mg/l. Especially if the seepage water remains a long time in the fill, concentration may grow up to problem values. The limit values of concentration vary from country to country. In the respect of steel and concrete structures, a content of sulfate of the water below 250 mg/l is considered tolerable.

11 SETTLEMENT DUE TO SATURATION

Most soils and rock fragments exhibit a certain behaviour, which is known as 'settlement due to saturation', 'collapse compaction', 'hydro consolidation' or 'hydro compaction'. This phenomenon mainly shows up, when relatively loose material is drowned the first time. The reasons differ depending on the kind of soil.

Fine grain soils show an aggregation of particles which collapses when the soil is saturated (Fig.7). Depending on the level of overburden stress at which it is wetted, this kind of soil exhibits a volume decrease (collapse) while sustaining essentially unchanged vertical stresses. Following Lawton et al. (1989) the maximum

253

amount of wetting-induced collapse occurs when the soil takes in water under an overburden stress equal to the compactive prestress. Kalachev et al. (1989) report this effect being underestimated because of the usual tests in the oedometer apparatus. Depending on the soil, real values exceed the test results by up to 30%.

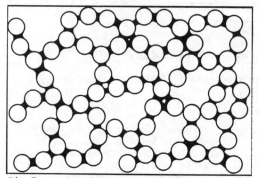

Fig.7: Collapsible soil structure

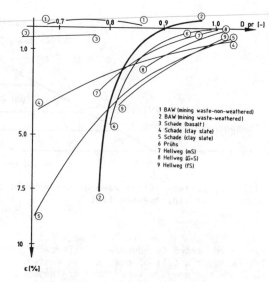

Fig.8: Settlement due to saturation of selected soils and mining wastes

1 BAW (mining waste-non-weathered)
2 BAW (mining waste-weathered)
3 Schade (basalt)
4 Schade (clay slate)
5 Schade (clay slate)
6 Prühs
7 Hellweg (mS)
8 Hellweg (G+S)
9 Hellweg (fS)

Coarse silt and fine sand develop an apparent cohesion which vanishes when the soil is saturated to about 80% and more.

Rock fragments suffer high edge stresses and the release of latent stress when drowned the first time.

Examples are given by: Kezdi (1973) for loessian soils, Lawton (1989) for clayey sand, Prühs (1982) for silty sand, Hellweg (1981) for fine sands, Schade (1984) for rock fragments.

Settlement caused by saturation may reach 35 % for loessian soils and up to 10% for sandy soils. Rock fragments (e.g. the fill of an abandoned open pit mine) show settlements up to 2 % (Charles et al., 1984). Fig.8 shows some examples of settlement due to saturation versus dry unit weight of the fill material.

As to coal mining wastes, this phenomenon is on one hand dependent on the grain size distribution, on the other hand on the mineral content. Tests had shown that the behaviour of the whole fill is governed by the behaviour of the fine grained fractions up to about 0.2 mm. Depending on the degree of compaction, mining waste fills suffer 5 - 10 % of settlement due to saturation (curve #2 in Fig.8)

Prevention against this kind of settlement may be taken by good compaction of the fill and a placement of the material preferably on the 'wet side' of proctor density. The tendency of the material towards settlement may be estimated by proctor tests. It has been shown that materials with significant curvature of

the proctor compaction curve also show significant settlement due to saturation (Hauss, 1990). Fig.9 shows the relation of the minimum radius of curvature of the settlement curve and the rising of the proctor compaction curve at an interval of $\Delta w = 3\%$. Until now it is only a qualitative method, but it gives a useful hint to initiate prevention measures when placing a fill.

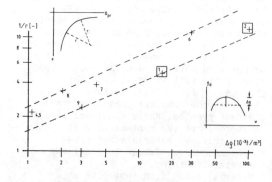

Fig.9: Radius of curvature of the settlement curves versus the rising of proctor compaction curves

254

12 CONCLUSIONS

Placement and compaction are of major importance concerning all relevant parameters of coal mining wastes used as fill material. There is direct influence on shear strength, compressibility, permeability and settlement due to saturation. The flow pattern of seepage water has to be taken into account either using rather coarse waste material containing few or no fines or performing stability calculations with a proper seepage surfce. German experience shows that coal mining wastes may be used successfully as earth dam fill, even in areas of mining subsidence, provided that there are no problems with the chemistry of the material or the seepage water. The water content should be on the 'wet side' of proctor density, thus preventing the possibility of settlement due to saturation.

Standard compaction rules ask for layers of 0.3 to 0.6 m, compacted by 4-6 compaction passes with a dynamic compactor. The mentioned range is due to the compactibility of the waste material. Layers of a thickness of 0.6 m are only successfully compacted using dynamic compactors with a service weight of 4-6 tons. With static compactors, layers of 0.4 m must not be exceeded.

This way minestone shows up as a fill material comparable to others, reliable and sufficiently consistent with environmental protection measures.

REFERENCES

Annen, G. & Stalmann, V. 1968. Die Eignung von Nebengesteinen des Kohlenbergbaus (Waschberge) zum Bau von Deichen und Dämmen. Technisch-wissenschaftliche Mitteilungen der Emschergenossenschaft und des Lippeverbandes, Heft 7.

BAST-E9 1971. Verwendbarkeit des Nebengesteins der Steinkohle als Schüttmaterial für den Straßenbau. Empfehlungen der Bundesanstalt für Straßenwesen.

Carp, H. 1952. Deichbau im Emschergebiet. In Vorträge der Baugrundtagung in Essen, Bautechnik Archiv, Heft 8

Charles, J.A.; Hughes, D.B.; Burford, D. 1984. The effect of a rise of water table on the settlement of backfill at Horsley restored opencast coal mining site. In Geddes, J.D. (ed.), Ground movements and structures, p.423-442. London, Pentech Press.

Hauss, K.-D., 1990. Settlement due to saturation. Mitteilungsheft der BAW. Karlsruhe: Bundesanstalt für Wasserbau. (to be published)

Hellweg, V. 1981. Ein Vorschlag zur Abschätzung des Setzungs- und Sackungsverhaltens nichtbindiger Böden bei Durchnässung. Mitt.Inst.f.Grundbau, Bodenmechanik und Energiewasserbau der Universität Hannover, Heft 17.

Kalachev, V.Y. et al. 1989. Study of cohesive soil deformability in a compression apparatus of different design. Bull. IAEG No.40, p.105-110.

Kezdi, A. 1973. Handbuch der Bodenmechanik, Bd.III. Verlin, VEB Verlag für Bauwesen.

Kuhn, R. 1985. Binnenverkehrswasserbau. Berlin, Ernst.

Lawton, E.D.; Fragaszy, R.J.; Hardcastle, J.H. 1989. Collapse of compacted clayey sand. Journal of Geotechnical Engineering, Vol.115, No.9, pp 1252-1265.

Prühs, H.; Stenzel, G.; Feile, W. 1982. Untersuchungen von Sackungen an Keupersanden. Vorträge der Baugrundtagung in Braunschweig. Essen, DGEG.

Schade, H. 1984. Die Sättigungssetzungen von Steinschüttungen. Dissertation Technische Hochschule Darmstadt (D 17).

Schmidbauer, J. 1952. Standsicherheit von Bergehalden. In Vorträge zur Baugrundtagung in Essen, Bautechnik Archiv, Heft 8

Schöne-Warnefeld, G. 1973. Geotechnische Probleme bei Lagerung und Verwendung von Gruben- und Waschbergen. Bund Deutscher Baumeister, Architekten und Ingenieure, Ortsgruppe Witten, Festschrift S.4-9.

Reclamation, Treatment and Utilization of Coal Mining Wastes, Rainbow (ed.) © 1990 Balkema, Rotterdam. ISBN 90 6191 154 0

An assessment of the marine application of minestone

P.A. Hart, Portsmouth Polytechnic, UK

ABSTRACT: Scenarios are given which highlight the potential for use of minestone in the marine environment, and also the problems which need to be solved before it can be widely accepted by the professions.

The paper concentrates on a review of existing knowledge regarding the utilisation of minestone in the marine and related environments, mineralogy and chemistry, physical and mechanical properties, and tests to assess these properties.

It is concluded that although use of minestone is becoming more widely accepted for use in this context, a sustained and concerted research and development effort is required for a definitive design procedure.

1 INTRODUCTION

Based on an analysis of global warming and its effects, a report by the United Nations Intergovernmental Panel on Climate Change concludes that redesign and reinforcement of many coasts and coastal structures will be necessary in order to withstand increased flooding, erosion, storm surges and wave attack (Brown, 1990).

Thus there would appear to be an enormous potential for use of minestone in the marine environment in context of helping to prevent coastal erosion and in the reclamation of land from the sea. Acceptance by the civil engineering professions that minestone is a suitable fill for marine applications will be based on results confirming its ability to perform as required. Graded minestone could be utilised as fill and possibly as core and filter layer material, for example in beach replenishment schemes, construction of sea walls, coastal harbours, breakwaters and offshore islands.

To its advantage minestone is readily available in large quantities, for example, for the Holderness Coast, which happens to be one of the areas most in need of protection. Minestone could also be provided for use in construction of the Severn

Barrage, and for numerous other smaller river mouth barrages that are planned. However, the Netherlands have shown that transportation of minestone over large distances by sea-going barges is cost-effective. This raises the question as to whether or not it would be the same if minestone were to be transported from source to more distant coastal areas in the UK. Certainly this should be one of the alternatives considered and investigated by the design engineer.

Higher levels of risk are associated with the design of a marine structure such as a breakwater or an offshore island, as compared with other land based structures, mainly due to the fact that 'there are few established rules, and designs are heavily dependent on the experience of the individual designer ...' (Maritime Engineering Group, ICE, 1985).

Lessons have been learned from past experiences in so much that whilst all coastal defences have provided some immediate conservation of land, not all have been completely inert nor have provided any real long term benefits. For instance, by protecting cliffs from erosion by means of stabilising the toe of the slope, the beach can be deprived of fresh material which in turn may result in its' own erosion. Komar (1976) has shown that direct interrelationships exist between wave energy, beach gradient and sediment size based on field data. Application of such relationships in an assessment of the suitability of particle sizes of minestone is important especially in context of sub-merged barriers and artificial islands. Despite the fact that erosion-sedimentation systems are constantly changing in any case, certain coastal areas have undergone remarkable and noticeable changes in beach and cliff profiles subsequent to completion of local coastal defences. An example of this is the evolution of the north Kent coastline, which has been studied by So (1966) and Sandall (1990). From the earliest record in 1872 to 1959, the cliff-line at Beltinge was eroded landwards by approximately 100 metres. Prior to 1953, sea wall defences were designed on the basis of a predicted maximum high tide of +2.8 metres compared with the Newlyn Datum. Over the period 31 January to 1 February 1953, a storm-fed maximum water level of +4.7 metres occurred with consequent breaching of sea walls and flooding over 25 square kilometres of the alluvial plane. Existing coastal defences include higher sea walls and cliff toe protection by use of randomly placed granite-gneiss armour blocks, apparently imported from Scandinavia. A study of plans has revealed that in places the water depth near shore is increasing due to shingle erosion. This has the consequent affects of reducing the distance between high and low water mark and threatens to weaken existing defences to the point of rendering them ineffective. Another way that erosion can occur is by dredging of the sea bed to supply the construction industry with aggregates. For example, the small fishing village at Hallsands in south Devon was repeatedly wrecked by storm waves, and in the end had to be abandoned in 1917. This tragedy happened as a result of dredging of a natural offshore shingle bank in Start Bay which undoubtedly unbalanced the erosion-sedimentation regime in the bay causing the natural protective beach at Hallsands to be completely eroded. Remains of the present day beach can be seen at nearly four metres lower than the original beach level.

Although the Code of Practice (BS6349: Part 1: 1984) gives specifications for natural materials for use in the marine environment, as a result of the findings of the Working Party of the Maritime Engineering Group of the Institution of Civil Engineers (1985), the formulation of

a committee was stimulated and subsequently formed to prepare a state of the art document entitled 'Guidance Notes on the Design and Construction of Breakwaters'. This report is eagerly awaited.

Minestone could be accepted as a suitable fill material for marine applications provided that quality control is easily maintainable. In effect this would mean that quality control would need to be achieved at the exploration stage by core analysis, and at the tip storage by the analysis of bulk samples. However, such control over quality is only possible provided that the following over-riding factors are satisfied:

1 Minestone meets the specifications laid down by the Code of Practice

2 Contract specifications for the marine structure are available

With regard to the first of these factors, an understanding of the geological and mining factors that effect the nature of minestone and thus its chemical and mechanical properties must facilitate long term planning of its utilisation. A great deal of work has been carried out by and under the auspices of Minestone Services in relation to gaining the acceptance of the professions for land based applications, but for marine applications much is yet to be discovered, and although not specifically relating to minestone, the recommendations made by the Maritime Engineering Group (1985) reflect the research work that is generally required:-

'Core and fill materials:-

(a) Controls and measurement of particle shape
(b) Durability, mechanisms, rates and effects of deterioration
(c) Geotechnical behaviour of clays modified by the marine environment
(d) Geotechnical behaviour of core and fill materials, stability and settlement
(e) Effect of segregation during placement

Filter layers:-

(a) As above
(b) As above
(c) Design of filter layers taking account of true particle shape
(d) Degradation of core and filter layers due to erosion and cyclic agitation by wave action involving permeability, internal friction and cohesion.'

Mindful of these recommendations, the remainder of this paper will highlight some of the relevant work that has or has not been carried out, as it is imperative that any research programme commences with a review of the state of the art knowledge.

UTILISATION

Minestone resources at the end of 1982 were estimated at over one thousand million cubic metres ($1 \times 10^9 m^3$) on the ground in England, Scotland and Wales, apportioned 730 million, 200 million and 80 million cubic metres respectively (Nutting, 1984). Examples of marine applications of minestone fill are relatively few, especially in the UK. Some of the important marine and inland waterway applications are summarised in Table 1.

South eastern England has been subjected to a relative rise in sea level resulting from a combination of global warming and glaciation influences. In an attempt to combat the effects of this near Deal in Kent, rolled colliery shale, obtained from the nearby Betteshangar Colliery tip was used as the bulk fill to construct a new 3 km long sea wall. Shale was preferred to sand, clay and chalk materials, mainly due to its local availability, its occurrence in sufficient quantities, and when compacted because of its relatively low permeability to water. Although more arenaceous discard in the Colliery tip was considered as a possible source of 'rip-rap', to its disadvantage it was deemed to be too difficult to define both the whereabouts, and quantity of such material. On this point, borehole to borehole and borehole to surface seismic geophysical surveys and tomographic analysis could prove useful in establishing the presence and extent of sizeable intact materials in tips. However, much work needs to be carried out to evaluate the potential of this technique for this purpose. Consequently, at Deal, rock armour was brought in from further afield – the seaward facing 'rip-rap' armour layer protecting the shale from wave erosion was constructed mainly of one tonne blocks of Kentish Ragstone (a limestone quarried in Kent) and Carboniferous Limestone from Somerset (Hamilton, 1984). Evidence of deterioration of parts of the seaward facing slope is now apparent in so much as removal of the primary armour and scouring of the fill is visible. Further study of this phenomenon is required. Again seismic surveying and/or radar profiling may prove to be suitable techniques in assessing physical changes that occur within such marine structures so that weak links in the structure could be identified and quickly strengthened.

Minestone fill obtained from the Ruhr Coalfield was tipped from a barge to construct a sill as part of a dam in the Netherlands (Laan, Van Westen and Batterink, 1984). Using fill of an effective bulk density of 1.56 tonnes per cubic metre, a friction angle of 30^0 and a $d50$ value of 7 to 15 mm (comparatively fine grade), slope gradients of between 1:2 and 1:2.5 were achieved with only a 25% loss of minestone.

Existing structures should be considered as large scale 'experiments'. Although monitoring of these 'experiments' can be time consuming, a wealth of important performance data can be generated, which can be used to benefit subsequent designs.

FACTORS AFFECTING THE USE OF MINESTONE

Initial shear strength properties will govern slope design of a marine structure, but some variation of shear strength is likely to occur with time, this being a function of the extent of the materials' deterioration in the structure due to marine and/or subaerial processes. Deterioration susceptibility can be assessed by conducting various durability tests, some being more appropriate than others in this context. With regard to the design of a structure there should always be provision for its future remediation. Remediation may prove necessary because slope failure occurs or due to removal of material by excessive storm conditions. Thus there are three main alternatives facing the design engineer at the outset: design the structure so that any anticipated variation in shear strength, for example, will have no effect; base the design on initial shear strength and provide an assurance that in the eventuality that future remediation is necessary, there will be sufficient quantities of suitable aggregate readily available; incorporate time dependent shear strength variation and surplus fill requirement into the design. Of course, the design engineer would opt for the latter whenever possible because of an increased factor of safety.

The critical factors effecting the shear strength and durability of minestone fill in a structure in the marine environment can be summarised under three broadly defined categories:-

TABLE 1

Structure	Slope Gradient	Quantity (m³)	Period
Deal, UK Marine wall Hamilton, 1984	1:3	85,000	1978–1979
Netherlands Marine dams, canal and river banks Laan et al, 1984	1:2;1:2.5	85,000+	1953 on
River Ouse, Yorkshire River banks Kluth, 1984	1:3	350,000	1983–1984
South Queensferry, Scotland Marine harbour Couper and Montgomery (in press)	-	-	-
Channel Tunnel project Marine reclamation Sleeman and Fry (in press)	-	-	-
F R G Waterways Haussand Mayer (in press) Helbaum and Hauss (in press)	-	-	-

1 Properties of the intact materials are determined by:
(a) Clasts and inter-granular cement (if present) mineralogies
(b) Clast size, shape and degree of inter-lock
(c) Structure of inter-granular cement
(d) Nature and persistence of microscopic and mesoscopic tectonic, sedimentary and mining induced discontinuities
(e) Natural moisture content, and hydraulic conductivity
(f) Weathering state
Dibb, Hughes and Poole (1983) identified most of these as critical factors in an assessment concerning the durability of rock armour in the marine environment, but equally they could apply to minestone fill.

2 Properties of the marine environment are determined by:
(a) Water depth
(b) Wave action, marine currents
(c) Chemistry of sea water
(d) Local weather conditions
(e) Seasonal changes

3 Properties of the marine structure:
(a) Shape and size of the structure
(b) Degrees of compaction and mass permea-bilities of the minestone and other materials used
(c) Protection offered by artificial tetrapod or natural rock armour systems and filter layers

Whereas the first two categories can be considered to contain independent factors, the design of the latter is obviously going to be dependent upon them.

Relevant to this paper is information regarding the first and third categories. As for the parameters in the second category, only sea water chemistry is relatively constant, and because of this, it is considered that discussion of the other parameters in this category is beyond the scope of this paper.

It is reasonable to assume that there are three types of marine environment: sub-marine (or submerged); intertidal; supra-tidal (or subaerial); in each of these environments minestone fill will be affected by the elements in different ways. Weathering processes that are effective in the coastal environment are wave buffeting, sea-spray, saltwater attack, rain and freeze-thaw in the intertidal and supra-tidal zones, and mainly saltwater attack in the submarine zone. These natural processes could cause deterioration by physical breakdown of the minestone possibly by chemical changes but mainly by pressures developed in microcracks due to swelling of sensitive clay minerals and crystal growths, voids caused by piping of fines and possibly seismic shock caused by wave impact. If protected by armour layers, and provided that adequate drainage exists as far as fill is concerned, the latter could be considered to be of relatively low investigation priority. Recent work has made progress towards an understanding of the deformational behaviour of minestone under variable water conditions (Skaryznska, Zawisza and Rainbow, in press).

MINERALOGICAL AND CHEMICAL CHARACTERISTICS

Several factors would effect chemical changes of minestone in a structure placed in a particular marine environment:-

1 Composition of minestone
2 Composition of sea water filtering through the structure
3 Position within the structure
4 Permeability of the structure
5 Temperature and pH in the structure
6 Wetting and drying cycles dictated by tides
7 Design lifetime of the structure

In the absence of visible discontinuities, mechanical behaviour of intact Coal Measures rocks have been found to be controlled by mineralogy (Briggs, 1979; Hart, 1986). Based on existing information a detailed account was given of properties and relationships between mineralogy, physical properties and mechanical behaviour of minestone from the UK (Taylor, 1984). Minestone being a broken material is non-cohesive and shear strength is essentially a function of bulk density, particle size distribution and particle shape characteristics. All of these are controlled to a great extent by mineralogy, porosity and discontinuities ranging from the microscopic to macroscopic in size. Shear strength can be raised of course by artificial compaction, but the degree to which the material can be compacted largely depends on the nature of the minestone, in addition to the compaction technique used. Compaction of fill is possible to raise the bulk density to a theoretical maximum controlled by the density of the intact rock. This has the effect of increasing the friction angle, and decreasing the coefficient of permeability (Michalski and Skarzynska, 1984). This is an option for engineers wishing to stabilize ground by increasing its bulk density by using either vibro-compaction or conventional roller compaction techniques, and should be borne in mind at both the initial design stage and at the remedial design stage.

Collins (1984) gives the main mineral species of these materials in order of decreasing abundance as being quartz, illite, mixed layer clay (illite and montmorillorite), Kaolinite, chlorite, allophane, siderite, calcite, ankerite, pyrite, jarosite, alunite, feldspar, with limonite, haematite, magnesite and gypsum as trace minerals. It is well known that the species and proportions of clays that are present in Coal Measures rocks influence their ability to swell in the presence of water. From this point of view, the most influential clay mineral is mixed layer clay. Montmorillonite within the latter type greatly influences water swell ability, reflected in its inter-lattice distance variation from 9.6A in dehydrated material to 21.4A when saturated. Although the ideal montmorillonite structure has no lattice positions for cations of Sodium, these elements are adsorbed onto the outside and between the lattice units due to their overall negative charge.

A reasonable direct relationship exists between swelling strain and a mineralogical ratio, consisting of the minerals, illite plus kaolinite contents divided by quartz plus siderite contents. The illite variety was deemed to contain montmorillonite in mixed layer form. Water saturation was also shown to significantly reduce various important mechanical parameter values − angle of internal friction, cohesion, Young's Modulus, uniaxial compressive strength − by 20% to 80% for values of the ratio between 0.5 and 1.0. Changes in mechanical parameter values are implied to be particularly sensitive to relatively small changes in mineralogy, especially in siltstones, muddy siltstones and silty mudstones (Hart, 1986). Other workers found direct relationships between 'mixed layer clay' content and strain rate (White, 1956; Rowlands, 1977; Kruszewski, 1970), quartz and total carbonate content and uniaxial compressive strength, triaxial stress factor, Young's Modulus and rebound hardness (Price, 1960; Taylor and Spears, 1970; Barbour, 1979). Conversely, there are inverse relationships between 'mixed layer clay' content and strength parameter values (Briggs, 1979; Barbour, 1979; Buffington, Stephenson and Rockaway, 1980), and also for quartz and carbonate content and strain rate. Based on analyses of minestone triaxial and shear box strengths, Taylor (1984) suggested a three fold grouping of minestone-generating areas: Group 1 − South Wales, Kent, North East and Scotland; Group 2 − North Yorkshire, Doncaster, Barnsley, South Yorkshire, South

261

Midlands; Group 3 - North Derbyshire, North and South Nottinghamshire, and Western. Bulk samples were found to be non-cohesive with average friction angles for Group 1 of 39^0 up to about 75 kN/m^2 effective normal stress to 34^0 above this stress level; average friction angles for Groups 2 and 3 were 33.5^0 up to about 100 kN/m^2 effective normal stress; above 100 kN/m^2 the average friction angle for Group 2 was 32^0, and for Group 3 was 29.5^0. A striking feature is that minestone had higher shear strengths where it was associated with higher rank coals and where minestone contained inert clay minerals.

Determination of the chemical composition of minestone is an important factor in an assessment of its chemical stability. Variations in chemistry exist for minestone from different parts of the UK; pH ranges from 2 to 10 with a slight skewed distribution towards higher pH values - the large ranges probably being mainly due to the variable presence of pyrite, its degree of oxidation, and a redox potential of between 434 mV and 603 mV at 20^0C; chloride contents range from 0.05% to 0.25% with a skewed distribution towards the lower values (Rainbow, 1984); total sulphur contents range from 0.03% to 12% the distribution being distinctly skewed towards the lower values (Rainbow and Sleeman, 1984; Collins, 1984). Add to this the fact that the concentrations of potassium and magnesium are 1.1% and 3.7% respectively in sea water of pH 8 (Mason, 1966) and that there is the similarity in crystal structures of montmorillonit-, kaolinite, illite and chlorite (Deer, Howie and Zussman, 1975). One of the most important factors influencing the stability of montmorillonite is the availability of magnesium. In submerged structures, sufficient quantities of this cation could help to preserve montmorillonite, whilst in subaerial and intertidal structures which are well drained, the magnesium may be leached resulting in the formation of a more stable clay mineral assemblage. These are feasible reactions for montmorillonite in mixed layer clay in minestone, which could occur in engineering timescales, and which ultimately could have a stabilising or destabilising effect. Note that in one instance 'chemical weathering' had the affect of reducing the shear strength of loosely tipped minestone by 21%, but a more representative value for old tips is around 10%. It was also reported that peak shear strength of minestone as

represented by the friction angle could be reduced by over half where comminution occurred due to additional movements on pre-existing shear surfaces (Taylor and Garrard, 1984).

This reinforces the concern expressed by the Maritime Engineering Group (1985) about the need for investigation into the chemical effects of sea water on clay minerals and shear strengths over engineering design time scales. However, with a low in place permeability of 10^{-6}m per second as experienced for example at the Selby-Wistow-Cawood flood barrier (Kluth, 1984), weathering and alteration of the materials would be expected to occur at a relatively slow rate. Recent progress in some of these areas of research has been made; the effect of petrology on swelling behaviour (Skarzynska, Kozielska-Sioka and Rainbow, in press); the effect of particle morphology on minestone behaviour (Maradas, Canibano and Alonso, in press); chemical composition changes in pore fluids (Szczepanska and Twardowska, in press).

Mason (1966) suggests four basic principles governing adsorption:-

1 A decrease in the grain size of the adsorbent increases the surface area of the grains and thus increases the potential for adsorption to occur

2 Favourable conditions occur when the adsorbent and adsorbate form a compound of low solubility

3 The amount of adsorbate is directly proportional to its concentration in solution

4 Ions with higher charges are adsorbed in preference to those with a lower charge

The first of these items is of direct relevance to minestone utilisation, in so much that for minestone containing significant montmorillonite it may be desirable to extract the fine fraction minestone prior to emplacement to effect a higher initial shear strength, but taking a long term point of view, the separation of fines may turn out to be superfluous for structures above the level of water.

DISCONTINUITIES

Generally, an increase in the number of sedimentary and/or tectonic discontinuities per unit volume decreases the values of the

mechanical parameters and this also
influences the initial particle morphology
and the weathering breakdown of minestone
aggregates. Stratification includes lami-
nation, graded bedding, cross stratification, ripple marks, and erosion features
such as washouts, and macroscopic-
mesoscopic post-depositional features
such as cleat in carbonaceous matter,
joints and faults including bedding plane
shears and associated microscopic cracks
(Saleby, Money, Dearman, 1977; Hassani,
Whittaker, and Scoble, 1979; Briggs, 1979;
Jeremic, 1980). In certain circumstances
sedimentary and tectonic discontinuities
are related, as for example bedding plane
shears are often confined to weaker mudrocks, and differential compaction faults
are associated with washouts.

TESTS AND INVESTIGATIONS

Laan, Van Westen and Batterink (1984)
provided a classification for assessing
the sensitivity of minestone to weathering
based on sampling and screening of samples
taken from sites, and a quick laboratory
test utilising a simple process of drying-
soaking in tap water-drying-screening to
assess aggregate disintegration. Within
the Ruhr area, investigations proved that
for samples originating from further east,
their sensitivity to weathering decreased.
Evidence suggests that weathering of
minestone in a tip is limited to the few
metres adjacent to the atmosphere and that
'minestone which had not been dried
beforehand showed no signs of disintegration after immersion in sea water for two
years' (Laan, Van Westen and Batterink,
1984). Proof of the assumption that
illite was responsible for the chemical
stability of the minestone immersed in
sea water is difficult without accurate
mineralogical analyses to identify the
nature of the illite present.

Particle morphology of rock fragments in
marine construction has been given some
considerable attention by Fookes and Poole
(1981). Whilst standard BS1377 grading
analysis gives quantitative information
about particle volumes, the test provides
nothing useful regarding particle shapes.
A given sieve will allow rod-like particles
of variable volume to pass through whilst
preventing the passage of tabular
particles. Results of grading analysis
obtained from the BS1377 technique were
compared to those obtained by using a
measurement technique based on weight/
shape measurements of non-prismatic
particles. Fairly consistent differences

were apparent in so much that the latter
produced a coarser grading profile for
potentially suitable fills, filters and
core materials of non-Coal Measures
origin. Even BS812 shape classification
procedure has led to over-simplification
of shape for elongate and flaky materials,
and Fookes and Poole (1981) suggest using
a much more rigorous classification of
shape in order to achieve a basis by which
the efficiency of packing of the materials
can be better assessed.

The coarsest minestone could prove to be
suitable for core and filter materials,
and finer minestone as fill. Attainment
of highest bulk density, due to efficient
packing of particles is desirable to
achieve initial design shear strength for
core and fill, and to achieve effective
barrier properties and shear strengths for
filter layers. Filter layer design is
very much dependent on the nature of the
core or fill. A filter can be designed,
for example, by using Terzaghi's rule of
filters (Terzaghi and Peck, 1968) to
protect the underlayer from phenomena such
as piping of the core or fill into the
armour and/or to prevent puncturing of the
underlayer by the armour layers. Without
grading most minestone fills would contain
a large proportion of fines (gravel size
or less) and as such a fair amount of
attention would need to be devoted to the
design of filter layers.

Based on a review and investigations into
rock armour durability for marine breakwaters (Dibbs, Hughes and Poole, 1983)
crystallisation pressures and hydration
pressure salts formed in cracks and pores
were suggested as being the main destructive weathering mechanisms. Referring to
work by Winkler and Singer (1972) and
Correns (1949), Dibb, Hughes and Poole
(1983) quoted an equation to calculate
crystallisation pressures:

$$P_c = \frac{RT}{V_s} \ln \frac{C}{C_s} \qquad (1)$$

Where P_c = pressure exerted by crystal
growth; R = gas constant; T = absolute
temperature; V_s = molar volume of solid
salt; C = solute concentrations; C_s =
concentration of solute at saturation
point. Since halite has no hydrate above
0^0C, it must exert pressures due solely to
crystallisation, and thus the sulphate
soundness test in which the degradation of
the sample depends principally on the
formation of hydration pressures, is
inappropriate in this case. A theoretical
derivation of hydration pressure first

given by Winkler and Wilhelm (1970) was
used:-

$$P_h = \frac{(nRT)}{(\overline{V}_h - \overline{V}_a)} \quad \ln \frac{Pw}{Pw^1} \qquad (2)$$

Where n = number of moles of water gained
during hydration; V_h = molecular weight of
hydrated salt/density; V_a = molecular
weight of original salt/density; Pw =
atmospheric vapour pressure at T; Pw^1 =
vapour pressure of hydrated salt at T. It
was noted that diurnal hydration of sodium
sulphate to its heptahydrate and further
to its decahydrate produces a volumetric
expansion of 314% with sufficient resulting
pressure to cause tensile failure of normal
porous building materials (Bonnell and
Nottage, 1939).

SUMMARY OF DESIGN PROCEDURE

Fookes and Poole (1981) suggest the following steps in designing filter systems for a
breakwater:

1 Wave climate will dictate the design of
   the outer primary armour
2 The secondary armour is then designed to
   compliment the primary armour
3 Determine the shape and grading profiles
   of the intended core materials, and for
   most minestone materials, filters would
   be necessary
4 Design the filter so that it is both
   compatible with the secondary armour and
   the underlayers, and it may be necessary
   to design further transitional filter
   layers to separate the core/fill and
   armour materials to provide a smooth
   grading profile across the structure

Unlike with conventional fill materials,
there are few precedents involving minestone. Thus for any proposed marine
structure with the potential to involve
minestone as a fill, core and filters, the
following design procedure is recommended,
as suggested by Baird and Hall (1984) in
context of design of rubble mound breakwaters:

1 Define the environmental conditions to an
   acceptable accuracy
2 Appraise the geotechnical properties of
   the site and that of the available
   materials
3 Review construction procedures and
   available equipment
4 Construct physical and theoretical models
5 Review of results gleaned in stages 1 to
   4, but in particular pay attention to:
   (a) optimum usage of construction
       materials

(b) ease of construction
(c) foundation and slope stabilities,
    overtopping, scour due to wave action
    and currents
(d) mode of failure if anticipated design
    conditions are exceeded
(e) costs
(f) acceptance by local community

CONCLUSIONS

To date use of minestone in marine and
inland waterway applications has been
successful in so much that the aggregates
have provided an inexpensive and geotech-
nically suitable alternative as compared
with what may have been previously
considered as 'conventional' aggregates.
Although the use of minestone in this
context in the UK has been somewhat
limited, it has long been established as a
viable construction material for marine
purposes by other northern European
countries.

Whilst information has been reviewed with
respect to some of the problems confronting
the engineer concerned with the design of
marine defences, a definitive approach is
yet some way ahead. As a result of this
assessment it would appear to behove
concerned organisations to ameliorate their
resources in order to sustain research and
development programmes in this area with a
view to gaining wide acceptance of minestone by the civil engineering profession,
for both gentle and harsh marine
environments.

ACKNOWLEDGEMENTS

Thanks are due to Miss J Johnson,
Portsmouth Polytechnic, for typing the
manuscript.

REFERENCES

Baird, W.F. and Hall, K.R. 1984. The
   design of armour systems for the protec-
   tion of rubble mound breakwaters.
   Breakwaters - Design and Construction.
   Thomas Telford Ltd, 107-119.
Barbour, T.G. 1979. Relationships of
   mechanical index and mineralogic
   properties of Coal Measures rock. Pre-
   print of 20th US Symp. on Rock Mechanics,
   Austin, Texas.
Bonnell, D.G.R. and Nottage, M.E. 1939.
   Studies in porous materials with special
   reference to building materials. I. The
   Crystallisation of Salts in Porous
   Materials. Journal of the Society of
   Chemical Industry, 58A, 16-21.

Briggs, D. 1979. An investigation of mineralogical and mechanical properties of Coal Measure rock. Unpublished PhD thesis, University of Wales, Cardiff.

Brown, P. 1990. UN warns of global time-bomb. Guardian, 22 May 1990, pp1.

BS1377. Methods of Test for Soil for Civil Engineering Purposes. British Standards Institution.

BS812. Methods for sampling and testing of mineral aggregates, sands and filters. Part 1: Sampling, size, shape and classification. Part 2: Physical properties. British Standards Institution.

BS6349. Part 1: Construction Materials. British Standards Institution.

Buffington, D. Stephenson, R.W. and Rockaway, J.D. 1980. Residual strength of underlay. Preprint of 21st US Symp. on Rock Mechanics, University of Missouri-Rolla.

Collins, R.J. 1984. Manufacture of aggregates from colliery spoil. 2nd International Symp. on the Reclamation, Treatment and Utilisation of Coal Mining Wastes, Durham, September 1984, 14.1-14.17.

Correns, C.W. 1949. Growth and dissolution of crystals under linear pressure. Disc. Faraday Society, 5, 267-271.

Couper, A.S. and Montgomery, H. 1990. Minestone for a maritime village development at Port Edgar, South Queensferry. 3rd International Symp. on the Reclamation Treatment and Utilisation of Coal Mining Wastes, Glasgow 1990, paper 48 (in press)

Deer, W.A., Howie, R.A. and Zussman. 1975. An introduction to the rock forming minerals. Longman, 528 pp.

Dibb, T.E., Hughes, D.W. and Poole, A.B. 1983. The identification of critical factors affecting rock durability in marine environments. Quarterly Journal of Engineering Geology, 16, 149-161.

Fookes, P.G. and Poole, A.B. 1981. Some preliminary considerations on the selection and durability of rock and concrete materials for breakwaters and coastal protection works. Quarterly Journal of Engineering Geology, 14, 97-128.

Hamilton, W.A.H. 1984. A colliery shale sea wall at Deal. ibid, 45.1-45.17.

Hart, P.A.H. 1986. Investigations into the role of groundwater in promoting floor heave in coal mine gateroads. From Cripps, J.C., Bell, F.G. and Culshaw, M.G. (eds), 1986, Groundwater in Engineering Geology, Geological Society Engineering Geology Special Publication No 3, 115-126.

Hassani, F.P., Whittaker, B.N. and Scoble, M.J. 1979. Strength characteristics of rocks associated with opencast coal mining in UK. Preprint of 20th US Symp. on Rock Mechanics, Austin, Texas.

Hauss, K.D. and Mayer, B.K. 1990. Tests concerning reclamation, treatment and utilisation of coal mining wastes in waterway engineering. ibid. paper 31 (in press)

Helbaum, M.H. and Hauss, K.D. 1990. Minestone in waterway engineering. ibid, paper 32 (in press).

Jeremic, M.L. 1980. Influence of shear deformation structure in coal on selected methods of mining. Rock Mechanics 13, 23-38.

Kluth, D.J. 1984. Use of minespoil in the Selby-Wistow-Cawood barrier bank. ibid, 46.1-46.14.

Komar, P.D. 1976. Beach processes and sedimentation. Prentice-Hall, N.J.

Krusewski, T. 1970. Influence of mineral composition of clay rocks on stability of slopes from lignite pits in Poland. Proc. 2nd International Congress of the International Association of Engineering Geologists, Sao Paulo, VI.1-VI.6

Laan, G., Van Westen, J.M. and Batterink, L. 1984. Minestone in hydraulic engineering application, deterioration and quality. ibid, 5.1-1.10.

Maritime Engineering Group, Institution of Civil Engineers, 1985. Coastal Engineering Research. Reports prepared by the Working Parties on Beaches and Sea Walls; Siltation, Dredging and Dispersion; and Coastal Harbours, Breakwaters and Off-Shore Islands. Thomas Telford, London 123 pp.

Mason, B. 1966. Principles of Geochemistry. John Wiley & Sons, 329 pp.

Michalski, P. and Skarzynska, K.M. 1984. Compactibility of coal mining wastes as a fill material. ibid, 15.1-15.13.

Moradas, M.R.G., Canibano, J.G. and Alonso, M.T. 1990. The effect of particle morphology on coal waste behaviour. ibid, paper 57 (in press).

Nutting, M. 1984. The use of minestone in conjunction with the controlled disposal of domestic and commercial waste. 2nd International Symp. on the Reclamation, Treatment and Utilisation of Coal Mining Wastes, Durham, September 1984, ibid. 48.1-48.11.

Price, N.J. 1960. The compressive strength of Coal Measures rocks. Colliery Engineering, July, 283.

Rainbow, A.K.M. 1984. Research into reinforced minestone in UK, ibid. 7.1-7.12.

Rainbow, A.K.M. and Sleeman, W. 1984. The effect of immersion in water on the strength of cement bound minestone. ibid, 55.1-55.15.

Rowlands, N. 1977. An examination into mineralogical and mechanical properties of Coal Measures strata. Unpublished PhD

Thesis, University of Wales, Cardiff.

Saleby, M.R., Money, M.S. and Dearman, W.R. 1977. The occurrence and engineering properties of intraformational shears in Carboniferous rocks. Preprint of the Conference on Rock Engineering, University of Newcastle upon Tyne, 1977.

Sandall, P.J. 1990. A geological and geotechnical assessment of the area of the Wantsum Channel in North East Kent for the proposed construction of a colliery waste sea wall. Unpublished CNAA BEng (Hons) Thesis. Portsmouth Polytechnic.

Skarzymska, K.M., Kozielska-Sroka, E., and Rainbow, A.K.M., 1990. Swelling of minestone in relation to its petrographic composition. ibid, paper 56, (in press).

Sleeman, W. and Fry, R.J. 1990. The transportation and utilisation of minestone for the Channel Tunnel Project. ibid. paper 49, (In press).

So, C.L. 1966. Coastal changes between Whitstable and Reculver. Geological Association Memoirs, 77, IV.

Szczepanskra, J. and Twardowska, I. 1990. Transformations of chemical composition of pore solutions in coal mining wastes. ibid. paper 24, (in press).

Taylor, R.K. 1984. Composition and engineering properties of British Colliery Discards. NCB Mining Department.

Taylor, R.K., Garrard, G.F.G. 1984. Design shear strengths for UK Coarse Colliery Discards. ibid, 34.1-34.14.

Taylor, R.K. and Spears, D.A. 1970. The breakdown of British Coal Measure rocks. International Journal of Rock Mechanics and Mining Sciences, 7, 481-501.

Terzaghin, K. and Peck, R.B. 1968. Soil Mechanics in Engineering Practice. John Wiley & Son.

Winkler, E.H. and Singer, P.C. 1972. Crystallisation pressure of salt in stone and concrete. Bulletin of the Geological Society of America, 83, 3509-3514.

Winkler, E.H. and Wilhelm, E.J. 1970. Salt burst by hydration pressures in architectural stone in urban atmospheres. Bulletin of the Geological Society of America, 81, 567-572.

White, A.W. 1956. Underclay squeezes in coal mines. Transactions of the American Institute of Mining Engineers, 205, 1024-1028.

Reclamation, Treatment and Utilization of Coal Mining Wastes, Rainbow (ed.) © 1990 Balkema, Rotterdam. ISBN 90 6191 154 0

Utilizing coarse refuse for tailings dam construction

M. Dunbavan
BHP Engineering, Sydney, N.S.W., Australia

ABSTRACT: The majority of Australian coal mines use storage dams for coal tailings disposal. Increased storage capacity is often gained by raising the crest level of existing embankments. The downstream method of construction is usually adopted. Road and rail facilities and a water supply pipeline prevented use of this method at the Peak Downs Mine. The feasibility of using upstream raising with coarse refuse as the construction material was determined by a comprehensive geotechnical investigation. The work included checking the integrity of the existing structure, determining geomechanical properties of the tailings, construction and monitoring a trial embankment. With favourable results from the investigation, the detailed design for staged upstream raising to increase the height by 6m was completed. The performance of the first stage raising met design expectations, and another raising is being planned.

1 INTRODUCTION

The majority of coal mined in the Bowen Basin (Queensland, Australia) is high quality coking coal destined for export markets. BHP-Utah Coal Ltd operates eight mines in the region, one of which is Peak Downs Mine. This mine is situated approximately 150km southwest of the coal port facilities at Hay Point near Mackay.

Peak Downs Mine produces 5.2 million tonnes of clean coal annually. Coal is won from several seams in the Moranbah coal measures by open strip mining. Overburden is removed by large walking draglines using either one – or two – pass methods with overburden fragmented initially by drilling and blasting. Coal is mined using electric shovels and/or front enloaders, placed in large belly-dump trucks and hauled to a central washing and storage facility on site.

2 WASTE DISPOSAL PRACTICE

2.1 Types of waste produced

The marketing constraints on the coal products demand that the raw coal be processed. Two types of coal waste are produced during processing; coarse refuse consisting of high ash coal and stone (maximum particle size 50mm), and tailings consisting of fine coal particles, stone fragments and clay. The coarse refuse is stored in bins or heaps and when appropriate is transported by truck to a remote dump. Occasionally, the coarse refuse is blended with the mined overburden during land rehabilitation work.

2.2 Disposal method

The tailings is produced as a slurry with no holding facilities at the coal washery. Thus, tailings must be piped to a disposal pond at its production rate. Water is not plentiful in the coal mining areas of the Bowen Basin, with a large excess of evaporation over precipitation. Two issues arise with tailings disposal; storage of the tailings and recovery of process water for reuse.

2.3 History of tailings disposal

The topography on the Peak Downs mining lease has low relief and the major land use (other than surface mining) is cattle grazing. The nearest population centres are townships established to house workers from the coal mines. There is no public conflict over the use of storage dams for tailings disposal.

The current disposal area is in a shallow valley approximately 1km south of the coal washery. This storage was created in two stages; the first stage was an earthfill embankment to 7.6m height (RL253.0) with a crest length of 800 metres. The second stage was another earthfill construction over the downstream face of the first embankment. The crest height varied from RL256 at the southern end to RL260 at the northern end. Figure 1 shows a plan of the area. Tailings was discharged along the northwestern edge of the storage area with the location of the discharge point being changed occasionally to promote even filling of the area. The earthfill embankment was abutted by the clayey fraction of the tailings and not the coarser fraction, as is usually the case when upstream raising is considered.

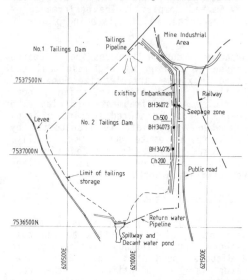

Figure 1. Locality Plan, No. 2 Tailings Dam

By late 1987, the need to plan for further tailings storage was addressed because less than 12 months storage capacity remained in the existing facility. Raising the existing dam crest level by downstream construction was not practical because of the proximity of a public road, railway and a water pipeline. Thus, two options were open; either raising the existing crest level using upstream construction, or locating a site to develop a new storage dam.

The cost of providing infrastructure for a new site and the time required for such a development made the upstream raising the preferred option. The possibility of using coarse refuse as fill material for raising the crest level was also preferred to importing earthen material from a remote quarry site. The existing embankment was classified as a "Referrable Structure" under Queensland government regulations. Any storage dam with a toe to crest height exceeding 10m and a storage volume exceeding 20000m^3 is required to meet engineering design standards approved by the Queensland Water Resources Commission. Approval for raising the crest level was required under these regulations.

3 APPROACH TO DESIGN

For the purposes of later comparison, the stability of the existing embankment was calculated. A programme of standard penetration testing was completed along the centreline of the embankment to assess the as constructed condition of the earth fill. Some variability was expected due to seepage observed in several discrete zones on the downstream face of the embankment. During the same programme, push tube samples were also taken for laboratory testing.

Using laboratory results for strength of coarse refuse material and results reported in technical literature for strength of tailings, the maximum acceptable height for a single stage raising was determined. The time to 90% consolidation under first stage loading was estimated. The stability of the final stage of raising was then assessed. The combination of these results gave profiles and a schedule for two stage raising to the specified crest level.

The values selected for strength and consolidation characteristics for the tailings were on the conservative side for the available information. To improve the reliability of the recommended design, especially regarding the tune to 90% consolidation, the construction and detailed monitoring of a trial embankment on the tailings was suggested. The trial was undertaken and showed a significantly shorter delay was required between stages than originally anticipated. Results from settlement measurements indicated the height of fill needed above final crest level to allow for tailings settlement so that long term maintenance could be minimised.

4 STABILITY OF EXISTING EMBANKMENT

4.1 Material properties

Geotechnical investigations of materials comprising the existing embankment included Standard Penetration Testing (SPT) rotary drilling with and undisturbed (tube) sampling, installation of standpipe piezometers and laboratory testing for index properties and shear strength. The location of the three SPT sites is shown on Figure 1 and undisturbed samples were taken from holes adjact to the SPT sites.

A cross-section of the existing embankment with the distribution of materials is shown in Figure 2. Undisturbed samples at different elevations were obtained using 73mm push tubes. Results from classification tests and consolidated undrained triaxial tests (with pore pressure measurement) are given in Table 1. The Emerson Crumb Test (AS1289:C8) was used to determine the susceptibility of the materials to slaking and/or dispersion which could result from changes in seepage rates when the tailings level is raised.

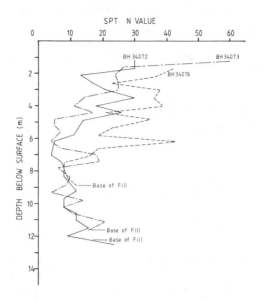

Figure 3. Results from standard penetration tests

4.2 Piezometric surface

Figure 2 shows the location of a drainage blanket across the base of the embankment. Seepage was observed through two areas of the downstream face; at the toe due to the drainage blanket, and at approximately mid-height along a 75m

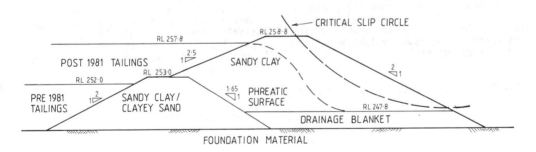

Figure 2. Cross-section of existing embankment

SPT results are for the three test locations are given in **Figure 3**. The variation in N values matched results from in situ density tests taken during construction. Layers of lower density were known to occur in the embankment and these influenced seepage paths through the embankment.

section near borehole BH34072 (see Figure 1). These observations and readings from the standpipe piezometers were used to determine the location of the piezometric surface.

Data indicated that the piezometric surface coincided with the top of the

269

Table 1 Properties for existing embankment

a) <u>Classification Tests</u>

Material	Classification	% Passing 0.075mm	Liquid Limit %	Plastic Limit %	Plasticity Index %
Gravelly Clayey Sand	SP-SC	30	25	14	11
Sandy Gravel, some silt & clay	SC-SW	17	29	17	12

b) <u>Shear Strength of Gravelly Clayey Sand</u>

Bulk Unit Weight 21 kN/m^3

Total Stresses:	Cohesion	18 kPa
	Friction	23°

Effective Stresses:	Cohesion	20 kPa
	Friction	26°

drainage blanket, inferring efficient operation of the drainage blanket and good drainage of the variably compacted sandy clay fill. The surface is shown in Figure 2. Particular attention was given to the mid-height seepage zone near borehole BH34072. This zone correlated with a poorly compacted layer and readings from a standpipe located in the layer showed the existance of a perched water table connected to the top of the tailings on the upstream side of the embankment. Laboratory tests on undisturbed samples from this layer showed lower shear strength compared with samples from other parts of the embankment. These lower values were used in design calculations.

4.3 Stability of existing embankment

In addition to the materials listed in Table 1, geomechanical properties were required for the tailings and the filter material in the drainage blanket so that an assessment of the existing embankments stability could be made. A summary of properties for all materials used for stability analyses is given in Table 2.

The likelihood and severity of possible earthquake loading at the tailings dam site was assessed. The nearest occurrence of any significant seismic event (ie reliably located) is approximately 200km to the east, offshore from Mackay. For a 1 in 50 year

Table 2 Properties used in stability analyses

Material Type	Unit Weight (kN/m^3)	Total Stress Conditions		Effective Stress Conditions	
		Cohesion (kPa)	Friction (degrees)	Cohesion (kPa)	Friction (degrees)
Present Embankment	21	18	23	20	26
Drainage Layer	18	0	35	0	35
Foundation	21	55	19	43	20
* Tailings	6.4	10	0	10	0
New Embankment	16	0	30	0	30

* Note: The parameters relate to the initial 2.5m fill, analyses of subsequent raising to the final design height assumed consolidation of the tailings had occurred with cohesion = 20kPa, zero friction, unit weight = 12.3kN/m^3.

return period, the maximum likely seismic event would have a magnitude of ML 5.6 at an epicentral distance of200km. This corresponds to a peak ground acceleration for the tailings dam of 0.015g. Such a low figure would have an insignificant effect on embankment stability.

Stability analyses for the downstream face were carried out using Bishop's Simplified Method which forms part of BHP Engineering's computer package SLOPE. Two cross-sections at Chainages 200 and 500 (see Figure 1) were analysed using steady stage seepage (effective stress) conditions. The critical slip circle (lowest factor of safety) was found using an automatic grid search built into the computer package. The range of the grid is specified by the user as part of the input data. The minimum values for factors of safety were 3.01 (Ch200) and 1.89 (Ch500). These values were used as a basis for comparison of similar results for the raised crest level.

Analysis of the upstream face of the existing embankment was not relevant for comparison with any aspect of the proposed raising.

5 STABILITY OF UPSTREAM RAISING

5.1 Constraints on construction

The fill material would consist entirely of coarse refuse and would be placed in a single lift for the first stage of raising. A standard compaction test (AS69-11A) on the coarse refuse gave a maximum dry density of 1.63 tonnes/m^3 at an optimum moisture content of 11%. Shear strength values were measured using both direct shear and triaxial tests with values selected for design calculations shown in Table 2.

The result of 49% from an Aggregate Crushing test on a compacted sample of coarse refuse indicated that material handling and compaction during construction could cause breakdown of the refuse to produce a more suitably graded fill. This would enable higher compacted densities and lower permeabilities to be achieved than indicated by laboratory tests.

The shear strength of the tailings was the other characteristic which had a significant influence on the pattern of

upstream raising. After a review of relevant literature and discussion with research staff (Dr David Williams, University of Queensland), a value of 10kPa was selected for the undrained shear strength of the tailings. Because placement of the first layer of fill would consolidate the tailings, the undrained strength was increased to 20kPa for subsequent stages of raising. The values selected were conservative at this stage of design because the purpose of stability analyses was to indicate the feasibility of the upstream construction method rather than providing information for the detailed design of the raising.

Thus, the constraints on upstream construction were the use of coarse refuse as fill material, the maximum height of the first stage due to the low shear strength of the tailings and the time required for consolidation of the tailings before the next stage of raising could commence.

5.2 Height of construction stages

Two distinct periods of tailings deposition are indicated in Figure 2. The lower zone has been consolidating under a higher pressure for a longer period than the upper zone. Thus, the upper zone should have lower undrained shear strength and be susceptible to failure under rapid loading due to placement of fill. The possible occurrence of a deep circular slip through the upper layer of tailings was assumed and Bishop's Simplified Method was used for stability analyses.

The shape of the fill was derived from the proposed final raised profile which had a crest width of 10m, an upstream side slope of 2.5h to 1v, and a downstream side slope of 2h to 1v. A sensitivity analysis for factor of safety versus height of raising was completed over a range of 1 to 5m. The minimum factor of safety (acceptable to mine management) was 1.25. For an initial fill height of 2.5m, the factor of safety was 1.27 at the end of construction and improved to 1.77 when 90% consolidation was achieved. If one lift was used to raise the crest level to its final height (6m above current level) when 90% consolidation was achieved for the first stage, the factor of safety would be 1.31 at the end of construction and 1.65 when 90%

consolidation was achieved. The location of the critical slip circle at the end of construction is shown in Figure 4.

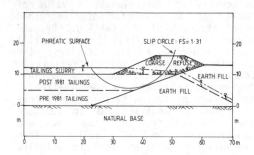

Figure 4. Proposed raised section.

5.3 Timing of construction stages

Prediction of consolidation times in geotechnical design is notoriously unreliable. Laboratory or field tests to determine consolidation characteristics had not been carried out for the tailings at Peak Downs Mine. However, laboratory test results were available for tailings at Riverside Mine (also operated by BHP-Utah Coal Ltd) located approximately 50km north, but recovering essentially the same coal seam. These results were adopted for this analysis and values were:

Compression Index, C_c = 0.162
Coefficient of Volume Decrease, m_v
= 0.47m^2/MN
Coefficient of Consolidation, c_v
= 2.0x10^{-7} m^2/s

Assuming one-way vertical drainage with a path length of 2.5m, 90% consolidation would be achieved in approximately 12 months. The settlement of the fill would be 180mm. The method of tailings deposition is likely to create a series of horizontal layers of higher permeability which would reduce the drainage path and hence the time to reach 90% consolidation. The estimate of 12 months is an upper bound to the actual time expected for field conditions.

5.4 Data for detailed design

In many cases, the only way to obtain reliable data for geotechnical design is to build a large scale model. Once the feasibility of upstream construction was demonstrated, the construction of a trial embankment with associated site investigation and performance monitoring

was recommended. The management at Peak Downs Mine adopted the recommendation and a major geotechnical study was undertaken by BHP Engineering with Uniquest (on behalf of the Department of Civil Engineering, University of Queensland) as sub-contractors.

6 TRIAL EMBANKMENT CONSTRUCTION AND MONITORING

Staff from the University of Queensland undertook the majority of the field and laboratory work associated with the trial embankment. A summary of the methods and results is presented in this section.

6.1 Preliminary investigations

To provide a valid basis for comparison, the undrained shear strength profile down through the tailings before placement of any fill was measured using a vane shear device. Both peak and remoulded values were measured. Block samples were recovered from near the surface for laboratory testing. Figure 5 shows the average undrained shear strength profile prior to construction. These values were used in the design of a trial embankment with the purpose of determining the rate of consolidation, magnitude of settlement and increase in strength of tailings due to consolidation. The embankment was not designed to induce failure in the tailings because this would have rendered the study useless in terms of predicting actual embankment performance.

6.2 Design and instrumentation

A rectangular embankment having plan dimensions of 26m by 100m with a height of 2.5m was selected. The long axis was parallel to the centreline of the existing embankment and approximately 100m away. The trial section was linked to the existing embankment by a narrow causeway built from coarse refuse.

Instrumentation consisted of settlement plates at depths of up to 3m below the surface of the tailings along the centre-line and edges of the trial section, standpipe piezometers and access tubes for a settlement profiler (Geokon Model 4651) placed on top of the tailings from one longer side of the section to the other. The location of instruments and the trial sections is shown in Figure 6.

272

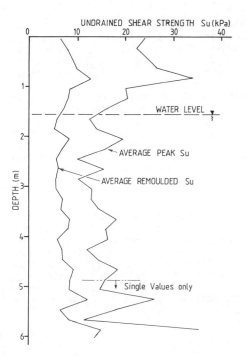

Figure 5. Undrained strength of tailings

6.3 Construction

Over a period of 6 hours, a 1.48m lift of coarse refuse was placed over the target area. This was monitored for three days (including the placement period) after which a further 1.19m of fill was added over a period of 9.5 hours. Monitoring settlement and water levels continuedthrough to 38 days after construction commenced. A second series of in situ undrained shear strength measurements were made before another 2.07m of fill was placed in 6.5 hours. Monitoring settlement and water levels continued for another 59 days. After that time, the majority of measurement points became inaccessible due to inundation by fresh tailings.

The total compacted volume of coarse refuse in the trial section was approximately 6050m^3. At an average compacted bulk density of 19 kN/m^3, this results in an applied vertical stress at the centre of the trial section of 90kPa.

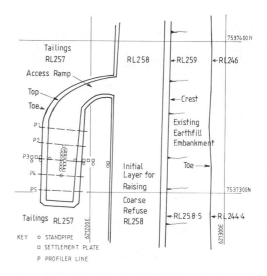

Figure 6. Plan of trial embankment

6.4 Settlement measurements

Results from the five settlement profile lines showed different values due to different construction loadings caused by the passage of earthmoving equipment around the instruments along the centreline of the section. The readings from individual surface settlement plates all agreed closely with measurements from the settlement profiler. However, the profiler gave a much clearer indication of the effects of construction than could be inferred from settlement plates. Figure 7 shows the settlement profiler results for line P5 (see Figure 6) and Figure 8 shows the variation in standpipe level during the monitoring period.

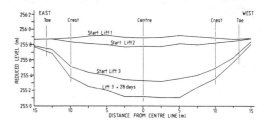

Figure 7. Settlement history, profiler line P5

273

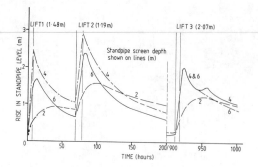

Figure 8. Standpipe level response.

7 DESIGN OF THE PROTOTYPE EMBANKMENT

7.1 Results from trial section

The trial section demonstrated that staged construction for an upstream raising is feasible within the design constraints of maximum height for each stage and the minimum time between stages. The use of simple measuring devices such as settlement plates and standpipes provides a reliable indication of degree of consolidation within the tailings.

The rate of pore pressure dissipation was significantly faster than predicted. Complete consolidation was observed within four weeks of the placement of the second and third lifts of the trial section, representing an order of magnitude faster than the 12 months initially anticipated. Increase in strength of tailings due to consolidation would have also occurred more rapidly, giving a shorter time for the lower factor of safety immediately after construction.

7.2 Design for raising

The two stage construction plan with lifts of 2.5m and 3.5m initially anticipated was retained. Using values for tailings strength measured in the field (undrained values) and laboratory(drained values), the factors of safety for the final crest level (6m raising) were 1.78 immediately after construction and 1.86 for the longer term.

With the placement of the first 2.5m of fill, the maximum settlement of the raised section was expected to be 150mm. A period of 4 months between construction stages was recommended. With the second stage of 3.5m fill, the maximum settlement was expected to be 350mm.

7.3 Performance and future work

Although detailed settlement measurements were not obtained for the raised section, no departure from expected behaviour for the raised section was reported by mine staff. Interest in a further upstream raising has been expressed by staff at Peak Downs Mine. Geotechnical investigations are being made into a possible 10m raising for the No. 2 Tailings Dam.

8 ACKNOWLEDGEMENTS

The author expresses his appreciation to BHP–Utah Coal Ltd for permission to publish this paper. The company's contribution of data on the existing embankment was very valuable. The efforts of Mr Dennis McManus (BUCL Technical Services) in obtaining longer term settlement readings are acknowledged. The significant contribution of Dr. David Williams, Department of Civil Engineering, University of Queensland is acknowledged. Dr. Williams and his staff completed most of the difficult and "dirty" tasks associated with site investigations and instrumentation for the trial embankment.

Lastly, the author acknowledges the contributions of his former colleagues at BHP Engineering, Messrs. Alex Eadie and Steve Rosin, for their assistance during the project and the assistance of Miss Dawn Anderson in preparing the typed version of this paper.

Reclamation, Treatment and Utilization of Coal Mining Wastes, Rainbow (ed.) © 1990 Balkema, Rotterdam. ISBN 90 6191 154 0

Seepage phenomena and control of phreatic surface within two lagoon embankments, constructed of coarse coal waste

B. E. Broś
Agriculture Academy of Wrocław, Poland

ABSTRACT: The paper presents two case studies regarding coarse coal waste embankments of fine coal refuse impoundments in Lower Silesia coalfield, in Poland. The consideration is based on 2 years field measurements of movement and phreatic surface within the dams. The field measurements of phreatic surface are compared with the results of model tests of seepage flow. The differences between fine coal waste impoundment embankments and conventional water storage dams are emphasized.

1 INTRODUCTION

Coal-mine refuse impoundment are engineering structures that involve two aspects of public concern. One is the structural stability of the embankments and the possible release, if failure occured, of large volume of water and semi-fluid deposits. Such an event cause extensive downstream pollution and pose a threat to life and property. The other aspect of public concern is the possibility of pollution, under normal operation, in which polluted effluent might escape through waste embankment into the streams or groundwater of the area. Instability of existing coarse coal waste embankments is usually associated with supernatant water seepage through these embankments. The phreatic surface governs to a large degree the overall stability of the embankment under loading conditions, in addition to influencing the susceptibility of the embankment to seepage-induced failure.

With respect to seepage control in refuse impoundment embankments it should be noted that unlike conventional dams, fine refuse dams are constructed for the purpose of storing a slurry of solids and water and the problems associated with seepage control for such

embankments differ from those encountered in the design of conventional water storage dams.

In August, 1969, two years after Aberfan / Wales / disaster, came to the failure of coal-mine refuse impoundment of GENERAL ZAWADZKI coal mine, in Upper Silesia, in Poland. About 90 thousands m^3 of water and semi-fluid deposits rushed into mine galleries intercepting 92 miners retreat.

Prior to 1969 some properly planned waste impoundment systems were designed in Poland using earth dam design practice. However the majority of the refuse systems were developed by the operators in the field, using techniques evolved over the years for small operation with a low production rate. When the operation grew in size and rate of production of refuse, no change occured in disposal practice other than to handle greater quantities of materials, without any design or construction control.

Following the GENERAL ZAWADZKI fine refuse impoundment disaster the Polish Ministry of Mines and Energy established a mandatory procedure for the technical and operational management of coal-mine refuse impoundments. According to this procedure all coal-mine refuse impoundments should be treated as civil engineering earthwork, with the aim of disposing of coal-mine refuse safely and economically.

The paper presents two case studies regarding upstream-type and downstream-type coarse coal waste embankments of fine coal waste impoundments, in the Lower Silesia coalfield, in Poland, based on field measurements of water levels within the embankments and embankments movement.

2 COAL-MINE REFUSE IMPOUNDMENT OF WAŁBRZYCH COLLIERY

Coal-mine refuse impoundment of WAŁBRZYCH-colliery in the city of Wałbrzych, in Lower Silesia, is typical side-hill refuse disposal structure.

The embankment of the impoundment is a homogeneous downstream-type dam, 22 m high and 695 m in length, errected in three stages, from compacted fresh coarse coal waste, mostly shale. In the downstream side of the embankment system of three internal, longitudinal drains have been installed. The drains of trapezoidal cross-section, double--layered, from broken stone and gravel, have been located on three different levels (Fig. 1).

The site of the coal refuse impoundment is underlain by practically impermeable carboniferous sandstones and conglomerates.

Fine coal waste is pumped in a slurry form into impoundment at the opposite side of the embankment.

2.1 Description of materials

Some physical properties of the fresh coal coarse waste, mostly shale, placed and compacted into embankment, are as follows:

Grain size range. - Fresh coarse coal waste, with more than 90% of shale and mudstone and a very small amount of sandstone and low quality coal, showed a range in particle size from gravel / 40 to 60% / to sand / 30 to 40% /.

The coarse coal waste mostly consisted of plate-shaped, elongated particles of shale, which were responsible for the poor engineering properties as well, as mechanical and chemical weathering.

Initially, as the coal waste is dumped on the surface of the embankment it is loose with a low unit weight and a large permeability. However, under compaction and as the effective overburden pressures increase with the raising of the embankment crest, elongated particles of shale break producing a particle matrix with larger amounts of fines and denser particles arrangement. This also results in a decrease in the friction angle and permeability.

Water content. - Water content of fresh coarse coal waste from spoil tips varied over a large range from 4,5 % to 18,9 %, generally about 6 to 9 %.

Optimum water content. - Optimum water content of fresh coarse coal waste varied from 8,4 to 10,3 %, generally about 8 to 9 %.

Total unit weight. - In-place densities of coarse coal waste, after spreading in thin layers of 20 cm and compacting with a bulldozer are lower and variable than in natural soils because of their lower specific gravities and large variations in their petrographic properties, particle size distribution, and degree of compaction.

The range of total unit weights was found to vary between 1,22 to 2,46 tonne/m³, most often between 1,80 to 2,0 tonne/m³

Because of the low unit weight the saturation of coarse coal waste in embankment due to a rise in the phreatic surface leads to a significant loss of effective unit weight.

Permeability. - Laboratory test on samples compacted to Proctor density have shown coefficients of permeability varying:

from $1,3 \times 10^{-3}$ cm/s to $1,9 \times 10^{-4}$ cm/s

The permeability of coarse coal waste material can vary over a large range because of the large variability in the gradation and in-place density. Fresh coal waste with no compaction and breakdown of particles would have a much larger permeability, where as compacted coal waste after some weathering would have a lower permeability than indicated by this range.

Shear strength. - The shear strength of the coarse coal waste

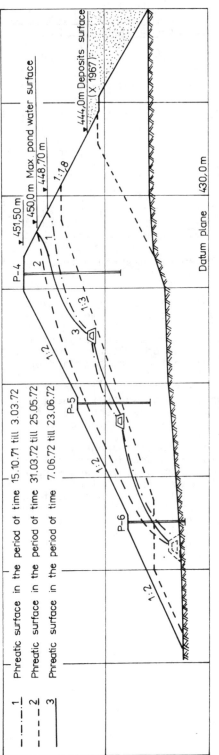

1 ——·——·—— Phreatic surface in the period of time 15.10.71 till 3.03.72

2 —— —— —— Phreatic surface in the period of time 31.03.72 till 25.05.72

3 ———————— Phreatic surface in the period of time 7.06.72 till 23.06.72

Fig. 1 Phreatic surfaces measured in piezometers

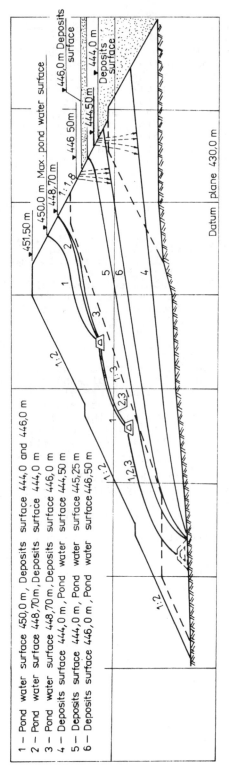

1 – Pond water surface 450,0 m, Deposits surface 444,0 and 446,0 m
2 – Pond water surface 448,70m, Deposits surface 444,0 m
3 – Pond water surface 448,70 m, Deposits surface 446,0 m
4 – Deposits surface 444,0 m, Pond water surface 444,50 m
5 – Deposits surface 444,0 m, Pond water surface 445,25 m
6 – Deposits surface 446,0 m, Pond water surface 446,50 m

Fig. 2 Model tests of seepage flow employing viscous fluid

277

was found to vary with gradation, unit weight and mineralogical composition.

Shear strength tests have been carried out in the direct shear test apparatus. The lower limit of effective friction angle represents a value of about 20° for weathered coal waste and the upper limit about 44° for fresh coal waste. The tests indicated considerable decrease of the friction angle with an increase of the content of fines and of water content.

2.2 Control of phreatic surface

The location of the phreatic surface, or internal water level, within an embankment exerts a fundamental influence on its behaviour, and control of phreatic surface is of primary importance in embankment design, construction and exploitation. The objective of prime importance is to keep the phreatic surface as low as possible in the vicinity of the embankment face.

After completion of the construction of the last stage of embankment in 1969, eleven piezometers have been installed to measure water levels within the dam. Three piezometers were placed in each of three transverse sections. Two piezometers were placed on the longitudinal axis of the dam.

Tailings ponds are usually raised slowly and take many years to reach full design height. The rise of the phreatic surface is caused by the rise in the impoundment pond level due to gradual filling of the reservoir.

In the space of time from October 1967 to the end of the year 1972 pond water level raised 6 m, from 444,0 m to the extreme level 450,0 m / Fig.1 /.

The first readings started from September 1970 and until the end of 1972 were taken every week. During the whole year 1971 pond water level was constant on the level of 448,70 m. All this time, till February 1972, water levels in piezometers were much the same, without any considerable variations. On the turn of 1971 pond water level raised 1,30 m, to the level of

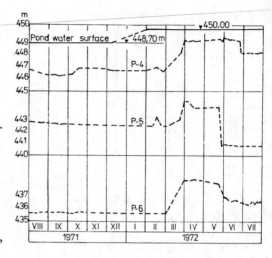

Fig. 3 Phreatic surface measurement

450,0 m. This last water elevation was reflected in piezometers reading with the delay of about 2 months, causing a considerable rise of water levels in March 1972 and remain of this culmination in April and May 1972 / Fig. 3 /.

In this space of time the effect of upper drains was very small and the top flow line rised above the zone of these drains / Fig. 1, line 2 /.

In the time of seepage water culmination pronounced horizontal seepage has been observed and seepage flow lines emerging out of the downstream slope of the embankment tended to slough off the slope because of the hydrodynamic pressure.

In the month of May 1972 the water level behind embankment was higher than normal due to heavy rainfalls. In the same time some small slides occured in the downstream slope of embankment.

After some months, because of swelling of shale, the water leakage in the dam stoped and the phreatic surface dropped, causing.steady seepage flow and all the drains permitted discharge of the seepage water / Fig. 1 , line 3 / .

In the central, elevated part of the embankment, of about 200 m in length, the phreatic surface in the time of the seepage flow culmination was situated very high, being

278

uneven and undulated and showing
differences of some meters in its
height. The downstream slope surface
was in different places very wet
because of the horizontal seepage.

Since a compacted coarse coal
waste embankment is placed and
rolled in horizontal layers, it is
to some degree stratified regardless
on the care taken in the construction control. Because of the stratification, it bekomes more pervious
in the horizontal direction than
in the vertical.

Almost all the additional flow
occurs through the face of the dam
above the slimes, where water is
in direct contact with the embankment.

A seepage flow problem can be solved by constructing a scaled model
and analyzing seepage flow in the
model. Soil models are of limited
use mainly because of the difficulties caused by capillarity.

Model employing viscous fluid was
used to study seepage flow in the
embankment, since the viscous material will follow the same flow
laws as water in a soil.

A model in the scale of 1 : 75
with transparent glass plates spaced
closely together and filled with
grease oil was constructed. The results of the research work are
shown in Fig. 2.

In the case, when all three drains
have been functioned:
- when the pond water have had
  great thickness the different
  levels of deposits surfaces
  influenced only in a very small
  degree the shape of flow lines
  between the drains, for different levels of pond water

- when the thickness of ponded
  water is smaller than 1,25 m
  takes place the interruption of
  the top flow line and seepage
  from the disposal site produces
  a wetting front that moves vertically downward, toward the
  water table in saturated zone
  of the embankment.

- only in the case, when the
  thickness of ponded water is
  greater than 1,25 m, the top
  flow line in embankment is
  connected with the surface of
  pond water as a continuous line.

2.3 Movements in embankment due to reservoir filling

Measurements made in many earth and
earth-rock dams have shown that
large settlements, horizontal movements, and cracking are frequently
caused by reservoir filling. These
complex movements may be understood
as the combined effect of two coun-
teracting effects of reservoir
filling, namely:
- the water load on the embankment
- the softening and weakening of
  the fill material due to wetting

The measuring devices have been
installed in three piezometric
cross-sections. Postconstruction
embankment movements were measured
in ten points during the period
of two years: 1970 - 1972.

The vertical postconstruction
settlements of the embankment in
the period of two years varied
      from 1,6 cm to 6,9 cm

The measured settlements were
proportional to the heights of
embankment in the corresponding
cross-sections. The greatest settlement of the crest of embankment was
6,9 cm and it makes 0,37 % of the
height of embankment in adequate
cross-section.

Cumulative postconstruction
compression of the USBR dams, in USA,
from all cases averaged less than
0,2 % in the first 3 years after
completion and less than 0,4 % in
period up to 14 years / Sherard
et al., 1963 /.

The measurements of the horizontal
movement were made during the same
period of 2 years of operation.
The crest of the embankment have
moved downstream and the horizontal
movements varied
      from 1,4 cm to 6,1 cm
an amount a little under 100 %
of the vertical settlement of the
crest.

In the period of time from March
72 to May 72, have appeared the
biggest horizontal downstream movement in the range 1,9 cm to 5,3 cm,
which reflected high level of the
water in reservoir and seepage flow
culmination in this time.

Therefore the downstream movements
resulting from the water load are
greatest during the later stages
of reservoir filling.

The measurements of the embankment
vertical movement carried out every

month on several benchmarks ,
distributed along the crest and
downstream slope, showed that the
settlement of the embankment was
three times interrupted by process
of the dam expansion, as a result
of seepage water penetration from
the reservoir into embankment, causing a swelling of the shale.

In the period of time from 1 Aug.
70 to 1 Sept. 70 all benchmarks
heaved from 1 mm to 13 mm, then
in the space of time from 21 Sept.
70 to 29 Oct. 70 they heaved from
1 mm to 5 mm, and finally from 27
May 72 to 13 June 72 the heave was
from 1 mm to 10 mm.

Neither was any indication in
Europe or in USA of embankment
swelling, because none of the dams
was constructed of soils having a
large percentage of fines with high
plasticity. However swelling of well
constructed earth dams of highly
plastic clay has been reliably measured. At two dams in Israel built
of clay with liquid limit of about
80, the upper 6 to 7,5 m of both embankments swelled when the reservoir
filled, and the crests heaved several centimeters / Sherard et al, 1963/

Annen and Stalmann /1963/ observed swelling of carboniferous shales
in colliery spoil tips. According
to them the embankments constructed
from shale can expand because of
swelling, and in result the crest
of these embankments may heave of
about 2 to 3 cm.

3   COAL-MINE REFUSE AND FLY ASH
    IMPOUNDMENT OF VICTORIA COLLIERY

Coal-mine refuse and fly ash impoundment of VICTORIA colliery, situated
in the city of Wałbrzych is typical
cross-valley impoundment, confined
by a dam extending from one valley
wall to another, located near the
head of the drainage basin to minimize flood inflows.

The embankment 34 m high, is an
upstream-type dam from locally
available natural material as well,
as from coarse and fine coal wastes
and fly ash, constructed in the
very long period of time, often
uncontrolled, and heightened in
many stages / Fig. 4 / .

The oldest initial dike / zone „1"
in Fig.4 / is probably constructed

from coarse mining rock waste with
some local gravel and sand.
Subsequent dike has been built up
with compacted coarse coal waste,
mainly shale / zone „2" in Fig.4/.
For the last two stages, compacted
fly ash and fine coal refuse have
been used with outer slopes constructed from fresh coarse coal
waste.

The initial area of the reservoir
is divided into two lagoons, which
allows a sequential filling operation. The fly ash and fine coal waste
are pumped in a slurry form into
storage lagoons, by spigotting off
the top of the dike. This forms a
gently sloping beach, where the
coarsest fraction settles near the
point of discharge and the fine
fraction, mostly silt, is deposited
progressively toward the centre of
the disposal area. Every next dike
is underlain by deposited fly ash
and fine coal refuse / zone „3" in
Fig. 4 / .

In the downstream side a big two-
-layer graded filter drain at the
toe of the embankment as well, as
internal drains at every toe of
subsequent dikes were installed.

3.1 Description of materials

Two sorts of materials, placed in
the embankment, were investigated,
namely fly ash mixed with fine coal
refuse and coarse coal waste.

Some properties of fly ash mixed
with fine coal waste are as follows:
according to grain size distribution
investigated material can be mostly
classified as silt or sandy silt,
with water content ranging
from 21,1 to 87,3 %
    Specific gravity 2,08 tonne/m^3
    Total unit weight:
        1,03 to 1,88 tonne/m^3
    Coefficient of permeability:
    $1,7 \times 10^{-6}$ to $7,4 \times 10^{-7}$ cm/s

Some properties of coarse coal waste
are as follows:
    according to gradation material
can be mostly classified as gravel
and sandy gravel with water content
ranging from 5 to 20 %.
    Total unit weight:
        1,8 to 2,0 tonne/m^3
Coefficient of permeability:
    2×10^{-5} to 2×10^{-3} cm/s

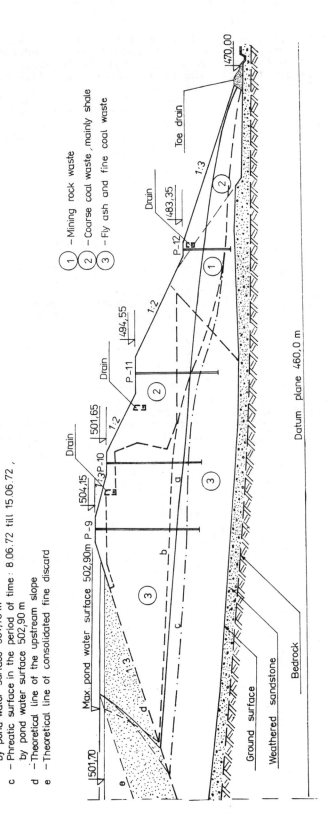

a – Theoretical phreatic surface
b – Phreatic surface in the period of time 1.05.70 till 8.07.70,
    by pond water surface 501,70 m
c – Phreatic surface in the period of time: 8.06.72 till 15.06.72 ,
    by pond water surface 502,90 m
d – Theoretical line of the upstream slope
e – Theoretical line of consolidated fine discard

① – Mining rock waste
② – Coarse coal waste , mainly shale
③ – Fly ash and fine coal waste

Fig. 4   Phreatic surfaces within embankment according to measurements in piezometers

281

Unknown in details is material placed in the initial dike / zone „1" in Fig. 4 /. Probably it consists from mining rock waste mixed with sand and gravel. Its coefficient of permeability may be estimated as

$$1 \times 10^{-1} \text{ to } 1 \times 10^{-2} \text{ cm/s}$$

3.2 Control of phreatic surface

In the period of years 1968-1969 16 piezometers have been installed in the embankment for purpose to measure phreatic surface within the dam.

The mode of discharge of fly ash and fine coal refuse into lagoons caused the formation on the upstream slope a zone of consolidated deposits which thickness can be estimated as t = 30 m and its coefficient of permeability as $k = 10^{-4}$ cm/s

The embankment is not homogeneous and because of that for determining the location of the theoretical top flow line the zone of consolidated fine refuse was theoretically replaced for material, from which the embankment was mainly constructed and which coefficient of permeability is $k_i = 10^{-3}$ cm/s, according to proportion:

$$\frac{k}{k_i} = \frac{T}{t}$$

where: T is a substitute thickness of the zone of consolidated deposits with the coefficient of permeability $k_i = 10^{-3}$ cm/s.

In Fig. 4 we can see:
- theoretical phreatic surface „a" which is stated by the assumption that the cross-section of embankment has been formed as an semi-
-impervious upstream zone and much more pervious downstream zone
-observed in 1970 phreatic surface „b", which was situated very low
- observed in 1972 phreatic surface „c", which was situated much lower than the phreatic surface „b" , despite of much higher position of pond water in 1972 than in 1970. This phenomenon may be explained by the effect of sealing up the bottom of lagoon by fines in time.

Because of the presence of a slimes zone, considerable headloss occurs within these low-permeability fine refuse and seepage flow becomes directed horizontally outward toward the embankment face. The resulting phreatic surface is considerably low and distinctly draw down within the zone of initial dike with higher permeability.

4 CONCLUSIONS

The main difference between coarse coal waste embankments and common earth or rock embankments is, that the coarse coal wastes have a low buoyant weight on submergence, because their specific gravity is significantly lower than common soil and rocks, and therefore have a low shear resistance. Hence, the rise of the phreatic surface and seepage parallel to the slope results in a major loss of effective weight and stability of embankment.

Embankments constructed from different sorts of coal waste are not isotropic. Because they are compacted in thin lifts, the horizontal permeability is usually higher than the vertical one.

Phreatic surface location within the embankment is in a high degree influenced by pond location, variations of the spigotted fine waste and anisotropic permeability of the fine refuse deposits.

REFERENCES

Annen,G.,Stalmann, V. 1969. Waschberge im Deich- und Dammbau. Glückauf, nr 26, pp. 1336-1343.
Broś,B., Kowalski,J., Haszto,B. 1974. Kształtowanie się filtracji w zaporze ziemnej zbiornika osadowego na drobnoziarniste odpady po przeróbce węgla kamiennego. P.T.M.T.S., Gdańsk.
Mittal,H.K., Morgenstern,N.R. 1976. Seepage control in tailings dams. Can. Geotech. J. Vol. 13
Sherard,J.L., Woodward, R.J., Gizienski,St.F., Clevenger,W.A. 1963. Earth and earth-rock dams. J.Wiley, New York.

Reclamation, Treatment and Utilization of Coal Mining Wastes, Rainbow (ed.) © 1990 Balkema, Rotterdam. ISBN 90 6191 154 0

The settlement of constructions founded on uncompacted fills

J. Fiedler
Inst. Nabeul, Tunisia (Formerly: KMP Prague, Czechoslovakia)

ABSTRACT: The behaviour of 2 industrial halls founded on the mining waste 12 years old of about 20 m thickness is described. Site investigations were carried out by the combination of laboratory and in situ testing/penetration, pressurometer /. Lateral nonhomogeneity of subsoil compressibility was evaluated for the calcul of soil-structure interaction. The rough estimate of total and differential settlement of the fill due to the own weight was based on an analogy with the behaviour of similar fills in the region because the settlement on the site was not measured. These values of settlement were also introduced in the calcul on a computer.

The halls were founded on a heavily reinforced strip foundation. The settlement of the first one corresponds approximately to the prediction. The realisation of the foundations for the second hall was carried out in the very difficult winter conditions. This influenced the properties of the fill. A very large settlement appeared during the erection of the superstructure. One part of the hall had to be reinforced. During a couple of years the settlement of every column of the hall was measured. Some results of these measurements will be presented as well as the evaluation of experience gained during the project and realisation.

1 INTRODUCTION

Shortage of suitable sites in areas of intensive opencast coal mining led to attempts to build even on uncompacted mining wastes which form very non homogenous and compressible underground. It brings a lot of problems during site investigations, design and construction. The problems of site investigations of uncompated fills are described in (1).

Structural engineer needs not only the information about the compressibility of the subsoil (especially about the non homogeneity and the behaviour of the soil after the contact with water) but also a forecast of the future settlement of the fill due to its own weight. This is a difficult task particularly if no measurements of settlement on the site are available. Then only a rough estimate of the total and differential settlement based on an analogy with behaviour of other fills is possible.

· The analysis of soil-foundation interaction on a computer enables to consider lateral non-homogeneity as well as differential settlement of the subsoil due to own weight. Incertitude concerning the soil data oblige to make a couple of variants. Thus a simplified method of interaction analysis is preferable than too sophisticated one. The influence of non-homogeneity was treated in (2).

Writer participated for nearly 10 years on a site investigation, project, realisation and subsequent observation of one industrial area founded on a mining waste. The behaviour of one heavy object was satisfactory and corresponded approximatively to the prediction (based on methods mentioned above). The description of the design and the settlement of this object is in (3). The settlement of the other hall was excessive already during the erection of the superstructure. This called for partial strengthening of the object. A detailed observation of the information. Some of them are presented in this contribution. Some restrictions in the presentation are due to the fact that the writer does not participate on the job now.

2 GEOTECHNICAL CONDITIONS AND DESIGN OF FOUNDATIONS

The subsoil was created by backfilling of the old opencast mine with overburden from the nearby area about 12 years before the start of the construction activity. The thickness of the fill is about 12-20 m. The fill is composed mainly of clay ($1p$ = 25-35% L = 50-60%, activity by Skempton A<0,7). Consistency of soil samples (1_c = 0,85-1,22). There was mainly a hard consistency in the boreholes and excavations without visable macro-pores. The underground water was not detected during site investigations but due to the position of the fill, the influence of water on the behaviour of the fill was taken into account. Some small springs were detected during the excavations.

From the beginning of the design stage some points for settlement measurements were installed on a surface of the fill. Their settlement was about 1-2 cm year.

If the objects were founded on piles the negative friction would be important. The settlement of the fill around the object could damage the underground conduits. In addition the owner (who was the contractor as well) had not his own piling department. Therefore heavy reinforced foundation strips lying on a thick layer of a compacted sand were chosen.

The dimensions of the object were about 60x60 m, (sketchy figure of the hall is in fig 1). The schematic section of the foundation is in fig 2. The pressure on the top of the sand layer due to the construction was 117 kPa. The thickness of the fill was about 20 m.

The estimation of the fill settlement due to the own weight was based on an analogy with the behaviour described in English and Czech literature. Differential settlement of 0,09 m was adopted for the calcul. The different variants of lateral non homogeneity of the compressabilty were analysed (the ratio E_{max} / E_{min} =2,2 was considered). Then the maximum bending moment Max = 3,8 MNm was chosen. The reinforcement placed "up and down", in every point of the strip was calculated

for this bending moment.

The quantity of steel was considerable but this was judged necessary due to incertainties in the fill behaviour. The foundation strip with opening in the centre would be preferable to reduce the weight of the structure but the technology suitable for the contractor was adopted.

3 REALISATION

The first object was successfully founded during the summer and autumn but there was a delay in the start of the construction of this object. Thus it was decided to realise the foundation during winter period. Unfortunately that winter was very unfavourable. The rainfall was twice more than the long term average. There was a couple of periods of frost and thaw. The realisation of the foundations in these conditions was difficult and slow. In spite of some measures taken by contractor the protection of the subsoil turned up to be insufficient and the fill properties were changed. As soon as during the erectio of the superstructure the excessive settlement suddenly occured with apparition of fissures in the concrete construction.

Comparison of the projected and real position of the different parts of the structure enable to estimate roughly the settlement (see fig 3).

After a detailed analysis of the state of the construction it was decided to strengthen it only partially. The brick wall (not yet realised) enter the columns of the row B was replaced by a reinforced concrete wall of small thickness. The further construction activity was permitted under condition that the settlement of every column of the hall and some new points around the building would be regularl measured.

Writer was charged by evaluation and the forecast fo the future behaviour from the geotechnical point of view after every series of measurement in the period 1981-1985. The measurements and observation of the hall continue even today. The results presented concern mainly the behaviour before 1985, however some new datas are included.

There was a difference in the behaviour of each row of columns. After the strengthi of the row B the settlement was regular there. The situation in the row D was not so favourable. The settlement of this row is in fig 4. There is a continual increase in the deflection there, however with the small velocity. As yet no strengthening was undertaken there.

The settlement of the fill due to the own

weight continues even up today with different intensity. For example enter 3.1985 and 10.1988 one point settled 4,9 mm only but the other 35,1 mm (the settlement of this point since the start of construction activity about 7 years ago is 110 mm). During the period 10.86 - 10.88 the settlement of the columns of the hall (47 points) was between 5 and 48 mm.

The great risk of the construction on the uncompacted fills is relied to the fact that the settlement due to the later change of water regime may appear even when the initial behaviour is satisfactory. This can be illustrated on the settlement of the row A. In the critical period of the consruction the settlement and deflection was minimal. However after 2 years the differential settlement due to gradual change of the soil properties in the corner of the hall appeared (see fig 5). Since that time this row continues to turn in the other direction than the row D which is defavourable for the superstructure. Thus the behaviour of the hall will have to be observed also in the future.

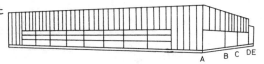

Fig. 1 Sketchy figure of the hall

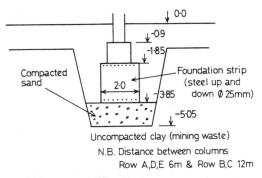

Fig. 2 Foundation of the hall

4 CONCLUSION

It is probably impossible to exclude the construction on uncompacted fills in the areas of intensive opencast mining. However it is necessary that all participants are well informed about inherent risks of the construction under these conditions. Prudence is recommended during the design and realisation. A long term observation of finished objects is desirable. The described construction was one of the first of these dimensions realised in the region. Thus not always the optimal solution was achieved. However the experience gained could be useful for those will realise the constructions under similar conditions.

(1) Fiedler J. Site Investigations and foundations on uncompacted fills, VII Konf. Mech. gruntov 1 fundamentowanija, Poznan, 1984.

(2) Fiedler J. The influence of the subsoil model on the behaviour of a strip foundation of an industrial hall (in Czech) Konference Zakladanistaveb, Brno 1984.

(3) Feidler J. Berechnungen zur Grundung Einer Industriehalle auf einer Heterogenen Kippe Bauplannung-Bautechnik 1984, Nr.1,22-23

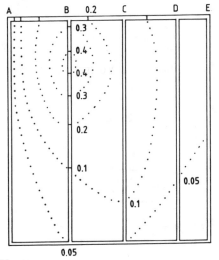

NOTE
THE POINTS ON THE FOUND STRIPS SHOW THE SETTLEMENT OF THE STRIP

THE DOTTED LINES ARE PRESENTED FOR VISUALISATION ONLY (THEY DO NOT CORRESPOND TO THE SETTLEMENT OF THE SOIL SURFACE BETWEEN THE STRIPS)

THE VALUES OF SETTLEMENT ARE IN METERS

FIG.3 SETTLEMENT AFTER SUDDEN DISPLACEMENT DURING THE ERECTION OF THE SUPERSTRUCTURE IN 1981

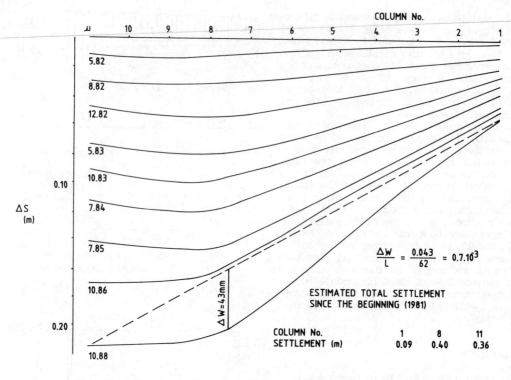

FIG.4   MEASURED SETTLEMENT IN ROW Ⓓ SINCE 1982

$$\frac{\Delta W}{L} = \frac{0.043}{62} = 0.7.10^3$$

ESTIMATED TOTAL SETTLEMENT
SINCE THE BEGINNING (1981)

COLUMN No.	1	8	11
SETTLEMENT (m)	0.09	0.40	0.36

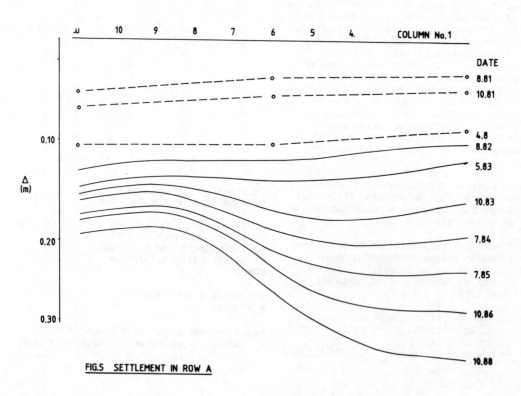

FIG.5   SETTLEMENT IN ROW A

Reclamation, Treatment and Utilization of Coal Mining Wastes, Rainbow (ed.) © 1990 Balkema, Rotterdam. ISBN 90 6191 154 0

Mechanism and performance of intensified ramming on coal refuse foundation

Li Shuzhi & Lu Taihe
Mine Surveying Research Institute, Central Coal Mining Research Institute, Tangshan Branch, People's Republic of China

ABSTRACT: More and more ground back filled with loose coal refuse is used as construction sites in China. How to reinforce the refuse foundation is a problem requiring immediate solution. It proves that intensified ramming is the best approach for reinforcing the loose refuse foundation. This paper goes on to elaborate on analysis of the laws governing the settlement of refuse at different depth, the stress of the refuse foundation and the variation of particle size, density and moisture content of refuse caused by intensified ramming. Reference is also made of the reinforcement mechanism and results obtained. Discussion is made on the correction coefficient for the depth of compaction resulted from intensified ramming when using the Menard's emiprical formula. It is therefore of vital significance for evaluating the results obtained with this method and the design of the intensified ramming operation.

INTRODUCTION

Reuse has been made in China of the waste land reclaimed from subsided areas in coal mines. More and more subsided areas restored by means of backfilling with coal refuse and industrial waste materials will be used for construction sites. How to reinforce the newly packed loose foundation has become a problem requiring immediate solution.

Intensive ramming is an effective approach for reinforcing loose foundations, though rarely applied for treating refuse ground. Little is known so far of the mechanism of intensified ramming in this case, and no in-depth investigations into the reinforced depth, selection & determination of opitmal parameters for intensified ramming operations have been conducted. Analysis is made in the paper of the mechanism of intensified ramming operation and the reinforced depth obtained based on test results gained in this field.

GENERAL DESTRIPTION

The site selected for conducting the experiment was originally a flat farmland which had become subsided and waterlogged due to coal mining. The site with a subsidence value of 5m and water depth of 3m had been backfilled with coal refuse 2 years ago.

The backfilled materials were composed of sandstone, shale and free-ash coal which were mostly in flaky and pointed form. It was therefore a loose ground filled up with soft rock pieces.

The experiment was conducted in the following sequence: excavation of the subsided pit layer by layer → backfilling on layer by layer basis and laying of sensing elements and settlement indicator layer ⟶ intensified ramming & observation ⟶ re-excavation & inspection.

The rammer used was made of reinforced concrete with an undersurfase area of 2 x 2m and a total weight of 9.2t. The hoisting equipment was a 25 ton capacity rubbertired hydraulic crane. The hammer was released at a height of about 11m under the control of an automatic unhooking unit. The ramming operations were continuously carried out for 11 times with a total ramming energy of 1100 t/m.

MECHANISM OF COMPACTION OF REFUSE FOUNDATION WITH INTENSIFIED RAMMING

1. Analysis of deformation of refuse foundation treated by intensified ramming

The refuse foundation after compaction with intensified ramming has been found to experience great deformations. The magni-

tute of settlement as observed was shown
in Fig. 1.

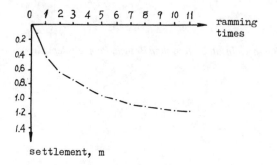

settlement, m

Fig. 1 Settlement of refuse foundation
treated with intensified ramming operation

As can be seen from Fig.1, the deformations
to which the newly-filled loose foundation
was subjected involve 3 stages:

Compaction stage: In this stage the newly
filled refuse foundation changes from loose
state to a relatively compacted stage. It
features that there appears a noticeable
vertical compression yet with little breakage of refuse. The initial 2 times of ramming will bring the foundation into the
compression stage. The magnitude of settlement after these 2 times of ramming usually
amounts to 637 mm which shares about 52% of
the total. The number of ramming times in
this stage depends mainly on bulk density
of the refuse layer and also the energy of
each ramming applied. As for the refuse
ground backfilled over years or already
compacted, there may not exist such a stage
when intensified ramming operation is
applied.

Solidation stage: In this stage the refuse
ground becomes further compacted or even
highly compacted – a stage in which the
load-bearing capacity of the foundation is
remarkably enhanced. It features that the
refuse particles are subject to breakage,
resulting in variation of size consist and
closer contact of skeleton particles. The
deformation observed is mainly in the form
of vertical deformation. This stage appears
when 3–7 times of ramming are applied. The
settlement value is found to be 429 mm,
sharing 35% of the total.

Reconsolidation stage: In this stage the
reinforcing effect brought about by intensified ramming is less pronounced and
instead there appears a greater lateral
compression deformation along with a settlement rate tending to be equal in speed. No
more ramming operations should be carried
out in this stage. If the compactness of

of the ground is not up to the design
standards, the impact energy applied by
each ramming should be increased to avoid
any possible waste of labour and adverse
effect on the foundation.

There is no rigid demarcation lines
between the 3 stages as described above
and in some cases it is very hard to make
a clear distinction. But it is of no real
importance. In actual operation, the designing standard is to be met through ramming
on a trial basis.

2. The additional stress to which the
   refuse ground is subjected during
   intensified ramming。

Shown in Fig. 2 are the additional stresses
observed with pressure cells.

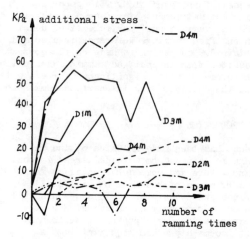

Legend
—— Vertical additional stress below
   the rammer
—·— Vertical additional stress at 2.5m
   apart from centreline
——— Horizontal additional stress at
   2.5m apart from centreline

Fig. 2 Additional stress vs number of
ramming times

As evidenced by observation results, the
additional stress experienced by the refuse
foundation increases rapidly at the initial
stage. However after 2–3 times of ramming
fluctuation occurs as a result of looseness
of refuse layers, uniformity of particle
size, degradation of size caused by intensified mining and redistribution of stress.
Under the effect of impact energy produced
by intensified ramming, the stress in the
refuse foundation increases continuously

and finally a new balance is produced due to redistribution of internal stress caused by the degradation of skeleton coarser particles because the breaking limit of refuse is reached. With further ramming operations, the skeleton refuse particles will be subjected to further degration in size, and a new balance will be reached due to redistribution of stress. So with continuous ramming, the additional stress will fluctuate in value.

The pressure cell imbedded in refuse foundations backfilled under normal conditions is subjected to uneven loads. When the compression member of the pressure cell is located at the place where the coarse skeleton refuse particle are subjected to compression, the cell would indicate a larger value, and the reverse is true when the compression member is located in the pressure protected area in the skeleton coarse refuse particles. Moreover the pressure cell which showed a larger stress value before intensified ramming would indicate a larger additional stress during intensified ramming, and vice versa.

The additional vertical stress induced by intensified ramming operation increases rapidly during the initial 5 times of ramming and after that it increases slowly. While the contrary is true of the additional horizontal stress under the same conditions. View as a whole, the additional horizontal stress is comparatively small. It indicates that, 1) The deformation of refuse foundation induced by intensified ramming is mainly in the form of vertical deformation, and the additional horizontal stress is relatively small; 2) The initial 5 times of ramming can produce the most pronounced reinforcing effect.

3. Variation in size, density & moisture content of refuse particles

Size compositions of the refuse before & after intensified ramming are listed in Table 1. Intensified ramming causes intensive degradation of refuse particles, especially those in the upper layers. The remarked decrease in size after ramming not only changes the size composition of particles in each layer and also increases the nonuniformity coefficient of refuse, thus resulting in a much improved size consist.

The densities of refuse measured both before and after intensified ramming are listed in Table 2.

The refuse foundation freshly filled up in a natural state is very loose with a density of only 1.541 g/cm³ which will increase to a large extent after 2 years of natural settlement. However it varies slightly with increase of depth, being 1.685 g/cm³ in average. When intensified ramming is applied, the density greatly increases. The average density of the 5 refuse layers observed through excavation is 1.822 g/cm³, about 18.2% higher than that of freshly refuse foundation.

Before intensified ramming the moisture content of the refuse fillings are high in the upper layers and low in the lower layers due to the effect of natural precipitation. The moisture content is remarkably reduced after intensified ramming. The moisture is found to be low in the upper layers and high in the lower layers, the very reverse of the situation before intensified ramming. It indicates that during intensified ramming, dewatering is effected, beneficial to the further consolidation of refuse foundation with continuous ramming.

EFFECT OF REINFORCEMENT OF REFUSE FOUNDATION PRODUCED BY INTENSIFIED RAMMING

The reinforcing effect produced is mainly ascortained by the depth that can be reinforced. Reinforcing depth is a key parameter for the optimal design of intensified ramming operations, which is generally estimated by using the Menard's empirical formula: i.e.

$$h = \sqrt{QA}$$

Where h – Reinforcing depth, m
Q – Weight of rammer, t
A – Dropping height of rammer, m

Is Menard formula applicable for direct calculation of the reinforcing depth effected by intensified ramming in refuse foundation? Determination of the applicability of Menard formula is therefore made through study and analysis on the relation of settlement, additional stress and density of refuse foundation to depth under conditions when intensified ramming is applied.

1. Relation of settlement of foundation to depth

Shown in Fig.3 is the relation of settlement of refuse foundation to depth when intensified ramming is applied.

It is evident from Fig. 3 that at shallow depth, the settlement of refuse foundation assumes approximately a linear relationship with depth, and the straight section of the curve has a large gradient.

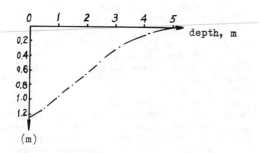

Fig. 3 Settlement of refuse foundation vs depth

The settlement is found to decrease rapidly with increase of depth. Thus it can be ascertained that at deeper depth the settlement will rapidly decrease with the increase in depth, i.e. a steep rather than flat curve is observed. Therefore the following conclusions may be reached: Under the effect of an impact energy of 100 t-m per ramming in the test, the ramming-induced deformation of refuse foundation usually takes place within 5 m in depth. The refuse foundation down to a depth of 5 m can get reinforced because reinforcing effect is generally produced at places deformation has taken place.

2. Relation of additional vertical stress to depth

When intensified ramming operation is applid, the relation of additional vertical stress to depth is shown in Fig.4. The additional vertical stress will increase with the increase of depth, but when a certain depth is reached, it will decrease with the increase of depth with a decreasing rate greater than the increasing rate. With increase of number of ramming times, additional stress values at different depths tend to increase and extend to further depth. It can seen from variations of additional stress with depth that the results obtained by intensified ramming of coal refuse foundation vary with the depth. The reinforcing effect in the upper layers is getting better and better with increase of depth, and an optimal value is reached at a certain depth. Further below the reinforcing effect deteriorates with decrease of depth. It can be seen from Fig.4, under the effect of ramming energy in the test that the maximum additional stress takes place at a depth of about 3 m with a transferring depth of 4.5-5 m. In consideration of the actual number of ramming times, the trans-

ferring depth of the additional stress is estimated to be about 5 m, i.e. the effective reinforcing depth is about 5 m.

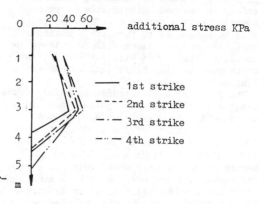

Fig. 4 Additional vertical stress vs depth

3. Relation of coal refuse density to depth

Shown in Fig. 5 is the variation of density of refuse foundation with depth.

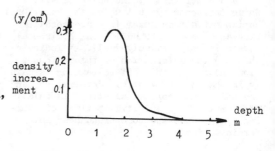

Fig. 5 Density increament vs depth

From Fig.5 it can be seen that the density increases remarkably with increase of depth at the initial stage and then rapidly reduces and then gradually changes to zero. The presence of the maximun value may be attributed to the loosening of surface layer caused by the vertical component of shock waves rather than the density increament of the uppermost layers. So the increase of density of refuse foundation due to intensified ramming, contributes to reinforcement of foundation. It can be seen from Fig. 5 that the influence depth caused by density variation due to intensified ramming is about 5 m.

In summary, in the tests a refuse foundation which has experienced 2 years of natural settlement was treated with a rammer with an undersurface area of 2x2 m^2 and a total weight of 9.2 t. The dropping height of the hammer was about 11 m, the ramming operations were repeated for 11 times with a total ramming energy of 1100 t-m. Analysis made on the relation of settlement additional stress and density increament experienced by the refuse foundation revealed to depth that the reinforced depth of refuse foundation by intensified ramming is about 5 m much less than the figure of 10 m obtained by Menard's empirical formula (h =$\sqrt{QA}$).

CONCLUSION

1. The mechanism of reinforcement and consolidation of refuse foundation can be described as follows: Under the effect of great impact energy, shock waves are produced in the refuse layers, resulting in severse deformation and stress between the refuse particles, consequently the following effects are produced: Degradation and rearrangement of refuse particles; continuous contact between skeleton particles; increase of density of particles due to compression of voids; further consolidation of refuse particles due to reduction of free moisture; great enhancement of bearing capacity.

2. The value obtained by using Menard formula (h =$\sqrt{QA}$) for direct calculation of the reinforcing depth effected by intencient of 0.4 - 0.7$\sqrt{QA}$

3. With a number of ramming times less than 7, the reinforcing effect brought about is less pronounced. If the reinforcing effect and depth are to be enhanced the impact energy applied by each ramming should be increased. However to rely solely on the increase of ramming times is infeasible both economically and technically.

4. Deformation of refuse foundation induced by intensified ramming is mainly in the form of vertical deformation and horizontal deformation is relatively small. The reinforcing range in the lateral direction is less than static-pressure angle of 45.

Table 1. Size composition of refuse particles before & after intensified ramming

Layers in descending order	Limiting diameter		Mean diameter		Effective diameter		Nonuniformity coefficient		Remarks
	Before	After	Before	After	Before	After	Before	After	
1	100	25	63	13	5	3	20	8.3	Freshly filled
2	48	19	31	11	4	3	12	6.3	Freshly filled
3	45	24	29	13	7	3	6.4	8	
4	31	38	22	23	7	3	4.4	12.6	
5	40	39	30	26	7	3	5.7	13	
6	44		34		8		5.5		
7	92		68		9		10.2		

Table 2. Density of refuse measured before & after intensified ramming

Unit: g/cm^3

Layers in descending order	1st	2nd	3rd	4th	5th	6th	Remarks
Before	1.541	1.541	1.694	1.665	1.719	1.661	Layer 1 & 2 are freshly filled
After	1.802	1.849	1.945	1.789			

Reclamation, Treatment and Utilization of Coal Mining Wastes, Rainbow (ed.) © 1990 Balkema, Rotterdam. ISBN 90 6191 154 0

Compaction of opencast backfill beneath highways and associated developments

J.Thompson
British Coal Opencast Executive, Stoke on Trent, UK

J.M.W.Holden & M.Yilmaz
Scott Wilson Kirkpatrick Geotechnics, UK

ABSTRACT

The paper describes the controlled backfilling that was undertaken at two opencast sites prior to the construction of the new A42 Birmingham to Nottingham Trunk Road near Ashby-de-la-Zouch in Leicestershire, UK. It presents the processes involved in obtaining agreement to opencast operations in advance of highway construction, the specification employed for backfilling, the levels of technical supervision employed, the standards of compaction achieved and the results of a backfill settlement monitoring programme.

1. INTRODUCTION

The construction of highways over coalfields can prevent the extraction of significant economic shallow reserves of coal. Opencast mining with controlled backfilling in advance of highway construction is a way of overcoming the problem and has been undertaken on a limited number of sites to date, generally under minor roads. The proposed alignment of the A42 Trunk Road in the vicinity of Ashby-de-la-Zouch in Leicestershire potentially sterilised significant coal reserves within two adjacent areas which were subsequently to become the Flagstaff and Lounge Sites. This dual two-lane carriageway road forms part of the important new M42/A42 route between Birmingham and Nottingham. Therefore, minimal post-construction differential settlements of the backfill were required to ensure a satisfactory ride quality and efficient functioning of the highway drainage system. To achieve this the British Coal Opencast Executive, the Department of Transport (DTp) and their Consultants undertook detailed studies of case histories available at the time to develop a suitable compaction specification.

Controlled compaction has been undertaken beneath a total length of 3.5 km of the route to depths of up to 38m. Compaction has been closely supervised and extensive testing of the backfill has been correlated with the methods of compaction which were adopted. In addition, a large number of instruments were installed in the backfill and monitored over a period to date of up to about four years to establish how the pattern of surface movements and strains within the backfill vary with time, the rise in the groundwater table and with rainfall. Post-construction settlements of the highway to date have been small.

This paper describes the processes involved in drawing up an agreement between the Opencast Executive and the DTp, the specification which was developed and the experience of compaction operations and also presents and comments on the control testing and movement monitoring results.

2. PLANNING AND DEVELOPMENT OF THE OPENCAST SITES

2.1 General

There has been significant debate concerning the alignment of the A42 Birmingham-Nottingham route in Leicestershire. Figure 1 indicates the line which was finally adopted in the area of Ashby-de-la-Zouch. It will be noted that the route is divided for the purposes of construction into the A42 Measham and

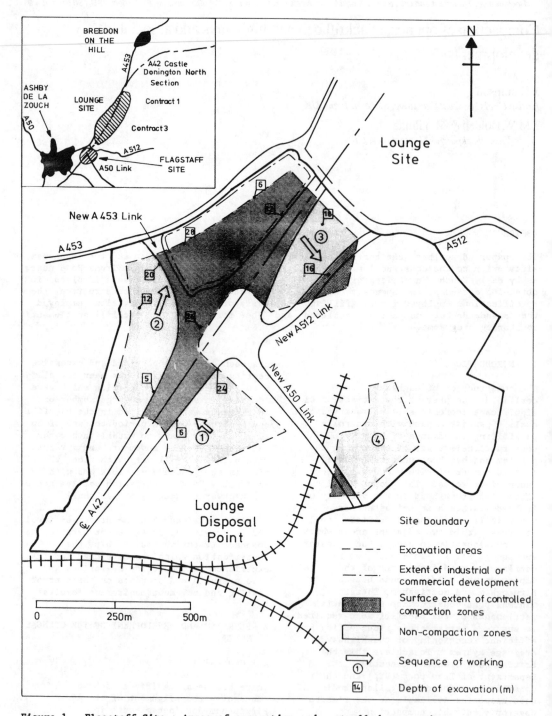

Figure 1. Flagstaff Site : Areas of excavation and controlled compaction.

Ashby Bypass and Contracts 1 and 3 of the A42 Castle Donington North Section. The Opencast Executive became involved in the debate concerning the road alignment because it did not wish to see significant reserves of low cost coal sterilised by this major road development. This view was supported by Minerals Planning Guidance Note No 3 and the Leicestershire Minerals Local Plan. Consequently the Executive's proposals for the recovery of this coal were dealt with quickly and were accepted without recourse to local public inquiries.

2.2 Flagstaff Site

In October 1983, the DTp confirmed the line of the proposed A42 Measham and Ashby Bypass. The northern section of the bypass crossed proven opencast coal reserves, known as the Flagstaff Site. Consequently the Executive submitted a planning application for the Flagstaff Site in March 1984 and permission was granted in October 1984. Site operations commenced in July 1985 and the land was backfilled ready for soils replacement by November 1986 with 244,500 tonnes of coal having been recovered from two seams. Construction of the A42 commenced in August 1987 and the road was opened in August 1989. The road incorporates a continuously reinforced concrete roadbase.

Flagstaff Site, illustrated in Figure 1, was approximately 126 ha in extent with three excavation areas, totalling about 35 ha. The gross excavation volume was approximately 4.4 Mm^3 with an average weekly excavation rate of about 55,000 m^3. The depth of excavation varied from a few metres at the outcrop up to about 30m.

In addition to the main A42 alignment, an interchange and associated link roads to the A453, A512 and A50 roads have been constructed within the Flagstaff Site. Controlled compaction was undertaken beneath all the roads and beneath an area to the north east of the interchange which it is anticipated will be designated for industrial or commercial development. The controlled compaction zones are shown on Figure 1 of the opencast area. The depth of backfill within these zones varies from 4 to 29m.

The methods of working both Flagstaff and Lounge Sites have been chosen to minimise settlements of the completed A42 road. At Flagstaff Site excavation and subsequent backfilling were advanced northwards in Area A before being advanced eastwards and southwards in Area B. Excavation and backfilling within Area C was undertaken last. This sequence ensured early completion of backfilling beneath the section of the A42 which was built first (in the A42 Measham and Ashby Bypass contract).

2.3 Lounge Site

In July 1985, the DTp published the draft orders for the A42 Castle Donington North Section. The southern part of this route crosses proven opencast coal reserves known as the Lounge Site. Whilst the Executive had initially objected to the road proposals, the objections were withdrawn on the basis that it was possible to agree a programme of opencast working and road construction which would avoid the sterilisation of coal reserves. The agreement between the Executive and the DTp included the controlled compaction of opencast backfill beneath the road line, limiting the gradients of excavation batters beneath the road to less than 1v:1h overall in order to reduce differential settlement, and the possible construction of a temporary link road between the A453 road and the southern extent of Contract 1 if coaling in the southern part of the Lounge Site prevented concurrent opening of Contracts 1 and 3 (see Figure 2).

The Lounge Site planning application was submitted in December 1985 and planning permission was granted in March 1986. Site operations for the recovery of 3.75M tonnes of coal from eight seams commenced in September 1986 and will be completed by July 1990.

The Executive's agreement with the DTp included the restoration of the corridor beneath Contract 1 by April 1989. This was achieved and Contract 1, for the construction of the central part of the A42 Castle Donington North Section, was awarded in July 1989. Highway construction work is currently in progress and is scheduled for completion in April 1991. The Executive also anticipates restoring the corridor beneath Contract 3 by July 1990 so that this 2.7 km section of the road, linking the A42 Measham and Ashby Bypass with the southern end of Contract 1, can be opened concurrently with Contract 1. This will avoid the construction of the temporary link road to the A453.

Lounge Site, illustrated in Figure 2, is approximately 231 ha in extent, with two excavation areas totalling 136 ha. The

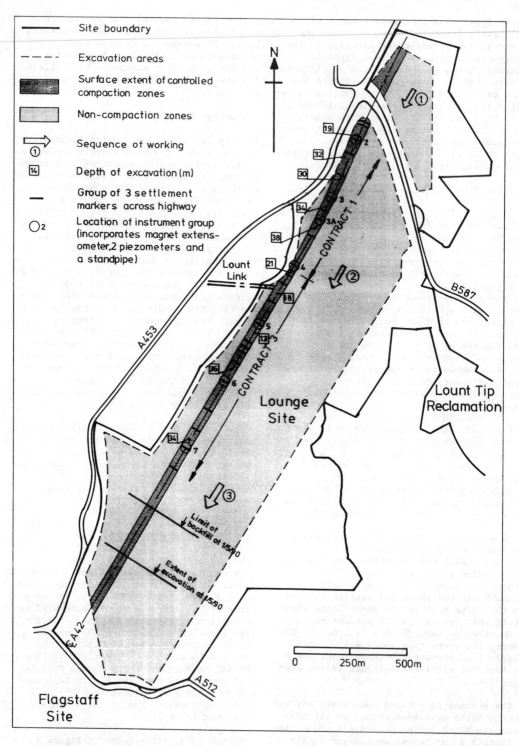

Figure 2. Lounge Site : Areas of excavation and controlled compaction and location of instrumentation.

gross excavation volume is approximately 47.7Mm³ with an average weekly excavation rate of about 250,000 m³. The depth of excavation varies from a few metres at outcrop in the south up to about 76 metres in the north.

Controlled compaction has been undertaken across the full width of the highway corridor, typically a width of 45 to 60m. The total length of the A42 to be built over backfill within Lounge Site is 2.7 km. The depth of backfill beneath the highway varies between 10 and 38 m (see Figure 2). The total volume of backfill subjected to controlled compaction is approximately 3.7M m³. The site has been excavated and backfilled from the north-east towards the south-west so that the period of time between completion of backfilling and commencement of highway drainage and pavement construction is greater in the deeper parts of the site.

3. GEOLOGY

3.1 Flagstaff Site

The Sherwood Sandstone Series occurred to depths of 20m in the higher parts of the site, resting unconformably on the Lower Coal Measures which outcropped over the remainder of the site. There was a thin covering of Glacial clays on the Sherwood Sandstone strata in places.

A feature of the site was the relatively large volume of material excavated down to depths of typically 6m in both the Sherwood Sandstone Series and the Lower Coal Measures which had weathered to sands or silty clays. A considerable proportion of this material had high natural moisture contents or became wet due to high rainfall during overburden excavation.

At depth the Sherwood Sandstone Series comprised very weak to weak moderately weathered clayey, silty, sandstones. In the Lower Coal Measures, beneath the completely weathered mudstones described above, there were highly weathered, very weak mudstones and siltstones which became moderately weathered to fresh and weak to moderately weak with depth. Within the mudstones, moderately strong to strong thickly bedded, sandstone horizons up to 5m thick were encountered. Two coal seams varying in thickness from 0.2 to 1.1m were worked.

A series of perched water tables existed in the Lower Coal Measures prior to the excavation with the phreatic surface within 5m of the ground surface. A perched water table also occurred in the Sherwood Sandstone Series. However, seepage into the excavation was very limited.

3.2 Lounge Site

Lower Coal Measures strata are present beneath the site, overlain in the southern third of the site by up to 10m of Sherwood Sandstone. Glacial Till overlies the outcrop of the Sherwood Sandstone and the Coal Measures. Typically the Till is less than 3m thick but in places channels have been incised through the Sherwood Sandstone into the Lower Coal Measures and infilled with up to 12m of Glacial Till.

The Glacial Till is typically a very stiff silty sandy clay containing gravel size erratics. The descriptions of the Sherwood Sandstone Series and the Lower Coal Measures rocks are similar to those given for the Flagstaff Site. The only major difference is that in places near the base of the excavation the Lower Coal Measures mudstones were moderately weak to moderately strong and required blasting. Furthermore, the moderately strong siltstone and sandstone horizons in the Lower Coal Measures were only up to 1m thick. Eight coal seams were worked at Lounge Site.

Perched water tables in the Lower Coal Measures and the Sherwood Sandstone Series were indicated by limited seepage from more permeable horizons in the excavation walls. Generally the site was very dry.

4. COMPACTION SPECIFICATION

4.1 Form of Specification for Compaction

To minimise settlements of backfill its air voids content after compaction must be limited. Practicable targets yielding acceptable settlements would appear to be 12% air voids in materials drier than the BS 1377 : Test 12 optimum moisture content (hereinafter abbreviated to omc) and 5% air voids in materials wetter than omc. What form of compaction specification should be used to achieve this standard of backfill? The two main types are outlined below. Whichever is adopted, it must allow unacceptable backfill to be quickly identified so remedial action can be taken before the material is overtipped.

4.2 Method Specification for Compaction

Method specifications which specify combinations of the number of passes of plant of various types and the maximum layer thickness for different categories of backfill have been used commonly in highway and opencast earthworks in recent years. They are based on the DTp Specification for Road and Bridgeworks 1976 (The Blue Book) as at Lounge Site (for materials with moisture contents greater than omc), and at Flagstaff Site. The more recent specifications are based on the newer DTp Specification for Highway Works 1986 (The Brown Book). Both specifications were developed for the materials encountered in the relatively shallow cuttings for highways and for the types of plant and working methods used in constructing highway embankments. Consequently for opencast operations they have been modified to some extent to take account of the faster rates of earthmoving and the drier more competent materials of larger size from large capacity hydraulic excavators. However, with the current limited state of knowledge it is perhaps inevitable that changes to the method specification need to be made on site to achieve the limits on air voids proposed earlier.

Fewer problems will be encountered if a method specification is developed specifically for use on opencast contracts. However, the development of such a specification must await the accumulation of experience in the compaction and subsequent performance of different types of opencast backfill. It should be possible to predict the likely performance of backfills compacted to this opencast specification on the basis of simple tests for moisture content, compaction, liquid and plastic limits and strength (as measured in soaked 10% fines value tests or undrained triaxial tests). These tests could be carried out on the cores obtained during the proving of coal reserves. In addition a knowledge of the hydrogeology of the site is required and this could be obtained from installation of piezometers during the early investigations.

One of the major drawbacks of this type of Specification is that it relies on a high level of site supervision and places the onus on the Client to ensure testing is carried out and the desired level of compaction achieved.

Good results can be achieved with a method specification provided amongst other things, that the full time compaction supervisors (both on the Contractor's and Consultant's staff) ensure that the rate of deposition of fill and the capacity of the compaction plant are consistent with the specified number of passes.

4.3 Performance Specification for Compaction

Performance specifications require the Contractor to demonstrate the adequacy of compaction by undertaking in situ testing, generally of dry densities. Numerical values of minimum target dry densities to achieve acceptable air voids contents can be specified if sufficient testing of the various materials on the site has been undertaken in advance of the preparation of the Contract Documents. This was done at the Lounge Site where a target minimum dry density of 2.0 Mg/m^3 was specified for materials with a moisture content less than omc. More commonly, target minimum dry densities are specified as a percentage of the maximum dry density obtained in standard compaction tests, for example, 95% of the maximum dry density in a BS 1377 : Test 12 standard compaction test. However, in this case there should be provision for early testing in order to fix the numerical limits on the dry densities to be achieved with the various material types. This will allow field testing to quickly establish if the degree of compaction is acceptable and whether specified limits on moisture content require any refinements.

The alternative procedure of comparing the field densities with the maximum dry density obtained in a laboratory compaction test on a sample taken from the layer under investigation may take several days. In this period the suspect layer may have been overfilled making rectification difficult and resulting in operational problems.

Maximum air voids values can be specified directly. However, where this is done, related values of maximum dry density may be established for each material type early in the contract. This will reduce the requirement for very frequent assessment of specific gravity values (which are required with values of dry density to allow air voids to be calculated).

The major argument in favour of adopting a performance specification for compaction is that a Contractor is best qualified to assess the optimum way of achieving the desired state of compaction with the

298

available plant at tender stage. Furthermore, by quantifying the limits of acceptable compaction the onus is upon the Contractor to provide the test results to indicate the necessary standard of compaction has been achieved. Provided the foregoing recommendations are followed a performance specification for compaction can produce good results. It is particularly suited to sites where the properties of the backfill are reasonably consistent. If a performance specification is used it is important to incorporate the required maximum thickness of layers and the frequency of control testing.

Disadvantages of performance specifications other than those already referred to include:-

a) the potential for Contractors to underestimate the required amount of compaction plant at tender stage, resulting in contractual conflict later;

b) the requirement for more control testing by the Contractor resulting in increased costs.

4.4 Extent of Compacted Backfill

The specified standards of compaction at Flagstaff and Lounge Sites (which are summarised in Table 1) were extended through the full depth of backfill. The lateral extent of controlled compaction zones relative to the highway or development boundaries are shown on Figure 3. Lateral support to the compacted material was ensured by requiring that uncompacted material should be placed at the same rate up to lines descending at 1v:3h from the surface extent of the controlled compaction zone. The steepness of the pit walls beneath the highway was restricted to an overall maximum gradient of 1:1 with local gradients no greater than $1^1/4(v):1(h)$. These limits were fixed to ensure that differential settlements of up to 250mm of the highway pavement and associated drainage occurring after construction could be tolerated.

4.5 Specification of Moisture Contents

The moisture contents of controlled backfill material at Lounge and Flagstaff sites were required to lie between the limits indicated in Table 2. Although originally excluded by the Specification, a significant volume of mudstone with moisture contents lower than omc - 2% had to be employed at Flagstaff and the compactive effort was increased to maintain air voids at acceptable levels. In the specification for the adjacent Lounge Site it was recognised that similar dry material would need to be incorporated in the controlled backfill and the minimum acceptable moisture content was reduced to omc - 4%. As indicated previously air voids contents were limited to less than 12% by requiring all material drier than omc to be compacted to a dry density of 2.0 Mg/m^3.

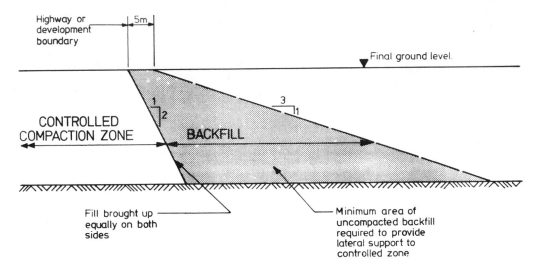

Figure 3. Lateral extent of controlled compaction zones.

Table 1. Specified Standards of Compaction.

Site	Moisture contents (%)	Specification	
		Type	Detail
Flagstaff	omc - 4 to omc - 2	Performance	Minimum dry density = 1.85 Mg/m^3
	exceeding omc - 2	Method	DTp Specification for Roadworks and Bridgeworks
Lounge	omc - 4 to omc	Performance	Minimum dry density = 2.0 Mg/m3
	exceeding omc	Method	DTp Specification for Roadworks and Bridgeworks

Note : Table applies to cohesive, granular and dry cohesive materials

A limited amount of spraying of the dry mudstone with water bowsers was carried out in some areas on Flagstaff Site. However, it was found that improvements in compaction only occurred for relatively small increases in moisture content. If excessive amounts of water were applied it entered the voids between the dry mudstone fragments and the material became difficult to compact behaving as an unacceptable granular material at a moisture content well in excess of its optimum moisture content. If time had been available it should have been possible to increase the moisture content of the lumps of dry

Table 2. Limits on Moisture Content of Backfill.

Material	Moisture Content %	
	Maximum	Minimum
Cohesive	1.0 x PL	omc - 4%
Granular and dry cohesive	omc + 2%	

Note : omc determined from BS1377 : Test 12

cohesive material, and not only the matrix moisture content, thereby rendering the material acceptable as a "cohesive material" with an upper limit on moisture content acceptability of 1.0 x Plastic Limit. However, operational constraints generally precluded this.

On the basis of this observation it is now considered that the upper limit on moisture content for dry cohesive soils should be about omc + 2%.

Only a small volume (7-8%) of the total excavation volume will be compacted upon replacement at Lounge Site. This has allowed the Contractor a good deal of flexibility in material selection, such that the acceptability of material has not been a significant issue. At Flagstaff Site the situation was similar. However, when the percentage of backfill to be compacted is high it is important in finalising the site working method to examine the overburden movements in detail to ensure that when very dry or wet materials are being excavated there are zones available outside the compaction area. At sites where all the backfill undergoes controlled compaction, additional compaction of dry materials and interlayering of wet material with dry cohesive or granular material may be necessary.

300

4.6 Technical Supervision

At Flagstaff and Lounge Sites and the seven other sites within the Central West Region where controlled compaction has recently been completed or is currently being undertaken, the Opencast Executive employs Consultants to:-

i) advise on the formulation of the compaction specification;

ii) provide technical (not contractual) supervision of placement and compaction;

iii) design, install and monitor a system to assess the behaviour of the backfill, and

iv) prepare a Geotechnical Certification Report on the compacted state of the backfill incorporating instrumentation monitoring.

The main method of checking field dry densities is the nuclear density gauge in conjunction with laboratory testing of moisture contents because moisture content values from the nuclear gauge cannot be relied on. The gauge readings are checked regularly against sand replacement density tests and infrequently against large water replacement density tests.

At sites where major compaction works take place, the Contractor is required to provide an on-site laboratory to a specification agreed between the Executive and its Consultant. The laboratory is used principally to check moisture contents, plastic limits, compaction characteristics, specific gravities and gradings.

Levelling of surface settlement markers at regular intervals is the main source of information on how the backfill is performing. However, magnetic extensometers are installed with standpipes and piezometers to measure variations in strain with depth and the way it varies as the groundwater rises. The instrumentation is installed in phases so that it is possible to commence monitoring shortly after backfilling is complete in each area of the site.

To undertake the site supervision of compaction the Executive's Consultant normally employs an Inspecting Engineer or Inspector and a Laboratory Technician with part-time support from a surveyor and a project geotechnical engineer.

5. COMPACTION PERFORMANCE

The descriptions of the main backfill materials after compaction at Flagstaff and Lounge Sites are summarised in Table 3. The predominant compacted backfill at both sites was the Lower Coal Measures mudstone.

At Flagstaff Site compaction to the DTp Specification for Highways and Bridgeworks was specified. For the Stothert and Pitt 72T vibrating rollers towed by D7 dozers proposed by the Contractor, four passes on a layer thickness of 200mm complied with the Specification for most of the materials. The Contractor also proposed the use of Caterpillar 825 B and C tamping rollers. The requirements of the DTp Specification were based on a much lighter tamping roller and as a result compaction trials were held. These established that a combination of four passes of these rollers on a layer thickness of 275mm achieved an equivalent standard of compaction.

At Flagstaff Site, for mudstones with moisture contents in the range omc - 4 to omc - 2%, which were precluded by the original Specification, the Contractor was required to ensure an absolute minimum value of dry density of 1.85 Mg/m^3 to limit the air voids content and to avoid the associated strains on saturation. This was achieved with typically 6 passes of a Caterpillar 825C on a layer thickness of 300 to 400mm.

At Flagstaff material was compacted using a Caterpillar 825C tamping roller with a blade to spread and compact in 275mm layers. Alternatively, Stothert and Pitt 72T rollers were substituted and the layer thickness reduced to 200mm. The feet on the 825C tamping roller were effective at breaking down the larger particles of mudstone. Particularly good standards of compaction were achieved when the tamping and vibrating rollers were employed together.

At Lounge Site minimum dry densities of 2.0 Mg/m^3 were specified for dry material with moisture contents in the range omc -4 to omc. Material with moisture contents exceeding omc were specified to be compacted to the methods in the DTp Specification for Highways and Bridgeworks. The vast majority of the material has been dry and has been compacted in accordance with the former requirement.

On both sites, where sandstone met the requirements of the DTp Specification for

Table 3 : Properties of Compacted Backfill

SITE	STRATUM	DESCRIPTION OF COMPACTED BACKFILL	LAYER THICKNESS (mm)	NO OF PASSES	PLANT	w (%)	PL (%)	BS1377:Test 12 omc (%)	mdd (Mg/m³) (1)	field γd (Mg/m³) (2)	Relative Compaction % (2) (1)	SG	Va
FLAGSTAFF	WEAK COAL MEASURES w > omc - 2	Dark grey silty CLAY with weak fine to medium mudstone fragments	275 - 200	4 / 4	CAT 825C or B D7 + 72T	11 - 14 (12)	22 (22)	13 (13)	1.93 - 1.95 (1.94)	1.92 - 1.98 (1.94)	99- 102 (100)	2.73 - 2.75 (2.73)	2- 7 (5)
	WEAK COAL MEASURES w < omc - 2	Dry grey silty CLAY with weak fine to coarse mudstone fragments	300 - 400	6	CAT 825C	10- 11 (10.3)	22 (22)	13 (13)	1.94 (1.94)	1.89 - 1.98 (1.93)	97- 103 (100)	2.73 - (2.73)	7- 11 (10)
	VERY WEAK TO WEAK SHERWOOD SANDSTONE SERIES	Clayey silt with pockets of sand or silty clay with pockets of sand	275 - 200	4 / 4	CAT 825C or B D7 + 72T	11- 15 (13)	20- 22 (20)	12 (12)	1.90- 1.93 (1.91)	1.85- 1.98 (1.89)	97- 104 (99)	2.71 - 2.74 (2.72)	2- 9 (6)
	COAL MEASURES AND SHERWOOD SANDSTONE MIXED	Mixtures of the two materials	275 - 200	4 / 4	CAT 825C or B D7 + 72T	11- 14 (13)	20- 22 (20)	12- 13 (12)	1.90 - 1.95 (1.92)	1.85 - 1.93 (1.90)	96- 106 (100)	2.71 - 2.73 (2.72)	4- 8 (6)
LOUNGE	WEAK COAL MEASURES	Dry grey silty clay with weak fine to coarse mudstone fragments	300 - 400	6 - 8	CAT 825C	7-11 (9)	18 -26 (23)	10 - 11 (11)	1.9 - 2.0 (1.95)	2.0 - 2.1 (2.03)	100 - 109 (104)	2.58 - 2.75 (2.68)	2 - 8 (6)
	VERY WEAK TO WEAK SHERWOOD SANDSTONE SERIES	Clayey sandy silt with fine to coarse sandstone fragments	300 - 400	6 - 8	CAT 825C	10- 12	15	12- 14 (12)	1.93 - 2.0 (1.98)	Insufficient test results	Insufficient test results	2.61 - 2.68 (2.64)	Insuf' test result (2)
	GLACIAL TILL	Very stiff silty CLAY with fine to coarse gravel size fragments	300 - 400	6 - 8	CAT 825C	10 - 12 (11)	21	12 - 14 (12)	1.91 - 1.99 (1.96)	2.01 - 2.07 (2.03)	103 - 105 (103)	2.67	1 - 3 (2)

Note: Typical ranges quoted with typical average values in brackets

Highway and Bridgeworks for rock fill, and in particular where the soaked ten percent fines value tests showed it would not break down on saturation, it was allowed to be placed at depths in excess of 15m provided it was compacted as rock fill in accordance with the requirements of Clause 609 of that Specification, ie 12 passes of a grid roller or a heavy vibrating roller on layers not exceeding 400mm thick. Where the sandstone was of excessive size, deposition was permitted in a single layer up to one metre thick at the base of the excavation at depths in excess of 15m.

Typical ranges and average values of moisture content, plastic limit, omc and maximum dry density (to BS 1377 : Test 12), field dry densities (obtained generally using the nuclear gauge), relative compaction, specific gravity (SG) and air voids content (Va) are summarised in Table 3 for the various backfill materials employed.

At Flagstaff Site field moisture contents generally ranged from 11 to 16% with an average value within 1% of the omc value of 12 to 13%. However, some dry mudstones had moisture contents in the range 10 to 11%. Field dry densities ranged from 96 to 106% of BS 1377 : Test 12 maximum dry density with an average value of 100%. Air voids values typically varied between 2 and 9% with an average value of about 5%. However, the dry Coal Measures mudstones (with moisture contents in the range omc-4 to omc-2%), which were compacted to achieve a minimum dry density of 1.85 Mg/m^3, had a typical average air voids value of 9%.

At Lounge Site the moisture contents of the Coal Measures mudstones, the predominant backfill material, were comparatively lower, varying between 7 to 11% with an average value of 9% compared with the omc value of 10 to 11%. This is consistent with a slightly stronger material being derived from greater depths. Field dry densities ranged typically from 2.0 to 2.1 Mg/m^3 with an average value of 2.03 Mg/m^3. These figures are 100, 109 and 104%, respectively, of the maximum dry density (to BS 1377 : Test 12) value. The reported air voids values of 2 to 8% (average 6%) are low, especially if the low moisture content of the material is taken into account. By way of comparison the dry mudstone at Flagstaff had an average air voids content of 10%. The higher values of relative compaction and the lower air voids contents reflect the higher compactive effort at Lounge Site compared with Flagstaff Site.

6. MONITORING

6.1 General

Comprehensive monitoring of the surface settlement and strains within the backfill and restoration of the groundwater table has been undertaken at both sites using the layout of instruments shown in Figures 2 and 4. The monitoring programme has quantified the magnitude and rate of creep movements in the backfill as they have settled under their own weight and also the magnitude of movements associated with saturation.

6.2 Flagstaff Site

The results of monitoring over a maximum period of 41 months are presented for depths of backfill in the range 10 to 28m.

Figure 5 shows lines depicting the best fit to the variation with time of average compressive strains (ie surface settlements expressed as a percentage of backfill depth) beneath each surface settlement marker. It also sums up some of the difficulties of trying to establish lines of "best fit" through results at backfill sites where scatter can be considerable. There is a tendency for strains to be approximately linear on the log time plot and the variation of strain with time can be defined by the creep compression rate parameter α, which is the strain occurring in a log cycle of time. An average α value of 0.15% may be inferred.

Figure 6 shows a similar plot using data from the extensometers. A similar average α value of 0.16% may be inferred.

The variation of average strains with time at Flagstaff Site has been compared with the results from other compacted backfill sites in Figure 7. Also presented are the standards of compaction and the lithology of the backfills since these affect behaviour. It can be seen that the strains are very low particularly in comparison with the results from the area of uncompacted backfill at the Holly Bank Site. (Results from the compacted area of this site are discussed later).

Figure 8 shows a typical variation with depth of the strains measured between adjacent magnets in all the extensometer installations within controlled backfill some 8 months after the start of monitoring. It is typical of the data from

303

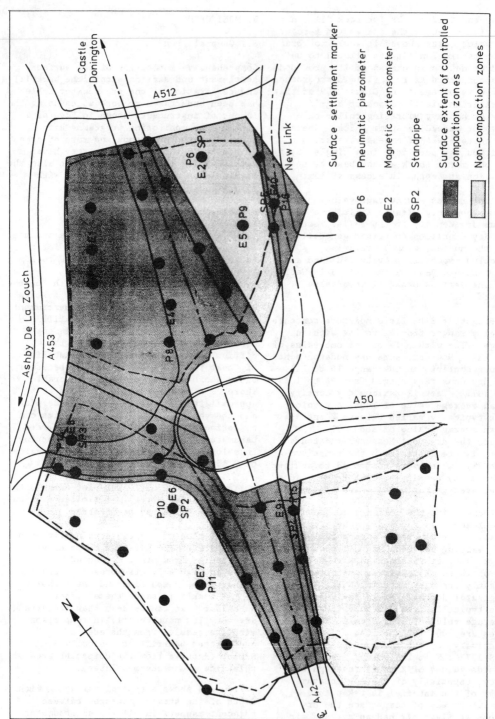

Figure 4. Flagstaff Site : Location of instrumentation.

304

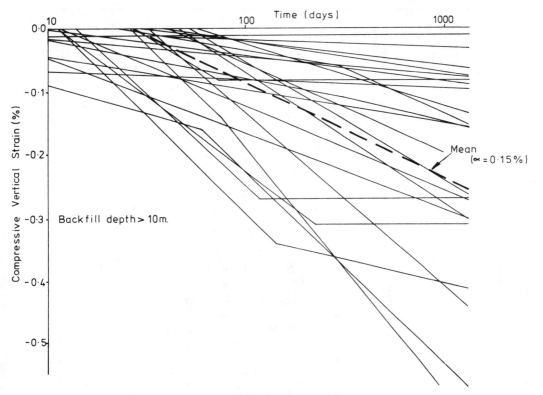

Figure 5. Flagstaff Site : Variation with time of strain over full depth of backfill from surface settlement markers.

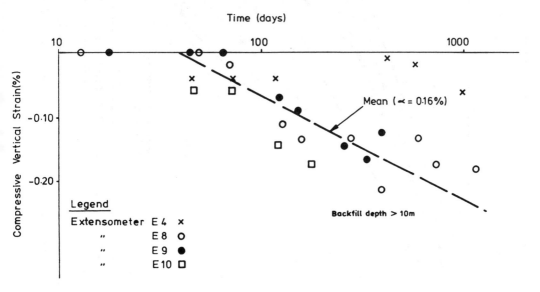

Figure 6. Flagstaff Site : Variation with time of strain over full depth of backfill from extensometers.

305

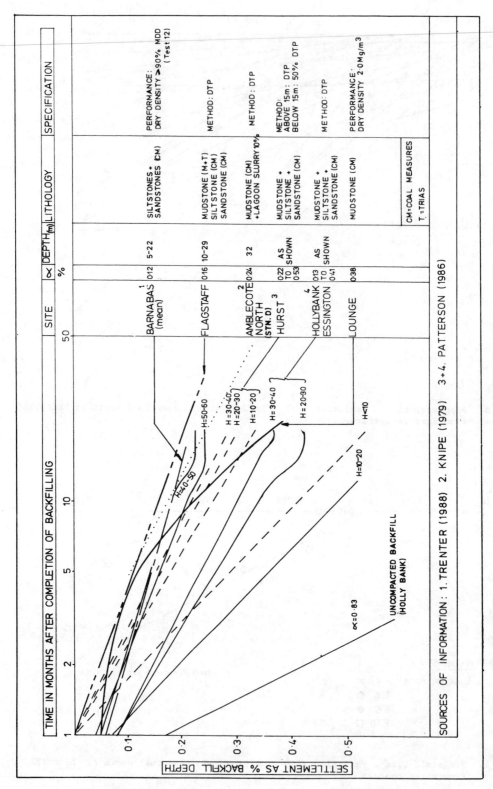

Figure 7. Summary of rates of settlement of compacted backfill.

306

this site in two respects:

(a) significant expansive strains are evident within the top 4m of backfill; and

(b) average compressive strains below 5m depth appear to be fairly constant.

The expansive strains at this site appear mainly to arise from swelling of the backfill near the surface where effective stresses are low and the mudstones absorb rainwater seeping downwards.

Examination of the extensometer readings in conjunction with the water level records for adjacent piezometers or standpipes has allowed strains on saturation by groundwater to be assessed. Strains on saturation are strongly influenced by the air voids content and total overburden stresses. Results from the site are presented on a plot of strain against air voids prior to saturation in Figure 9.

For overburden stresses in the range 300 to 500 kN/m^2 compressive strains due to saturation range between 0 and 0.4% with a mean value of 0.2%. Compressive strains are small because good compaction and the avoidance of very large or dry material ensured air voids were less than 10%. For overburden stresses in the range 40 to 100 kN/m^2 expansive strains of near surface backfill on saturation are up to 0.9% two years after completion of backfilling. These values compare well with other published results for similar materials, air voids contents and overburden stresses. The expansive strains are on the low side of the range predicted on the basis of work by Cox (1978) on Mercia Mudstone.

The results of monitoring are summarised in Table 4. The projected average creep strain over the full backfill depth at the end of the 25 year pavement design life is only 0.39% which corresponds to settlements of only 40 mm for 10m of backfill and 110mm for 28m of backfill.

6.3 Lounge Site

The backfill movements have only been monitored over a relatively short period of up to 21 months at this site.

Figure 10 shows plots of the variation of average surface settlements with log time for various depths of backfill. Initial strains over about the first year were characterised by a compression creep rate parameter, α, of about 0.1%. Higher

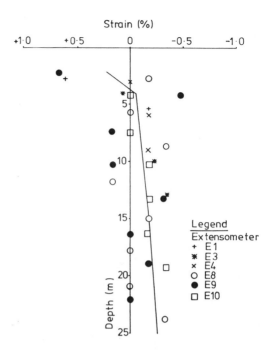

Figure 8.   Flagstaff Site : Variation of strain with depth in backfill 8 months after instruments installed.

movement rates were subsequently recorded (and are discussed later) but the most recent results suggest the movements are levelling off. Average compressive strains over the 21 month period to date vary between 0.2 and 0.6% corresponding to average settlements between 40 and 180mm. These trends are also apparent in the results from the magnet extensometers which are presented in Figure 11. The average compressive strains from the extensometers are compared with values from other compacted sites in Figure 7. Although the values are greater than those recorded at Flagstaff Site they fall in the middle of the range of other published results.

Localised higher strains occurred in an area where the zone of controlled backfilling was located immediately adjacent to an oversteepened section of the side wall of the pit. These higher strains are characterised by the results from extensometer 4.

307

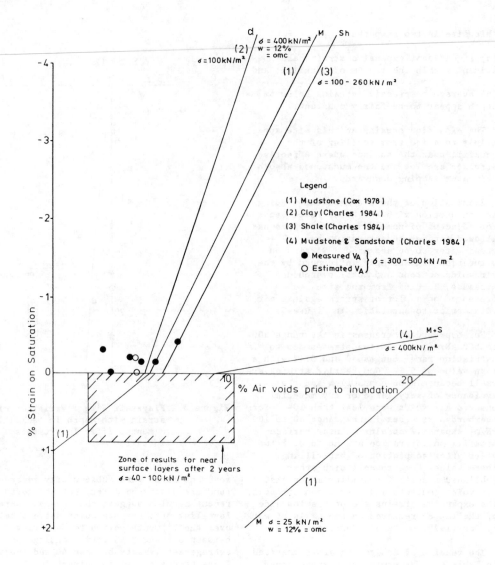

Figure 9. Comparison of strains on saturation with air voids of backfill.

Table 4. Flagstaff Site :
          Summary of Measured and Projected Movements of Backfill.

	Average	Range
Depth of backfill (m)	15.5	10 to 28
Surface settlements after 2 years (mm) Extensometers Surface settlement markers	40 40	12 to 52 2 to 90
Av. compressive strain over full depth of backfill after 2 years (%) Extensometers Surface settlement markers	- 0.20 - 0.22	-0.02 to -0.21 -0.01 to -0.53
Creep compression rate α (Strain per log cycle in days)	0.16	0 to 0.4
Projected av compressive strain over full depth of backfill after 25 years (%)	- 0.39	-
Projected average surface settlement after 25 years (mm)	-	40 to 110 (10m) (28m)

Note: Times quoted from completion of backfilling

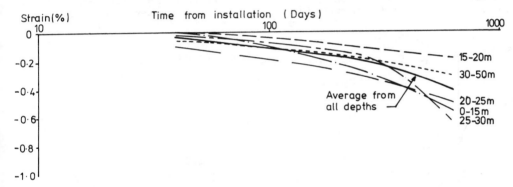

Figure 10. Lounge Site : Plot of variation with time of average strain for various
           depths of backfill from settlement markers.

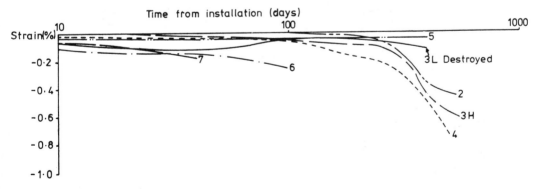

Figure 11. Lounge Site : Variation with time of strain over full depth of backfill from
           extensometers.

309

Variation of strains within the backfill 500 days after installation of the extensometers (except in extensometer 5 where the period is 330 days) are presented in Figure 12. Trends are difficult to discern but the results for extensometer 3H in particular, confirm the trend at Flagstaff site for strains to be fairly constant with depth. This conflicts with the trend of results at the Holly Bank and Hurst Sites (shown in Figure 7) for overall strains (surface settlements expressed as a percentage of backfill depth) to decrease with increasing depth of backfill. Such a trend suggests that strains, occurring after backfilling is completed, decrease with depth.

Expansive strains near the surface were only observed in one extensometer at Lounge. This may reflect the fact that the highest magnets were below the zone in which most of the expansive strain was likely to be occurring.

Figure 13 is of interest because it shows that strains within the backfill, as characterised by the results from extensometer 3H, remained very small for a period of eighteen months over the dry summers of 1988 and 1989 and the intervening dry winter. The start of a period of increased strain rates in October 1989 coincided exactly with the start of a period of heavier rainfall. It is considered that the movements occurring over the 1989/90 winter are due to saturation of the backfill by downward percolating rainfall and lateral groundwater flow from the adjacent intact strata. Inundation of the backfill has not occurred yet to any great extent at Lounge. The maximum depth of water at the base of the backfill, as registered by the piezometers, is only 2m. The component of strain on saturation recorded by extensometer 3H to date is about 0.6% which correlates well with the values of strain suggested by the work of Charles (1984) (as presented in Figure 9) for a typical air voids value in the backfill of 6%.

It is interesting to speculate on why strains at Lounge Site are greater than those at the Flagstaff Site only a short distance away. The generally greater depth of the excavation at Lounge Site led to a significantly greater proportion of the

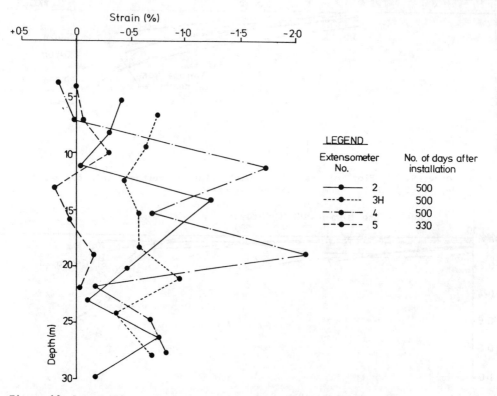

Figure 12. Lounge Site : Variation of strain with depth in backfill.

Lounge Rainfall Record

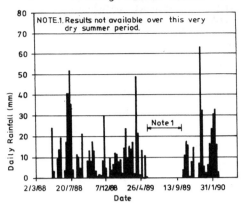

Figure 13. Lounge Site : Comparison of the rate of strain in extensometer 3H with rainfall.

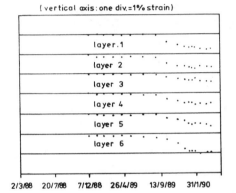

backfill being very dry (moisture contents less than omc - 2%) and somewhat stronger. This could have resulted in a backfill with a greater proportion of air voids which was more susceptible to settlement on saturation. However, the air voids contents quoted in Table 3 do not support this explanation. The fact that the largest strains have been recorded where the side wall of the excavation is closest to the highway corridor and that the whole length of the compacted backfill zone on Lounge Site lies close to the side wall of the excavation suggests another explanation. The shear stresses developed between the backfill and the side wall may have caused arching within the fill and a commensurate slowing of the initial rate of settlement. As a result, more settlement

may have been recorded following the installation of instrumentation 2 to 3 months after the completion of filling in any area.

7. CONCLUSIONS

Settlements of highways on opencast backfill must be anticipated and, due to the variability of the backfill materials and the time variation between deposition in different areas, differential settlements are likely. Differential settlements will also occur where the highway crosses excavation walls, whether these be external or internal buried walls. Effects can be mitigated by adopting slack wall gradients, providing steeper falls on drainage, providing more surface water drainage gullies than normal and avoiding water entering the backfill in localised areas from unlined ponds or drainage systems. Movements and groundwater levels should be monitored for at least a year after completion of backfilling before decisions on the timing of subsequent development are taken. If movement rates at this time are slow then development may be feasible. However, the results from Lounge site suggest that careful consideration needs to be given to the possibility of delayed settlements, particularly if rainfall in this period is low. With these precautions the effects of backfill settlement are likely to be no worse than settlements of highway embankments on soft natural ground and on the basis of movements observed at Flagstaff Site are likely to be considerably better.

It is important that monitoring of further backfill sites is undertaken and monitoring of existing sites is continued over long periods to establish patterns of movement behaviour. These should be correlated with careful records of backfill materials and control testing. From these records it should be possible to develop variable standards of compaction which are related to the lithology of the backfill, the sensitivity of the proposed development to movements and the period between completion of backfilling and development of the site.

8. ACKNOWLEDGEMENTS

The contracts for the working of the Flagstaff and Lounge Sites were awarded by British Coal Opencast Executive, after competitive tendering, to A F Budge

(Mining) Limited. The A42 Measham and Ashby Bypass and Contract 1 of the A42 Castle Donington North Section have both been undertaken by A F Budge (Contractors) Limited. At the time of writing Contract 3 of the A42 Castle Donington North Section has not been awarded. The design and construction supervision of the roads and auditing of compaction operations within the opencast sites has been undertaken by Scott Wilson Kirkpatrick on behalf of the East Midlands Regional Office of DTp (Director, Mr S Rose CEng FICE, FIHT, dipTE). Scott Wilson Kirkpatrick have been also employed by the Opencast Executive to compile the compaction specifications, undertake the technical supervision of site compaction operations, monitor the backfill settlements, and prepare the Geotechnical Certification Reports for the Flagstaff and Lounge Sites.

The authors are grateful to their colleagues, and particularly Mr S J Hodgetts, Mr R J Hodgson, Mr P M Staten and Mr J M Saunders for assistance in the preparation of this paper and to British Coal and the DTp for permission to publish. Any opinions expressed are those of the authors and not necessarily those of British Coal or the DTp.

9. REFERENCES

Charles J A (1984) : Some geotechnical problems associated with the use of coal mining wastes as fill materials. Symp on the Reclamation, Treatment and Utilization of Coal Mining Wastes. NCB Minestone Executive, Durham, Paper No 50.

Cox D W (1978) : Volume change of compacted clay fill. Proc Conference on Clay Fills pp 79-86. Institute of Civil Engineers, London.

Knipe C (1979) : Comparison of settlement rates on backfilled opencast mining sites. Proc Symp on the Engineering Behaviour of Industrial and Urban Fill pp E81-E98. University of Birmingham.

Patterson D A (1986) : Settlement of fill with particular reference to opencast coal mining sites. MPhil Thesis. University of Aston.

Scott Wilson Kirkpatrick & Partners (1989) : Flagstaff Opencast Coal Site Geotechnical Certification Report. Unpublished.

Trenter N A (1988) : Compaction procedure and settlement monitoring at the Barnabas opencast site, near Chesterfield. Proc of the 2nd International Conference on Construction in Areas of Abandoned Mineworkings. June 1988 University of Edinburgh.

Reclamation, Treatment and Utilization of Coal Mining Wastes, Rainbow (ed.) © 1990 Balkema, Rotterdam. ISBN 90 6191 154 0

Long term settlement of opencast mine backfills – Case studies from the North East of England

S. M. Reed
British Coal Opencast Executive, Stoke on Trent, UK

D. B. Hughes
British Coal Opencast Executive, Newcastle upon Tyne, UK

ABSTRACT: The paper presents the results of four investigations relating to long term uncompacted backfill settlement in the North East of England. The research which covers a period from 1974 to 1988 is aimed at providing a better understanding of the complex deformation processes which backfill materials are subject to, and consequently contribute to the design of compacted backfill placement for restored sites destined for structural development. The results indicate the interaction of mine geometry, fill properties, method of working and groundwater on long term deformations.

1. INTRODUCTION

Observations related to backfill settlement in the North East of England commenced in the 1960's, when Kilkenny, (1968), investigated the suitability of restored opencast mine sites for structural development. A detailed investigation using surface levelling stations and borehole instruments was conducted on the Horsley site monitoring the effects of groundwater recovery and the related induced collapse settlement of backfill materials, Charles et al (1977), (1984). In 1983, British Coal Opencast Executive commissioned research conducted by Nottingham University, Department of Mining Engineering into groundwater recovery problems associated with opencast mine backfills, Reed, (1986). This project which was conducted on a national basis of field work received a substantial input from monitored sites in the North East of England, and presented results of monitoring from 1975 to 1986 on a variety of sites. This paper extends the findings of this work.

1.1 Uncompacted and compacted backfills

The results presented in this paper are applicable to backfill masses which have been conventionally replaced by plant without specific compaction treatment.

Where structural development of a site is known prior to operations, e.g. for road construction or light structural development, then the backfill is compacted in accordance with a designed specification. The specification ensures that future construction is not adversely affected by movements in the foundation fill. The behaviour of uncompacted fills is however important for the following reasons;

a). Development of an uncompacted site several years after completion which was not previously envisaged at site design stage.

b). Construction of services, water courses etc. on fills not warranting a high degree of compaction.

c). Understanding of the settlement properties of uncompacted fills is a prerequisite to the efficient design of a compaction operation.

1.2 Factors affecting the rate and magnitude of backfill settlements

The most commonly recognised factors affecting the settlement process of an uncompacted backfill are as follows;

a). Method of fill placement.
b). Fill materials
c). Influence of groundwater.
d). Mine geometry.

Broken Coal Measures rock strata can exhibit considerable volume changes when used as a backfilling material. Silts and plastic clays expand when wet, whilst soft organic materials such as peat can compress on dumping. It is where bulkage is high that the possibility of significant settlement exists. In the course of normal operations, mining machinery can give complimentary albeit indiscriminate compaction to fill materials. The initial bulkage is related to initial particle size, which in turn can be dictated by mining method. End-tipping by dump trucks or dragline spoiling can lead to a gradation of particle sizes with depth in the fill. The larger blocks of material will tend to fall further with gravity and consequently the lower layers of a fill may be significantly less dense than nearer surface. The use of scraper plant gives greater compaction in its method of layer replacement and by the fact that the smaller blocks handled produced a fill of greater initial density. The advent of large hydraulic shovels has also enabled larger block volumes of overburden to be excavated with consequently increased bulkage.

Traditional settlement theory postulates a logarithmic decay of fill settlements with time. This however assumes the self weight of the fill to be the controlling mechanism and does not consider additional effects such as the impact of water table fluctuations on fill stability. The stresses exerted by the action of water can have a great effect on the individual blocks or particles within a backfill mass, owing to the ability of water to weaken rock strength. If the rock strength is considered to be proportional to its surface energy, then this is reduced in the presence of water, with a resulting reduction in compressive strength, figure 1.

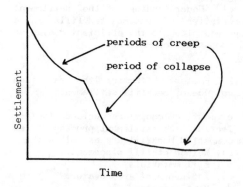

Fig. 1 Traditional Settlement theory

A water table which rises gradually within a fill, either as a consequence of natural recovery of levels after mining or by external influences, (e.g. cessation at adjacent pumping sites or of deep mine pumping), will have an affect of accelerating the physical and chemical degradation of fill materials. Several authors have produced research to show the close correlation between collapse, (or sudden accelerated settlement), and fluctuations in the water table within the fill. The following case studies demonstrate the effect of water recovery on such backfill movements, and indicate the timescales which may elapse prior to sudden collapse settlements uperimposing on conventional creep movements.

2 INTRODUCTION TO CASE STUDIES

The four case studies, (Radcliffe, Coldrife, Radar and Sisters sites), relate to opencast mines in a relatively small area of Northumberland close to the coast. The positions of these mines is indicated in figure 2. The geology of the area is typically Coal Measure mudstones, sandstones, coals and seatearths under a covering of glacial drift materials up to 30 m thick. A common ratio of overburden excavated on a site to coal won is 20:1. Typical working methods for sites involve loose tipping draglines, shovel and truck, as well as motor scraper plant. The groundwater over much of the area is affected by regional deep well pumping immediately adjacent to the former Hauxley Colliery shaft and by nearby active opencast workings.

3 CASE STUDY 1. RADCLIFFE SITE

3.1 Observations

Radcliffe site worked from 1971 to 1976. The instrumentation scheme which consisted of surface levelling stations commenced in 1975 over a traverse crossing an excavated and backfilled slope as detailed in figure 3. The results of the monitoring to date is presented in figure 4.

The following observations can be made from inspection of the above two figures;

a). Settlement profiles are similar for all stations, showing;

o A slow period of settlement from late 1974 to mid 1976.

o A period of collapse settlement

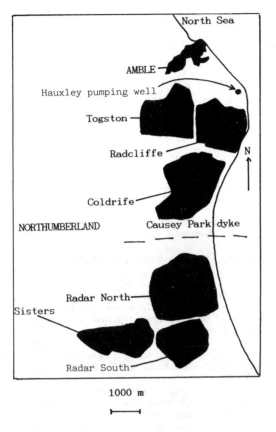

Fig. 1. Site location plan

Surface levelling stations

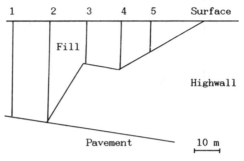

Fig. 2 Radcliffe site. Monitoring scheme.

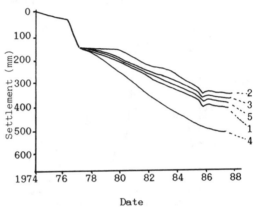

Date

Fig. 3 Settlement results, Radcliffe site

between mid 1976 to late 1976.

o Continued long term settlement from late 1976 to mid 1985.

o A small period of collapse settlement from mid 1985 to early 1986.

o A period of settlement recovery (heave) followed by continued settlement from early 1986 to mid 1986.

o Continued long term settlements to mid 1987.

b). The magnitudes of settlements are not related to fill depth. The magnitudes of backfill settlement are up to 3% of the fill depth.

Table   1 details magnitudes of settlement relevant to the study.

No piezometers have been retained actually within the Radcliffe site boundaries, however fluctuations in water levels can be inferred from piezometers on nearby sites and water levels recorded in the Hauxley Colliery shaft over the period of monitoring.

Total settlement 1975 - mid 1987

Stn	Fill Depth (m)	Settlement (mm)	Settlement %fill depth
1	34	399	1.17
2	36.5	332	0.91
3	15	348	2.32
4	18	504	2.80
5	12.5	373	2.98
6	0	+2	-
7	0	+6	-
8	0	+3	-

Table 1 Settlement; Radcliffe site

Figures   5 and 6 show water levels from 1975   to mid 1985 for the colliery   shaft

315

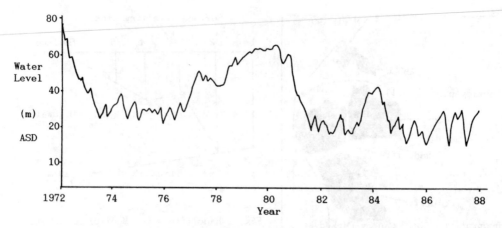

ASD; above site datum = 100 m below ordnance datum

Figure 4. Water levels recorded in Hauxley Colliery pumping well

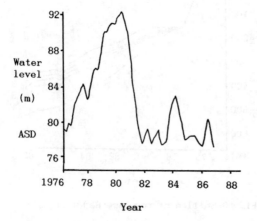

ASD = above site datum = 100 above OD.

Fig. 5   Water levels in piezometer,
          Togston site

some 1000m distant and a piezometer on the adjacent Togston site 2000 m distant. The influence of the Hauxley pumping can be seen to have significant impact on the water levels on the Togston site and hence may be used to subjectively consider water levels affecting the monitoring exercise on the Radcliffe site.

Water movements in the colliery shaft controlled by well pumping operations can be summarised as follows;

o Minor fluctuations during 1975 and 1976.

o Rising water levels, initially at a high rising rate from the beginning of 1977 to April 1987. Then a steady climb of

water levels to around March 1980

o A sharp drop in water levels then occurred between from the end of 1980 to April 1981.

o From April 1981 onwards, occasional fluctuations in water level with a prominent peak from April 1983 to July 1984.

The collapse settlement in the Radcliffe fill occurring between August 1976 and November 1986. In this period the water levels in the shaft were falling until October 1986 when the general period of rising water levels commenced. With reference to the piezometer on the adjacent Togston site, water levels commenced to rise from mid 1976 onwards to early 1980. It is apparent that these rising water trends are also applicable to the Radcliffe site, however the sudden collapse settlement was only restricted to a few months rather than the approximate three years of recovery recorded on the Togston piezometer.

3.2 Collapse versus creep settlement

Of more importance than the period of collapse settlement on the Radcliffe site, is the extensive creep settlement that followed. To the last date of monitoring available, it is apparent that the fill has been settling since the period of collapse at more or less a uniform rate. The period since the completion of collapse settlement is of the order of 10.5 years. The uniform rate, (of up to 38 mm/year on station 4), has been slightly disturbed by a

316

sudden collapse movement towards the end
of 1976. On this occasion, collapse
settlement was noted on all stations
except station 5 situated in the
shallowest fill of 12.5 m deep. (Closest
to the highwall). The period is associated
with rise of water levels, however as
figures 4 and 5 show, other water level
rises previous to this occasion and after
the initial period of collapse settlement
did not result in further collapse
settlement of the fill. The most
significant comment lies in the fact that
the fill was capable of undergoing a
significant further collapse settlement
some 10 years after an initial period of
collapse settlement, and after a prolonged
period of creep settlement.

3.3 Magnitude of settlements

Previously mentioned was the fact that the
overall magnitude of settlements recorded
on each station was independent of depth.
This is particularly relevant for station
4 situated within a bench of the highwall.
The most likely reasons for this are
twofold;
   o The degree of compaction afforded to
the fill in close proximity to the
excavated wall may be reduced as plant
have difficult access to the interface of
the highwall and the backfill.
   o The interface between the solid and
backfill may act as a conduit for
surface water, thus lubricating the fill
in this area more than in the main
body of the fill, e.g. station 1, this
resulting in an increased rate of
settlement occurring through gradual
infiltration and percolation of
rainfall/runoff.
   A third reason may be that there were
significant material differences between
the fill in this area to that in other
parts of the traverse. This cannot be
proved, other than by site investigation,
and is most likely to be insignificant as
compared with the two hypotheses
postulated above.

3.4 Summary of findings; Radcliffe site

The following summary observations can be
made from the investigation on the
Radcliffe site.
   o The potential effect of long term
settlements, (in this case over 10 years).
   o The inter-relationship between ground
water fluctuations and collapse backfill
settlement.

4 CASE STUDY 2. COLDRIFE SITE

The site worked from 1966 to 1971, and
was restored to form a country park. The
restoration involved the construction of
a lake with a concrete lined channel
leading to the sea. Channel excavation
was commenced in April 1983 and the
installation was completed by July 1983.
In June 1985, it was reported that a 45
m length of the channel had subsided.
The channel and subsidence zone is
illustrated in figure 7, along with the
working limits of the two opencast
seams. The occurrence involves delayed
settlement within a fill, approximately
30 m deep some 14 years old.

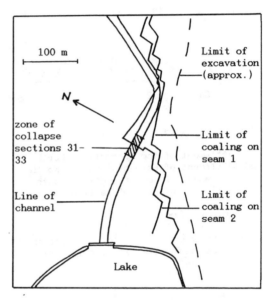

Fig. 7 Coldrife site, area of subsidence

4.1 Investigation

Sections of the concrete channel had been
periodically levelled prior to and after
the collapse. In addition to this
monitoring, after the collapse of the
channel, three boreholes were drilled and
the nature of the backfill investigated. A
piezometer was installed in one of these
boreholes.

4.2 Magnitude of settlement

The sections levelled are situated at 5.5
m intervals along the line of the channel.
Figure 8 details the total magnitudes of
settlement over the available monitoring

period. Figure 9 details the settlement records against time for the section of greatest subsidence, number 34. These curves are representative of the other sections in the collapse area.

-59	-46	-113	NM	-162	-114
	31		32		33
-57	-47	-99	-69	-130	-104

-163	-135	-152	-128	-118	-123
	34		35		36
-138	-105	-131	-112	-107	-104

-78	-107	NM	-86	NT	ST
	37		38		X
-78	-84	-32	-53	NB	SB

NM; not measured.   X: section no.

NT Settlement of top of section, North
NB Settlement of base of section, North
ST Settlement of top of section, South
SB Settlement of base of section, South

Settlement in mm.

Fig 8. Maximum recorded total settlements
Coldrife site, 11.9.84–12.5.87

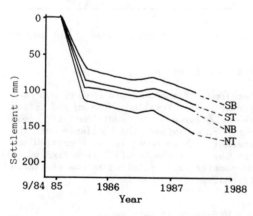

Fig. 9 Subsidence in section 34, Coldrife
site

Examination of the above movements show the general slumping and tilt of

the channel towards the height of the depression recorded on sections 33 and 34. Some degree of rotational movement is apparent. Records of settlement on either side of the collapsed zone indicate magnitudes of the order of 10 mm movement.

4.3 Groundwater behaviour

The behaviour of the groundwater on this site is again controlled to a large extent by the Hauxley pumping, but also by dewatering operations on the East Chevington site to the South. Reference is made to figure 5 in Case study 1.

The nature of the water within the fill was subsequently investigated by piezometer installation adjacent to the channel. The piezometer was installed to 4 m below the backfill pavement, acting purely as a stand-pipe within the fill. Results taken over the period of late 1985 to late 1989 show an almost constant water level of approximately 5 metres above the base of the backfill. Inspection of the water levels within the shaft and on the Togston piezometer show for the period of collapse settlement, (between November 1984 and June 1985), that water levels were dropping after a peak recorded around early 1984. With the Coldrife site being at a greater distance from the Hauxley shaft than the Togston site, it may be that the subsidence was related to this peak in water levels, which had taken a longer time to manifest within the Coldrife fill owing to the distance between the site and the control of groundwater. Previous peaks in water levels prior to channel construction indicate that collapse settlements may have actually occurred previously within the fill, but were not recorded.

4.4 Subsurface investigation

Three boreholes drilled to investigate the subsurface conditions were also gamma ray logged. Interpretation of the logs enabled the following conclusions to be drawn;
o   Water levels identified as commensurate with piezometer readings.
o   Broken mudstone and sandstone fill materials with greatest degree of cavitation towards the base of the fill. Occasionally cavities are described as large.

4.5 Summary

This case study complements that of the Radcliffe site, by indicating that significant long term backfill movement can occur in uncompacted fills. Also demonstrated is the effect of water recovery on fills of this age can still result in collapse settlement. The scenario of collapse or increased settlement rates close to the interface beween fill and the excavated rock slopes of the sites is also apparent, as in the Radcliffe case. Whilst this is not purely a differential settlement at the fill/rock interface, it is an indication that backfill is naturally less compacted in such locations. The observation that the backfill contains a greater proportion of cavities towards its base, is again a reflection of the loose tipping method of working in conjunction with the difficulty associated with backfilling close to the excavation boundary.

5 CASE STUDY 3. RADAR SITE

The site worked between 1957 and 1973. On completion of backfilling, three surface levelling traverses were installed. Figure 10 details the layout of the traverses. The period of monitoring has been from the end of 1975 until mid 1987. Settlement movements can be related to a piezometer close to the excavation.

5.1 Traverse A

Traverse A passes over both the Radar site and the previously mined Radar South site which completed coaling in 1956. Settlement curves and water levels are presented in figures 11 and 12. The water levels from piezometer 611 indicates that the general trend of water recovery continued from the start of monitoring until approximately mid 1981. After this period only minor fluctuations in water level were observed. In this particular area the effects of dewatering at Hauxley can be discounted owing to the presence of an impermeable dyke to the North of the site. Total settlements recorded on the traverse are recorded in table 2.

The settlement curves thus show a good correlation between the rate of water recovery and settlement of the fill, although the settlement curves themselves appear to have a more classical logarithmic decay curve, rather than a sudden collapse in level associated with a sudden groundwater rise. The slow steady recovery of groundwater within the fill may explain this phenomenon. Settlement magnitudes are again not directly related to fill depth. Stations A2 and A3 are

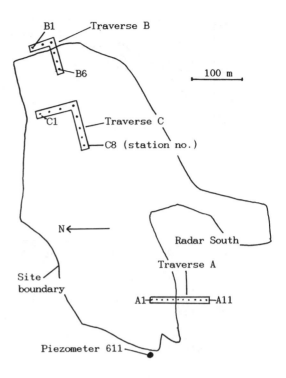

Fig. 10 Levelling traverses, Radar North

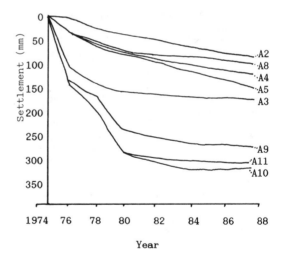

Refer table 2 for total settlements.

Fig. 11 Radar North, Traverse A, settlement results, 1974 – 1988

ones situated in the newest fill of the traverse and thus greater magnitudes of settlement have been recorded as might be expected.

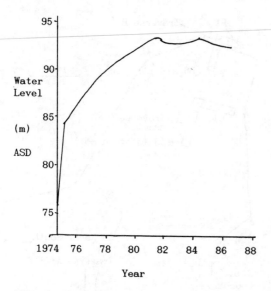

ASD; above site datum = 100 m below
Ordnance datum

Fig. 12 Piezometer water levels, No. 611
Radar North site

traverse A, although a period of
monitoring was omitted between 1976 and
1983, resulting in a significant loss of
information. Since 1983 however there
has been very little movement on the
stations, and this may be taken to
correspond with the groundwater
observations on piezometer 611.

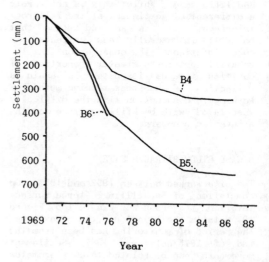

Fig. 13 Radar North, Traverse B settlement

Traverse A, Monitoring from May 1974

Stn	Settle-ment mid1981 (mm)	Settle-ment mid1987 (mm)	Fill Depth (m)	Settle-ment %fill depth
A1	+6	+27	6	–
A2	-58	-89	7	1.27
A3	-172	-174	7	2.49
A4	-91	-123	18	0.68
A5	-111	-157	35	0.95
A6	-11	-6	boundary of	
A7	+4	+15	two sites	
A8	-85	-102	21	0.49
A9	-261	-272	24	1.62
A10	-308	-322	21	1.53
A11	-300	-309	18	1.72

Table 2.   Magnitudes   of   backfill
settlement; Radar Site

Traverse B, Monitoring from 1969

Stn	Settle-ment mid1987 (mm)	Fill Depth (m)	Settle-ment %fill depth
B1	+19	0	–
B2	+17	0	–
B3	+17	0	–
B4	-336	35	0.96
B5	-644	55	1.17
B6*	-402	55	0.73

* defunct after 11.2.76

Table 3. Magnitudes of backfill
settlement; Radar Site

5.2 Traverse B

Settlement results for traverse B are
presented in figure 13 and table 3. The
traverse location on figure 10 indicates
that three of the stations were sited on
solid ground, the traverse crossing a
highwall with three stations on fill.
The trend of the instrument movements
are similar to those recorded on

5.3 Traverse C

Eight levelling stations have been
monitored from 1983 onwards. Small
displacements have been observed over this
period in the range +17 to -63 mm. With
such a small range of displacements no
trends were identified. It is presumed
that a great deal of settlement had been
completed prior to monitoring.

5.4 Summary

This case reinforces the concept of long term movements of significant magnitude. The role of the slowly recoverying groundwater table has been indicated as the main direct controlling factor for both traverses monitored. In this case however after the completion of water recovery, the settlement rate, (i.e creep settlement), has been practically nil.

6 CASE STUDY 4. SISTERS SITE

The project consisted of the monitoring of the settlement of a sewage pipe line constructed on backfill restored in 1974. The project commenced in 1982, and the scheme is presented in figure 14. Instrumentation consisted of three magnetic extensometers/ piezometers in the fill complemented by 3 sets of 10 surface levelling stations. Seven manhole covers also acted as surface levelling points. The design of the pipeline was such that some degree of settlement could be tolerated by the introduction of flexible joints between the individual pipe sections.

6.1 Surface displacements

The time elapse between the date of backfilling and the commencement of monitoring was some 7.5 years. In the period of monitoring only small movements have been noted, either settlement or heaving movements. No distinct trends in settlement have been observed. The fill varies from 16 to 32 m deep. Table 4 details the total displacements recorded on all stations between 1982 and late 1988.

By inspection the fill can be seen to be fairly stable. All measurments are well within the tolerable design limits of the flexible sewer, and only one result, MHE, (77mm), can be said to have displaced with any significant magnitude.

6.2 Groundwater observations

The piezometers installed as part of the extensometer instruments have recorded fluctuating water levels within the fill over the monitoring period. On instrument E3 there is the indication that the fill has been saturated and dry in the upper 16 m. No evidence of collapse settlement has been observed at any time.

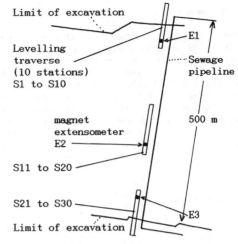

a) Plan of instrumentation

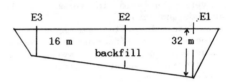

b) Section along line of sewer

Fig. 14 Sisters site instrumentation

Extensometers, (Total settlement)

| E1 | −23 | E2 | +7 | E3 | +9 |

Manhole covers

| MHE | −77 | MHF | +5 | MHG | +8 |
| MHH | +5 | MHJ | −5 | MHK | 0 |

(situated at equal spacings along sewer MHE close to E1)

Surface levelling stations

S1	+3	S2	+2	S3	−8	S4	−7
S5	−7	S6	−19	S7	−21	S8	−23
S9	−23	S10	+10	S11	+10	S12	+10
S13	+10	S14	+7	S15	+6	S16	+7
S17	+5	S18	+4	S19	+6	S20	+6
S21	+6	S22	+7	S24	+7	S25	+5
S26	+7	S27	+6	S28	+6	S29	−9
S30	0						

settlements in mm
results not available for S23

Table 4. Sisters Site, settlement results 1982-1988

6.3 Summary

The Sisters site illustrates that stability may be achieved in uncompacted mine backfills, and in this case a period of at most 7.5 years after backfilling provided a fill exhibiting very little in the way of movement. The delay in monitoring meant that it was inevitable that the early behaviour of the fill was lost. The case however does prove that uncompacted backfills can ultimately form a stable foundation for such structures.

7 CONCLUSIONS

The following conclusions can be drawn from this work.

a). The settlement rates of uncompacted backfills have been observed to be very variable in terms of both magnitude and rate.

b). Both the rates and magnitudes of displacement are affecting by a complex interaction of mining method, mine configuration, material properties and groundwater.

c). Long term backfill movements of over 10 years have been observed. These movements do not show, at this moment in time, any signs of retarding.

d). The siting of structures, even of the most simplistic nature, e.g. water channels, require detailed site investigation when siting on backfill materials regardless of age. The nature of the entire fill depth requires assessment, in order to establish material properties, voids and groundwater behaviour. The investigation should consider external influences, e.g. remote pumping and their likely effect on backfill stability.

ACKNOWLEDGEMENTS

The authors wish to thank British Coal Opencast Executive for permission to publish this paper. The views expressed are those of the authors, and do not necessarily reflect the views of British Coal.

REFERENCES

Charles J.A. Naismith W.A. and Burford D. 1977. Settlement of backfill at Horsley restored opencast coal mining site. Proc. Conference on Large Ground Movements and Structures, UWIST, Cardiff.

Charles J.A. Hughes D.B. and Burford D. 1984. The effect of a rise of water table on the settlement of backfill at Horsley restored opencast coal mining site. Proc. Conference on Large Ground Movements and Structures, UWIST, Cardiff.

Kilkenny. W.M. 1968. A study of the settlement of rstored opencast coal mining sites and their suitability for building development. Bulletin No. 38. Dept of Civil Engineering, University of Newcastle upon Tyne.

Reed S. M. 1986. Groundwater recovery problems associated with opencast mine backfills. University of Nottingham, Ph.D thesis.

Reclamation, Treatment and Utilization of Coal Mining Wastes, Rainbow (ed.) © 1990 Balkema, Rotterdam. ISBN 90 6191 154 0

Investigation on the crushing behaviour of minestone and coal preparation waste with regard to their utilisation

J. Leonhard & Th. Schieder
DMT, Institute of Raw Materials and Preparation, Essen, FR Germany

ABSTRACT: The investigations were aimed at producing, by means of appropriate reduction and classifying methods, mine fractions susceptible of being used in the construction industry as an aggregate. The basic test for dissociation and removal of carbonaceous and clayey components as well as reduction of waste were run on crushing machines of various design and mine waste from collieries selected under regional aspects. Based on the results obtained a concept was developed working on a two-step comminution of mine wast by impact crushers. The concept makes it possible to dissociate, in a first step (preliminary crushing) a sometimes quite high proportion of organic and soft clayey material by taking advantage of selective reduction effect and remove that material classification . The second step (secondary-crushing) then yields a high-grade material from the coarse feed. The products thus obtained and used as aggregates in lime bricks and shaped concrete blocks complied with the commercial requirements of the construction sector.

1 INTRODUCTION AND OBJECTIVES

In connection with the rationalisation and advanced mechanization of underground coal mining and roadheading activities the waste content of ROMcoal has considerably increased. The majority of this waste is disposed on dumps, whereas waste stowing underground was more and more reduced in the Federal Republic of Germany. Although a sizeable proportion of waste is already used for the construction of roads and embankments as well as for relandscaping, efforts continue to enhance the sale of waste to other markets or open up new applications.

The objective of the studies was therefore to produce, by means of appropriate methods of comminution and classification size categories susceptible of beeing accepted by the construction industry as an aggregate for wall bricks, pavement bricks and concrete. Those aggregates have to meet certain standard specifications as far as grain shape, grain resistance and contents of noxious organic and clayey components is concerned, in order to ensure their safe use.

2 SELECTION OF APPROPRIATE CRUSHING MACHINES

2.1 Experimental set-ups and materials

Comminution takes place essentially by compression (slow and fast) or by impact. For the present investigation on waste reduction we selected therefore those machines exerting predominantly said loads. Figure 1 is a schematic drawing of the equipment examined.

The AUBEMA 11665/50 type two roll crusher had drums of 650 mm diameter and 500 mm width. To influence on the final grain size by modifying the gap widths and also to equalize wear action one of the two drums has been laid out as a freely movable element. At 2 x 15 kW drive capacity the rated speed of rotation amounts to 1000 rpm which results in a peripheral speed of 2,9 m/s. The gap width at the base of the crushing teeth was 10 mm. The feed between the drums is subjected to compressive and shear loads. The tooth profiles support the crushing forces so that it is possible to handle also big and hard lumps. The reduction rate which should not be beyond 4 or 5 was set in such a way that the upper grain size of the comminution product amounted to approx. 40 mm.

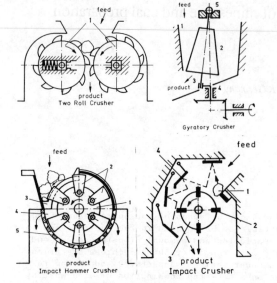

Fig. 1 Different equipment for mine waste reduction

The AUBEMA 1586/GR III 2730 H type granulator is a gyratory whose downward tapering crusher bowl is formed by a hollow cone crusher shell and a crushing cone. The crusher cone axis is slightly inclined in relationship to the shell axis and its top supported in such a way that it is capable of exerting tumbling movements. The lower end of the cone axis is received by an excentric bushing. The feed is reduced by compressive and impact load. The rotational speed of the excentric shaft is 320 rpm, the gap width is 25 mm and the drive capacity 90 kW.

The AUBEMA 1706/4 type impact crusher functions by the feed material beeing flung by the impact ridges of the operating rotor against the impact wings where comminution is brought about. The material is spalled by mere impact load at the natural separating faces. The diameter of the rotor was 600 mm and its width 400 mm. The impact crusher is equipped with 4 impact ridges of 90 mm height. The peripheral speed of the rotor amounted 30 m/s, at a lower gap width of 30 mm.

The 1208/4 AUBEMA type impact hammer crusher is a device wherein the impact arms, movably suspended around the rotor periphery, are brought into a radially extended position once they have attained their full peripheral velocity. Upon entering the impact area of the rotor the feed material is seized by the impact heads, flung against the impact wall by

the rapid succession of blows and subjected to a preliminary reduction by the impact action itself. Secondary reduction is brought about in the lower section between rotor and impact trajectory. The rotor diameter is 800 mm and its width 400 mm. 15 impact arms are arranged on the rotor in three axial rows. Peripheral speed of the rotor was 30 m/s, gap width 40 mm.

As experimental material for the tests in the different machines served washery refuse and minestone from the hardcoal preparation plant at Heinrich Robert colliery. The feed (120 to 45 mm) was obtained by screening the size smaller than 45 mm from the overall size range.

Machine parameters were selected in such a way that the upper grain size produced was approx. 40 mm which made the results comparable.

2.2 Experimental results

The reduced material was examined in terms of grain size distribution, proportion of cubes and degree of selective comminution. We defined those grains as cubes whose length to width ratio was 3 : 1. Table 1 is a summary of the results.

Table 1. Results of preliminary crushing as a function of the load type

Preliminary Crushing

type of crusher	two-roll crusher	gyratory crusher	impact crusher	imp. hammer crusher
type of strain	compression	compression/ impact	impact	impact
rate of reduction	4 - 8	5 - 20	-20	10 - 20
rate of selektive reduction	insufficient	low	good	very good
degree of cubature (%)	53	66	62	70
grain size fraction > 16mm(%)	72	51	51	28
grain size fraction < 4 mm (%)	9,6	18,2	20,0	25,3

As can be seen the two-roll crusher turned out to be particularly effective in terms of a gently reduction. The proportion of >16 mm was 72 %. On the other hand 53 % only of the size ranges examined between 4 and 40 mm were of a cube shape. We found furthermore that selective comminution was dissatisfying. Coarse coal lumps and coal/mineral intergrowths were observed in the fractions up to 16 mm. The situation did not improve very much during the subsequent tests with narrower gap widths and increased rotational speeds.

The gyratory (granulator) excelling by its high throughput and reducing capacities did not function up to the expectations either. The portion of cube shapes at the product was, as expected, within the limits of gap width setting. The most inconvenient drawback was that with dimishing grain size the proportion of cubes decreased disproportionly. Dissociation was dissatisfactory, too, since even in the fraction beyond 4 mm uncrushed, carbanaceous particles were present.

The product yielded by the impact crusher, on the other hand, complied with most of the requirements. The size proportion >16 mm amounted to 50 %; the proportion of cubes in the 8 to 4 mm fraction was beyond 60 %. The carbonaceous content was noticeably enriched in the < 4 mm fraction.

The impact hammer crusher, finally, met all of the specifications. The proportion of cube shapes amounted to roughly 70 % in all of the fractions. No organic matter was found in the fraction < 4 mm. In addition, this tape of mill offers more flexibility since by adjusting its rotor rotation and/or rotor/impact plate spacing, there is a means of accomodating to varying constitutions of the waste material.

It appeared by and large that compressive load conveys the necessary energy to a feed consisting of layered sedimentary rock of varying hardness, to bring about some selective reduction. Impact reduction in an impact hammer mill turned out to be especially appropriate in terms of controlled yield of desirable proportions of defined size ranges, of improved grain shape and of selective effect.

3 IMPACT REDUCTION OF THE WASTE FROM DIFFERENT COLLIERIES

3.1 Experimental set-up and programme

Reduction experiments were carried out in a single-rotor impact crusher of Hazemag AP SO 403 type. The rotor diameter was 460 mm, at 340 mm width. The waste was impacted at three different peripheral rotor velocities, i.e. at 9.08, 14.75 and 21.76 m/s. The feed rate was approx 1.5 t/h. Washery refuse minestone pre-screened to 40 mm were used as test materials. Their origin, feed size and carbon content are summaried in Table 2.

Table 2. Test material

Type of waste		Grain size mm	Coal (waf) % by wt.
VB	Prosper II	140 - 40	2.1
GB	Heinr.Robert	130 - 40	5.7
VB	Lohberg	200 - 40	2.0
VB	Sophia-Jacoba	120 - 40	3.6
VB	Ibbenbüren	150 - 40	3.7
GWB	Niederberg	150 - 40	8.2

VB = Material from predestoning machines
GB = Minestone
GWB = Coarse washery refuse

3.2 Experimental results

The result of selective impact reduction of waste material depends on the raw material properties and impact energy. Our experiments were aimed at characterizing the dissociation of the carbonaceous portion in the waste and its enrichment in the < 0,5 mm, < 1 mm, < 2 mm and < 4 mm fractions, as a function of the peripheral rotor velocity (impact energy), by the respective mass proportions of coal contained in those fractions. The carbon proportion allows, at the same time, to draw some conclusions on the enrichment of other soft or brittle rock proportions, e. g. slaty clays, in the smalls fractions.

A first assessment of the reduction behaviour of the waste feed can be derived from the grindability curves shown on Figure 2. The mean grain diameter has been

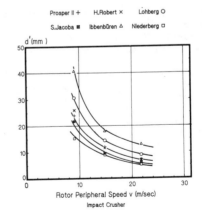

Fig. 2 Grindability curves of different wastes

plotted as a function of peripheral rotor velocity. According to this figure the wast colliery ougth to be the "softest" and that

from Ibbenbüren the "hardest" material.
The hardness scale thus resulting does,
however, not correlate at all with the
carbon contents of the size fractions.

If the various waste fractions are to be
used for construction purposes it may be
purposeful or even necessary to remove,
after reduction the lower fractions in
order to make the residual carbon content
of the useful product comply with the
standard requirements. Said removal makes
only sense, however, if the to be discarded remainder is not excessively high.

The results are illustrated by the reduction of waste from Sophia-Jacoba
colliery. The respective proportions of
the size ranges <4 mm, <2 mm, <1 mm and
<0,5 mm have been plotted on the peri-
pheral rotor velocity (Figure 3). These
proportions go up with increasing energy
supply.

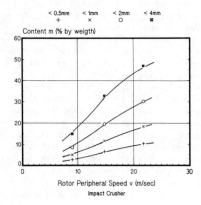

Fig. 3. Proportions of the different size
fractions as a function of peripheral
rotor velocity (material from predestoning
machine, Sophia-Jacoba coal mine)

Figure 4 shows the standardized residual
carbon content in the useful product if
the respective lower fractions of <4 mm,
<2 mm, <1 mm or <0,5 mm are removed.
Here, again, one has a characteristic
curve behaviour depending on the raw material properties of the feed. As can be
seen, it is impossible to completely tran
fer the carbon proportions to the lower
size ranges.

If the carbon content in the waste from
Sophia-Jacoba were to be reduced from
initially 3.6 down to 3.0 % by wt. in the
final product, one would have to select
7.5 m/s of peripheral rotor velocity and
discard the proportion <4 mm amounting to
approx. 10 %.

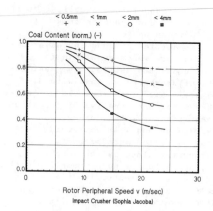

Fig. 4. Standardized carbon content in the
product fractions as a function of peripheral rotor velocity for removal of the
proportion of <0,5, <1, <2, <4 mm

A similar residual carbon content would be
obtained if the peripheral rotor velocity
were 9.0 m/s and the proportion of <2 mm of
roughly 8 % removed.

Table 3 contains the required peripheral
rotor velocities and to be removed volumes
of waste feed if the residual carbon content is to be reduced to 3 % by wt. in the
final product.

Table 3. Required peripheral rotor velocity and removal of <2 mm viz. <4 mm size
fractions in order to obtain a residual
carbon content of 3 % by wt. in the
cominuted product

Colliery	SJ	Ibb.	Ni.
Initial material carbon proportion (waf) in % by wt.	3.6	3.7	8.2
Required percentual reduction or residual carbon proportion of 3 % in the reduction product, % by wt.	17	19	63
Required v for removal of the <2 mm material, m/s	9	13	30,5
Proportion <2 mm % by wt.	8	9	63
Required peripheral rotor speed v during removal of the <4 mm material m/s	7,5	10	22
Proportion <4 mm % by wt.	10	9	55

Dissociation of carbonaceous constituents during impact reduction and enrichment of these constituents in de lower size ranges is characteristic for any and all of the waste materials involved. Assuming a high carbon content of the initial material one would have to remove, however, large quantities (for the Niederberg waste e. g. the fraction <4 mm up to 55 % by wt. viz. < 2 mm up to approx. 30 % by wt. of the total waste) in order to effectively diminish the carbon content in the useful products. This does, however, not always make sense from an economic point of view.

4 TWO-STEP REDUCTION OF WASTE

4.1 Experimental set-up and implementation

We developed a concept of two-step reduction of waste material, Figure 5, under due consideration both of the results represented in chapter 2 as well as of the varying raw material properties of the material and of the experience gained by the industry of stones and earthes.

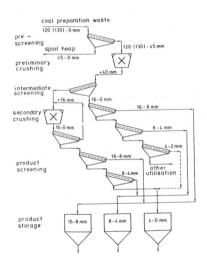

Fig. 5. Flowsheet of the two-step reduction of washery refuse and minestone

The first reduction step (preliminary reduction) is meant to dissociate an as high as possible proportion of organic and clayey material by utilizing the

effect of selective reduction; that proportion is removed by subsequent classification. The feed material (130 - 45 mm) leaves the process, crushed to an upper size of 40 mm, with a desirable proportion of the size beyond 16 mm of 40 % by wt. of the product yield. This proportion is impact-ground in a second step (secondary reduction) to <16 mm.

To make the results of two-step and single-step grinding comparable we used the plant furthermore for reducing test material to the smaller than 4 mm range, under recirculation of the oversize. We had impact hammer mills installed which were run at 30 m/s in the prereduction and at 55 m/s in the secondary reduction step. The test material was waste from Heinrich Robert, Prosper II and Lohberg collieries. Mineralogic constitution of these waste types is presented on Figure 6. Said

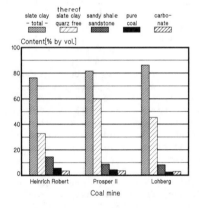

Fig. 6. Distribution of rock types and coal in the waste

waste types differ specifically by their proportion of non-quartzeous slaty clay (HR 32.5 %, P II 60.1 %, LO 46 %), sandstone plus sandy slate HR 14.5 %, P II 8.8 %, LO 8.2 %) and of free carbon (HR 5.7 %, P II 4.4 %), Lo 2.56 %).

4.2 Experimental results

Figure 7 shows the results of two-step reduction for the waste material from Heinrich Robert colliery.

Noticeable shifting can be seen to have happened between the different reduction steps as far as rock constitu-

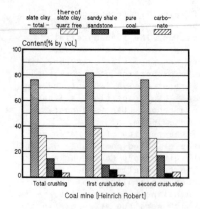

The suitability test on reduction products for constructional purposes was
done on the fraction 4 - 0 mm from preliminary and secondary reduction as
well as on the smaller than 4 mm material of a total reduction product
prepared for comparative purposes.
Testing was done for the suitability as
calcareous waste blocks manufactured by
the sandlime-brick-process. Concrete
shapes were manufactured from the 8 -
4 mm fraction of preliminary and secondary reduction according to a uniform
recipe.

Fig. 7 Distribution of rock types and
carbon in the comminuted products from
Heinrich Robert colliery

5.1 Manufacturing of calcareous waste blocks

tion and carbon content are concerned.
The proportion of non-quartzeous slaty
clay e. g. diminishes between prereduction via total reduction throught to
secondary reduction from initially 38.6 %
by vol. via 32.5 % by vol. down to 30.5 %
by vol. Taking the different size categories of each reduction product one ob-
serves an increase in non-quartzeous
slaty clay content along with reducing
grain size. This is particularly striking
for the waste from secondary reduction.

The proportion of sandy slate and sandstone in the products on the other hand
gos down from 17.2 % by vol. (secondary
reduction) via 14.5 % by vol. (total reduction) to 9.9 % by vol. (prereduction).
It was furthermore found that the rock
types higher in quartz, e.g. sandy slate
and sandstone, are enriched in the coarser size ranges of all the reduction products. This applies, again, specifically
to the material from secondary reduction.
The carbon content of waste products
will, in general, go up along with increasing fineness of the material and
amount to between 2.5 and 32 % by vol.
(average: 6.3 % by vol.) for the prereduced product, between 1.0 and 14 % by
vol. (average: 3.5 % by vol.) for the
product of secondary reduction and,
finally, between 1.0 and 18 % by vol.
(average: 5.7 %) for total reduction.

Single-step total reduction below 4 mm
requires much energy which is translated
by the high ultra-fine proportion yielded. A removal of undesirable mineral
proportions would require screening of
all of the material.

Calcareous sandstone blocks are manufactured from limestone and predominantly quartzeous aggregates and, after
intimate blending and compaction,
shaped and steam pressure-hardened
(hydrothermal process). The binder
structure thus formed and consisting of
so-called calcium/silicate hydrates, is
essentially determiniative of the properties of the final product (compressive strength, thermal conductivity,
weathering resistance).

To asses the product quality we manufactured hydrothermally bound blocks
from the size range 4 - 0 mm yielded
by the tests on preliminary, secondary
and total reduction. In this case the
quartz sand usually serving as an
aggregate in the calcareous sandstone
process, was completely replaced by
waste sand. Comparison of the compressive strengths and apparant densities
achieved was to contribute to the
assessment of the different sand types
in terms of their lime-binding and
compaction properties. Calcinated lime
was used as a binder. To all of the
tests 6.6 % by wt. of lime was added.
Pressing humidity was adjusted to 3.2 %
by the addition of water. The carbon
contents of the sands from crushed
waste, after preliminary, secondary and
total reduction of Heinrich Robert.
Prosper and Lohberg waste materials,
can be taken from Figure 8.

Table 4 contains the characteristics
of wall blocks manufactured from the
different waste sands. Figure 9 illustrates the compressive strengths of
the calcareous waste blocks, obviously

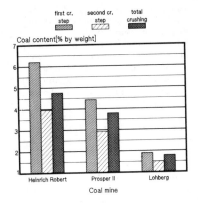

Fig. 8. Proportion of free carbon in the fraction < 4 mm

Calcareous Waste Blocks

aggregate: waste sand < 4mm, binder: lime 6,6 %
block size: 2 DF (240 × 114 × 112 mm)

constructional features	coal mine		
	H. Robert	Prosper II	Lohberg
compressive strength (N/mm²)			
① from - to	12,4-16,7	12,4-13,7	13,3 18,1
average	13,6	12,8	15,6
apparent density (kg/dm³)	1,57	1,78	1,81
compressive strength (N/mm²)			
② from - to	19,1-19,4	13,8-16,9	13,5-18,5
average	19,2	15,1	15,6
apparent density (kg/dm³)	1,78	1,83	1,81
compressive strength (N/mm²)			
③ from - to	17,2-19,0	14,0-17,0	14,1-19,6
average	18,4	14,8	17,2
apparent density (kg/dm³)	1,78	1,81	1,81

aggregate from: ① preliminary crushing ② secondary crush.
③ total crushing

Table 4. Characteristics of calcareous waste blocks for constructional purposes

any waste material from secondary reduction meets the required standards for compressive strength class 12, (i. e. average compressive strength required 15.0 N/mm² and lowest individual level 12.0 N/mm² pursuant to DIN 106, calcareous sandstone blocks. The blocks made from prereduction material contained high proportions of non-quartzeous slaty clay (Prosper) or free carbon (Heinrich Robert) both of which behaved inert in the calcareous sandstone process and therefore did not yield adequate compressive strengths.

If the quartz content of the feed waste is sufficient total reduction of the material below 4 mm will likewise yield a waste material suitable for the calca-

reous sandstone process; this appears from the results of block production using Heinrich Robert and Lohberg waste.

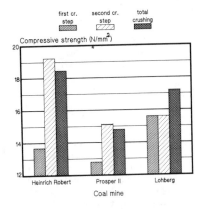

Fig. 9. Compressive strength of calcareous waste blocks (2 DF size)

5.2 Manufacturing of concrete shapes

The properties of the waste fraction between 8 and 4 mm yielded by preliminary and secondary reduction of Heinrich Robert, Prosper and Lohberg waste were checked by using the respective materials as an aggregate in normal concrete. To achieve a size blend of steady screen characteristics we replaced the lacking 4 to 0 mm range by natural sand. The aggregate blend thus consisted of 55 % by wt. waste chips and 45 % by wt. sand of < 4 mm. The proportion of cement binder referred to the aggregate quantity amounted to 20 % by wt. The water/cement ratio was defined at 0.3.

The characteristics obtained from determination of the compressive strength of cubes (after 28 days) and of the apparent density of concrete shapes of 150 × 150 × 150 mm sample size have been summarized in Table 5.

Judging by the preceeding table, concrete blocks made from the secondary reduction product exhibit by better compressiv strengths than blocks produced from the prereduced material; furthermore their scatter of individual values is closer and their apparent density higher. This is obviously due to the improved grain shape and higher strength of the individual grain in the material from secondary reduction. The

329

Table 5. Characteristics of concrete blocks for constructional purposes

Concrete Blocks
aggregate: waste chips 8-4mm, 45% naturalsand <4mm
binder : 20% cement

constructional features	coal mine		
	H. Robert	Prosper II	Lohberg
cement addition (%)	20 (PZ 55)	20 (PZ 35 F)	20 (PZ 35 F)
① compressive strength (N/mm²) from - to	35,8-78,1	41,2-48,1	53,8-55,7
average	63,8	46,6	54,9
apparent density (kg/dm³)	2,31	1,90	2,07
cement addition (%)	20 (PZ 55)	20 (PZ 35 F)	20 (PZ 35 F)
② compressive strength (N/mm²) from - to	62,6-68,2	47,0-50,8	56,2-57,3
average	64,1	49,3	56,7
apparent density (kg/dm³)	2,43	1,92	2,12

aggregate from
① preliminary crushing ② secondary crushing

strength differences observed e. g. when confronting the results of Lohberg and Prosper waste, both for the preliminary as well as the secondary reduction materials, may be attributable to a higher proportion of non-quartzeous slaty clay in the aggregate, as far as the Prosper waste is concerned.

6 LAYOUT CONCEPT OF A PLANT FOR THE TREATMENT OF WASHERY REFUSE AND MINESTONE

6.1 Flowsheet and weight distribution

Starting from the experimental results we elaborated a flowsheet of a technical plant for the manufacturing of sand and chip fractions from waste (Figure 10).

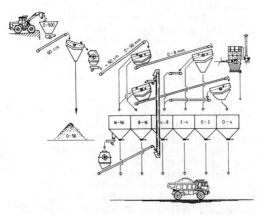

Fig. 10. Flow sheet of a two-step waste reduction plant

The initial feed (0 to 500 mm) is transported, via bucket wheel loader, to a feed hopper whence it passes over a discharge chute and conveyor belt to a 56 mm prescreening unit. It is planned to remove bigger foreign matter-bodies (timber, metal etc.) manually during that passage. The material > 56 mm is then reduced in a first impact crusher (preliminary reduction) to be subsequently screened at 8 and 16 mm. The < 8 mm material is subjected to further classification into the fractions 0 to 2 mm, 2 to 4 mm and 4 to 8 mm. As a function of the properties of the feed material, the fractions < 2 mm viz. < 4 mm could be discarded due to their high contents of carbon viz. clayey constituents. The material yielded from screening to 4 - 8 mm and 8 - 16 mm can be considered as a final product. The > 16 mm material is feed to an impact hammer crusher (secondary reduction) where it is reduced to < 16 mm, with oversize recycling, and then screened into the size ranges 0 - 4 mm, 4 - 8 mm and 8 - 16 mm. The reduction products are stored separately. A schematic representation of the weight distribution of the above process is contained in Table 6.

Table 6. Weight distribution of two-step waste reduction

Weighted Flowsheet					impact hammer crusher at face end		impact hammer crusher - sieve circuit -		product -final grain size -	
grain size	feed-raw material -		impact crusher						Σ	Σ
mm	%	t/h	%	t/h	%	t/h	%	t/h	%	t/h
0- 2			14,0	7,0	25,0	6,0	39,0	9,4	32,8	16,4
2 - 4			8,0	4,0	13,0	3,1	14,0	3,4	14,8	7,4
4 - 8	20,0	12,5	10,0	5,0	19,0	4,6	23,0	5,5	21,0	10,5
8 -16			20,0	10,0	25,0	6,0	24,0	5,7	31,4	15,7
16 - 56			48,0	24,0	18,0	4,3	–	–	–	–
56 - 500	80,0	50,0	–	–	–	–	–	–	–	–
	100,0	62,5	100,0	50,0	100,0	24,0	100,0	24,0	100,0	50,0

We started from a feed rate of 62.5 t/h, with 50 t/h being fed to the first and approx. 24 t/h to the second reduction equipment, after preliminary screening. The final product represented by the sand fraction < 4 mm is approx. 24 t/h viz. by the chips fraction 4 - 16 mm aprox. 26 t/h. If the material < 4 mm has to be subjected to preliminary reduction the sand proportion goes down by 11 t/h.

6.2 Investment and operational costs

A crushing capacity of 50 t/h was set as base for assessing the investment costs of a waste reduction plant. Said plant will have a yearly throughput of

100 000 t, for single shift operation.
Table 7 given a breakdown of the investment costs.

Table 7. Investment costs of a waste
reduction plant

Investment costs	TDM
1. Machinery and equipment	1 100
2. Electrics, including in- stallation	95
3. Steel constructions	230
4. Dedusting equipment	100
5. Covering	50
6. Installation	150
7. Assessory costs, 5 % of item 1	55
	1 780

Dust reduction and silencing items have
been included for environmental protection. The total investment thus amounts
to DM 1 780 000.

When estimating the operational cost we
assumend three technicians to be required
for plant operation. The installed electric capacity is 290 kW. Electricity
costs are based on a rate of 0.16 DM/kWh.
Global maintenance cost was entered at
5 % of investment costs. Wearing costs
appear here as the greatest cost factor.
Depreciation and capital service were
determined on the base of a capital re-
flux factor of 0.15 (10 years of operation, 9 % interests).
Table 8 gives the operational costs.

Table 8. Operational costs of a two-step
waste reduction plant

Operational cost	DM/
1. Labor	1.50
2. Material	
- Energy	0.93
- Maintenance	0.90
3. Capital costs	2.67
	6,00

Based on the above rough cost estimation
one has to expect operational costs of
approx. 6 DM/t which may go down along
with higher throughputs.

7   SUMMARY

1. Pre-screened waste >40 mm or separated minestone is, after treatment, a
quite suitable material for aggregates

2. Impact strain is the ideal way of
comminuting heterogeneous waste
material in order to dissociate its
carbonaceous and clayey constituents.
   The equipment and process parameters
should, however, be adapted to the specific type of feed. Impact strain as
exerted by a gyratory, will yield high
proportions of cubic product within a
narrow size range corresponding to the
discharge gap width of the equipment.
Said properties may be required for the
production of some types of chip material.

3. The material yielded from single-
step total reduction to smaller than
4 mm can be used for wall blocks with
calcareous binder (KS method) only if
the initial feed is rich in sandstone
and low in carbon.

4. Two-step-reduction offers a possibility to get rid of the "harmful" son-
stituents after preliminary reduction.
   In that case material from a secondary reduction step can be used for
cement-bond shapes of higher compressive strengths as well as of more moderate scatter of individual properties.
This is attributed above all to the
improved grain shape of the feed.
   For the blocks with calcareous binder
manufactured by the sandlime-process,
we were able to demonstrate that specifically enrichment in non-quartzeous
clay minerals, in combination with
enriched carbonaceous constituents,may
reduce the compressive strength of the
blocks so far that they will no longer
comply with the standards.

5. The properties of aggregates recovered from a secondary reduction step
are considerably less dependent on the
origin of the waste and yield good product characteristics for constructional
purposes.

6. The investment costs of a two-step
100 000 annual tons through-put waste
reduction plant for the manufacturing
of sand and chip fractions amount to
roughly 1.780 000 DM. A rough estimation of operational costs gave DM 6/t.

ACKNOWLEDGEMENTS:The investigations
were carried out under the sponsorship
of the Commission of the European
Communities and of the Ministry of
Economy, Trade and Technology of North-
rhine-Westphalia, FRG

Reclamation, Treatment and Utilization of Coal Mining Wastes, Rainbow (ed.) © 1990 Balkema, Rotterdam. ISBN 90 6191 154 0

Full-scale compaction trials with coal mining wastes

J.A. Hinojosa
MOPU, Dirección General de Carreteras, Madrid, Spain

J.González Cañibano & J.A.F.Valcarce
HUNOSA, Dirección Técnica, Oviedo, Spain

A.Falcón & J.L.Ibarzábal
MOPU, Demarcación de Carreteras del Estado, Oviedo, Spain

J.M.Rodriguez Ortiz
E.A.T., S.A., Madrid, Spain

ABSTRACT: This paper describes and records the results obtained in full-scale compaction trials carried out using coal mining wastes so as to study influence of the different variables: compaction energy, vibrating and non-vibrating roller, number of roller passes, layer thickness, type of waste, waste particle size distribution, etc. The data obtained from the aforementioned tests were applied in the construction of a trial embankment using coal wastes, subjected to artificially generated traffic.

1 INTRODUCTION

Earthworks are one of the most interesting uses of coal mining wastes as they allow great quantities of material to be employed without, in the mayority of cases, special treatment.

In the Federal Republic of Germany, France, England, etc., numerous studies carried out have led to the placement of millions of tons of coal wastes in highway and road embankments. In Spain, however, few of any are the studies designed to determine the potential for such coal waste use and, in addition, there are not standards to regulate its utilisation in mentioned field, the use of coal waste is scarce. Studies are however under way aimed at introducing these materials in a systematic fashion in earthworks, which should lead to the employment of far greater amounts in the next years.

Full-scale compaction trials using coal-mining wastes (both spoil-heap and washery) have been carried out to study the influence of several variables: compaction energy, vibrating and non-vibrating roller, number of roller passes, layer thickness, type of waste, waste particle size distribution, etc..

These data are the basis for the construction of a trial embankment using coal wastes.

2 WASTES TESTED

Four types of wastes were selected for the full-scale compacting trials (three from Asturias –North of Spain–, one from Leòn –Centre of Spain–, three washery, one spoil heap), considered to be representative of the wastes existing.

Graph 1 shows the particle size distribution of the different wastes tested and Table I sets out various geothecnical characteristics of the same.

3 TESTING PROCEDURE

To carry out the full-scale compacting trials, once the surface of the chosen place was prepared, a transition layer of the same type of waste to be used in the trial was spread to avoid possible influences of material from the place selected on the wastes.

This transition layer, 50 m in length, 35 m in width and 30 cm thick was compacted by 6 passes of 10.2 t static weight roller (the first non-vibrating, the rest vibrating). In situ density being determined

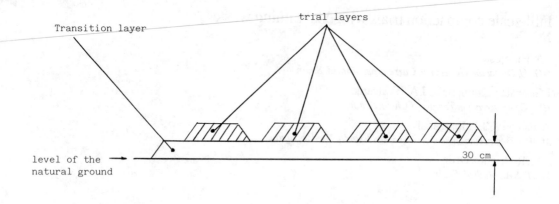

Transition layer

trial layers

level of the
natural ground

30 cm

by sand method and load tests with a 30 cm diameter plate have been carried out.

The scheme below shows the way in which trial layers were spread over the transition one.

The trial layers, 35 m long and 5 m wide at the upper level, were compacted as follows:

. 1st trial layer:

A 10.2 t static weight roller, non-vibrating, was passed over, using sand method in situ density control. The same 10.2 t non-vibrating roller was then used twice more with in situ density control by sand method.

A further two passes of 10.2 t roller were performed with sand method density determination until at least 98% of the dry maximum density obtained in Proctor Normal tests on laboratory wastes coal samples, or constant density, was achieved.

. 2nd trial layer. Procedure was exactly the same as for the first, but in this case a 16.9 t static weight roller was employed.

. 3rd trial layer. Procedure was the same as for the 1st layer but:

- the first pass was with a non-vibrating roller
- the second and subsequent passes were with a vibrating roller

. 4th trial layer. Procedure was exactly the same as for the third layer, but this time the 16.9 t roller was used.

In all the trial layers, after the final pass, load tests were

performed using a 30 cm diameter plate and samples were taken of the coal wastes to determine particle size distribution after compacting.

Load trials were carried out at 0.5 kg/cm^2 intervals up to 3 kp/cm^2 for two load cycles.

The deformation module determined by

$$E_2 = 0.75 \ D \ \frac{\Delta p}{\Delta s}$$

being

E_2 : deformation module

D : plate diameter (30 cm)

Δp : increase in pressure (normally between 0.5 and 1.5 kp/cm^2)

Δs : settling

should be $E_2 \geqslant 500$ kp/cm^2 for the second load cycle.

Trial layers of 30 - 50 and 70 cm thickness were employed.

4 RESULTS AND DISCUSSION

Graphs 2 and 3 show density with respect to number of passes of 10.2 and 16.9 t static weight rollers in the case of León fresh coal washery wastes. It can be seen from these that:

. For a roller in identical conditions and with identical layer thickness, density increases with the number of roller passes up to a certain maximum. The density remains constant,

334

or even decreases, with subsequent roller passes. This is due to the fact that increased energy produces particle disorder and thus diminished density.

. For the same roller and layer thickness, density achieved is greater for the vibrating than for the non-vibrating roller.

. The greater the layer thickness, the more the number of passes required to achieve maximum density. This is normally reached after 5 vibrating or non-vibrating passes. More passes may be necessary at 70 cm thickness layer.

. For the same number of passes and the same roller in identical conditions (whether vibrating or not) greater densities as layer thickness diminishes are obtained.

. It appears from a comparison of densities obtained from the different rollers used, with or without vibration, that for the same tickness:

-The same densities are obtained at 30 cm layer thickness for the same number of roller passes, whether vibrating or not, with 10.2 and 16.9 t static weight rollers.

-At 50 cm layer thickness, using at non-vibrating roller, densities obtained with the 10.2 t roller are rather higher than those given by the 16.9 t one.

-At 50 cm thickness with vibrating roller and 70 cm thickness with vibrating or non-vibrating rollers -with few passes- greater densities are obtained with the 16.9 roller than the 10.2 t static weight one, but the results are reversed once the number of passes exceeds seven.

Table II sets out the deformation modules obtained in the load trials carried out, from which it can be seen that modules in excess of 1000 kp/cm^2 are generally achieved, far above the 500 kp/cm^2 required.

Graphs 4 - 5 and 6 show density with respect to the number of passes for the 16.9 t static weight roller in samples 3 - 4 and 2 respectively.

The conclusions drawn being similar for those for sample 1.

The deformation modules obtained for samples 2 - 3 and 4 are presented in Table III, from which it can

be seen that they are above the 500 kp/cm^2 required and correspond in general to the densities achieved, greater deformation modules being obtained with a vibrating roller than a non-vibrating one for the same number of passes and the same thickness.

All the coal wastes tested suffer degradation on compacting, as previously observed in laboratory tests, as can be seen in Graph 7, which shows the particle size distribution before and after compacting in the case of spoil heap wastes, sample 3. This degradation is, however, small, and does not affect the quality of the material.

5 CONCLUSIONS

It may be deduced from these full-scale compaction trials that good results are obtained (100% dry maximum density) with 5 passes of 10.2 t roller or 7 passes of 16.9 t roller (static weight), vibration being applied in both cases.

There is little appreciable difference, in most types of coal wastes, in density whether 10.2 or 16.9 t static weight rollers are used.

Furthermore, it would seem unadvisable to use layers of more than 50 cm thickness.

The compaction process does, however, produce a certain degradation in the wastes, increasing the number of fines particles ($<$ 0.080 mm) by 2-3%.

ACKNOWLEDGEMENTS

The authors would like to express their gratitude to MOPU (Direcciòn General de Carreteras), HUNOSA and OCICARBON for the facilities afforded them in the elaboration of this paper and for financing the project "The use of coal mining wastes in earthworks", to Doña Josefina Grela Vàzquez for the typing and to Don Ceferino Fernàndez Cuetos for the drawings.

335

Table I. Geotechnical characteristics of the tested coal wastes

Type	Liquid limit	Plastic limit	Plasticity index	Normal Proctor dry maximum density (g/cm^3)	optimum humidity (%)
Sample 1	20	18	2	1.93	8.0
Sample 2	18	14	4	1.99	9.0
Sample 3	25	16	9	1.87	12.0
Sample 4	18	12	6	2.10	7.0

Table II. Deformation modules in coal waste load trials in sample 1

No. of passes	layer thickness cm	roller type	compacting method	deformation module
9	30	10.2 t	non-vibrating	1323
9	30	10.2 t	vibrating	1406
9	30	16.9 t	non-vibrating	1184
9	30	16.9 t	vibrating	1812
9	50	10.2 t	non-vibrating	1023
9	50	10.2 t	vibrating	1607
9	50	16.9 t	non-vibrating	1025
9	50	16.9 t	vibrating	1406
9	70	10.2 t	non-vibrating	1125
9	70	10.2 t	vibrating	1125
9	70	16.9 t	non-vibrating	937
9	70	16.9 t	vibrating	1250

Table III. Deformation modules in load trials for coal waste samples 2 - 3 and 4

Waste	No. of passes	layer thickness cm	roller type	compacting method	deformation module E_2 (kp/cm^2)
2	5	30	10.2 t	non-vibrating	1023
2	5	30	10.2 t	vibrating	1250
2	5	30	16.9 t	non-vibrating	938
2	7	30	16.9 t	vibrating	1250
2	9	50	16.9 t	non-vibrating	750
2	9	50	16.9 t	vibrating	1500
3	9	30	10.2 t	non-vibrating	1125
3	9	30	10.2 t	vibrating	1406
3	9	30	16.9 t	non-vibrating	1250
3	9	30	16.9 t	vibrating	1184
3	9	50	16.9 t	non-vibrating	1125
3	9	50	16.9 t	vibrating	1323
3	9	70	16.9 t	vibrating	1125
4	9	30	10.2 t	non-vibrating	1250
4	9	30	16.9 t	non-vibrating	1184
4	9	30	16.9 t	vibrating	1323
4	9	50	16.9 t	non-vibrating	1071
4	9	50	16.9 t	vibrating	1250
4	9	70	16.9 t	vibrating	1184

Sample 2 Fresh washery coal wastes. Asturias
Sample 1 Fresh washery coal wastes. León
Sample 3 Spoil heap wastes.
Sample 4 Fresh washery coal wastes. Asturias

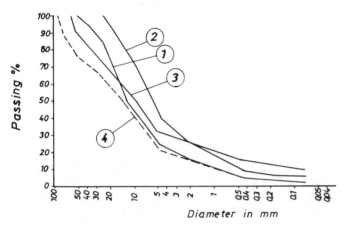

Graph 1. Particle size distribution of coal wastes employed in
full-scale compacting trials.

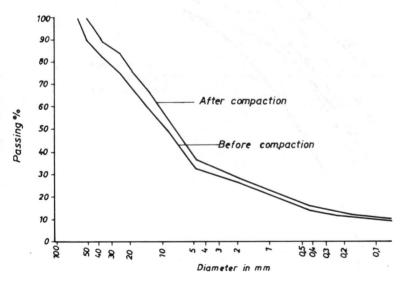

Graph 7. Particle size distribution before and after full-scale compaction trials. Sample 3.

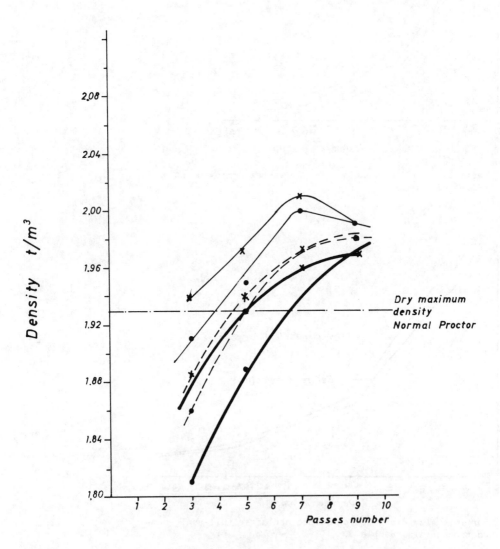

Graph. 2 Density versus roller passes. Roller of 10.2 t static weight. Fresh washery coal wastes. Leòn. (Sample 1).

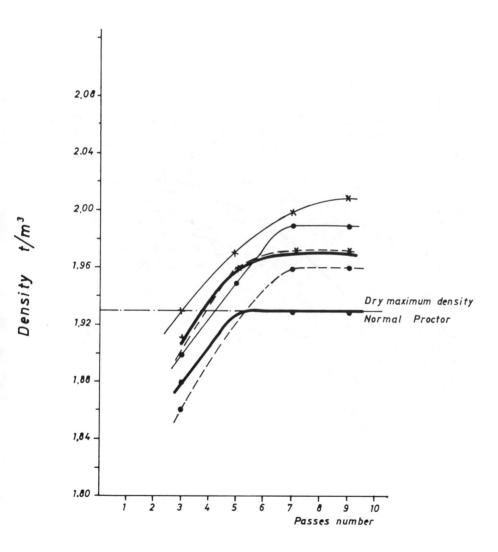

Graph . 3 Density versus roller passes. Roller of 16.9 t static weight. Fresh washery coal wastes. Leòn (Sample 1).

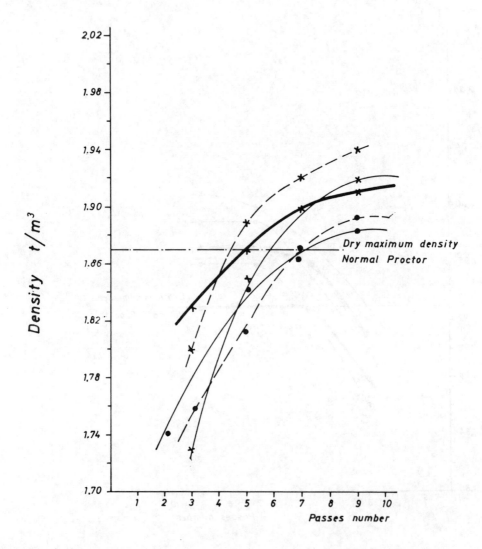

Graph. 4 Density versus roller passes. Roller of 16.9 t static weight. Minestone (Sample 3)

340

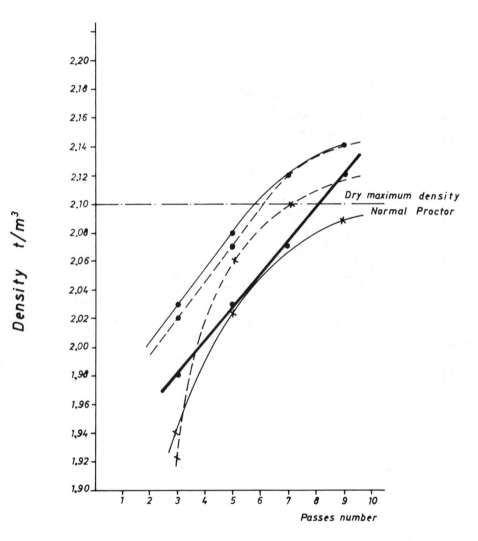

Graph 5   Density versus roller passes. Roller of 16.9 static weight.
Fresh washery coal wastes. Asturias (Sample 4)

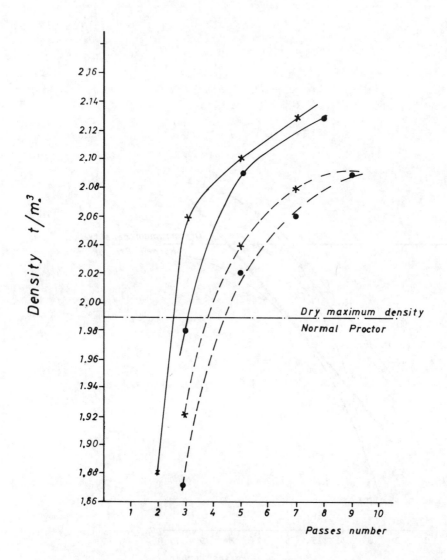

Graph. 6 Density versus roller passes. Roller of 16.9 t static weight.
Fresh washery coal wastes. Asturias. (Sample 2)

Reclamation, Treatment and Utilization of Coal Mining Wastes, Rainbow (ed.) © 1990 Balkema, Rotterdam. ISBN 90 6191 154 0

A combined anchor and geotextile system of reinforcement for Singrauli minestone utilization

R.B.Singh
Institute of Technology, Banaras Hindu University, Varanasi, India

ABSTRACT: The performance of a combined anchor and geotextile system of earth reinforcement, using semi-Z shaped mild steel model anchors alternating with a heavy grade non-woven synthetic geotextile in depth, was studied in a test box using Singrauli minestone soil finer than 20 mm as the fill soil. Comparative studies for fully geotextile reinforced and fully anchor reinforced systems were also carried out. The end results of the tests presented in the paper show that the combined system is the most efficient one and is characterized by a fairly high rigidity and high holding capacity.

1 INTRODUCTION

Anchored earth is one of the most recent, simple and versatile system of soil reinforcement. The holding capacity of the reinforcing system is a consequence of the passive resistance developed by standard round mild steel bar anchors, having the effective end bent into a triangular or Z-shape, pulling out of the compacted soil fill. The first ever application of anchored earth was reported by Jones, et. al. (1985), and the first Indian application has been reported recently by Singh (1989). Researches by the author at the University of Glasgow and Banaras Hindu University spanning over the last six years, using minestone waste soils as fill material in model experimentation, have given confidence in the use of this technique for such soils (Singh 1987, Singh and Siavoshnia 1988a). Performance studies of triangular-shaped polymeric anchor of special design were also carried out using a range of British minestones (Singh, et. al. 1987).

In reinforced soil, geotextiles have been used as reinforcing elements (Mitchell 1981) to a limited degree. A combined anchored earth and geotextile system (nicknamed "combined system"), using anchors alternating with synthetic geotextiles over the height or depth of the structure was proposed by Singh and Finlay (1986) as offering a more advantageous system, which could find applications in embankment situations for all types of soils. The end results of a model study on the application of the abovementioned combined system using semi-Z shaped round mild steel bar anchors and a non-woven permeable synthetic geotextile in a compacted fill of minestone finer than 20 mm size were presented recently by Singh and Siavoshnia (1988a, 1988b). Considerable improvement in the holding capacity of the combined system was noted, as compared to the system using anchors alone or using geotextile alone. The important end results of the more recent model tests are presented in this paper, reinforcing the abovementioned conclusions.

The growing awareness of the environmental and ecological problems associated with large open-cast coal mining areas in India, particularly those of the Singrauli coal belt area of eastern Uttar Pradesh and northern Madhya Pradesh, prompted these studies to be continued further. Anchored earth using semi-Z shaped anchors, and the combined semi-Z shaped anchor and synthetic non-woven geotextile system, are both proposed as suitable systems for mitigating the problem of economic utilisation or acceptable disposal of the waste coalmine overburden soils in the construction of useful civil engineering structures, like haul roads, highway earthworks and embankments.

2 THE MINESTONE SOIL

The minestone soil used in experimentation was the finer than 20 mm fraction of a colliery overburden soil, not subjected to any heat alteration, obtained from the Singrauli coal belt area. Table 1 gives the engineering properties of the soil.

Table 1. Engineering properties of the minestone soil.

1. Natural moisture content	4.5%
2. Specific gravity of solids	2.41
3. Particle size distribution (Particles coarser than 20 mm were removed from the minestone soil used)	
finer than 0.002 mm	**6**.5%
0.002 mm - 0.06 mm	14.5%
0.06 mm - 2.00 mm	73%
2.00 mm - 20.00 mm	**6**%
4. Proctor test	
maximum dry density	1.79g/cm^3
optimum moisture content	11%
5. Index tests (0.0063 mm fraction)	
Liquid limit	19%
Plastic limit	14%
Plasticity index	5
6. Direct shear test at slow rate (0.020 mm/min), to simulate drained condition for sample prepared at the test condition having dry density of 1.70 g/cm^3 and moisture content of 13%	
$c' =$	0.03 kg/cm^2
$\emptyset' =$	28°

3 THE GEOTEXTILE

An Indian manufactured 100% polypropylene, needle-punched, white, non-woven, heavy-grade, geotextile with bonding one side on to an open-mesh white nylon scrim was used in the studies. The addition of the scrim gave the geofabric some added strength as well as good surface bonding property for soil reinforcement. The tested characteristic physical properties of the geotextile are given in Table 2.

Table 2. Characteristics of the synthetic geotextile.

Sl. No.	Physical Property	Value
1.	Weight (g/m^2)	213
2.	Thickness (mm)	2.8
3.	Grab tensile test Breaking strength (kg force)	
	(a) warp way	51
	(b) weft way	48
4.	Wing tear test Tear strength (kg force)	
	(a) warp way	13.5
	(b) weft way	11.7

4 THE SEMI-Z ANCHOR

Details of the semi-Z shaped with rounded corner anchor used in the studies are given in Fig. 1. It also shows the observed slip zone boundaries in the plane of the anchor corresponding to the three-dimensional soil wedges that develop on the straight and inclined arms of the anchor during pull-out. A theoretical approach for predicting the pull-out capacity of

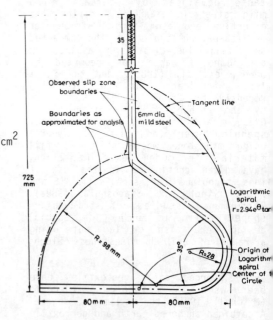

Fig.1 Semi-Z shaped anchor and soil wedge boundaries in the plane of the anchor in Singrauli minestone fill.

fully anchored systems has been given by Singh, et. al. (1985), and has been applied for such a purpose by Rai (1989).

5 THE TEST BOX

Model tests were performed in a wooden test box, reinforced with heavy steel members, with internal dimensions of 735 mm length by 545 mm width and 415 mm height. An opening was provided at the front face to accommodate five wooden interlocking facing panels measuring to an assembled overall dimension of 540 mm width and 340 mm height. The anchor or the geotextile, as appropriate, and the minestone compacted fill was brought up in appropriate lifts to fill the test box. A steel loading frame and hydraulic jack were used for applying reaction loading to the top of the test box in order to conduct the stress-controlled loading tests. The applied loading, horizontal displacements of the five movable front panels, and the drop in the top level surface of the compacted fill ('settlement') were recorded suitably as the test was continued to failure or near failure condition.

Details of the model set-up have been given by Singh and Siavoshnia (1988b) and Rai (1989).

6 THE TESTS

Three sets of tests were carried out, one each corresponding to the fully geotextile reinforced, fully anchor reinforced, and combined anchor and geotextile reinforced system. In each test the compacted minestone fill was prepared to a dry density of 1.70 g/cm3 and a moisture content of .13 per cent. The respective compaction procedure was established by trial to obtain compacted fill with the abovementioned dry density and moisture content.

The fully geotextile reinforced and the fully anchor reinforced systems had three layers of geotextile and three layers of anchors, respectively, installed in the middle of each of the first (I), third (III) and fifth(V) layers of the compacted fill (the five layers of compacted fill correspond with the five movable interlocking facing panels - panel V being the topmost). In the combined system

test, the soil sample in the test box was reinforced with three layers of anchors (I, III and V panels) and two layers of geotextile in between (II and IV panels).

Geotextile was cut to the dimensions of 635 mm by 540 mm for installation in the test box, and was connected to the respective facing panel by first folding one side of it around a light weight aluminium flat strip, which was then connected by steel screws to the inside face of the facing panel. A simple nut connection was used for the screwed end of the anchor for connecting to the respective panel. Each layer of anchor comprised of two anchor elements positioned 280 mm apart centrally in the box.

Load testing of the three systems was done by controlled application of uniformly applied stress increments through reaction loading, and recording data, as mentioned in section 5.

Applied vertical stress versus 'settlement', and lateral displacement of panel V versus applied vertical stress curves, for each of the three tests described, are shown in Figs. 2 and 3, respectively. The curves for other panels were similar though showing smaller displacements than for panel V.

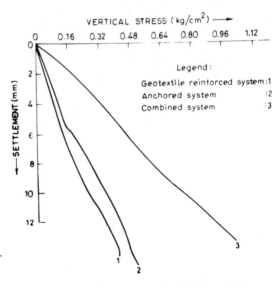

Fig. 2 Applied vertical stress versus 'settlement'

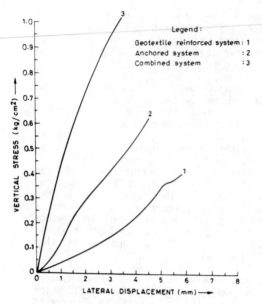

Fig. 3. Lateral displacement of panel V versus applied vertical stress

The programme of investigation on the combined system has so far covered a silty clay soil and a fly ash soil from a thermal power plant, in addition to the minestone soil. The trend of results has been found to be similar for all the soils tested.

7 DISCUSSION

Fig. 2 shows that the 'settlement' (drop in the top level surface of the compacted fill) exhibited by the combined system for any given vertical stress applied is much less than that exhibited by the other two systems. Likewise, Fig. 3 shows that the lateral displacement of panel V exhibited by the combined system for any given vertical stress applied is much less than that exhibited by the other two systems. The system behaviour shown by these two figures, clearly indicates that the combined system is characterized by a fairly high rigidity and high holding capacity, and is the most effective of the three soil reinforcing systems studied. Further, it indicates that the use of geotextiles singly as reinforcement may not be very useful for embankment situations.

These figures give an indication, that in the increasing order of their holding capacities, the three systems could be arranged as the fully geotextile reinforced system, the fully anchor reinforced system and the combined anchor and geotextile reinforced system. Detailed studies (Rai 1989) have shown this conclusion to be true.

8 CONCLUSIONS

A study of the relative effectiveness of the three soil reinforcing systems, viz., synthetic geotextile reinforcing system, anchored earth system using semi-Z shaped mild steel anchors, and the combined anchor and geotextile reinforced system, was attempted using Singrauli minestone finer than 20 mm as fill. It was shown that the combined system is the most efficient, followed in the descending order of effectiveness by the fully anchored system and the fully geotextile reinforced system. It is proposed that coalmine waste soils, like Singrauli minestone, reinforced with the combined system, could be conveniently used as bulk fill for economic civil engineering uses, such as in embankment situations.

9 ACKNOWLEDGEMENT

The author gratefully acknowledges the financial support, the facilities and opportunities for research on Anchored Earth provided by Banaras Hindu University, University of Glasgow and Minestone Services of British Coal. He expresses his deep gratitude to Professor Hugh B. Sutherland for his guidance in the early stages of research, and thanks his ex-research students Mr. Mehdi Siavoshnia and Mr. D.K. Rai for conducting the tests on the Singrauli minestone soil.

10 REFERENCES

Jones,C.J.F.P., Murray, R.T., Temporal, J. and R.J. Mair 1985. First Application of Anchored Earth. Proc. 11th Intern. Conf. of Soil Mech. and Found. Engg., San Francisco 3: 213-218.

Mitchell, J.K. 1981. Soil Improvement - State of the Art Report. Proc. 10th Intern. Conf. of Soil Mech. and Found.Engg., Stockholm 4: 509-565.

Rai,D.K. 1989. A Comparative Study of Anchored, Geotextile, and Combined Anchor and Geotextile Systems of Soil Reinforcement. M. Tech. thesis, Department of Civil Engineering, Banaras Hindu University, Varanasi.

Singh,R.B., Finlay, T.W. and H.B. Sutherland 1985. Fabric Studies in Model Investigations of Anchored Earth. Proc. 7th Intern. Working Meeting on Soil Micromorphology, Paris. Elsevier (1987): 545-551.

Singh, R.B. and T.W. Finlay 1986. Response paper. Proc. Indian Geotech. Conf. (IGC-86), Delhi 2: 211-213.

Singh,R.B. 1987. Anchored Earth Study using semi-Z Shaped Model Anchors. Proc. Indian Geotech. Conf. (IGC-87), Bangalore 1: 373-375.

Singh,R.B., Finlay, T.W. and A.K.M. Rainbow 1987. Reinforced Minestone Using Special Designed Reinforcement. Proc. 2nd Intern. Conf. on the Reclamation, Treatment and Utilization of Coal Mining Wastes, Nottingham: 583-586.

Singh,R.B. and M. Siavoshnia 1988(a). Field Characterisation of an Indian Geotextile and Its Application in Soil Reinforcement. Proc. Indian Geotech. Conf. (IGC-88), Allahabad 1: 195-198.

Singh,R.B. and M. Siavoshnia 1988(b). Performance Study of a Combined Anchored Earth and Geotextile System of Reinforced Soil. Proc. First Indian Geotextiles Conf. on Reinforced Soil and Geotextiles, Bombay: F.31-F.37.

Singh, R.B. 1989. First Application of Anchored Earth for the Approach Embankment of a Road Culvert. Proc. Indian Geotech. Conf. (IGC-89), Visakhapatnam 1: 339-342.

347

Reclamation, Treatment and Utilization of Coal Mining Wastes, Rainbow (ed.) © 1990 Balkema, Rotterdam. ISBN 90 6191 154 0

Friction characteristics of polypropylene reinforcing straps in various fills

A. Bouazza, T.W. Finlay, N. Hytiris & M.J. Wei
Glasgow University, UK

ABSTRACT: Large shear box, laboratory pull-out, and field pull-out tests have been carried out on two types of polypropylene strap designated A and B up to 4m long embedded in five different fills, viz two minestones, one burnt shale, a sand, and pulverised fuel ash (p.f.a.).

The tests produced values of friction co-efficient, pull-out $\underline{v}$ displacement, and maximum pull-out force for the various combinations of test type and strap.

The results are presented and indicate that strap type B offers a higher resistance to pull-out than type A in all the fill materials. They also show that the friction coefficients derived from the large shear box tests are generally higher than from the laboratory pull out tests for the two minestones at all values of overburden pressure, unlike the sand $\underline{v}$ strap B where the opposite is the case. The p.f.a. and burnt shale show higher values from the shear box at overburden pressures greater than about 40 and 80 kN/m^2 respectively. Finally, the friction coefficients from field pull-out tests are generally lower than those from both the shear box and the laboratory pull-out tests, with the exception of the burnt shale.

The results and their significance in design are discussed.

1 INTRODUCTION

Reinforced earth as a construction technique in civil engineering was introduced in 1966 and involves the amalgamation of reinforcing elements into a mass of fill. Many reinforced earth structures have been erected world-wide and normally comprise metallic reinforcement in a cohesionless soil. More recently, attention has turned to the use of non-metallic reinforcing elements in an attempt to overcome the problem of corrosion, and to the use of fills other than selected granular fills currently specified. The research described in this paper investigates the friction characteristics of polypropylene reinforcing straps in five different fill materials and the work formed part of a larger research programme sponsored by British Coal's Minestone Services division and carried out in the Department of Civil Engineering at Glasgow University.

Laboratory shear box and pull-out tests, and full-scale field pull-out tests have been performed using two types of polypropylene strap embedded in five different fill materials, viz two minestones, one burnt shale, a sand, and pulverised fuel ash.

The friction characteristics obtained from the tests are presented and their significance in design is discussed.

2 MATERIALS

2.1 Reinforcing straps.

Samples of polypropylene strap, designated A and B were supplied by I.C.I. Both types comprised bundles of continuous, aligned, high tenacity polyester fibres enclosed in a durable polypropylene sheath. Strap A, beige in colour, had a cross section 85 mm x 2.5mm and a reported maximum tensile strength of 30 kN. Strap B had a cross-section 92mm x 3.5mm, a reported maximum tensile strength of 50 kN, and was black in colour to resist ultra-violet radiation and bacterial attack. Both types of strap had a textured surface to enhance the friction against the surrounding soil.

2.2 Fill materials.

The various fill materials used were

1. Wardley minestone
2. Wearmouth minestone
3. Horden red shale
4. Loudon Hill sand
5. Methil p.f. ash

The properties of the fill materials were determined from tests carried out by British Coal's own laboratories, and included particle size distribution, and dry density/moisture content relationship, the results of which are presented in Table 1.

The friction co-efficient, μ, obtained from the shear box test can be expressed as

$$\mu = c_a/\sigma_v + \tan \delta$$

where c_a and δ are the adhesion and friction angle measured, and σ_v is the normal stress or overburden pressure. Obviously, in a cohesionless fill, $\mu = \tan \delta$.

3.2 Laboratory pull-out test

A steel box 2m long by 0.4m wide by 0.5m deep was used. The fill was compacted to mid-height in the box, at the moisture content and density given by the field tests, and a single reinforcing strap 1.5m long was placed in position

Table 1   Properties of fill materials.

	Wardley minestone	Wearmouth minestone	Horden red shale	Loudon Hill sand	Methil p.f.a.
Particle size distribution %					
Cobbles > 60 mm	30	10	0	0	0
Gravel 2-60 mm	51	78	64	23	0
Sand 60μm-2 mm	15	9	27	75	36
Silt/clay <60μm	4	3	9	2	64
Dry density/m.c. $\lceil \gamma$dmax kN/m^3	18.2	17.7	18.6	18.1	11.1
2.5 kg rammer $\lfloor$ o.m.c. %	10.0	8.0	15.0	15.0	36.0
Test conditions $\lceil \gamma$d kN/m^3	17.8	17.7	17.7	16.5	11.6
$\lfloor$ m.c. %	9.5	5.7	12.5	7.0	27.0

3   TESTS PERFORMED

3.1 Shear box test.

A large, 300 mm x 300 mm, shear box was used to determine the shear strength characteristics of each of the fill materials at their natural moisture content and at a density corresponding to that obtained from the full-scale field tests. The frictional characteristics of the reinforcing straps were found by fitting a timber block into the bottom half of the shear box and fixing a continuous 300 mm square layer of the strap material to the surface of the block, level with the join between the upper and lower halves. The upper half of the box was then filled with compacted fill material, and the shear test done after applying a vertical load to give a normal stress.

centrally and fed through a slot in the front face. Further fill was placed and compacted level with the top of the box. A steel cover plate with a rubber membrane below was then attached to the top of the box. Air pressure introduced between the cover plate and the membrane developed a normal stress (or overburden pressure) on the fill, and the pull-out was applied to the strap, the pull-out force and displacement being continuously monitored during the test. The apparent frciition co-efficient, f*, was obtained from the expression

$$f^* = T/2BL \ \sigma_v$$

where T = maximum pull-out force
    B = width of strap
    L = length of strap
    σ_v = normal stress or overburden pressure

3.3 Field pull-out tests.

The field tests were carried out at two locations, at Wardley colliery in N.E. England and at Barony colliery in Ayrshire. At each location, a large open-topped box with one open end, was constructed from rolled steel sections and timber railway sleepers. The boxes were each 5m long by 4.5m high, that at Wardley being 2m wide, and at Barony 3.1m wide. During use, the box was filled with compacted fill material, and at specific levels reinforcing straps were laid which passed through the front face of the box. When the box was full, a specially designed pull-out jack, supplied by Minestone Services, was used to pull out individual straps with a continuous record of load $\underline{v}$ displacement. The friction co-efficient f^{**} from these tests was obtained using the same expression as for the laboratory pull-out tests, and taking the overburden pressure as a function of the fill height above the strap.

4 RESULTS

The shear box results are shown in Table 2. To facilitate the comparison between shear box and pull-out tests, the fill/strap friction co-efficients, μ, have been calculated and are given in Table 3.

Table 4 shows the friction co-efficients f^* derived from the laboratory pull-out tests.

The field pull-out test friction co-efficients f^{**} are detailed in Table 5. It should be noted that the highest overburden pressure available was restricted by the height of fill to 70 kN/m^2 compared with the 120 kN/m^2 for the laboratory pull-out test.

5 DISCUSSION

The frictional resistance of the strap surface as given by the shear box test is shown in Fig. 1 where it can be seen that strap B is superior to strap A in all the fill materials except Loudon Hill sand at low overburden pressure.

Table 2 Shear box test results.

		Wardley minestone	Wearmouth minestone	Horden red shale	Loudon Hill sand	Methil p.f.a.
Fill alone	c kN/m²	14	8	12	0	3
	Φ deg	33.0	37.0	41.5	31.0	36.5
Fill/strap A	c_a kN/m²	4	3	3	3.5	5
	δ deg	18.0	18.0	23.5	20.0	21.0
Fill/strap B	c_a kN/m²	4	6	2.5	0.5	7
	δ deg	20.0	21.0	27.5	22.5	24.0

Table 3 Friction co-efficients μ from shear box tests.

Overburden pressure, σ_v kN/m²	Wardley minestone strap		Wearmouth minestone strap		Horden red shale strap		Loudon Hill sand strap		Methil p.f.a. strap	
	A	B	A	B	A	B	A	B	A	B
20	0.525	0.564	0.475	0.684	0.585	0.646	0.539	0.439	0.634	0.795
60	0.392	0.431	0.375	0.484	0.485	0.562	0.422	0.423	0.467	0.562
120	0.358	0.397	0.350	0.434	0.460	0.541	0.393	0.418	0.426	0.504

Table 4 - Friction co-efficients f* derived from laboratory pull-out tests.

Overburden pressure, σ_v kN/m²	Wardley minestone strap		Wearmouth minestone strap		Horden red shale strap		Loudon Hill sand strap		Methil p.f.a. strap	
	A	B	A	B	A	B	A	B	A	B
20	0.650	0.303	0.418	0.459	1.078	1.492	0.636	1.165	0.725	0.918
60	0.280	0.389	0.230	0.254	0.396	0.626	0.350	0.644	0.353	0.477
120	0.200	0.173	0.198	0.262	0.289	0.498	0.314	0.415	0.350	0.434

Table 5   Friction coefficients f** from field pull-out tests.

Overburden pressure, σ_v kN/m²	Wardley minestone strap		Wearmouth minestone strap		Horden red shale strap		Loudon Hill sand strap		Methil p.f.a. strap	
	A	B	A	B	A	B	A	B	A	B
4			0.956	0.543	3.566	4.518				
7										
8	0.313	0.374							0.740	1.315
10.6							0.485	0.717		
12.2									0.522	0.310
15			0.451	0.489	0.897	1.449				
17.4									0.397	0.367
18.6							0.296	0.693		
19	0.503	0.644								
20			0.519		0.625					
22.6									0.315	0.499
26				0.393		0.503				
26.5							0.344	0.568		
27.8									0.243	0.291
30	0.315									
32			0.354	0.450	0.303	0.493				
33									0.214	0.269
34.4							0.257	0.270		
38.3									0.196	0.292
41	0.237	0.219								
42.4							0.212	0.304		
43				0.262		0.561				
43.5									0.192	0.244
49			0.222		0.426					
50.3							0.127	0.320		
52		0.239								
54			0.245	0.213		0.438				
58.3							0.260	0.417		
63	0.180	0.182								
65			0.151	0.229		0.401				
66.2							0.198	0.252		
70			0.244	0.186		0.307				

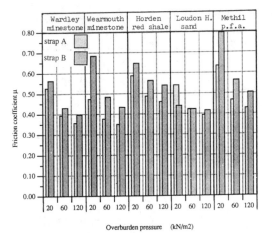

Fig 1   Comparison of straps A & B - shear box

When the friction co-efficients obtained from the shear box and the laboratory pull-out tests are compared it can be seen from Table 6 that the shear box derived values are generally greater than the laboratory pull-out values except at low overburden pressure in the non-minestone materials.

To compare values of field pull-out friction co-efficient f^{**} with the values of f^{*} resort was made to interpolation from Table 5 to determine f^{**} for strap A at overburden pressures of 20 and 60 kN/m^{2}, and Table 7 presents the comparison which shows that the friction co-efficients from field pull-out tests are generally lower than those from the laboratory pull-out tests.

The tests have shown that the fill materials can be ranked generally in descending order of friction co-efficient, regardless of how this is measured, as Horden red shale, Methil p.f.a., Loudon Hill sand, Wardley minestone, and Wearmouth minestone, and that strap B was superior to strap A.

The laboratory pull-out tests, Fig. 2, indicate that strap B is also better than strap A except for Wardley minestone at low and high overburden pressures.

Of more significance perhaps is the fact that the three different methods of measuring the friction co-efficient gave very different results, with the order from highest to lowest being shear box, laboratory pull-out, and field pull-out. Since the friction co-efficient is used in design, a problem obviously arises as to which value to use. It is the authors' opinion, based on research work carried out by Wei (1988) that although the shear box test gives the frictional characteristics of the reinforcement surface it does not take into account the effect of strap flexibility as it is pulled out of the soil under real conditions. This effect is reproduced in the laboratory and field pull-out tests, both of which give lower values of friction co-efficient which would be more suitable for use in design.

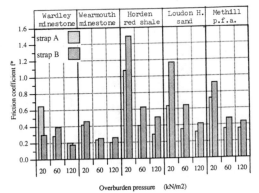

Fig 2   Comparison of straps A & B - Laboratory pull-out

Table 6   Ratio of shear box μ to laboratory pull-out f^{*}.

Overburden pressure, σ_v kN/m²	Wardley minestone strap A	Wardley minestone strap B	Wearmouth minestone strap A	Wearmouth minestone strap B	Horden red shale strap A	Horden red shale strap B	Loudon Hill sand strap A	Loudon Hill sand strap B	Methil p.f.a. strap A	Methil p.f.a. strap B
20	0.81	1.86	1.14	1.49	0.54	0.43	0.85	0.38	0.87	0.87
60	1.40	1.11	1.63	1.91	1.22	0.90	1.21	0.66	1.32	1.18
120	1.79	2.29	1.77	1.66	1.59	1.09	1.25	0.93	1.22	1.16

Table 7   Ratio of laboratory pull-out f^* to field pull-out f^{**}.

Overburden press. σ_v kN/m^2	Wardley minestone	Wearmouth minestone	Horden red shale	Loudon Hill sand	Methi p.f.a.
20	1.38	0.80	1.71	2.19	2.13
60	1.56	1.21	–	1.19	1.96

6 ACKNOWLEDGMENTS

The role of British Coal's Minestone Services Division in this work was invaluable, particularly in the field tests at Wardley and Barony Collieries.
These could not have been done without the assistance of the Director, Dr. A.K. M. Rainbow, Geotechnical Engineer, Stephen Barnett, and the on-site staff.

7 REFERENCE

Finlay, T.W., Wei, M.J. & Hytiris, N. 1988   "Friction characteristics of polypropylene straps in reinforced minestone".   Proc. Int. Geotechnical Symp. on Theory and Practice of Earth Reinforcement, Fukuoka, Japan pp 87-92.

Reclamation, Treatment and Utilization of Coal Mining Wastes, Rainbow (ed.) © 1990 Balkema, Rotterdam. ISBN 90 6191 154 0

Waste dump management – With particular reference to slope design

O. P. Upadhyay, D. K. Sharma & D. P. Singh
Department of Mining Engineering, Institute of Technology, Banaras Hindu University, Varanasi, India

ABSTRACT : Opencast mining operations involve huge volumes of overburden removal, dumping and back filling in excavated areas. Exhorbitant increase in rate of growth of waste dumps in recent years, resulting in dumps of greater height in order to occupy minimum ground area has given rise to danger of dump failures. Knowledge of physico-mechanical properties of overburden material, especially where it is not hard and compact but virtually friable, is helpful in slope stability. Consolidation and permeability properties of waste material and volume to weight relationship of the excavated material are of importance. The present paper discusses various engineering properties involved in waste dump slope design and remedial measures necessary to make the slope more stable.

1 INTRODUCTION

Various waste materials resulting from mining operations include soils from strip and placer mining, overburden, excavated rock and ground ore in slurry form and fine wet materials such as filter cake from coal treatment plants. These may be used for backfilling mined out spaces or dumped/ponded, depending on their characteristics (according to their geological origin, particle size, processing and water content), volume and availability of storage spaces.

Our study is confined only to waste material placed on ground surface and embankments necessary to support them. Waste heaps are increasing in volume and height with the increase in mining activity and trend towards mining of low grade ores. Increase in rate of water distribution, height and rate of growth of dumps, affect the stability and have prompted investigators to study these aspects so as to minimise possibilities of failures.

2 TYPES OF DUMP FAILURES

Different types of failure patterns include cylindrical failure, multiple block plane wedge failure and flow slides.

2.1 Cylindrical failure

Deep seated failure (Fig. 1a) occurs when there are no planes of weakness whereas shallow failure (Fig. 1b) results when it is restricted by weak planes or other discontinuities.

2.2 Multiple block plane wedge failure

A two-block wedge failure consisting of an active wedge and a toe wedge is shown in Fig. 2a. wedge may be conceived to be consisting of an active block, a middle block and a passive block (Fig. 2b). This is common mode of failure when dump is placed on a weak layer.

2.3 Flow slides

These include debris slide, mud flow, rocky mud flow, mud slide, earth flow, and mud spate. Flow slides occur in a wide variety of situations commonly under wholly or partly subaqueous conditions (Hutchinson 1986; Koppejan 1948; Casagrande 1965; Anderson 1967; Bjerrum 1971; Torrey 1984). Sub-aerial flow slides in waste tips are becoming more frequent as the number and size of these increase (Bishop 1973; Hutchinson 1986; Dorby and Alvarez, 1967; Davies 1968;

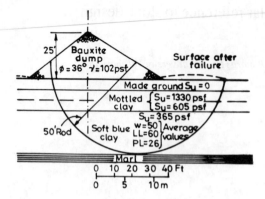

Fig. 1a Failure of a Bauxite dump at Newport. Cylindrical failure surface (after Bishop 1973)

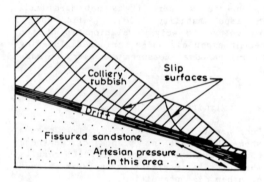

Fig. 1b Failure of a colliery waste dump. Cylindrical failure surface (after Bishop 1973)

Blight 1977; Campbell 1978). A typical flowslide of limestone waste is shown in Fig. 3a. Excellent reviews of the flowslides and debris flow have been reported by Rouse (1984) and Johnson

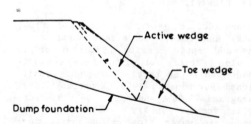

Fig. 2a Illustration of active wedge and the toe wedge. Plane failure surface. (after Campbell 1986)

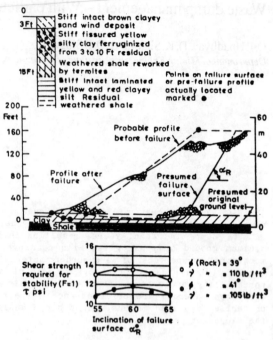

Fig. 2b Details of first Vlakfontein slide plane failure surface (after Bishop 1973)

and Rodine (1984) respectively. An important mechanism in explaining catastrophic flowslide has been presented by Shreve (1986) and is shown in Fig. 3b. Speed

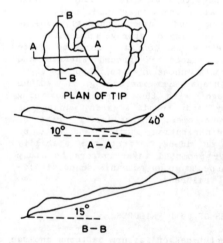

Fig. 3a Plan and sections of flowslide of limestone waste, Derbyshire (after Bishop 1973)

356

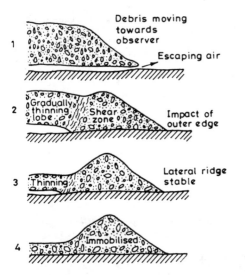

1 Debris moving towards observer
Escaping air

2 Gradually thinning lobe | Shear zone
Impact of outer edge

3 Thinning
Lateral ridge stable

4 Immobilised

Fig. 3b Air-layer lubrication and the formation of lateral ridges (Shreve, 1968) (after Rouse 1984)

and magnitude of flowslides, however, depends on soil type, stress history and change of stress systems prior to failure (Bishop 1973).

Compressed air cushion as suggested by Shreve (1968) is also a likely mechanism to explain very high speed obtained by the flowing debris.

3 WASTE DUMP MANAGEMENT

For proper management of waste dumps, effect of various geological, geometrical, hydraulic factors and physico-mechanical properties should be minimised, so that there are fewer incidents of dump failures.

3.1 Geological factors

Geological investigations should include regional geological reports, maps, aerial photographs, borehole petrography, presence of burried channels, weak formations, etc. Gold (1986) suggested use of nearby valleys as waste dump to be safe and economical. Spoil instability at Pain-tearth's Mine resulted from the combination of weak floor conditions and undercutting of spoil heaps (Hebil 1986). Weathering of waste material after dumping has impact on stability. Location of pit and geologic

nature of reserves combine to introduce factors which can adversely affect waste dump stability e.g. steep underlying terrain, variable waste material type, variable particle size range, surface water, snow and frost, and waste deposition rates (Tassie 1988).

3.2 Geometrical factors

Geometry of dump is governed by shear strength of waste materials in the dump and in the foundation. Areal extent of waste pile is controlled by volume of waste, height of pile and permissible slope at the perimeter. Method of dumping and nature of disposal area also regulate permissible height of waste dump in accordance with physical and mechanical properties of waste material (Melnikov and Chesnokov 1969).

3.3 Hydraulic factors

Water content is one of the most important index properties for a given soil. Natural water content provides a quantitative assessment of soil compressibility and sensitivity of the preconsolidation stage. Water content depends on void ratio, particle size of minerals, organic contents of dump material and ground water condition. Moisture in the spoil pile controls the stabiligy through its influence on pore-water pressure and shear strength (Richards 1980). An increase in the void ratio reduces stability as the available space for water in the dump increases. Similarly, moisture content of the dump can be correlated with the angle of internal friction of dump material. It also affects maximum compaction limit of dump material (Reymond et al. 1980). Seepage through dump material depends not only on the permeability but also on local discontinuities such as fissures, joints, etc. Greater the permeability, easier will be water flow within a dump and hence reduced stability.

3.4 Physico-mechanical properties

Grain size parameter has a prominent influence on natural slopes affecting the height of the dump. Despite the fact that factors like velocity at the time of discharge and soil-water ratio are unfavourable for steep slope, coarser materials of dump can withstand steeper slopes. Well graded materials have a

357

higher density than poorly graded mixtures (Coates et al. 1986). Stability of waste piles composed of clayey soil could be increased by increasing sandy component. Pure sands are more stable than poor clayey loams and sandy loams. Slope angle is directly proportional to grain size and coarser the dump material, steeper would be the slope.

Critical height of dump is higher for sand than for clay size particles while mixture of the two has minimum critical height. Stability of waste dump which rely on frictional strength alone is not independent of height but a function of both height and rate of dumping (Caldwell et al. 1986).

Shear strength plays important role in the stability and for coal wastes it is generally low. The effective cohesion and effective angle of internal friction which form two main components of shearing stress, vary between wide limits e.g. effective cohesion C' varies from 0 to 1488 Pa and ϕ varies from 22 to 84 degrees. Shear strength of foundation should be appreciably higher than that of dump material. Maximum height and permissible slope are governed by shear strength of dump material. In Cannon Mine at Wenatchee, WA, embankment was divided into four zones as shown in Fig. 4 and properties of various zones are given in Table 1 (Davies 1968).

Table 1 Average soil properties (after Davies 1968)

	Zone 1	Zone 2	Zone 3	Zone 4
Gradation				
% Clay	10	-	-	-
% Silt	50	5	-	-
% Sand	40	25	90	10
% Gravel	-	70	10	90
Limits				
Liquid Limit (LL) (%)	30	30	-	-
Plastic Limit (PL) (%)	8	4	-	-
Strengths				
Cohesion (K Pa)	20	-	-	-
Friction Angle (Deg)	29	37	37	37
Hydraulic conductivity	10^{-8}	*	10^{-4}	*

* - not measured

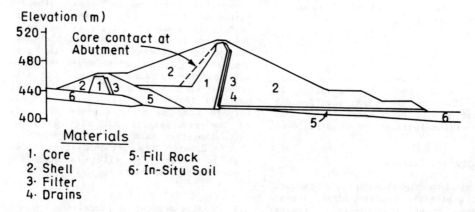

Fig. 4 Embankment cross section (Caldwell et al. 1986)

4 SLOPE DESIGN

It is imperative that stability of mine waste embankment must be assured, as instability could endanger persons working on or near them. Methods of design commonly used for tailing embankments have been discussed elsewhere (Sharma et al. 1977) and only those for waste piles are given here.

Stability analysis is a procedure of successive trials. A potential sliding surface is chosen and Factor of Safety (F.S.) against sliding along that surface is determined. Different surfaces are selected and the analysis is repeated until the surface having lowest F.S., known as critical surface, is found and the calculated F.S. against sliding along this surface is the indicated F.S. for the slope. This critical surface may lie within embankment or totally outside or at any position between these limits. These calculations should be based on effective stress analysis for which knowledge of pore water pressure is essential.

The critical surface for embankments is assumed to be cylinder whose axis is oriented parallel to strike of the slope, which in two dimensional cross-section is a circular arc. The shape of the surface can often be non-circular also. For trial sliding surface, F.S. calculations should account for changes in shear strengths and varying pore water pressure. This can be taken into account by method of slices in which a trial surface is chosen and potential sliding mass is divided into a number of vertical slices. Each slice is acted upon by its own weight producing shearing and normal forces on its vertical boundary and shearing and normal forces along its base.

In infinite slices method, a circular trial sliding surface is selected and stability of whole mass considered. Stability analysis using "Simplified Bishop Method" is a variation of method of slices and limited to circular surface only. Factor of safety is expressed by the equation :

$$F.S. = \Sigma (c'b\ sec\alpha + N'tan\ \phi')/\Sigma W_o\ sin\alpha$$

$$\ldots\ (1)$$

where,

F.S. = Factor of safety,
W_o = weight of material within the slice
c' = effective cohesion for the soil,
b = width of the slice,
α = angle of inclination at the centre of the base of the slice,
N' = effective normal force on the base of the slice,
ϕ' = effective angle of internal friction.

For Bishops Simplified Method, N' is the sum of forces in the vertical direction :

$$N' = \frac{[W_o - b\ sec\alpha\{u\ Cos\alpha + (c'/F.S.)\ Sin\alpha\}]}{[Cos\alpha + (tan\ \phi'\ Sin\alpha/F.S.)]}$$

$$\ldots\ (2)$$

where, u' = pore pressure.
Therefore,

$$F.S. = 1/\Sigma W_o\ Sin\alpha\ .$$

$$x[\ \frac{(c'b + (W_o - ub)\ tan\ \phi')\ Sec\alpha}{1 + (\ tan\ \phi'\ tan\alpha/F.S.\ )}\ ]\ \ldots(3)$$

The equation is solved by successive approximation, since F.S. appears on both the sides.

Morgenstern and Price (1965) have presented a method for calculating F.S. for non-circular surfaces of sliding using a computer for a number of iterative steps. Wedge analysis for F.S. calculation is used where configuration of trial sliding surface conforms approximately to two ormore intersecting tangents.

Where soft foundation structure exists beneath the embankment, F.S. against horizontal translation should be checked with trial sliding surface passing through soft foundation layer. Where an embankment or slope is subject to earthquake, ground motions produce stress fluctuations so that dynamic shearing stresses are alternately higher and lower than static shearing stress.

Numerous charts solutions have been published in which Cousin's is the most comprehensive one and is preferred.

Techniques of Finite Element Method (FEM) or the Limiting Equilibrium Method (LEM) were evolved during the last 50 years for a quick and accurate modelling of slopes by discretizing whole mass into tiny geometrical elements to which various properties are assigned. Fig. 5 exhibits an overstressed zone indicated by FEM analysis for the spoil, weak plane and floor (Choudhury et al. 1986).

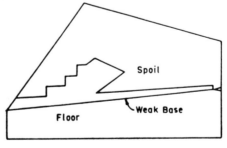

Fig. 5 Overstressed zone indicated by Finite Element Analysis (Choudhary et al. 1986)

The method employed for placing dry waste will depend on :

- location of waste pile in relation to mine,
- topography of waste disposal area,
- foundation conditions for the waste pile,
- nature of waste materials,
- maximum height of the waste dump, and
- type of material handling equipment available.

Dumping could be from mobile haulage units or conveyors and pile constructed from bottom up or from top down. When trucks are used to dump waste from the top, it is imperative to dump short of the crest and then doze the material over the crest. A series of slides may occur as a result of crest being extended in horizontal direction creating angle steeper than angle of repose and breaking from time to time. In case of conveyor, the main conveyor can be extended by a short movable discharge conveyor incorporating a flinger. The method of soil placement and compaction can be categorised into three major areas (Khandelwal et al. 1987).

1 Dumped and graded : It includes general spoil piles, material on the outslope, or any material placed in lifts greater than 1.2 m. This gives minimum compaction and density.

2 Truck haul and graded : It includes hollow fill and valley fill. Generally, truck places the spoil in piles which are graded with dozers. This gives inter-mediate compaction and density.

3 Thin lift and graded : Typically done by dozers pushing their lifts ahead of the blade and tracking the entire surface area before placing the next lift. This gives maximum compaction and density.

Angle of friction and in turn strength of spoil depend much on the degree of compaction. It has been reported that an increase from 90 per cent to 100 per cent of standard proctor compaction increased angle of internal friction in spoil sample from 27.7° to 30.6° (Paul et al. 1984). Density of spoil also correlates with the cohesion, but correlation is not as good as moisture content – angle of internal friction relationship.

Shear strength and other characteristics of wastes and foundations are often not adequately known in advance. Improvement in shear strength through compaction can be investigated on trial embankment which is constructed, samples taken and tested in laboratory.

A high concentration of fines is detri-mental to stability and should therefore be directed to a stable, unused portion of a dump. A stable crest is selected to minimise risk of instability of dump during disposal of a concentration of fines. Coarse material should be sent to area of possible instability as it has better drainage characteristics and will also assist in bringing down blocky material hung up in excess fines near the crest. Modification of profile of a slope by cut and (or) a fill is a commonly used stabilization measure. Toe of an actual or potential landslide should be loaded and head unloaded, geometry of such earthwork is decided on experience basis and on results obtained by succession of trial stability analysis (Hutchinson 1977 and 1984). Haul distance should be kept to a minimum at all times.

Instrumentation are installed in the embankment and/or its foundation to monitor changes occurring in respect of piazometric levels, seepage flows, embankment movements and total pressure.

The areal extent of waste pile will be governed by volume of waste, height of pile and permissible slopes at its perimeter. Stability of waste pile is governed by shear strength characteristics of the materials comprising waste pile, shear strength of in-situ foundation materials, pore water pressures within waste pile and foundation, slope of ground surface, height of waste pile and slopes at the perimeter of pile.

Height of pile and slopes surrounding its perimeter are independent, both governed by shear strength of waste material and that of foundation. Height can be increa-sed if slope is flattened and conversely slope can be steepened if height is reduced.

If shear strength of foundation is higher than that of wate material then maximum height of pile and maximum slope will be governed by shear strength of waste material. Where shear strength of foundation is lower than that of waste material then maximum height and slope will be governed by shear strengths of the foundation. Maximum permissible height and slope must be determined by stability analysis.

Various approaches in determining the permissible heights and slopes of piles of frictional wastes, cohesive or degrading wastes are depicted in Figs. 6 and 7 (Coates 1977) respectively.

Piles of free draining, frictional wastes:

Coarse frictional waste on competent foundation : pile can be constructed to any height providing $\beta \leq$ angle of repose (Fig. 6a).

Frictional waste on shallow weak foundation: h and β governed by strength of weak stratum. Probable mode of failure is horizontal translation. Determine permissible combination of h, β by stability analysis (Fig.6b).

Weak stratum at shallow depths:
h and β governed by depth and strength of weak stratum, probable mode of failure is horizontal translation. Determine permissible combinations of h and β by stability analysis (Fig. 6c).

Weak foundation extending to considerable depth :
h and β governed by strength of foundation. Mode of failure may be horizontal translationor deep seated shear failure on competent surface of sliding. Determine permissible combination of h and β by stability analysis (Fig. 6d).

Cohesive or degrading waste piles :

Competent foundation :
h and β governed by strength of waste material , probable mode of failure is along circular or non-circular surface within waste material. Determine permissible combination of h and β by stability analysis (Fig. 7a).

Shallow weak foundation :
h and β governed by strength of weak stratum. Probable mode of failure is horizontal translation (Spreading). Determine permissible combination of h and β by stability analysis (Fig. 7b).

Weak stratum at shallow depth :
h and β governed by strength of waste material or by strengths of weak stratum. Critical failure surface may be circular or non-circular through waste, or may be composite surface passing through weak stratum. Determine combination of h and β by stability analysis (Fig. 7c).

Weak foundation extending to considerable depth :
h and β governed by strength of foundation. Mode of failure may be horizontal translation or deep seated shear failure on composite surface of sliding. Determine permissible combinations of h and β by stability analysis (Fig. 7d).

6 CASE STUDY

The authors recently took up investigation of waste dumps of two large opencast coal mines, Mine A and Mine B, in India. First of all, the dumps were surveyed (Fig. 8a and b) and formation of input parameters was done. Based on the samples collected, some parameters were determined in laboratory and are tabulated in Table 2.

Table 2   Some parameters of waste dumps

Properties Determined	Mine A	Mine B
1. Max. density of spoil dump	1.90*	1.94*
2. Optimum moisture content	11.9%	11.5%
3. Liquid Limit (LL)	14%	13%
4. Plastic Limit (PL)	11%	11%
5. Plasticity index	3	3
6. Field Density	1.39*	1.28*
7. Grain Size Analysis	Mostly sand	Mostly sand

* - gm/cc

Triaxial test was performed from which cohesion and angle of internal friction (Fig. 9) were determined. The results are tabulated in Table 3.

Table 3   Results of triaxial test of waste dumps

Properties	Mine A	Mine B
Cohesion (kg/cm^2)	0.65	0.85
Angle of Internal friction	11°	9°

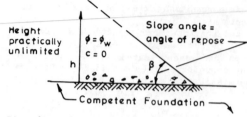

Fig. 6a Coarse frictional waste on competent foundation

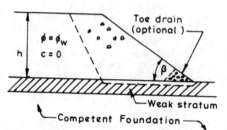

Fig. 6b Frictional waste on shallow weak foundation

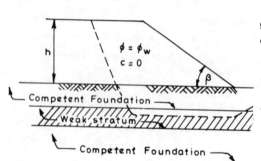

Fig. 6c Weak stratum at shallow depth

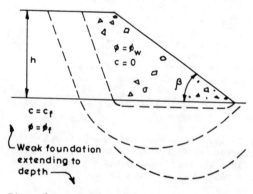

Fig. 6d Weak foundation extending to considerable depth

Fig. 6 Determination of permissible heights and slopes for piles of free draining, frictional wastes (after Coates et al. 1977).

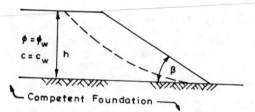

Fig. 7a Competent foundation

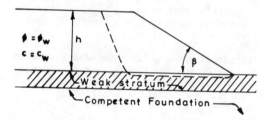

Fig. 7b Shallow weak foundation

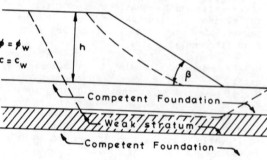

Fig. 7c Weak stratum at shallow depth

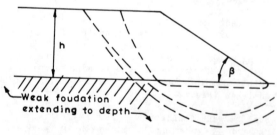

Fig. 7d Weak foundation extending to considerable depth

Fig. 7 Determination of permissible heights and slopes for cohesive or degrading waste piles (after Coates et al. 1977).

362

Fig. 8a Spoil heap at Mine A

Fig. 8b Spoil heap failure at Mine A

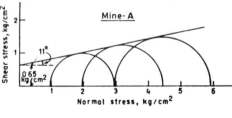

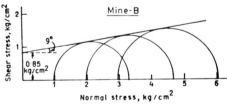

Fig. 9   Mohr's circle for determination
of cohesion and angle of internal friction
in mine A & B

Shear strength of the soil will be
determined by field tests as well as
laboratory tests. The field tests include

- Standard penetration test,
- Static cone penetration test,
- Vane shear test.

The common laboratory tests are :

- direct shear test,
- triaxial compression test.

7 CONCLUSIONS

The above study leads us to the conclusion
that operational strategies for safe
and efficient use of waste dumps is necessary. Dump failures can not be entirely
eliminated but its occurrance can be
minimised with adherence to dump procedures.
Dump failures are invariably preceeded
by extensive development of tension cracks
along the crest of the dumps. Dump failures
can be predicted 4 to 8 days in advance
with early indications of instability
and sign of imminent failure is clearly
evident 1 to 1.5 days in advance.
    Various geological, hydraulic and strength
parameters of spoil heap material as
well as base, need to be properly investigated and considered at the design stage
for economical and safe design.

REFERENCES

Anderson, A. and Bjerrum, L. 1967. Slides
    in subaqueous slopes in loose sand and
    silt. Norwegian Institute Publication
    No. 81.
Bishop, A.W. 1973.   The stability of
    tips and spoil dumps. Quarterly Journal
    Engineering Geology, Vol. 6, 335-376.
Bjerrum, L. 1971. Subaqueous slope failures
    in Norwegian fjord. Norwegian Geotechnical
    Institute Publication No. 88, 1-8.
Blight, G. 1977. Slopes in industrial
    waste. Proceedings of Ninth International
    Conference on Soil Mechanics and Founda-
    tion Engineering, Vol.2, Tokyo, 600-604.
Caldwell, J. Moore, D. Hutchinson, I.
    and Sluder, T. 1986. Design and cons-
    truction outlined of the Cannon mine
    tailing impoundment. Mining Engineering,
    Vo.38,No.8, 813-815.
Campbell, D.B. and Shaw, W.H. 1978.
    Performance of a waste rock dump on
    moderate to steeply sloping foundations.
    Proceedings First International Symposium
    on Stability in Coal Mines, Vancouver,
    395-405.

Campbell,D.B.1986.Stability and performance of waste dumps on steeply sloping terrain. International symposium on Geotechnical Stability in Surface Mining, Calgary, R.K.Singhal (ed.), Balkema.

Casagrande, A. 1965. Role of calculated risk in earth work and foundation engineering. ASCE Jl. of Soil Mechanics and Foundations Engineering Division, 91 (SM4), 1-40.

Chowdhary, R.N., Nguyen, V.V. and Nemeik, J.A. 1986. Slope stability considering progressive failure. Jl. Mining Science and Technology, Vol. 3, 127-139.

Coates, D.F. and Yu, Y.S. (eds.) 1977. Pit Slope Manual, Chapter 9 - Waste Embankments, CANMET (Canada Centre for Mineral and Energy Technology) Report 77-1, 129 p.

Davies, W.E. 1968. Coal waste bank stability. Mining Congress Journal, 54(7), 19-24.

Dorby, R. and Alvarez, L. Seismic failure in Chilean tailing dam. ASCE Jl. of Soil Mechanics and Foundation Engineering, 93(SM6), 237-260.

Gold, R.D. 1986. Performance and operations of waste dumps in steeply sloping terrains case at Fording Coal. International Symposium on Geotechnical Stability in Surface Mining, Balkema.

Habil, K.E. 1986. Spoil pile stabilization at the Paintearth's Dam Forestburg, Alberta. International Symposium on Geotechnical Stability in Surface Mining, Balkema.

Hutchinson, J.N. 1977. Assessment of effectiveness of corrective measures in relation to geological conditions and type of slope movement. General Report Theme 3, Symposium on Landslides and other Mass Movement, Prague.

Hutchinson, J.N. 1984. An influence line approach to the stabilization of slopes by cut and fills, Canadian geotech. Jl., Vol. 21, 363-370.

Hutchinson, J.N. 1986. A sliding - consolidation model for flow slides, Canadian Geotech. Jl. Vol. 23, 115-126.

Johnson, A.M. and Rodine, J.R. 1984. "Debris Flow" slope instability. In Brusden and D.B.Prior (eds.). John Wiley & Sons Ltd.

Khandelwal, N.K. and Mozumdar, B.K. 1987. Stability of overburden dumps. Jl. of Mines, Metals and Fuels, Special Number on Surface Mining, 253-260.

Koppejan, A.W., VanWamelen, B.M. and Weinberg, L.J.H. 1948. Coastal flow slides in Dutch Province of Zeeland. Proceedings Second International Conference on Soil Mech. & Foundation Engineering, Rotterdam, Vol. 5, 89-96.

Morgenstern, M.R. and Price, C.E. 1965. The analysis of the stability of general slip surface. Geotechnique, Vol. 15, 79-93.

Melnikov, N. and Chesnokov, M. 1969. Safety in opencast mining. MIR Publishers, Moscow.

Paul, G.S. Christopher, J.G. and John, S.J. 1984. Surface coal mine spoil study. Vol. I, USBM, DFR 52(1)-84.

Pernichele, A.D. and Kahle, M.B. 1971. Stability of waste dumps at Kennecotta Bingham Canyon Mine. Transactions Society of Mining Engineers, AIME, Vol. 25, 363-367.

Reymonds, P.M., Peter, H.D., Robert, A.R., David, A.R. and William, L.T. 1980. Surface Mine Slope Stability Evaluation. Vol. I, USBM, DFR 78(1)-80.

Richards, B.G. 1980. Finite element analysis of spoil pile failure at Goonyella Mines, CSIRO Dvn., APP. Geomech., Australia, Tech. Report 76.

Rouse, W.G. 1984. "Flow Slides" slope instability. In Brusden and D.B.Prior, John Wiley & Sons, Ltd.

Sharma, D.K., Ratan, S. and Dhar, B.B. 1977. Role of soil mechanics in mining with special reference to mine tailing disposal. Jl. Mines, Metals and Fuels, Vol. 25, No. 9, 261-263.

Sherve, R.L. 1968. The Blackhawk Landslide. Geological Society Amer. Spec., Paper 108, 47 p.

Tassie, W.P. 1988. Waste dump management at Quintette Coal Ltd. CIM Bull. Vol.81, No. 9, 35-39.

Torrey, V.H. and Weaver, F.J. 1984. Flow failures in Mississippi River banks. Proceedings Fourth International Symposium on Landslides, Toronto, Vol. 2, 355-360.

Yudhbir and Basudhar, P.K. 1989. Mechanics of spoil dump instability and relevant geotechnical design aspects, Proceedings Third National Seminar in Surface Mines, Dhanbad, 5.5.1-5.5.9.

Reclamation, Treatment and Utilization of Coal Mining Wastes, Rainbow (ed.) © 1990 Balkema, Rotterdam. ISBN 90 6191 154 0

Survey of bituminous coal waste treatment and utilization in the Ostrava-Karviná Coalfield

M. Hlavatá-Sikorová
Czechoslovak Academy of Sciences, Institute of Industrial Landscape Ecology, Ostrava, Czechoslovakia

M. Vítek
General Headquarters of the Ostrava-Karviná Collieries, Department of Coal Preparation and Coking Plants, Ostrava, Czechoslovakia

ABSTRACT: The Ostrava-Karviná Coalfield produces apart from 22.4 million tons of run-of-mine bituminous coal 18 million tons of coal wastes and about 1.5 - 1.8 million tons of coal slurries and flotation wastes. The waste materials from coal mining and preparation are utilized in the following spheres: stowing of worked-out underground workings, HALDEX waste treatment technology, reclamation, production of building materials for civil engineering, fluid combustion of wastes.

1 INTRODUCTION

The Ostrava-Karviná Coalfield (OKR) represents that part of the Upper Silesian bituminous coal basin which by its minor part (approximately 1600 km^2) penetrates from Poland into northeastern Moravia in Czechoslovakia. Bituminous, mostly coking coal is mined by 15 mine enterprises associated in the concern of the Ostrava-Karviná Collieries. The annual output of about 22.4 million tons of run-of-mine coal (in 1988) is prepared by 15 preparation plants. In addition to bituminous coal a great volume of wastes is produced: about 18 million tons of mine wastes and tailings from coal preparation plants, 25 million m^3 of waste water from coal preparation plants with about 1.5 to 1.8 million tons of coal slurries mixed with flotation wastes and of flotation wastes. Waste water from coal preparation plants is cleaned in large outdoor lagoons (in the OKR approx. 800 ha). The material contained in all the lagoons cannot be considered as wastes. It represents mineral raw material reserves which are recovered and used as a low quality fuel.

2 CHARACTERISTICS OF CARBONIFEROUS WASTES RECOVERED FROM COAL MINING AND PREPARATION IN THE OKR

The characteristics of wastes recovered from coal mining and preparation in the OKR can be seen from different views. One of them includes also the following classification:

Mine wastes
adjacent rocks hoisted to the surface with non-differentiated size
- wastes from drift development – course-grained with great sandstone contents
- wastes from in-seam development – course-grained with contents of coal substance, pelites and siltstones

Wastes recovered by preparation processing
- wastes from gravity preparation with a size greater than 0.5 mm to 1 mm and smaller than 50, 80, 120 or 200 mm
- flotation wastes with a size smaller than 0.5 to 1 mm; they are pumped into lagoons
- manually recovered wastes with a size above 120 mm (at present – minimum volume)

Wastes recovered from spoil banks mine wastes or tailings from a preparation process stored on a spoil bank.
A total area of about 2900 ha is covered by spoil banks. More than 250 million tons of wastes are stored on 80 spoil banks. In many cases the spoil banks are used for ground levelling on undermined territories and are already reclaimed.

3 INDUSTRIAL UTILIZATION OF CARBONIFEROUS COAL WASTES IN THE OKR

The priorities applied to waste utilization in the OKR include:
1) Stowing of mined out areas
2) HALDEX waste treatment technology
3) Technical reclamation
4) Treatment for the production of building materials
5) Utilization in civil engineering
6) Fluid combustion

ad 1)
Stowing of underground mined-out areas
The total volume of stowed areas has been decreasing continuously. At present only one million ton of coal waste is stowed underground. As the price for stowing of one ton of wastes is about 40 to 60 crowns and costs for storing one tone of waste on a spoil bank are about 15 crowns (Kčs), the question of efficiency of stowing minedout areas is always decisive.
The stowing and its favourable effects upon cost reduction for damages caused by undermining and mining process itself (reduction of pressure in adjacent mine workings, prevention of spontaneous caving, self-ignitions, improvement of climatic conditions underground, etc.) are not yet fully appreciated.
A certain increase of stowing volume can be expected by introducing new technical processes for utilization of flotation waste and utility fly-ash mixture. This technical procedure has been planned so far only for two mine enterprises with an operation started in 1990.

ad 2)
HALDEX waste treatment technology
The Haldex Ostrava, a Czechoslovak-Hungarian mining joint-venture, is aimed not only at recovery of residual coal substance from coal preparation wastes but also at application of secondary wastes from Haldex technology in building industry.
The Haldex technology is based on separating the input raw material in hydrocyclones in a medium of waste input itself for separating coal from wastes (0.5 - - 31.5 mm) and of Haldex own raw slurries functioning as a heavy medium of the system. The coal slurries of 0 - 0.5 mm size are treated by flotation. The Haldex plant processes annually about 250 000 tons of wastes and other 250 000 tons of law quality fuels which are not saleable in energy industry, and recovers about 70 000 - 90 000 tons of coal. The secondary wastes are predominantly stored on spoil banks and partly used by building organizations for civil engineering projects; flotation waste have been operationally tested for the production of brickware.

ad 3)
Technical reclamation
Wastes have been traditionally used for rehabilitation and reclamation operations by which territories affected by underground mining are reconstructed and corrected. These operations in the OKR are of highly specific nature. The surface areas subject to undermining are repeatedly affected, i.e. several times in a period of some decades. The greatest undermining effects can be seen in the Karviná part, but the largest waste production per ton of run-of-mine coal is found in the Ostrava part of the OKR (in the Karviná part the waste production amounts to about 0.25 - 0.5 ton per ton of saleable coal, and in the Ostrava part 1 - 1.9 ton of waste per one ton of saleable coal). This proportion is a result of a structure of coal reserves worked, of mining concentration and configuration of undermined territory. The rehabilitation and reclamation of the Karviná part territory by waste

materials from Ostrava collieries is costly and is feasible only thanks to the existing railway network of the OKD-Doprava (Transport) providing for a cheaper and ecologically sounder transport than in case of vehicle transport. Costs per 1 t km by railway of OKD-Doprava was 0.68 Kčs in 1988 and by road transport 2.41 Kčs. At present about 8 - 10 million tons of wastes are used annually for rehabilitation and reclamation.

ad 4)
Treatment of wastes for the production of building materials
Brickware production
Since 1981 the North Moravia brickworks at Hranice have been utilizing flotation wastes from the Paskov Mine. These wastes have an ash content of 65 - 70 %A^d, a heating value of 4.18 - 5.02 MJ.kg^{-1} and a size of 20 - 25 % W^r_t. The wastes are added to brick raw materials in the weight proportion of 10 - 15 %. The production tests of bricks made from a mixture of flotation wastes and brick clays have proved good brick properties. Flotation wastes have a function of grog in brick production, decreasing a sensitivity of products against drying and burning, enriching clays with Al_2O_3, increasing a sintering interval, improving product quality and reducing rejection rate. By their carbon contents the wastes mainly reduce the consumption of process fuel funcioning as a fuel within products. In addition, tests of bricks produced from ground wastes have been performed. They have been proved to comply with the ČSN Standard requirements.

Due to a shortage of investment means the construction of brick plant in the vicinity of the Paskov mine has not yet been implemented.

At present approximately 20 000 tons of flotation wastes from the Paskov and Dukla Mines are utilized for the brick production.
Utilization of wastes in cement clinker production
The utilization of wastes for the cement clinker production is limited by a chemical composition of ash and harmful substance contents, i.e. Fe_2O_3, MgO, MnO, TiO_2, P_2O_5, alkalis, heavy metals and chlorides. Flotation wastes from the Paskov Mine and 1.máj Karviná coal preparation plants have been used with a heating value below 17 MJ.kg^{-1} and ash contents exceeding 40 %. The significance of utilizing high ash content slurries for the cement clinker production consists in their complete waste-free utilization, i.e. utilization of both heating value and ash components as a fuel and a raw material component for clinker burning.

Pilot plant technical tests of cement and chemical analysis of clinker proved a suitability to utilize high ash content slurries also in terms of clinker quality. This method enables 20 % savings of fuel oil as against a traditional burning.
Production of expanded aggregates
Technical expanding tests made for the utilization of ground Carboniferous wastes with the size bellow 2 mm at a temperature 1000 °C for the production of expanded aggregate proved that wastes are slightly expandable. The strength of ceramsite which is made from wastes is three times greater at compression in a cylinder than it is set in the standard. The preparations of the ceramsite production from wastes were not continued because of a nearby aggloporite production plant which shoud have produced aggloporite from ash.
Production of moulded bricks
Moulded bricks were manufactured from burnt shales, also these wastes of 0 - 5 mm size are suitable for the production of cements and a 5 - 15 mm size for the production of porous concrete. Such utilization of wastes is presently irrelevant as large reserves of burnt shales are found in reclaimed spoil banks.

ad 5)
Utilization of wastes in civil engineering
The utilization of wastes in civil engineering is in the OKR widely spread because wastes have mostly a favourable granulometric composition which does not require any other treatment. In 1980 a derictive for waste utilization in civil engineering was elaborated. This directive defines the basic properties of waste and specifies

a suitability of its application for earth structures, structural and foundation structures, water structures, embankments and back fills below premises of earth and industrial structures.

The spoil bank fill material made up of Carboniferous rocks of 0 to 200 mm size (with untreated fill material up to 500 mm size) consists of sandstones, siltstones and claystones, with sporadic occurrence of conglomerates. Admixture of combustible substances varies between 15 and 25 %.

In this way 2 - 4 million tons of wastes per year are presently utilized within the OKR. The wastes are increasingly used (approximately 200 000 tons per year) in construction of roads within the OKR but also by other enterprises.

ad 6)
Fluid combustion of wastes
The fluid combustion appeared to be a very optimistic alternative for the utilization of coal substance residues in the OKR. At present, after 25 years of research, only one fluid combustion generator is operating in the OKR area. This generator burns in a fluid bed a mixture of wastes (the so called "fluid fuel") and liquid tar and oil wastes.

The fluid oil is a specially prepared mixture of wastes of 0 - 30 mm size with a heating value of 5.0 - 5.4 $MJ.kg^{-1}$. The solid fuel covers 25 % generator capacity. In 1989 the price of fluid fuel increased by 30 %, i.e. 36 $Kčs.t^{-1}$. Under this condition the waste combustion is inefficient.

4 CONCLUSION

The existing model of national economy in Czechoslovakia has not created sufficiently stimulating conditions for a complex utilization of Carboniferous wastes both with the waste producer and with a potential consumer. A number of research tasks completed with successful pilot plant tests or even by operational technical tests has not been materialized as the present price policy has placed many

a time the progressive technology at a disadvantage.

REFERENCES

Hlavatý, A. 1989. Dosavadní výsledky a další záměry ve využití hlušin v čsl.-maďarském podniku Haldex Ostrava. Racionalizace hospodaření s hlušinami v OKR. Ostrava: ČSVTS GŘ OKD.
Směrnice pro využití hlušin v inženýrském stavitelství. 1981. Bratislava: Výskumný ústav inžinierskych stavieb.
Sikorová, M. 1989. Příspěvek k řešení problematiky jemnozrnných odpadů po úpravě uhlí v OKR. Ostrava: VŠB.
Vejpustková, J. 1986. Studie možností využití méněhodnotných paliv z OKR pro výrobu cementářského slínku. Ostrava.
Vítek, M. 1985. Ekonomika zakládání, náměty na zlepšení základky a její využití v OKR. In Zakládání vyrubaných prostorů v OKR. Ostrava.
Vítek, M. 1989. Současný stav a perspektivy v hospodaření s hlušinami na úseku bezpečného ukládání a využití hlušin. In Racionalizace hospodaření s hlušinami v OKR. Ostrava: ČSVTS GŘ OKD.

Reclamation, Treatment and Utilization of Coal Mining Wastes, Rainbow (ed.) © 1990 Balkema, Rotterdam. ISBN 90 6191 154 0

A study on utilization of coal mining wastes in Korea

Kim Bok-Youn & Kim In-Ki
Korea Institute of Energy and Resources, Seoul, South Korea

Coal mining in Korea has been the most active and prosperous sector among the local mining industry owing to its important role as an indigenous fossil fuel for domestic heating.

Currently, 18 million tons of waste rocks are being produced every year while producing over 20 million tons of coal, and most of them are being disposed to the mountain slopes or fields without any treatment. These wastes sometimes cause environmental pollution or natural disasters in the rainy season.

This paper presents the general status of coal mining waste covering the volumetric distribution, geotechnological properties of the wastes and the future prospects on the utilization of them abandoned useless.

Outline of the conclusions are as follows:
- The wastes are generally composed of 60% of sandstone, 30% of shale and 10% of the other rocks, and their size distribution reveals that 4 mesh oversize takes more than 70%
- Considering the geotechnological properties such as specific gravity, liquid limit, plasticity index, permeability, compaction characteristics and California bearing ratio, the wastes are classified to GP or GM by the unified soil classification system, and interpreted as an available filling material for road construction, embankment and reclamations.
- The future prospects on utilization of these wastes are very bright considering a lot of civil work projects are intensively being undertaken as part of a national development scheme.

1 INTRODUCTION

The Coal mining industry in Korea is the most active and the largest sector of local mining industry because of its important role as an only one domestic fossil fuel resources.

Currently, annual coal production over 20 million tons and about 18 million tons of waste rocks are generated while producing such an amount of coal. All the coal mine wastes are disposed at the slopes or fields without any special measure, which sometimes they cause air and water pollution, especially in the rainy season and the residential area located down stream used to be damaged by the flush of the wastes.

This paper presents the art of the state on Korean coal mine wastes ranging from the volumetric distribution to the physical and chemical properties to figure out the future prospects for the utilization of the wastes.

Judging from the physical properties, wastes rocks can be used for aggregate, light aggregate or fireproof material, but it would not have economical merit at the present situation.

Taking into account geotechnical properties it is believed that they are available for banking of civil works or base material for road construction.

2 STATUS OF KOREAN COAL MINES

Most of Korean coal belongs to the anthracite, 85% of which is embedded in Permian and the other 15% in Jurrasic formations respectively. There are also some lignite deposits in Tertiary formation but it is almost negligible. Figure 1 shows the distribution of coal fields in Korea.

The structure of the local coal seams is so much disturbed by tectonic movements that dip, strike and seam thickness are frequently varying and the average inclination appears more or less 40 degrees. Table 1 presents the deposits and production states by coalfields.

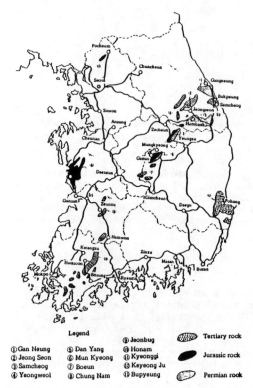

Legend

① Gan Neung	⑤ Dan Yang	⑨ Jeonbug	Tertiary rock
② Jeong Seon	⑥ Mun Kyeong	⑩ Honam	
③ Samcheog	⑦ Boeun	⑪ Kyeonggi	Jurassic rock
④ Yeongweol	⑧ Chung Nam	⑫ Keyeong Ju	
		⑬ Bupyeung	Permian rock

Fig. 1 Location of Coal Fields

These kind of unfavourable seam conditions make it very hard to adopt mechanized mining methods such as longwall or room and pillar methods which are predominant in Europe and American coal mines. The most popular coal mining method in Korea is "slanted chute caving method", which is a kind of block caving method, and daily production from one face is only 36 tons in average. Table 2 shows the status of local coal mining methods.

Currently over 20 million tons are produced from more than 300 coal mines and 90% of the coal are consumed for house heating.

3 SOURCE OF WASTE ROCKS

As mentioned above, the large number of mining faces necessarily requires long extensions of roadways which naturally

generate large volume of waste rocks. Statistics say that approximately 50,000 metres of roadways are developed every year at 2,000 heading faces all around the local coal mines. In other words, 2.61 m of roadways are driven for every 100 tons of coal production.

Assuming the average roadway excavation profile area is around 10 m² taking into account usual overbreaks, waste rocks only from the heading faces can be estimated to be 13,500,000 tons per year.

Mechanized equipment such as road-header are not applicable due to the higher strength of the rocks than that of European coal bearing formations. Accordingly, most of the faces employ blasting method. Drilling density and specific charge range from 0.3-0.4 m/hole and 1.7-2.9 kg/m respectively.

As the second largest sources, about 3,000,000 tons of wastes which take about 15% of ROM coal are generated from the coal preparation plant, another 1.5 million tons from repair works and the other miscellaneous underground works (Table 3 & 4).

4 PROCESS OF WASTE DISPOSAL

All of the wastes have to be hauled up to the surface, because the local coal mining methods do not need backfill at all. The hauled-up wastes used to be disposed along the mountain slope or field without any special treatment.

The wastes from the heading faces are normally irregular in shape and size. These wastes are loaded on mine cars by rocker shovel and hauled up by winder to the surface. The wastes from preparation plant are rather regular in size and mostly comprised of shale and coaly shales, which are picked up by hand after screened by 50 mm meshes.

All of these wastes are disposed at mountain slopes or fields usually located at the upper stream of residential areas as shown in photograph 1. This kind of waste heap sometimes causes disasters by sliding down, especially in the rainy season. The only measure to prevent such a disaster is to install concrete or stone retaining walls as shown in Photograph 2.

5 GEOTECHNICAL CHARACTERISTICS OF THE WASTES

5.1 Size distribution

Because of over 75% of the wastes are produced from rock heading faces, there are lots of large-size and hard rocks. Fig. 2

Photograph 1   Waste Heap Upstream of Residential Area

Photograph 2   Waste heap protected by stone wall

shows the size distributions of the wastes tested with the under-size of 75 mm screen for convenience.

5.2   Kinds of rocks and their physical
       properties

Most of the forming rocks of the wastes are sandstone, shale and coaly shale and their portions are shown in Table 5.
   Physical properties and chemical compositions of the rocks are presented in Table 6 and 7 respectively.

5.3   Shape of waste rocks

Shape classification was carried out with the wastes ranging from 20 to 75 mm in

diameter the results are shown in Table 8.

5.4   Atterberg limit

The test carried out by D.S.Han in 1988 shows that the wastes are non-plastic materials as presented below:

   liquid limit :   20.8 %
   plastic limit : 19.0 %
   plastic index :   1.8 %

5.5   Compaction characteristics

Typical test results are shown in Fig. 3. Maximum dry density appeared to be 2.168 when the moisture content is 7.26 %

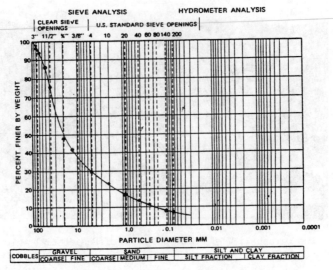

Figure 2 Size distributions of waste rock

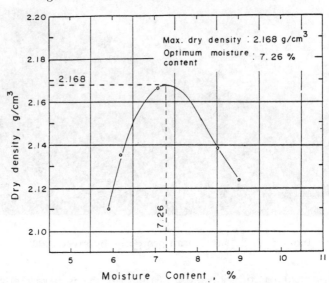

Figure 3 Dry density – moisture content
(after D.S.Han, 1988)

5.6 Geomechanical characteristics

Taking into account the aforementioned
size distributions and Atterberg limits,
the wastes can be classified as GP or GM
class based on Unified Classification
System.

As for the permeability, the wastes are
classified as medium class, of which
permeability coefficient is 0.64×10^5 m/
sec. Compaction characteristics were
already shown in Fig. 3 and it is noticeable that Korean wastes comprise lower
moisture contents with higher dry density
compared with the Tertiary wastes in western countries (Table 9).

Degradation rate under the compaction
turned out to be very low. Fig. 4
presents the differences of size distributions before and after the compaction.

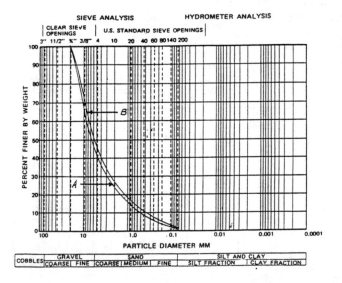

Figure 4   Size distribution before/after compaction
(after D.S.Han, 1988)

Judging from the California Bearing Ratio 43% in dry basis, the wastes are classified as "good" which is available for the base or sub-base of road and airstrip. When it is saturated by water, swelling rate rose to 0.94% and CBR number was reduced to 21%.

6   PROSPECTS FOR THE UTILIZATION OF THE WASTES

As mentioned above Korean coal mine wastes has its own unique characteristics in terms of the kinds of rocks and their physical or chemical properties. Hard sandstone occupies about 57% of the wastes and the shape of rocks are rough, angular and very irregular in size. Currently, some of the mines are using the hard sandstone as aggregate for their own uses after separation and crushing, but the quantity is almost negligible. Some of the shales generated from the preparation plant are supplied to the cement factory as the raw material for alumina, but the total quantity is only about 100,000 tons per year.

Theoretically, fire-proof material or light weight aggregate can be made with the waste if the appropriate process is provided but in view of economic efficiency, it is not feasible yet at this moment.

Considering the geomechanical properties of the wastes, the wastes are available for the construction of road and airstrip and also as general filling materials of civil works.

Taking into account the on-going nationwide development plan with the growth in national economy, coal mine wastes have a good potential for wide utilization in the future as the base materials of various construction works such as highway, dam, irrigation, reclamation and the other civil work sites.

7   CONCLUSION

- Korean waste rocks which is from the Palaeozoic formations are hard and irregular in size and angular shaped
- 75% of wastes are from rock-heading faces and the rest 15% from preparation plant
- Major forming rocks are sandstone and shale of which physical properties are much different from each other and the separation of them is practically difficult.
- It is found that the waste is available for filling material and base of road.
- Utilization as fireproof material or lightweight aggregate is not feasible at this moment.

Table 1. Deposit Condition and Production Scale

Coal Fields	Seam Thic. (m)	Dip	Calo. Value (kcal)	Reserves (1,000t)	Minable Reserves (1,000t)	Prod. Scale (1,000TPY)
Kangnung	0.9	20-40	4,117	66,475	42,358	888
Jungson	0.6	25-45	4,099	454,319	71,525	601
Samchok	2.4	30-60	5,319	548,781	293,503	14,042
Danyang	0.7	20-40	4,539	78,971	32,082	539
Munkyung	1.7	40-70	5,338	87,032	44,739	2,310
Boeun	1.1	40	4,895	24,658	16,661	272
Chungnam	0.6	40	3,883	308,705	88,521	1,747
Honam	0.8	25	4,642	76,615	49,301	888
Others			3,325	13,330	9,068	79
Total				1,658,888	674,758	21,370

Table 2. Coal Mining Methods

Mining Method	Portion (%)	Productivity (t/MS)	Daily Production per Face (t/d/face)
Slanted Chute Caving	82.1	2.86	36
Top Slicing	2.7	3.26	57
Sub-level Caving	14.3	3.2	55
Long Wall	0.5	3.4	75
Total	100	2.91	39

Table 3. Annual output of waste rocks

Source	Weight (1000)	Portion
Heading	13,500	75 %
Preparation	3,000	17 %
Others	1,5000	8 %
Total	18,000	100 %

Table 4. Waste production by Coal Fields

Coal Fields	Waste Production (1000 t/y)	Portion (%)
Kangnung	756	4.2
Jungson	504	2.8
Samchok	11,826	65.7
Danyang	450	2.5
Munkyung	1,944	10.8
Boeun	234	1.3
Chungnam	1,476	8.2
Honam	756	4.2
Others	54	0.3
Total	18,000	100

Table 5. Forming Rocks of the Waste

Forming Rocks	Portion (%)
Sandstone	57.51
Shale	25.15
Coaly Shale	1.70
Others	14.48
Total	100

Table 6. Physical Property of Rocks

	Sandstone	Shale
Specific Gravity	2.67	2.73
Porosity (%)	2.86	2.62
Absorbtion Ratio (%)	0.7-0.87	0.3-1.44
P-wave Velocity (m/sec)	4,200-4,900	3,200-4,670
Comp. Strength (kg/cm²)	700-2,200	300-1,000
Tensile Strength (kg/cm²)	145	150
Young's Modulus (x10^5 kg/cm²)	4.4	3.8
Poisson's Ratio	0.32	0.24

Table 7. Chemical Composition of Rocks

	Sandstone	Shale	Coaly Shale
SiO	72.06	57.84	53.67
Al_2O_3	18.87	35.97	41.04
Fe_2O_3	3.36	2.92	3.00
Ash Fusion Point (°C)	+1,500	+1,500	+1,500

Table 8. Shape Analysis of Waste Rocks (Weight %)

	Sandstone	Shale	Coaly Shale	Others	Coal	Total
Angular	31.14	7.86	0.44	7.14	0.49	47.07
Flaky	13.64	9.90	1.16	3.99	0.11	28.80
Flaky and elongated	3.08	0.92	–	1.2	–	5.2
Elongated	9.65	6.47	0.09	2.16	0.56	18.93
Total	57.51	25.15	1.69	14.48	1.16	100

(After D.S.Han, 1988)

Table 9. Compaction Characteristics

	Korean Waste	American Waste
Max. dry Density	2.168	1.920
Water Content (%)	7.26	10.9

Reclamation, Treatment and Utilization of Coal Mining Wastes, Rainbow (ed.) © 1990 Balkema, Rotterdam. ISBN 90 6191 154 0

Minestone fill for a maritime village development at Port Edgar, South Queensferry, Scotland

A.S.Couper & H.Montgomery
Lothian Regional Council, Edinburgh, UK

ABSTRACT: Port Edgar is a well established east coast marina and water sports centre with a magnificent setting beside the famous Forth Road and Rail Bridges.

The area exposed when the harbour dries out at low tide offers tremendous potential for redevelopment.

A feasibility study was undertaken to investigate the possibilities for redevelopment and the most effective way of reclaiming the harbour. Near by, lie many large tips, already identified for reclamation. The prospect of infilling the harbour with minestone could deal with two reclamation problems at the same time.

Examination and testing of the potential of the nearest tips of retorted oil shale waste, highlighted many design issues associated with the tidal action of the sea, on the long term settlement of the reclaimed land. It also highlighted an inter-relationship between performance and quality of the waste arising from the way it was originally tipped.

1 PORT EDGAR

Port Edgar is 16 kms west of Edinburgh, beside the town of South Queensferry. It lies right below the Forth Road Bridge and looks onto the famous Rail Bridge which is 100 years old this year. (Fig. 1) Within its horseshoe shape it has 270 fully serviced pontoon berths, accessible at all states of the tide. There is no marina like it on the east coast of the United Kingdom from Amble to Inverness.

It has 17ha of land, 9ha of which contain former naval base buildings, the rest being a steep heavily wooded escarpment.

Despite the fine views, the excellent covered boat storage, the commercial pier and the tenants, the harbour is silted up to an average depth of 2.5m, is vulnerable to winter storms, has a 6.3m tidal range, an access road that is too steep and narrow, and has inadequate services.

1.1 Marine History

From 1912 until 1978 Port Edgar was owned by the Admiralty. Prior to that by the North British Railway Company, who used a slipway and a few mooring posts at the northern end of the west breakwater, which was linked by a railway line to the shore. It dried out at low spring tides so must

have been of limited use then. In 1914 the Admiralty constructed five wooden piled pen structures within the harbour, to provide berths for up to sixty destroyers. To be able to use the harbour at all states of tide, major dredging work was carried out to give over 5.0 metres of freeboard in the entrance channel and at the berths. Ever since, maintenance dredging has had to be carried out. In 1939, modifications were made to the pen structures to accommodate minesweepers which are now the east and west piers. From 1978 it has operated as a public marina.

1.2 Recreational History

In 1978 Lothian Regional Council acquired Port Edgar from the Department of the Environment and installed pontoons for yachts off the east pier, as a first stage of a long term plan to develop it as a recreational complex. The appeal of the harbour to the Council was its uniqueness in having a large area of protected water with direct access to the River Forth and its extensive development land (Fig. 2). Unfortunately, the Council never embarked upon its development plan.

Although the marina had become a very successful yatching and water sports centre, constant dredging, the maintenance of old

Fig. 1 Port Edgar looking east 1988

buildings and repayment of interest on
capital, continued to absorb all income and
to create a substantial running deficit.
Without radical steps being taken there
seemed to be little prospect of improvement.

1.3 Key issue for the Regional Council

The key issue for its owners, Lothian
Regional Council, was how could the marina
be retained and public access be guaranteed
without spending more money, exacerbating
the deficit, when the Council wished the
deficit removed.
A jointly funded study with the Scottish
Development Agency was embarked upon, to
establish if a development package would be
financially viable and commercially
attractive.

1.4 Tourist potential

Knowing that the Forth Road Bridge carried
some 13.5M vehicles in 1987, it was apparent
that there was an enormous untapped
"impulse" visit market literally on the door
step of Port Edgar, besides the potential
day visit market offered by the 1.7m
residents within 1 hour's drive. On the

face of these figures, such a development
package ought to succeed (Fig. 3).

2.0 DEVELOPMENT OPPORTUNITY

The study brief had four key objectives to
investigate:-
1. The creation of a waterside village
incorporating housing, an extended marina
and a mix of other recreational and
visitor/tourist facilities,
2. The reduction of the annual deficit,
3. The cash return if such a development
were to proceed.
4. The long term management implications
there would be for the marina.

2.1 Market analysis and product mix

It was recoginsed from the outset that a
high quality, appealing atmosphere was vital
to attract people. The old coastal fishing
villages on the Firth of Forth seemed to
typify this warm, human appeal, but part of
that appeal was their unplanned nature. To
recreate such, seemed to be going against
the ideal of a planned concept. From
visiting new marina developments in the UK
and abroad, only the Dutch canalside

378

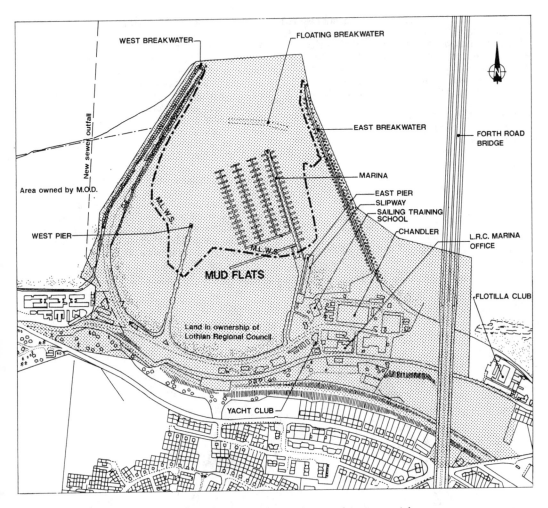

Fig. 2 Port Edgar Marina showing extent of Lothian Region ownership

townships managed to create that desired ambience. Shops, houses, cafes and small workshops there, all faced onto the waterfront. For a new development to achieve this ambience needed a clear understanding of the types of product that would fit in.

From an analysis of the market, four aspects stood out as important:-

1. Quality housing would create its own market appeal.

2. The sailing side of Port Edgar could benefit from better quality facilities for sailors, perhaps a club house with food and drink.

3. From the existing attractions within 1 hour's drive of Port Edgar, there was a potential day visit market to be tapped of around 22m trips.

4. Demand for quality office space in Edinburgh was high. As Port Edgar was only 20 minutes away from Edinburgh Airport, there was potential for high quality single solum or small professional offices.

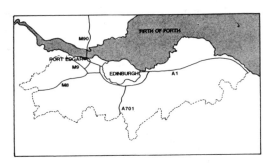

Fig. 3 Location of Port Edgar

379

For a new marina village to be successful it would have to combine these four aspects together into a harmonious whole, which would give the feeling that it was bustling with activity, was 'in contact' with the water, yet was commercially viable. To achieve this, it needs to have a mix of products as follows:-
- 320 yacht berths in the marina, plus 100 in the inner basin
- a sailing club
- a sea-life centre
- a naval interpretation centre
- a children's playground
- studio workshops
- a hotel
- a pub and bistro bar
- speciality shops
- offices
- housing (576 units, 35% town houses, 52% flats).

2.2 Emerging design philosophy

In parallel with the market analysis, studies of the physical characteristics of the harbour were undertaken. From individual studies investigating infill, silt, geology, infrastructure, water use, hydraulics and climate, a basic design philosophy emerged. It had the following objectives:-
1. To maximise the developable land available and minimise or preclude the areas of harbour which dry out at low tide.
2. To minimise the land reclamation costs within the context of objective 1.
3. To maximise the areas of land which would have water frontage with views across the water, so maximising total land value.
4. To take advantage of the unique setting by the Forth Bridges.

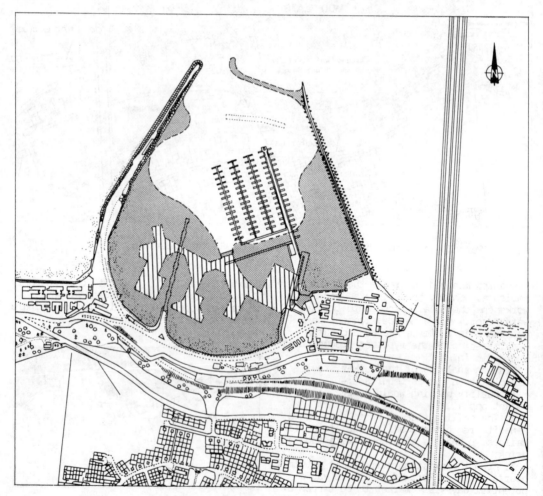

Fig. 4 Preferred option for redevelopment

5. To integrate the existing marina and boating areas into the scheme.

6. To minimise the effects of the high tidal range on the built form.

7. To provide a fully protected operational marina through reducing wave action and siltation.

8. To provide substantial weather independent sailing areas.

2.3 Final design solution

Many design solutions were tested by critically evaluating them against a common unit cost base which had been developed from the individual studies mentioned in 2.2.

The final design, the preferred option (Fig. 4), proposed the reclamation of the harbour to the low tide line yet preserved the existing marina pontoons. Its outer shape also took account of tidal action in the harbour. Within the development platform proposed, it was proposed to create an inner basin designed like a series of cannals for maximum water frontage. This inner basin would have a variable water level, controlled by a hydraulic flap (Fig. 5). The sense of close contact with the water could be achieved by building over it, a boardwalk, in much the same manner as in the Dutch waterside towns. In many ways

this was like the water association in the traditional harbours on the Firth of Forth.

Its advantages could be summarised as follows:-

1. It was the most appropriate development of land in relation to:
- tide
- mud flats
- infill requirements
- self cleansing action of the main harbour basin
- control of rough water
- ease of construction
- phasing of construction

2. The developmental land values in relation to land reclamation costs were optimised.

3. It was the most appropriate use of land in relation to:
- sunlight and shade
- views into and out of the development
- generation of a unique and special environment.

The total surface area created in this preferred option, including the inner basin, represented an increase of 130% in the area of developable land, without taking into account any adjacent land.

To evolve the design to this stage was not possible without extensive assessment of the bed of the harbour, where and what the infill might be, to what level would one

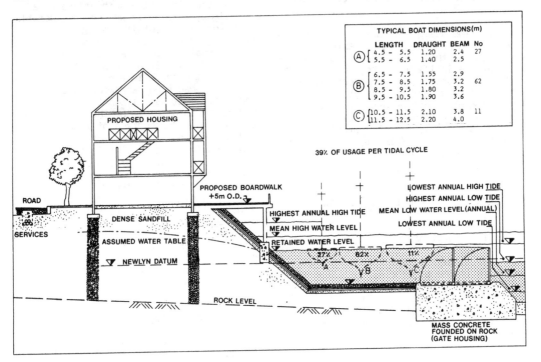

Fig. 5 Cross section through development platforms

construct a development platform and how one would protect it from the sea.

It represented the maximum development potential and could be summarised as follows:-

Table 1. Comparison of development area

	Developable Area Ha	Undevelopable Area Ha	Total Ha	Water Frontage Metres
Existing	8.9	8.1 (1)	17.0	1,200
Created	11.6	2.4 (2)	14.0	2,500
Total	20.5	10.5	31.0	2,500
Net Increase	130%	30%	82%	108%

Notes: (1) Escarpment Area   (2) Inner Water Basin Area

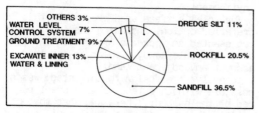

Fig. 6 Pie diagram of construction elements

To evolve the design to this conclusion was however, not possible without extensive assessment of the bed of the harbour, where and what the infill might be, to what level would one construct a development platform and how one would protect it from the sea. The subsequent sections describe that process and the evaluation of oil shale waste as a potential infill material. From the pie diagram (Fig. 6) one can see how important the infill element is to the overall scheme.

3 RECLAMATION OPPORTUNITIES AND PROBLEMS

3.1 Infill opportunities

Very preliminary estimates indicated that somewhere in the order of 1.3 million m^3 was required to create a platform on which new development could sit. The choice of potential infill (Fig. 7) within the region and Firth of Forth for reclamation was numerous:-

Retorted oil shale waste
Colliery waste

Demolition arisings
Dredged sand
Pulverised Fuel Ash
Rock

As the volume of material required was likely to be the largest single cost element, all of them were assessed in terms of availability, quality, quantity, haul distance, speed of delivery, cost, environmental impact and pollution.

3.2 Harbour problems

Having operated as harbour master for 12 years, knowledge had built up of some aspects of the harbour, in particular the accumulated silt, the wave conditions within the harbour and the lack of protection given by the breakwaters. All of these aspects needed examination and attention before any development could be considered. A site investigation was also necessary.

1. Accumulated silt

The historical maps and the Admiralty's hydrographic records gave a good indication of the extent of marine silt that had accumulated within the harbour over the past eighty years. It was of considerable depth and extent outwith the dredged channels, varying from 2.5 metres in the harbour basin to 6 metres at the harbour mouth (Fig. 8). It was apparent that substantial changes in accretion and deposition had been happening in the same timespan. What was also noted was a difficulty in determining the precise depth because the upper surface of it was so mobile and liquid. From this, it was estimated that the total volume of silt would be in the order of 365,000m^3.

Grading tests of samples taken at various depths indicated that the material was an amorphous, highly plastic silt, was not structurally sound and unlikely to improve, if drained or dried out. It would have to be removed to minimise settlement of the infill and dumped at sea, prior to placing any fill. Clearly it was a considerable restriction on development within the harbour basin.

2 Underlying geology

From the geological maps, the alluvium of the harbour bed formed the present beach which overlays a low raised beach. Nearby exposures of sandstones of the Upper Oil

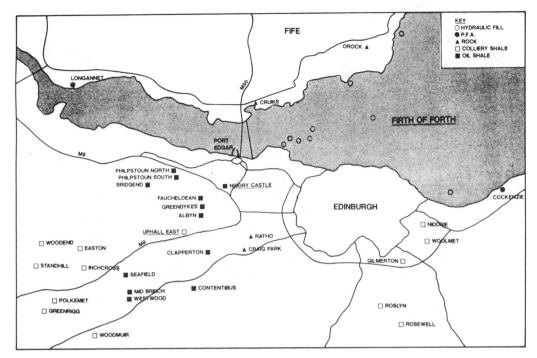

Fig. 7 Sources of infill in the region

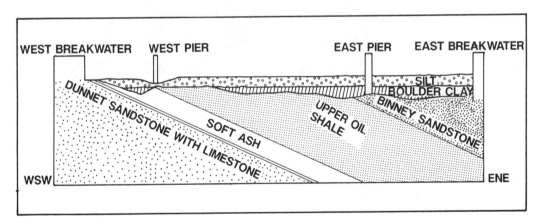

Fig. 8 Longitudinal section through harbour bed

Shale Group of carboniferous age were
evident, dipping at a shallow angle towards
the Firth of Forth (Fig. 8). Within this
was the Port Edgar Ash, a band of volcanic
agglomerate. The Burdiehouse limestone
forms the base of this rock group, which
was quarried on the south side of the
harbour and may have been mined under it.

It was essential to prove the harbour bed
geology and this was undertaken in parallel
with the design process. Forty two
boreholes were sunk by shell and auger
methods and five trial pits were excavated
to prove the geology and investigate the
superficial deposits, those within the
harbour being put down from a floating
pontoon. What this information indicated
was that within the harbour basin, bedrock
was between 2 and 10 metres below the
harbour bed level and overlain by a
generally stiff and very stiff glacial
sandy, silty clay, containing gravel in

383

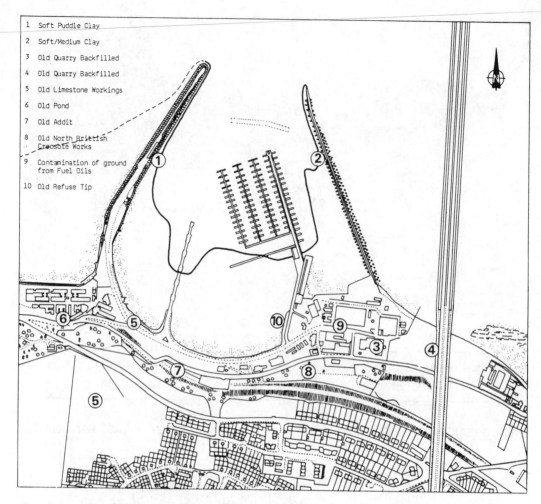

1 Soft Puddle Clay

2 Soft/Medium Clay

3 Old Quarry Backfilled

4 Old Quarry Backfilled

5 Old Limestone Workings

6 Old Pond

7 Old Addit

8 Old North Brittish
   Creosote Works

9 Contamination of ground
   from Fuel Oils

10 Old Refuse Tip

Fig. 9 Problem areas arising from site investigation

some areas towards the harbour mouth. However, it did show that there were lenses of soft laminated clay within the basin and around the southern shoreline. (Fig. 9)

It also indicated potential soil and mining problems in the landward area, where within the former naval base area and on the sloping land by the Forth Road Bridge, soft clays were present and deep mining of the Burdiehouse Limestone was evident. Additionally it showed up shafts and an adit within the mine workings, old backfilled quarries, an old refuse tip and some contamination of the ground by fuel oils. (Fig. 9)

Any plan of development would have to recognise these factors presented considerable restrictions on the design solution.

3. Wave conditions

Knowing more about the wave climate within the harbour was essential to be assured that any redevelopment would not make conditions worse within the harbour. The level of the development platform in relation to this was equally important. A computer model was set up to test the hydraulics fully in relation to the development options and in relation to future harbour management.

The tidal range in itself was a reclamation problem.
Existing ground level on the shoreline is +4.90m OD, 1.45 metres above the highest astronomical tide of +3.45m OD. The lowest astronomical tide is -2.85 OD. This gives a tidal range of 6.3 metres. As the top of both breakwaters is just over 2.0 metres

higher than the highest astronomical tide value, waves from spring tides frequently over-top the breakwaters.

From experience it was known that the wave climate within the harbour was far from ideal. The hydraulic study confirmed that wave heights at the marina pontoons were higher than recommended for berthing, and for over 70% of the harbour area. Further, the flood tide produced an anti-clockwise circulatory pattern within the harbour, with a maximum current of 0.3m/second, significantly more than the ebb tide. The flood tide was clearly the predominant vehicle for sediment transport into the harbour, which showed from samples taken to be mainly medium to fine silts. This was the prime reason for the extensive accumulation of silt within the harbour, the shape of which reflected the shoreline's influence on the anti-clockwise deposition. Regular dredging has been necessary ever since the harbour was built, and still is, to ensure clear freeboard under the pontoons. Beyond the breakwaters, in the River, the tides were shown to be in equilibrium.

From the hydraulic testing it was clear that the one year winter storm wave height would be over 2.2m, at a frequency of less than 6.0 seconds, within the harbour entrance without taking account of North Sea surge. This again confirmed what was known, that in winter, storm waves overtop the breakwater.

4 Lack of protection

What the hydraulic study established beyond doubt was the need for a new closure breakwater to minimise silting of the harbour and to reduce the wave climate within to the accepted standard for new marinas. Without it there would be no protection. This closure breakwater would link the ends of the existing breakwaters and run east/west to reduce the width of the harbour mouth and push the flood tide away from the entrance. From the hydraulic study, its alignment and length were refined to an extension of the east breakwater, leaving a new narrow mouth by the west breakwater.

No new northern breakwater would make sense if the existing rubble mound construction breakwaters were in poor condition. Apart from minor local repairs to the armouring on both harbour and seaward side, their structural condition was found to be sound. The west pier within the harbour was found not to be; it needed demolition.

The development platform level even with a new closure breakwater was considered an equally critical issue. Having established the need for the new northern breakwater, the hydraulic study showed the wave climate would, at the worst case, generate waves of 0.2 metres against the shoreline. With this protection, one could confidently design the development platform level relative to the highest high tide data. To ensure that foundations and services would be constructed above the highest high tide, the platform level for development was selected at +5.0m OD, giving a freeboard of 1.50m. The trend of rising sea level of about +3mm/year, is acknowledged but was considered best taken into account in the detail design stage (Fig. 10).

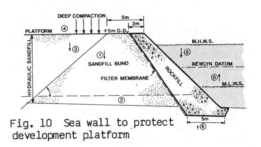

Fig. 10 Sea wall to protect development platform

4 EXAMINATION OF OIL SHALE

In Scotland, the Scottish Development Agency, under the Scottish Development Agency Act 1975 has the power to undertake land renewal schemes under Secions 7 and 8 of that Act. Since 1976 Lothian Regional Council has operated as Agent of the Scottish Development Agency implementing a major programme of rehabilitation of derelict land within Lothian Region.

In the national survey of 1988, 2041 hectares of land were recorded as derelict in Lothian Region. A large percentage of this dereliction consists of tips. With the priorities for rehabilitation concentrating on social and economic objectives, many of these tips have little prospect of extraction, as the demand for cheap infill materials has dwindled to a low level with the completion of the Central Belt motorway construction programme. For the oil shale tips of West Lothian, which represent 27% of the remaining derelict problem in Lothian, the potential for recycling as infill or construction materials represents a resource of over 100 million tonnes. In England and Wales colliery waste is referred to as shale, but in Scotland the word refers specifically to oil shale. At the present rates of extraction this resource could last until the 22 century.

Examination of the rehabilitation programme suggested there was the potential of cheap infill material close to Port Edgar. The advantages of removing a tip whilst forming land with development potential both at source and at destination made this an attractive concept.

4.1 Potential Sources

The nearest oil shale tips to Port Edgar, likely to obtain planning consent for extraction under the Town and Country Planning (Minerals) Act 1981 from the District Council, should the development of Port Edgar proceed, were Niddry Castle and Uphall East. Both have the benefit of being close to main roads and therefore lorry traffic could largely be routed to avoid or lessen the impact of going through communities en route to Port Edgar. It is estimated that Niddry Castle contains 21.6 million tonnes and Uphall East 3.2 million tonnes of retorted oil shale waste. The Niddry Castle source was within 9km and seemed to offer the possiblitity of transportation by overhead conveyor to the site. The cost of installing a conveyor was however over £1 million and effectively doubled the unit cost of oil shale. It was rejected as a transportation solution.

4.2 Origin of oil shale

Oil shale is found near the base of the carboniferous system. It was mined in West Lothian, primarily, for over a 100 years from 1860 to 1962. It was mined specifically to produce oil by a retorting process, although "downstream" product like paraffin wax and sulphate of ammonia were produced too. After retorting at a temperataure of over 700 c°, the spent waste was tipped. The pinky red colour and sheer bulk of the tips of spent oil shale are the only reminders of the once prolific but now defunct Scottish Oils Industry. The other unique feature of the tips are the "clinkers" which were due to fusing of the hot spent waste from the retorts.

4.3 Properties

A number of tests were carried on samples of retorted oil shale taken from Niddry Castle and Uphall East tips. To put these qualatative tests into context, the following descriptions of the properties of oil shale are given:

1. Retorted oil shale falls into five distinct colours or colour groups, which have different textures and moisture contents.

Table 2. Visual description, texture and moisture content.

	Colour	Texture	M.C.
1	Mixed red and yellow	Coarse	typ. 13%
2	Dark blue and red	Coarse	Often mixed with l.typ. 9%
3	Blue with some yellow	Coarse	In clinker form typ.5%
4	Salmon pink	Fine	typ. 24%
5	Dark blue with some yellow	Fine	typ. 24%

2. Particle size distribution of oil shale in its as found state falls within limits of Type 2 granular sub-base materials given in the DTp Specification for Road and Bridge Works, often very close to the requirements for Type 1 granular sub-base materials.

However, determination of the particle size distribution of oil shale, once subjected to compaction tests, shows that the particles are relatively soft and break down on compaction.

When crushed and tested for particle size, oil shale often shows a well graded silt size, with 10% clay size.

3. Extensive experience in Lothian and Central Scotland in the use of oil shale shows that the risk of spontaneous combustion is negligible, primarily due to the nature of the retorting process to extract oil and other products. A comparison of combustibility of colliery waste is given below.

Table 3. Combustibility in comparison with colliery waste by percentage.

Component	Retorted Oil Shale	Colliery Waste Burnt	Unburnt
Loss of Ignition	3.0	4.0	16.0

4. Compaction of oil shales from California Bearing Ratio tests will show the CBR values are high over a large range of moisture contents. The addition of water to oil shale does not appear to greatly affect the strength, as measured by the CBR, particularly under levels of compaction one would obtain in practice.

5. Oil shale is frost susceptible. Results showed most types of oil shales tested exceeded the maximum values permitted. Mixed with 5% cement however, the test showed results reduced from 50mm heave after 300 hrs, to 10mm.

6. The chemical composition of oil shale can be summarised below. Colliery waste has been included for comparison.

Table 4. Chemical composition in comparison with colliery waste by percentage

Component	Retorted Oil Shale	Colliery Waste Burnt	Colliery Waste Unburnt
$SiOm^2$	48.5	55.0	52.0
Alm^2Om^3	25.2	26.0	19.0
Fem^2Om^3	12.1	7.0	6.0
CaO	5.3	0.5	0.7
MgO	2.2	1.3	1.2
SOm^3	3.2	1.4	0.4
pH	5.4	-	-
Sulphate content of 1:1 shale-water suspensions (% as SO_3)	0.31	-	-

7. Specific gravity and water absorption results indicate the variability and heterogenity of the deposits in each tip source.

Table 5. Specific gravity and absorption

Test	Niddry Castle Sample	Niddry Castle Sample	Uphall East Sample	Uphall East Sample
Specific Gravity	2.76	2.66	2.39	2.25
Absorption (%)	20.5	16.9	8.8	7.4

8. Oil shale can contain soluble sulphates in potentially harmful concentrations. If used for reclamation, it would be advisable to separate concrete from the fill, by a layer of free draining inert material or an impermeable separation membrane. This would prevent sulphate bearing solutions being drawn into concrete by evaporation. Lothian Region's experience on another rehabilitation project with this problem indicates that the National House Builder's Council would insist on a separation layer.

5 ALTERNATIVES TO OIL SHALE

As mentioned in 3.1 the region offers a wide choice of infill materials. All were assessed on a unit cost and delivery rate established as follows:-

1. Colliery Waste

The significant numbers of colliery tips in the region could offer a major source of infill (Fig. 7). Laboratory tests indicated that there could be an influence of sea water causing long term problems of settlement. The unit price was estimated to be £2.20 - £4.70/m³ with a maximum delivery potential of 4000m³ day.

2. Demolition Arisings

The volume of clean demolition material arising in the region was unpredictable and of insufficient quantity annually to infill the harbour at a rate that could release new ground in the immediate or near future. Because of this and problems of control and compaction, it was rejected as an idea.

3. Dredged Sand

Sand existed in quantity on the bed of the Firth of Forth within an acceptable haul distance from the site (Fig. 7). It was in shallow water and was capable of being pumped ashore. Provided it was fine to medium grained with a low silt and clay content not exceeding 15%, it would be ideal for infilling large volumes. The unit price was estimated to be £1.60/m³ including royalty, with a maximum delivery potential of 12,000m³/day.

Dredged sand, hydraulically placed, was the most effective choice. It was the most economic and structurally effective choice for a large volume of fill and would eliminate major traffic pressure from the public roads. Unlike the other infill materials it avoided the complication of dewatering because it could be placed in the wet.

4. Pulverised Fuel Ash

Four power stations on the Firth of Forth offer PFA for sale but it was only available in extremely small quantities daily (Fig.7). It would also require a lengthy period of self settlement before it could even be worked on to improve its bearing capacity. By comparison with dredged sand, it would take 3000 days to fill with PFA whilst dredged sand would take 100 days. The indicative unit price was estimatated to be £3.20/m^3, with a maximum delivery potential of 400m^3/day.

5. Rock Fill

Rock fill for the whole volume was rejected as too expensive. However at least 130,000m^3 was required for a breakwater and armouring, to protect the development platform. Transport of this material to the marina by lorry would be difficult logistically and environmentally. Barging the rock from local loading facilities and dumping it by bottom opening barges, was preferred and would allow much speedier and economical construction (Fig. 7). Suitable stone of the Basalt Group, Quartz Dolerite, was obtainable, exactly the same as used on the Forth Road Bridge, 100 years ago.

6.0 NEED TO DETERMINE MORE

What was evident from general testing was a need to establish more about oil shale's hardness, toughness and durability, especially as it was proposed for infill in a tidal environment.

Two mechanical tests were used to measure hardness and. toughness and one a magnesium sulphate soundness test for durability assessment.

6.1. Hardness

A modified 10% fines test (BS812) was devised to measure crushing. This is a variation of the aggregate crushing test (BS812) in which resistance to crushing is measured by submitting a sample of 14 mm-10 mm chippings contained in an open ended steel cylinders to a load of 400 kN in a compression testing machine, the load being achieved in 10 minutes. The percentage by weight of fines (passing a 2.36 mm sieve)

formed in the test, relative to the initial weight, is known as the Aggregate Crushing Value.

Table 6. Tests to assess hardness, toughness and durability.

Test	Niddry Castle		Uphall East	
	Sample	Sample	Sample	Sample
10% Fines Value (Soaked) (kN)*	12	15	9	12
Aggregate Impact Value	51	51	62	52
Magnesium Sulphate Soundness Value		73.4		53.9

* 10% Fines Value tests were carried out on samples soaked for 48 hours.

However, in the standard aggregate crushing test, the fines formed in the early stages of the test tend to fill in the voids in the aggregate and thus restrict the effect of the later increases in load. Thus with weak aggregates, the test becomes progressively less sensitive to differences in the strength of tested aggregates. To meet this difficulty, a new form for the test was devised in which the load is increased only until 10 per cent fines have been formed. This can be estimated approximately from the distance moved by the platens. The conditions of test are otherwise the same as in the aggregate crushing test. The load in kN required to produce 10 per cent fines is known as the 10 Per Cent Fines Value.

In this case, the test was modified by soaking the aggregates for 48 hours before crushing.

6.2. Toughness

Aggregate impact tests (BS813) were undertaken to measure impact. Resistance to impact was measured by subjecting a bed of 14-10 mm chippings in a 102 mm diameter cylindrical steel cup to 15 blows from a 14 kg rammer, falling through 380 mm. The percentage by weight of fines (passing a 2.36 mm sieve) formed in the test, relative to the initial weight, is known as the Aggregate Impact Value.

6.3. Durability

A Magnesuim Sulphate soundness test (SDD Technical Memo 8/87 et al) was used to measure durability. This test simulates crystallization processes, which can be deleterious to some rocks with low tensile strengths.

In this test, samples of aggregate were subjected to alternate 48 hour cycles of immersion in a saturated solution of magnesuim sulphate, draining, oven drying and cooling. Crystallization in the pores of the aggregate may set up pressures which can exceed the tensile strength of the host material. The pressure may reduce the size of the aggregate by causing spalling or may even cause disintergration. The loss of material over a 2.36 mm sieve, expressed as a percentage of the original mass, after 5 cycles of immersion and drying, gave an estimate of the soundness and is known as the Magnesium Sulphate soundness value.

7 CONCLUSIONS

The use of retorted oil shale waste for the Port Edgar development project has not been entirely ruled-out.

The combination of high transport costs relative to those of dredged sand, the time needed to place and compact the oil shale, and the questions over its long term strength do move it well down the list of potential infill materials.

The mechanical tests to assess strength and impact resistance, together with the soundness results, suggest that oil shale is a weak material, possibly due to porosity and microfracturing. It is concluded that in respect of reclamation in a harbour, the material has poor durability and is sensitive to water. Testing of oil shale in other tips would confirm or reject this view.

Oil shale is generally suitable for use as a common embankment fill above ground water level. Apart from sulphates, the only constructional problem in Lothian Region's experience has been, in a few cases, instability of fill due to an increase in pore pressures. This was due mainly to over-handling and over-compaction and was resolved by reducing handling to a minimum, employing lighter compaction and ensuring even distribution of construction traffic

over the full width of the fill, together with some drainage measures within the embankment structure itself.

If relatively weak materials are used to reclaim areas in a tidal environment, breakdown may occur leading to re-orientation of particles and subsequent decrease in shear strength. Considerable settlement may then take place over a period of time depending on the loading, degree of breakdown, depth of fill, and the hydraulic gradient between ground water level in the fill and low tide. Whether the degree of settlement could be accelerated to overcome this tendency to breakdown, by preloading or applying deep compaction was not established. It would require the construction of a trial embankment to obtain reliable data.

Such a trial embankment could test whether stabilisation by lime and or cement would overcome this propensity to break down. It would add further to the costs of using oil shale but these could be offset by the benefits occuring from rehabilitating the tip site through extraction.

ACKNOWLEDGEMENTS

This paper represents the views of the author which are not necessarily those of Lothian Regional Council or the Scottish Development Agency.

The author wishes to thank the Director of Planning for allowing publication of this paper and to the staff in the Landscape Development Unit for their helpful assistance and comments.

REFERENCES

Webber, N.B. and Blain, W.R. 1989. Marinas; Planning and Feasibility. Couper, A.S., A feasibility study of Port Edgar, South Queensferry, Edinburgh p189-211. Southampton: Computational Mechanics Publications.

Lothian Regional Council Landscape Development Unit 1989. Port Edgar Development Phase II Study; Volume 11. Edinburgh: Lothian Regional Council.

Halcrow, Sir William and Partners Scotland Ltd 1988. Hydraulic Study. Glasgow: Halcrow

Wimpey Laboratories Limited 1988. Proposed Redevelopment of Port Edgar; Report on site investigation. Lab. ref. No S/25863. Broxburn: Wimpey.

Lothian Regional Council Department of Highways 1988. Laboratory testing results on oil shale samples. Edinburgh: Lothian Regional Council.

L & R Leisure Group in association with
CASCO, Kennedy, Cairns, Tozer Gallacher,
LRC Landscape Development Unit 1988. Port
Edgar Development Phase II; Volume 1.
Edinburgh: L & R Leisure Group.

Reclamation, Treatment and Utilization of Coal Mining Wastes, Rainbow (ed.) © 1990 Balkema, Rotterdam. ISBN 90 6191 154 0

The transportation and utilization of minestone for the Channel Tunnel project

W.Sleeman
Minestone Services, British Coal Corporation, Hebburn, UK

The Chunnel Tunnel project is the largest civil engineering work carried out in Europe. A major section of the project is the construction of the United Kingdom terminal near Folkestone in Kent. Large volumes of imported infill material were required in the construction of the terminal site, in addition to suitable fill materials recoverable from the tunnel excavation and from within the terminal site itself. Minestone from Snowdown Colliery in Kent was selected to contribute towards satisfying this imported fill requirement at the terminal site. Road transportation of bulk materials into the Tunnel project was not allowed, therefore rail loading facilities were set up at Snowdown colliery to service the scheme. This paper describes the assessment of the potential source sites and materials and comments on the establishment and operation of the rail facilities at both the Snowdown site and the receiving railhead at Ashford set up to handle the fill materials.

1 INTRODUCTION

The Channel Tunnel, planned to open 1993, is a major civil engineering project and one of the greatest ever undertaken by private enterprise. Out of a total of 31 miles, 24 miles will be below the Channel, making it the longest undersea tunnel in the world.

The tunnel's route will link the terminals to be built near Folkestone and Calais. Fig.1.1, 1.2.

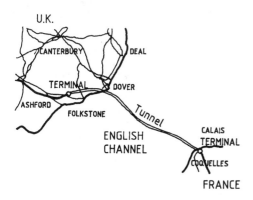

Fig.1.1. Route of Tunnel

Fig. 1.2. Location of Tunnel Project in U.K.

The exact line to be taken by the tunnel boring machines has been chosen after a careful study of numerous surveys and confirmed by extensive drilling operations.

Most of the tunnel's route will be bored through chalk marl, an impervious material generally considered ideal for tunnelling. The layer of chalk marl slopes like a shallow bowl, so the tunnel will have gentle gradients at each end, assisting acceleration and breaking of trains. Fissured chalk on the French side will be consolidated by injecting a mixture of cement and clay ahead of the boring machines. The tunnel will run up to 40m below the sea bed.

The fixed link that has tantalised far sighted engineers and businessmen for nearly 200 years is to be built by Eurotunnel, a private sector Anglo-French group.

The contractor to Eurotunnel is Transmanche-Link, TML for short. This is the joint venture organisation that has been awarded the contract to design, construct, test and commission the project for Eurotunnel.

TML brings together the experience and expertise of five British and five French engineering companies.

TML's British members - Balfour Beatty, Costain, Tarmac, Taylor Woodrow and Wimpey - are companies with first-class international reputations. So are the French counterparts - Bouygues, Dumez, Societe Auxilliaire d'Entreprises, Societe Generale d'Entreprises and Spie Batignolles. Major projects undertaken by TML's members around the world include motorways, railways, airports, power stations, dams, oil terminals, harbours, as well as tunnels such as the Mont Blanc.

TML's contract is divided into three parts:

1 Some 40%, including the terminals and approach roads, is at a fixed base price, adjustable for inflation.
2 About 10%, mainly shuttle rolling stock and other mechancial equipment, is procured by competitive tendering.
3 The remaining 50%, the main tunnelling work, is covered by a target price arrangement. TML shares in any savings, if it comes in below target, but it pays a penalty if the target figure is exceeded, because it has to share over-runs with Eurotunnel.

Careful and strict environmental management proposals were given a high priority in Eurotunnel's original proposal in 1985. Since then a continuous programme of consultation has been undertaken with statutory bodies, local planning authorities, individuals and more than 100 environmental and other groups. Eurotunnel have taken a number of significant steps to avoid or minimise unacceptable damage to the environment.

1 Design of terminal areas and access/ egress roads to minimise land take and 'enveloping' effect upon local communties
2 Comprehensive landscaping schemes.
3 Realignment of tunnel routing - in Holywell Coombe - to miss part of an important geological deposit.
4 Landscaping of the platform at Lower Shakespeare Cliff to create a new habitat of conservation interest, as well as a visually attractive area to be enjoyed by the public.

1.1 Spoil generated

The spoil generated by tunnelling is expected to comprise some 10 million cubic metres of predominantly chalk marl material.

In the U.K., handling of spoil is a combination of placement on the site of the terminal west of Castle Hill (some 1 million cubic metres) with up to 3,75 million cubic metres placed behind a sea wall at Lower Shakespeare Cliff.

In France the requirements for bulk spoil for the terminal site comprise some 3.25 million metres, with the balance currently being placed behind a dam sealing off the downstream side of a quarry and dry valley at Fond Pignon 1 km south west of the tunnelling sites at Sangatte. Spoil material generated from French tunnelling is wet and is pumped in semi liquid form by pipeline to behind the Fond Pignon dam.

1.2 Terminal facilities

The terminal areas for the transport system are of markedly contrasting sizes, the main U.K. facilty north of the M20 at Folkestone comprising some 140 hectare, while the French terminal between Frethun and the south west of Calais will be four times the size at 600 hectares in total though not all this area will be used for terminal facilities. The British facilities are at Dolland's Moor, where British Rail will marshall through international trains, an inland clearance depot for freight transfer at Ashford and an international passenger terminal adjacent to Ashford's existing British Rail station.

The main terminal functions are those of turning around shuttle trains within the enclosed loop system via a cut and cover tunnel on the U.K. side under the terminal buildings of over 1000 metres in length, unloading and loading vehicles for near continuous operation through the tunnels. The second key function is the reception and marshalling of road vehicle traffic inbound from the M20 British and A1 French motorway systems and the two national trunk road networks.

1.3 Environmental protection measures imposed

Issues relating to the environment constituted a significant part of the planning procedures and subsequent reports, together with those on safety of the system, commercial impact on competitive modes of transport and security conditions.

Overall, a total of 72 separate undertakings were made by Eurotunnel during and between the House of Commons and Lords Select Committee hearings. These varied widely in the form and content and in particular in terms of cost implications to Eurotunnel. For example Eurotunnel were required to accept, during the Commons Select Committee, the transport of all bulk fill and minestone materials to their various construction sites by rail, so far as this was possible, notwithstanding any cost benefits which might accrue from road transport. Another protective clause introduced at this time, permitted Dover Harbour Board to be consulted before design approval could be sought from Kent County Council for the sea wall structure behind which the total volume of spoil so placed.

1.4 Folkestone terminal

To construct the Folkestone terminal for the Channel Tunnel Fixed Link it was necessary to carry out extensive earthworks including the placement of approximately 2,400,000 cubic metres of imported fill to improve ground levels within the terminal area.

The general earthworks within the Channel Tunnel Folkestone terminal can be divided into five main elements:

1 Strip and stockpile topsoil;
2 Cut to fill operations;
2 a Placing of bulk material imported through the Ashford Railhead;
  b Placing of hydraulic marine sand;
  c Placing of land and sea drive tunnel spoil;

This paper is only concerned with one element of these earthworks, namely the import of minestone from Snowdown Colliery

through the Ashford Railhead as one source of bulk fill.

The fill requirement within the Folkestone terminal area amounted to approximately 3,400,000 cubic metres. The volume of suitable cut within the area was 1,000,000 cubic metres so it was estimated that the fill deficit was in the order of 2,400,000 cubic metres.

This deficit was to be made up in approximate quantities from the following sources:

1 1,000,000 cubic metres of hydraulic marine sand;
2 900,000 cubic metres of tunnel spoil;
3 500,000 cubic metres of minestone.

It was proposed that approximately 500,000 cubic metres of minestone be taken from Snowdown Colliery to be used as a selected fill material in areas where other potential alternatives, whether cut material or imported fill, were considered unsuitable. This was seen to be where a high degree of structural stability and compaction was required. In some cases minestone was to be reserved to form the final layer of fill below paved areas which were required to withstand high traffic loads.

2 SOURCES OF MINESTONE

When it was decided that considerable quantities of fill material was required for the Channel Tunnel project the minestone reserves in the Kent coalfield were investigated as potential sources of satisfying the demand.

Separated from the other coalfields by 150 miles the Kent coalfield lies in the area between Canterbury and Dover - but the three collieries were rarely seen by the holidaymakers on their way to and from the channel ports. The last new coalfield to be developed in Britain was important as it produced prime coking coals for the steel industry as well as supplying power stations and general industry. The seams were difficult to work, partly because of the conditions in which they were formed and subsequent geological events.

The first hint of coal under Kent came more than 150 years ago, but it was not until the mid-1880's that an abandonded Channel Tunnel boring was extended to prove the seams. Coal was struck at 353 m and in the next 30 years 29 successful borings followed the first strike.

In the 1920's miners moved to the new coalfield where virtually untouched reserves offered a lifetime's work, and they were followed by another influx during the depression of the 1930's.

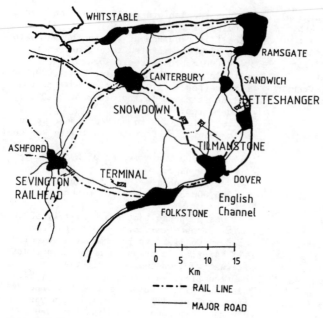

Fig. 2.1 Kent Area

Three collieries were operational in kent when the study began to find material for the Tunnel project.

Betteshanger, Snowdown and Tilmanstone collieries together employed over 3000 men producing 800 000 tonne of coal per year. All three collieries are now closed with Tilmanstone and Snowdown ceasing production immediately prior to the tunnel operations starting.

Figure 2.1 shows the location of the three mines in the Kent coalfield relative to the tunnel workings.

2.1 Origin of minestone

The modern mining process produces large amounts of waste rock material, unburnt colliery spoil, we chose to term Minestone.

Minestone is an admixture of those coal measures rocks associated with the coal seams.

In the early days of mining, coal was filled on a coaling shift and the development spoil was worked on a sparate shift. With the extension of mechanisation, coal production became multi-shift and the development spoil was reluctantly mixed in with the run-of-mine product. Also with mechanisation , the machine producing the coal was equally able to produce spoil from below, within or above the seam.

Indeed, in many instances by design, floor

spoil was taken to allow an additional section to be worked to suit the equipment or to suit the operators.

The coal preparation branch of mining technology developed techniques to cope with this deteriorating run-of-mine product. The washing process was initially an easy part of the production cycle. That picture contrasts greatly with some mines today, where the treatment of the run-of-mine is a major constraint on operations. The Coal Preparation plant presents the major operational problem for the mine, with the rate that the coal can be treated often dictating the mines output.

During 1987/88 British Coal Deep Mines produced 81.6 M tonnes of saleable coal, but with it 58 M tonnes of spoil was also produced.

The last comprehensive survey of the sources and quantity of dirt produced from British Coal mines took place in 1981 when the report of the Working Party on the disposal Disposal of Mine Waste published a table indicating spoil sources and quantities for the year 1979/80 Fig. 2.2 . This report indicated that 67 M tonnes of dirt were produced together with a total saleable output of 109 M tonnes representing a vend of 62%. The Working Party noted that some 52% of all spoil came from the coalface, that is from roof and floorcutting, in-seam spoil bands and geological disturbances. Other sources were rippings

SOURCE OF SPOIL	1979/80 MILLION TONNES	%	1987/88 MILLION TONNES	%
1. Coalface				
a) Roof, Seam, Floor and Geological Disturbances	35	52.2	33.7	58.1
b) Face Ends (Rippings)	10	14.9	4.7	8.1
2. Roadway Repairs, Back Ripping and Dinting	6	9	5.2	9
3. Ad-hoc Sources, Special Examinations	6	9	5.2	9
4. Development Drivages including surface drifts	10	14.9	9.2	15.8
	67	100	58	100

Fig. 2.2 Source and Estimated Quantity of Spoil

and face ends, (14.9%), roadway repairs, backripping and dinting (9%), special excavations and ad-hoc sources (9%) and major development drivages (14.9%). Recently Operational Research Executive have analysed the likely sources of spoil which make up the 58 M tonnes produced in 1987/88 Fig. 2.2 . It is fair to expect that whilst the coalface still represents the largest single source of spoil from underground there have been some variations in the percentages from other sources due to trends which have taken place over the 9 year period between surveys.

2.2 Uses of minestone

In general throughout the United Kingdom the major use of minestone, because very large quantities are often required, has been as imported fill. While the characteristics of available minestone vary from source to source, and sometimes between parts of a spoil heap, laboratory study and field experience have shown that the majority can be readily compacted into stable fills of high dry density. Minestone has been imported to sites for many purposes which include: the elimination of surface irregularities on building sites; construction of temporary haul roads; access roads and roadworks over low bearing ground; replacement of silts, peats, soft clays, water-logged and other unsuitable materials, to allow site development to proceed; back-filling of disused quarries, and gravel and clay pits to provide building or recreational land; raising of ground levels on low lying sites; blinding and covering of municipal tips; filling of disused canals and docks. Minestone from disused spoil heaps, properly handled - usually by spreading and compacting in layers as for highway earthworks - provides good stable ground, strong enough to support many types of structure on suitable foundations, easily trenched for services etc. Minestone is used in these ways and particularly in bringing derelict low grade land into better use. Land may be produced where urgently needed by using minestone as fill on foreshores.

The development and working of the Kent coalfield over the past 50 years or so has produced in the order of 13 million cubic metres of minestone in total. Some 6 million cubic metres are stockpiled at Betteshanger Colliery, some 3 million cubic metres at Tilmanstone and some 4 million at Snowdown.

Material from all three collieries has been utilized over many years for schemes in Kent. These include use in highways, as general fill to raise site levels for commercial and industrial developments and for sea defence works. In addition as a result of British Rail's assessment of

alternative supplies of suitable bulk-fill
materials, minestone from Betteshanger
Colliery, Kent was specified as the construction material for new railway embankments at the Gloucester Road Triangle,
Croydon. This scheme formed part of a
£120 M track and re-signalling scheme on
British Rail's Southern Region, Victoria
to Brighton main Line.

2.3 Suitability of source sites

The three sites identified were operational
at the time of the initial investigation
and rail facilities were available or
could be reinstated at all the collieries.
It was feasible to transport material from
any of the sites by road however the
commitment given by Eurotunnel resulted
in rail transportation being the only
option available. British Rail were
consulted on the feasibility of operating
from each of the collieries and the situation for the three source sites is given
below:-

a Tilmanstone - The Tilmanstone site
   was connected to the British Rail main
   line by a British Coal owned private
   rail line as shown in Fig. 2.3. The
   track and sidings at both the colliery
   and at Shepherdswell, where the line
   connected to the British Rail network
   were not in use and would require
   repairs and renovation to allow the
   high level of traffic necessary for
   the proposed movement of minestone.
   The line had been utilized for coal
   haulage by rail, however because of the
   alignment and condition of the track
   large trains with many wagons could
   not be accommodated. Trains would be
   required to be spilt into small sections
   of probably six wagons at Shepherdswell
   resulting in delays and slow turnround
   of trains. In addition only small
   2-axle wagons could be used. This
   accumulation of problems for operating
   a large movement of material by rail
   resulted in the Tilmanstone site being
   removed from the list of source options
   at an early stage.

b Betteshanger - Betteshanger minestone
   had been used for the construction of
   British Rail embankments in Croydon
   where material was transported by rail.
   A siding connected to the British Rail
   mainline by the colliery rail network
   was constructed, adjacent to the tip
   to serve that project and the track
   was still in place Fig. 2.4 .
   Examination of the siding revealed that

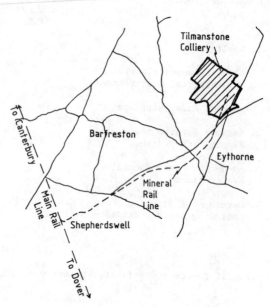

Fig. 2.3 Tilmanstone Colliery location

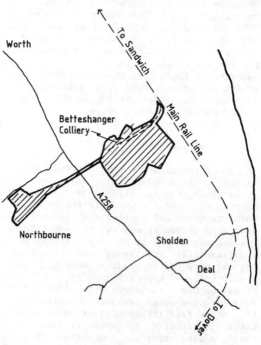

Fig. 2.4 Betteshanger Colliery location

396

substantial repair work was necessary to
bring it to a satidfactory standard for
the handling of the large quantity of
minestone for the tunnel project. Considerable amounts of coal were hauled by rail
from Betteshanger and any removal of minestone could only be accommodated if the
coal traffic was not affected.

As British Rail do not operate wagons
suitable for the transport of minestone,
wagons were required to be either purchased
or leased. Suitable 4-axle wagons were
available for hire from Tiger Rail, each
wagon being capable of carrying up to 75.5
tonnes of material Fig. 2.5 . Trains of
20 wagons, with a total minestone load of
1510 tonnes were proposed. Up to four
trains per day Monday to Friday being
envisaged. Further investigation revealed
the programme for Betteshanger to be optimistic due to the traffic restrictions
both on the mainline and the likelihood
of conflict with coal traffic at the
colliery. With the problems associated
with the existing siding and the traffic
difficulties the Betteshanger source was
reserved for further consideration should
problems occur at Snowdown.

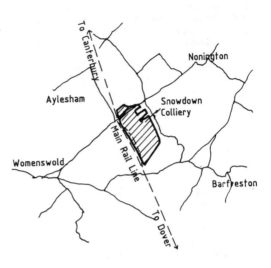

fig. 2.6 Snowdown Colliery location

Fig. 2.5 4-axle rail wagon

c Snowdown - Minestone was utilized by
road from Snowdown Colliery both as a
fill material and in cement bound form
for use as a paving material. A network of sidings remained on the site
adjacent to the British Rail mainline,
although the connection to the mainline
had been removed several years earlier
as coal was no longer moved from the
colliery, by rail. Fig. 2.6 . Much
of the track in the sidings was either

damaged, in a poor state of repair or
removed, however there were sections, which
although overgrown with vegetation, that
could be renovated satisfactorily. An
estimate of approximately £250,000 was
given by British Rail for the reconnection
of the siding to the mainline. The major
part of this cost involving signalling as
the mainline was heavily trafficked with
commuter services being the Canterbury to
Dover mainline. Trains of up to eighteen
4-axle wagons could be handled giving a
loaded weight of minestone of 13590 tonnes
per train. Up to four trains per day
Monday to Friday was envisaged. There was
no difficulty with loading trains or
operating from the siding as the works
would be well away from the colliery
activities. Snowdown was selected as the
prime source available and operations set
up to supply the tunnel project.

3 OPERATIONS AT SNOWDOWN AND SEVINGTON

Snowdown Colliery is situated in the East
Kent coalfield in an otherwise rural setting.
The spoil heap of approximately five to six
million tonnes of colliery spoil consists
of material with a proven record of
satisfactory performance in civil engineering
works. The spoil heap covers an area in
excess of 25 hectacres to the south of the
pithead works and to the east of the main

Canterbury to Dover railway line. The spoil heap and sidings and loading area is nearly a kilometre to the south of the Snowdown residential community and approximately half a kilometre from the nearest residence.

The minestone removed from Snowdown Colliery was unburnt colliery shale. Its suitability as a fill material results from the fact that it can be readily compacted in stable fills of high dry density. The moisture content of material which has been in a spoil heap for sometime usually stabilises to within about two per cent of the optimum for compaction. Because the material is comparatively weak the process of compaction causes fracture of larger particles so that local variations in size grading, densities and dry density/moisture content relationships are of less significance.

The chemical properties of minestone are such that the level of soluble sulphates are quite low and therefore the possibility of leachates contaminating watercourses is restricted. In the Folkestone terminal area minestone was only to be used in areas where the underlaying Folkestone sand beds and consequentially the underground aquifers are protected by the adequate thickness of impermeable Gault Clay as discussed and agreed with Southern Water Authority. Fig 3.1, 3.2, 3.3 .

There was one siding which remained extant at Snowdown Colliery. The connections north and south to the Britsh Rail mainline however had been removed as detailed earlier.

The track was generally in good condition although approximately 25% of the timber sleepers required renewal along with about 20% of the chair keys. Ballast was present in an acceptable condition and quantity although considerable clearing of undergrowth was needed both along the length of the siding and in adjoining areas used for loading.

Component:	(%)
SiO_2	37.8
Al_2O_3	18.5
Fe_2O_3	6.8
MgO	1.0
CaO	1.0
Na_2O	0.4
K_2O	3.1
TiO_2	0.7
SO_3	4.0
Loss on ignition	30.2

Fig. 3.1 Chemical Composition of Snowdown Minestone

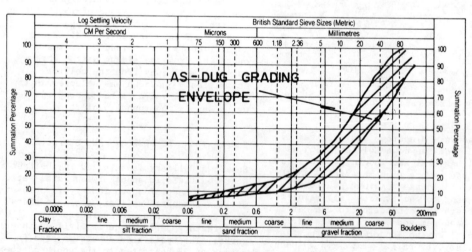

Fig. 3.2 Particle size distribution – Snowdown minestone

Natural Moisture Content (%) — 6.3

Dry Density/Moisture Content Relationship

Method	Maximum Dry Density (Mg/m³)	Optimum Moisture Content (%)	Voids Ratio (e)
Vibrating Hammer	2.05	8.9	0.25

Loose Bulk Density (Mg/m³) — 1.46

Atterburg Limits

Liquid Limit	44
Plastic Limit	25
Plasticity Index	19
Percentage passing 425 μm sieve	10

Sulphate, Sulphur and Chloride Contents

Acid Soluble Sulphate (% SO_3)	2.3
Water Soluble Sulphate (% SO_3 in 1:1 water extraction)	0.36
Pyritic Sulphur (% SO_3)	1.04
Total Sulphur (% SO_3)	4.03
Chloride (% Cl)	<0.1

pH Value — 4.7

CBR (%)

Mean CBR	Dry Density	Moisture Content
10.5	1.92	10.4

Fig. 3.3 Snowdown Minestone Characteristics

Fig. 3.4 Snowdown siding adjacent to B.R. Mainline

Existing gates and chain link fencing were
repaired or renewed along the boundary
between the siding and BR operational land.
The gates were provided across the turn-in
and turn-out connections re-instated by BR.

Two existing lighting towers, one at the
north and one at the south end of the
siding, were repaired and brought back into
operation including re-connection to a
power supply at the adjacent substation.
Lighting was to be provided for the loading
and moving of trains during the hours of
darkness.

Minestone was removed from those areas
of the spoil heap which offered the most
suitable material in terms of physical
consistency. The material was loaded in to
dumptrucks by backactors or loading shovels
and transferred to a stockpile area adjacent
to the rail sidings.   Fig. 3.5

Fig. 3.6 Loading train at Snowdown siding

Fig. 3.5 Excavation of minestone from spoil
heap

The 500 metre long siding was well suited
to a loading operation being on only a
light curve, almost level and with ample
room for manoeuvering loading equipment
adjacent to it. The stockpiled material
was loaded into rail wagons using rubber
tyred loading shovels. The stockpile
consisted of approxi .tely 20,000 cubic
metres which was initially loaded into
trains at the rate of approximately 700
cubic metres per train.   Fig. 3.6 .

Loading operations and timing of arrival
and departure of trains from the siding
was dependent on meeting the operating
hours at the temporary railhead at Ashford.
The railhead was operational between 0700
and 1900 and this entailed trains arriving
and departing from Snowdown during the
hours of darkness.

Two trains were to be used to supply a
maximum of four loads of minestone per day
to Ashford.

The selection of Snowdown as the source of
minestone was influenced by the better
gradient available on the rail link via
Dover to Ashford. The routing being Snow-
down/Dover/Ashford. The alternative of
Betteshanger as a possible source had the
disadvantage of being on the Canterbury/
Dover loop which is less suitable.

As was anticipated the proposed operations
had little environmental impact. Apart
from the initial arrival of plant and
machinery required for the minestone moving
operations there was no generation of lorry
traffic on roads as a result of this
proposal except for that between the Ashford
railhead and the Folkestone terminal.

During the operations arrangements for
the avoidance of environmental pollution
of the surrounding areas by dust or surface
water run-off were maintained.

Transmanche Link (TML) the contractors
for the works established a bulk aggregates
depot at Sevington near Ashford. This
facility was to enable aggregates for use
on the terminal site to be imported by
rail. Minestone was unloaded at Sevington
using hydraulic grabs Fig. 3.7 and loaded
from the stockpile ontoroad vehicles by
wheeled front end loading shovels. Fig. 3.8

From the railhead at Ashford the minestone
was taken by road along the M20 to junction
12 and then via the A20 and Danton Lane
site access to the Folkestone terminal area.

In addition to utilizing minestone on
the terminal area the material was used to
form the road embankment for a section of
the Ashford South Orbital road. This road
forms a major access route to the terminal
and the section involved passes directly
adjacent to the Sevington railhead.

Initially trains of eighteen wagons were
operated, however as the demand for minestone was reduced trains of nine to ten
wagons were employed carrying a total load

Fig. 3.7 Unloading facility at Sevington railhead

Fig. 3.8 Loading of minestone to lorries at Sevington

Fig. 3.9 Folkestone Terminal construction site

of 680-755 tonnes. Loading was carried out within one hour with the locomotive waiting with the train during loading. At the discharge terminal unloading was generally completed well within a two hour period.

4 CONCLUSION

In the period 1988-89 some 700 000 tonnes of minestone was transporated from Snowdown Colliery for use associated with the Channel Tunnel project. The material as utilized proved satisfactory for its applications and the recovery and transporation operations were carried out without difficulty.

Primarily because of the necessity to transfer the material from rail to public road vehicles to move the material between Sevington and the terminal site the economics proved costly for the major bulk movement initially considered. TML installed an onshore pipeline to carry sea dredged material recovered from offshore west of Folkestone. Whilst this operation was used successfully certain areas of the site were not considered environmentally suitable for the use of sea dredged material and minestone was used in these locations.

On a purely cost basis the supply of minestone could have been carried out using road haulage at a lower overall cost than either rail or the pipeline. The environmental cost to the communities in the Kent area of moving millions of tonnes of material by road was considered to be too high a price and therefore the additional cost for the rail movement was unavoidable in this instance.

Rail movement of large quantities of minestone has been shown to be environmentally acceptable and feasible and if appropriate terminal facilities can be provided it can also prove economical.

REFERENCES

CTG FM and TML. Construction Contract. 7 August 1986 et seq.
HMSO London. "The Channel Fixed Link" Dated 14 March 1986, Concession Agreement, April 1986 Cmnd. 9769.
Channel Tunnel Group - France Manche. Submission to Government for a Channel Tunnel Fixed Link, 30 October 1985, Eurotunnel.
W S Atkins & Partners with Sir William Halcrow & Partners. Main Audit Report of the Submission by the Channel Tunnel Group and France Manche. January 1986.
HMSO London. House of Commons Special Report from the Select Committee on the Channel Tunnel Bill, 18 November 1986.

HMSO London. House of Lords Special Report from the Select Committee on the Channel Tunnel Bill. 6 May 1987.

HMSO London. The Channel Tunnel Act 1987. Chapter 53, 23 July 1987.

Roberts M.H.P., The Channel Tunnel Project Its Statutory Framework and Measures for Environmental Protection in the U.K., Symposium on Mineral Extraction, Utilization and the Surface Environment, Minescpae 88, The Institute of Mining Engineers, Harrogate, 1988.

Barnett, S.A., Recent Case Histories of the use of Minestone in the Construction of Railway Embankments in the United Kingdom. Proceedings of the Symposium on the Reclamation, Treatment and Utilization of Coal Mining Wastes, Nottingham, 1987.

BS 6543:1985, Guide to use of Industrial By-Products and Waste materials in Building and Civil Engineering, British Standards Institution, London.

Blelloch, J.D., Waste Disposal and the Environment, Colliery Guardian, August, 1983.

Minestone Services, Information Sheets, British Coal, London, 1986.

ACKNOWLEDGEMENTS

The Author is grateful to British Coal for the support given in the preparation of this paper and for permission granted for its publication. The views expressed are those of the Author and not necessarily those of British Coal.

Reclamation, Treatment and Utilization of Coal Mining Wastes, Rainbow (ed.) © 1990 Balkema, Rotterdam. ISBN 90 6191 154 0

Reclamation and treatment of ultrafine coal wastes for direct firing in a circulating fluidized bed boiler

J. M. Brunello & M. Nominé
Cerchar, France

ABSTRACT: Sodelif, a joint venture between Houillères de Lorraine, Electricité de France and Stein Alsthom is building presently a 300 MWth Stein-Lurgi circulating fluidized bed boiler for electricity generation. One of the original features of this project, besides its size, is that high concentration slurries of ultrafine coal wastes will be used as feedstock. These coal wastes will be recovered from ponds where they have accumulated for years, slurried, pipelined and pumped to the boiler.

The present paper gives the characteristics of the coal wastes that are to be used, presents the pilot scale experiments that led to the definition of the reclamation and slurryfication process and discusses the rheological behaviour of the slurry with a view to full scale transport pipeline sizing.

1 INTRODUCTION

Houillères de Lorraine, the main coal producing Company in France, is also a significant power plant operator for electricity generation. The original features of these plants is that they are fired with by-products resulting from coal washing. In 1987, in order to maintain the generating capacity, it was decided to revamp a 125 MWe facility and to replace the old PC fired boiler by a Stein-Lurgi circulating fluidized bed boiler. In addition, the decision was taken also to use as a feedstock high concentration slurries prepared from coal wastes accumulated in ponds.

Cerchar was asked to provide support to the CdF Engineering Company "COREAL" for the design of the slurryfication and pipeline transport plants.

This paper is mainly dealing with the latter, with special emphasis on the rhelogical behaviour of the slurries.

2 CHARACTERISTICS OF THE PRODUCTS

Two types of products are considered for the preparation of the slurries: ponded wastes and freshly produced schlamms.

2.1 Ponded wastes

Recoverable resources are distributed into 12 ponds with an individual content ranging from 85 000 tonnes to 1,35 million tonnes and a cumulative content of 4,5 million tonnes. These wastes result from the treatment of schlamms by classifying cyclones with a 100μ m cut-point. Fine coal and clays contained in the cyclone overflow were let to settle in ponds where they have accumulated.

As some kind of hydraulic classifying occured during settling, the characteristics of the products vary from place to place. For example, near the outlet of the feeding pipe, one finds the coarsest products that settle rapidly: the average ash content is 40-45% and the mean particle size is 30-40μm. Layers of unusual products resulting from malfunctions in the washing plants operation may be found also: low ash (15%), coarse (up to 5 mm) coal, high density, high sulfur pyritic sands...

Near the banks of the ponds, products are generally higher in ash (up to 65%) and smaller in size (less than 10μm).

The moisture content of the products is generally in the order of 30% which makes them sticky and very difficult to handle.

All the laboratory and pilot scale trials have been carried out with a large sample extracted from one of the la Houve ponds which is one of the most important in terms of size and recoverable resource. The characteristics of this product are relatively constant particulary as regards the relationship between the ash content and the particle size distribution. The coarse fraction (50-315μm) is always low in ash (5-10%) whereas the fine fraction (0-50μm) is much higher (50-60%). On the average, an inverse relationship between ash content and mean diameter is found, which is of great importance for the slurryfication process.

2.2 Freshly produced schlamms

The operating washing plants of Freyming and de Vernejoul produce schlamms that are transported as low to medium concentration slurries (200 to 400 g/liter) to the preparation facility, on the premises of the Power Plant. They undergo concentration by cyclone and thickeners, moisture removal by vacuum filtration and thermal drying before being milled and fed into the boiler as pulverized fuel. These products exhibit fairly constant characteristics.
ash content 20-25%
moisture content (filter cakes) 20-25%
mean diameter 80-100μm

3 RHEOLOGICAL PROPERTIES OF THE SLURRIES

The objective was to meet two requirements simultaneously.
- high solid concentration (67-68%) in order to minimize heat losses in the boiler.
- low apparent viscosity allowing the transportation of 150 m3 /h over 650 m in a pipe of acceptable diameter (250-300 mm) with a pressure drop of less than 100 bars/km.
As these requirements may be hardly compatible, our problem was to define the best compromise.
The rheological behaviour of highly concentrated slurries, which are essentially non-newtonian fluids, has been extensively studied in relation with coal-water mixtures development (1,2,3). The main parameters that influence the rheological behaviour are :
- the solids content
- the particle size distribution and the mean diameter
- the surface properties of the solids that can be modified by suitable chemical additives.

3.1 Laboratory experiments

On the basis of our experience, we expected some difficulties in the formulation of highly concentrated slurries with ponded wastes due to their unfavourable properties: low mean diameter and hydrophilic surface.
This was confirmed by laboratory experiments performed with a Haake RV2 coaxial cylinder rheometer, according to a standard procedure defined by the Electric Power Research Institute (4). The rheological behaviour is depicted as a plot of shear stress (τ) against

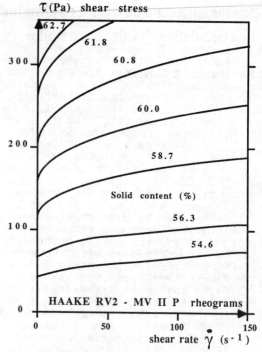

Fig.1 Typical rheograms of a ponded waste at different solid contents

shear rate ($\dot{\gamma}$). As it can be seen on figure 1, the shape of this plot generally conforms to the general equation.

$$\tau = \tau_0 + K\dot{\gamma}^n$$

where K is the flow consistency number, n the flow behaviour index and τ_0 the yield stress.

Fig 1 gives an example of typical rheograms of a ponded waste at different solid contents.

Another expression of the results is shown on fig 2 where τ_0 and K (average consistency determined between $\dot{\gamma}$ = o and $\dot{\gamma}$ = 30 s⁻¹) are plotted against the solid concentration for a suspension prepared with a La Houve waste.
It is quite obvious that the objective of solid concentration can not be met with this product, especially because of the strong increase in yield stress with the solid content: beyond 62%, the suspension looks like a thick paste that does not flow anymore. Attemps to decrease the consistency with chemical additives proved technically successful. Polyacrylates were found particulary effective but their cost was considered prohibitive for this special application.
Another possibility of improving the rheological behaviour is to alter the particle

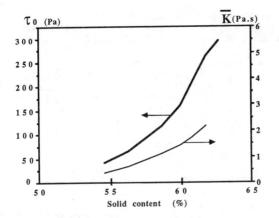

Fig.2 Evolution of yield stress and average consistency with the solid content

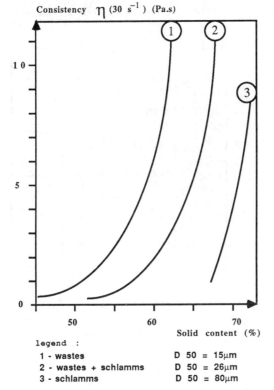

legend :

1 - wastes	D 50 = 15µm
2 - wastes + schlamms	D 50 = 26µm
3 - schlamms	D 50 = 80µm

Fig.3 Influence of the particle size distribution on the evolution of consistency with solid content

size distribution by widening it and increasing the mean diameter. This could be performed by mixing the fine wastes and the coarse, freshly produced schlamms.

Fig 3 where consistency (at $\dot{\gamma} = 30 \text{ s}^{-1}$) is plotted against the solid content for different mixing ratios shows that this method is fairly effective and markedly increases the solid content for a given consistency.

The results were confirmed by complementary tests in a capillary rheometer whereby the relationship between pressure drop Δp and flow rate Q can be established for various tube radius R. This relationship provides further information on the behaviour of the suspension when they are transported in pipes.

Fig 4 gives an example of the results. The experimental points have been plotted in terms of consistency variables.

$$\tau p(\text{tangential shear stress at pipe wall}) = R \frac{\Delta p}{2L}$$

$$\Gamma p(\text{apparent shear rate at pipe wall}) = 4 \frac{Q}{\pi R^3}$$

At this point, the problem is that the same slurry, tested in capillaries of different radii and at different length, exhibits a distinct flow curve (ie τp vs Γp) in each case. Attempts to interpret this rheological behaviour in terms of wall slippage and thixotropy failed so far. In effect, the isolation of the effect of each phenomenon would require that a

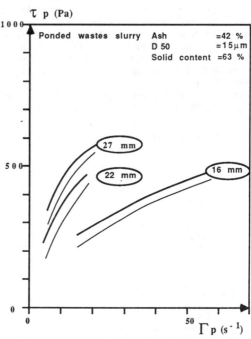

Fig.4 Typical set of capillary rheometer flow curves corresponding to different pipes geometries.

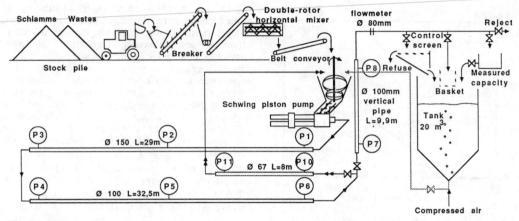

Fig.5 Pilot scale slurry preparation and pumping plant. Schematic drawing

permanent flow regime be attained, that is to say very long capillaries ans consequently an unattainable pressure discharge for the pump.

3.2 Pilot plant tests

As mentioned above, the unusual behaviour of the suspension makes it difficult to extrapolate the results of the laboratory tests for the calculation of the full scale pipe line size. It was thus considered necessary to build a pilot plant in order to minimize the uncertainty in extrapolation. Besides that, this pilot plant could give very useful indications on how to cope with problems of handling and mixing of the raw products.

The plant is represented on fig 5. Like with the capillary rheometer, the relationship between Δp and Q is established as a function of the characteristics of the suspension.
- type of solid: ponded wastes, freshly produced schlamms,
- solid content.

Some typical results are plotted on fig 6 in terms of consistency variables.

These results can be interpreted as follows. For each pipe diameter, an "operating area" is drawn up, limited by an horizontal line corresponding to the required maximum pressure drop (100 bars/km) and a vertical line corresponding to the nominal flow rate (150 m3/h). The flow curve of a given suspension must intercept the vertical line of the operating area.

For example, curve A corresponding to a 50% waste - 50% schlamms mixture at 67% solids can be readily transported in a 250 mm diameter pipe but not in a 200 mm pipe. Curve C, corresponding to a 100% waste at 63% solids

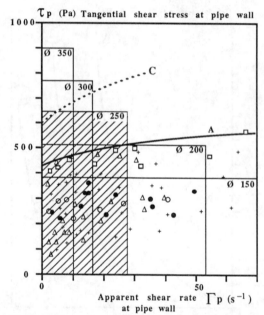

Fig.6 Plot of typical pilot tests results. Operational area of the commercial size slurry transport

will prove certainly dificult to transport, even in a 300 mm diameter pipe.

4 CONCLUSIONS

The study led to the conclusion that it is hardly possible to prepare concentrated slurries of fine coal wastes that meet simultaneously the requirements of combustion (high solid content, ash$\leq 45\%$) and those of transport in pipes ($\Delta p < 100$ bars/km for a 250 mm diameter pipe). The addition of freshly produced schlamms improves the characteristics of the slurry and makes it suitable for the transport and the feeding in a circulating fluidized bed boiler.

REFERENCES

CWM Development 3d to 8th International Symposia on Coal slurry Combustion and Technology.

J.M. Brunello, M. Elomrani, M. Nominé Préparation de mélanges charbon-eau: effet de la nature du charbon et amélioration de la qualité des suspensions par action sur les propriétés des particules. Powder Technology 57, 1989 pp 223 - 234

J.M. Brunello, M. Elomrani, M. Nominé On the extrapolation of wet ball milling process for CWM preparation. Third european conference on coal liquid mixtures 14-15 october 1987 Malmö (Sweden).

EPRI   Coal-water slurry evaluation EPRI CS - 3413, V1, R1 Jul. 1985.

Reclamation, Treatment and Utilization of Coal Mining Wastes, Rainbow (ed.) © 1990 Balkema, Rotterdam. ISBN 90 6191 154 0

Suggesting a temporary approach to the evaluation of strength characteristics for waste materials

H.B.Suchnicka
Technical University of Wrocław, Poland

ABSTRACT: Factors affecting the strength properties of waste material during excavation, transport and storage are described. The results of relevant laboratory tests are presented and interpreted. Most of the tests were performed on specimens modelled with a set of small lumps (a dozen or so millimeters in size) in direct shear apparatus. This kind of experiments was found to be of little utility. The approach suggested in this study makes use of some reducing factors which, together with the parameters of "undisturbed" soils, enable evaluation of the waste material characteristics. A method for estimating the reducing factors is shown.

1. INTRODUCTION

As the result of the ever increasing demand for raw materials, useful minerals are frequently extracted by opencast mining despite the considerable depth at which they occur. The overlying strata must be removed for storage on the nearby ground surface in the form of an external heap. Those structures (refered to as waste heaps or dumps) may extend to very large dimensions, thus creating serious hazards. This holds particularly for the initial open pit or other structures situated in the area adjacent to the dump in case of stability loss. To minimize the hazard, the design and the construction of the dump should meet the demands of engineering art. Thus, it is necessary to know at least some parameters of the waste materials, such as the density of soil and the strength characteristics. The problem of evaluating the material parameters for soils dumped onto heaps is extremely difficult. The various factors affecting the soil from the moment of extraction to the moment of storage not only change the material properties, but also contribute to the mixing of different types of soils (unless the selective system of dumping is introduced). All those factors are of a random nature.

It is obvious that an adequate solution to this problem may be obtained either by large-scale in-situ experiments or by using 1:1 models. These not only call for specialized equipment, but also take a very long time. That is why attempts have been made to find another solution—probably less accurate, but also less time-consuming. With these thoughts in mind a "semi-theoretical" approach has been adopted. The approach involves very careful analysis of all the elements that are expected to influence the properties of the waste material. On this basis, a quantification of this influence is attempted by introducing a number of reducing factors. They should express the decrease in soil shear resistance and in density due to excavation, transport and storage, as compared to the relevant values for "undisturbed" soil. This study has been limited to the waste materials consisting only of one type of soil. For a mixture of various types of soil it may be sufficient to make use of the approach suggested by Dmitruk and Suchnicka (1963).

2. FACTORS AFFECTING THE PROPERTIES OF WASTE MATERIALS

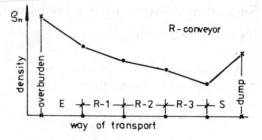

Fig.1. Variations of density

As a result of extraction for removal onto a waste heap, the overburden soil experiences grinding, loosening and destruction. The extent of these changes depends on the soil type and on the excavation method applied. In cohesionless soils loosening is dominant. Cohesive soils experience not only loosening, but also structural discontinuities. The structural discontinuities owe their origin to the occurrence of soil blocks which are formed due to excavation. This gives rise to the formation of a specific granular material which consists of compressible "grains". Dmitruk, (1965) was first to describe this soil behaviour. He defined that kind of material as a "granular medium of the second type". Soil looseness due to excavation is described by the loosening factor W_1. Hence, we can write

$$W_1 = \frac{\varrho}{\varrho^*} = \frac{V^*}{V} , \qquad (1)$$

where ϱ is density of intact soil, V is volume of intact soil, ϱ^* is density of waste material, and V^* is volume of waste material.

Following extraction, the overburden soil is transported for storage onto the waste heap. Transport is carried out by lorries, by train, or by belt conveyors. Lorries or train transport account for only insignificant changes in the material due to dynamic forces or mixing of different soil types. However, those changes become considerable in the case of conveyor transport, which is the basic transportation method in large opencast mines. For this reason, the influence of conveyor transport on the soil properties and waste materials behaviour cannot be neglected.

Conveyor transport produces dynamic loads of a random nature, which fall into two groups - vibrating loads and impact loads. Vibrating loads are generated by the vertical vibrations of the belt moving on the idlers and by the horizontal vibrations due to lateral whip. Impact loads originate from the non-

uniform distribution of the transported material along the belt, as well as from the malfunction of individual conveyor parts. Some impact loads are produced during running of the transported material from one conveyor to another. The process also accounts for further loosening, grinding and destruction of the soil. A schematic representation of density variations during conveyor transport is given in Fig.1. In rough estimates, density changes approach 7 to 12% and 30 to 50% in non-cohesive and cohesive soils, respectively.

During transport, cohesionless soils may either take the form of small lumps with diameters of up 20 mm or undergo liquefaction, depending on their water content. Cohesive soils, which initially have the form of large blocks, undergo grinding followed by smoothening and rounding of their surface. As a result of water outflow from the block, its surface becomes polished. This phenomenon was examined by Szafran, (1964) in direct shear tests. He found that there was a reorientation of soil particles in the zone adjacent to the boundary surface. Their position was becoming tighter and almost parallel. All those phenomena account for the decrease in the shear strength of the soil.

Another drawback of conveyor transport is the mixing of different types of soils which come from other conveyors. The precipitation water which is randomly received by the transported soil may also unfavourably affect its moisture.

The final stage of overburden removal is dumping into the soil area. Poured from a considerable

height, the soil mass experiences
an increase in density and compaction, but still retains a loose
"granular" texture. This is substantiated by the fairly steep cones
which form while the soil blocks are
falling down from the belt of the
spreader. The properties of the
waste materials are also influenced
by the dumping method, which is responsible for the spatial distribution of the soil blocks. The waste
material loses its "granular" texture with time, as well as under
the influence of the increasing
pressure exerted by the overlying
stratum. This leads to the increase
in density and shear resistance.
However, the net of discontinuities
between the previously formed soil
blocks remains unchanged despite
the compaction of the material.
Those surfaces should be regarded
as weakness zones. The differences
in the properties between "undisturbed" and waste soils are illustrated
by the lines of Fig.2, which show
the strength envelopes for the two
materials. Of course, the relations
between the lines should be described quantitatively - and this is
the objective of the present paper.

3. TEST RESULTS

The problem under study has rarely
been reported in specialized literature. Laboratory tests are usually
run on waste material specimens
modelled with a set of very small
soil lumps (a dozen or so millimeters in size). The strength of the
material is examined by direct
shear tests of 60x60 mm or 80x80 mm
samples. In some instances, the tests
were carried out in the triaxial
apparatus on 38 mm diameter and 76
mm high specimens which had been
prepared in a similar manner as
those for direct shear tests. A different approach was reported by
Szafran. To examine the variations
in the contact zone between two
blocks due to the dynamic forces
acting during transport, he performed a number of tests on specimens
with a fixed plane of dicontinui-
ties.
    For clarity, the interpretation
of the experimental results will involve the following symbols: DS =
direct shear stress; TC = triaxial

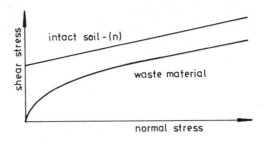

Fig.2. Shear envelopes for soil
and waste material

compression test; l = modelled lump
specimens; p = fixed shear plane
specimens, and n = "undisturbed"
soil specimens. The indices v and
w denote vibration and spraying
(with water), respectively.
    For the purpose of our tests it
has been assumed that the properties of the waste material are primarily influenced by the following
contributing factors: structural
failure or grinding (modelled by
lumps or by fixed shear plane),
dynamic loads (ground soil is vib-
rated prior to specimen formation),
and water content (either ground
soil is sprayed with water, or
ready samples are saturated under
laboratory conditions). Saturation
of soil samples is carried out to
represent the soil conditions that
might occur when the drainage system is gradually eliminated with
the increasing distance of the
working face.
    The physical characterization of
the experimental soil samples is
given in the table. Figures 3 and
4 show typical strength tests per-
formed on lump specimens. The $\varsigma - \bar{\varsigma}$
relation has been plotted for the
direct shear test only. Soil specimens for triaxial tests have been
prepared with a fixed density (approaching 2/3 of ς_n). The density
of the sprayed ground material reached its "natural" values at relatively low pressure (about 250
kPa). No such behaviour was
observed in non-sprayed specimens.
Under actual conditions, the density of the waste material is believed to reach the "natural"
value at pressures ranging between
800 and 1400 kPa. Thus, it is the
type and the consistency of the

411

Table. Physical characterization of experimental soils.

Soil	Parameter		DS	TS	Kind of test DS (according to Szafran)		
Type	Symbol		CE	CV	CH	ML	CLS
	w	%	27	28	32	13	11
		t/m³	1.95	1.96			
undisturbed	s	t/m³	2.71	2.82	2.70	2.67	2.66
	w_L	%	93	83	49	27	28
	w_P	%	31	39	27	16	18
	f_i	%	50	50	56	14	15
	f_p	%	15	20	6	13	54
waste	w^*	%	23-29	26-28	20-42	16-27	10-18
	ϱ^*	t/m³	0.94-0.98	1.34-1.37			
	w_w^*	%	24-39	34-41			
	ϱ_w^*	t/m³	0.78-0.81				

soil that influence the plot of the ϱ-σ relation.

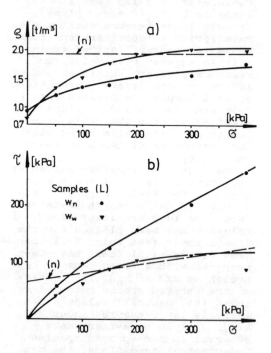

Fig.3. Direct shear test results,
(a) density versus normal stress,
(b) shear strength

The strength tests performed on soil lumps in the shear box yielded incredibly high shear resistance values. These should be attributed not only to the material properties, but also to the relation between the lump size and the specimen size (Fig.5). The influence of the lump size/specimen size relation is less distinct in triaxial tests. In spite of this, the utility of laboratory tests for waste materials raises objections. Those drawbacks cannot be eliminated by preparing specimens of small-size lumps, as this leads to the annihilation of the similarity in the response between actual and experimental soil material. The only remedy seems to lie in the application of such experimental systems that enable testing of large-size specimens.

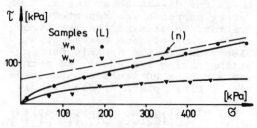

Fig.4. Triaxial test results

412

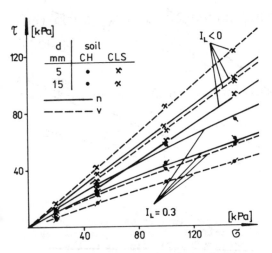

Fig.5. Effect of lump size on the shear strength of specimens

Being aware of the fact that the quantitative estimates obtained via conventional laboratory tests are of little utility for our purpose, we may ask whether or not they can provide information about the contribution of the forces acting during extraction, transport and dumping to the properties of the waste material. This holds especially for the time of vibration and for the water content. To answer this question, let us interpret the data reported by Szafran (1964). He made use of remoulded soil with controlled water content to prepare lump specimens, as well as specimens with a fixed shear plane. Unfortunately, Szafran fails to give the strength parameters for "undisturbed" soil. That is why, for the purpose of the present study, the lacking parameters have been determined approximately only (on the basis of grading and physical state).

The effect of lump size and vibration time on shear resistance is plotted in Figs.5 and 6. Thus, the nature of strength variations due to vibration depends both on the type and on the consistency of the soil. In low-plasticity soils, there is an increase of shear resistance during vibration. The increase is most intensive when water content approaches the plastic limit value.

High-plasticity soils experience a decrease of shear resistance in the course of vibration. Hence, we can see that neither lump specimens nor fixed shear plane specimens represent the waste material adequately. Only the results for fixed shear plane samples of high-plasticity soils can be regarded as the lower estimates for shear strength resistance. They can be of utility in the analysis of some stability problems (e.g., in determining the safety zone).

The difference in the strength envelopes for lump specimens, for specimens with fixed shear plane and for "undisturbed" soil with or without vibration is shown in Fig. 7. The influence of the type and consistency of the soil is plotted in Fig.8.

From engineering practice we know that the strength properties of the waste material are poorer than those of intact soil. And this is, again, an indication that laboratory tests performed via the above manner are insufficient to quantify the strength parameters of the waste materials.

The results of in-situ investigations confirm the disadvantageous

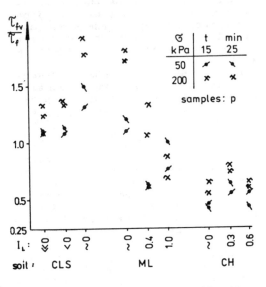

Fig.6. Effect of time of vibration on the shear strength of specimens with fixed shear plane

413

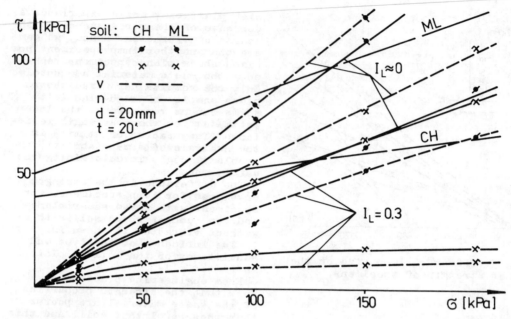

Fig.7. Direct shear test results

strength properties of the waste materials, especially at the initial stage of the dumping operation. Sounding has revealed that the compaction of the waste materials is very poor (relative density approaches 20%). This means that, irrespective of the soil type, the dynamic forces acting on the waste material during transport are responsible for the decrease in shear resistance.

All of the experiments have substantiated the non-linear relation between shear resistance and pressure. In practice, however, this nonlinearity is of importance to low normal stresses only. With the increasing pressure the relation can be described by a straight line with parameters $\tan \Phi^*$ and c^*. The tests results show that linear behaviour can be assumed to appear at pressures ranging between 100 and 200 kPa. The problem calls for site investigations.

4. ESTIMATIONS OF REDUCING FACTORS

Since the laboratory tests on lump specimens in a standard apparatus had failed to yield reliable results, it was necessary to find an-

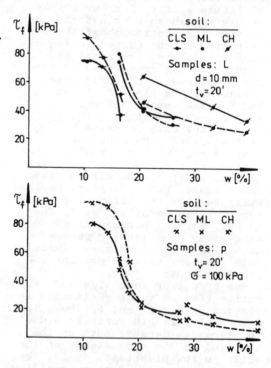

Fig.8. Shear strength variation with water content

other solution to the problem. The abundance of experimental data reported for both undisturbed and disturbed soils, and - on the other hand-some experience gained during observation of waste material behaviour have directed our attention to a semi-theoretical approach. In this approach, the unfavourable effects of extraction, transport and storage are established in terms of the reducing factors. Combined with the material parameters of undisturbed soil (x), the reducing factors describe the parameters of the waste material as follows:

$$x^* = \alpha_1 \alpha_2 \ldots \alpha_k x \qquad (2)$$

where α is the reducing factor, and k denotes the number of factors affecting the waste material properties.

The values calculated by the relation of (2) pertain to the initial stage of the dumping operation (after partial rebuilding of the macro-"grain" soil texture), when the most disadvantageous stability conditions occur.

It is advisable to evaluate the reducing factors as follows:
• For the loosening of the material

$$\alpha_\varsigma = \frac{1}{W_1} \qquad (3)$$

• For structure disturbance and grinding

$$\alpha_r = \frac{\tau_{r1}}{\tau_f} \quad \text{or} \quad \alpha_T = \frac{\tau_{r2}}{\tau_f} \qquad (4)$$

where τ_f is peak strength (TC), τ_{r1} is residual strength (multiplied shear method, DS), and τ_{r2} is residual strength (fixed shear plane method, DS).
• For dynamic loads

$$I_L \leqslant 0.25 \quad \alpha_d = 1$$
$$0.25 < I_L \leqslant 0.75 \quad \alpha_d \text{ from } (1-0.5)\,(5)$$
$$0.75 < I_L = 1 \quad \alpha_d \text{ from } (0-0.5)$$

The α_d factor is influenced by soil type and soil consistency. It has been anticipated that the soil type effect is incorporated in α_r. Hence, the conditions of (5) include the soil consistency effect only.

• For water content α_w.
Laboratory tests and engineering practice show that water content variations are of great importance to soil behaviour. That is why they must be considered with each change of conditions. The effect of atmospheric precipitation can be neglected.

The parameters of the waste material may be established via the following procedure:
• Density (irrespective of the soil type)

$$\varsigma^* = \alpha_\varsigma \varsigma \qquad (6)$$

• Internal shear resistance for cohesionless soil (with no liquefaction hazard)

$$\tan \varphi^* = \alpha_\varsigma \tan \varphi \qquad (7)$$

for cohesive soil

$$\tan \varphi^* = \alpha_r \alpha_d \tan \varphi \qquad (8)$$

or

$$\tan \varphi^* = \alpha_\varsigma \tan \varphi \qquad (7)$$

(with anticipation of a lower value).

• Cohesion resistance

$$c^* = \alpha_\varsigma \alpha_r c \qquad (9)$$

Apart from cohesionless soils, the strength parameters of undisturbed soil should be determined by TC. The value calculated in terms of (6) to (9) are of utility in typical stability analysis of slopes in opencast mining. To calculate the stability of the "safe" slope, it is necessary to substitute α_T for α_r in (8) and (9).

5. CONCLUDING COMMENTS

Laboratory tests in standard apparatus are of little utility (if at all) when determining the strength parameters of waste materials. The approach proposed in this paper is helpful in solving practical problems. The solution obtained is primarily based on engineering experience and intuition. Verification should be carried out by experimental investigations of actual objects and by the interpretation of the data sets obtained. We may seem to be playing too safe here, but this is always so in the case of an intuitive approach.

Irrespective of the approach, the values of the waste material parameters should be regarded as being of mere statistical nature. It is therefore convenient to relate them to the values of the undisturbed soil parameters. In this way, the number of the tests required for describing the material as a whole can be markedly reduced. Most of the data will be of utility in the design of the opencast and of the waste heap.

REFERENCES

Dmitruk, S. 1965. Soil mechanics and dimensioning of waste dumps. (In Polish), Zeszyty Nauk.Pol. Wrocł. Nr 21, Budownictwo XXV, Wrocław.

Dmitruk, S. and Suchnicka, H.B. 1963. On the calculation of admissible stresses for the foundations of structures on waste heaps. (In Polish), II Sesja Nauk. Wydz. Budow. Ląd. Pol. Wrocł.: 255-263.

Szafran, Z. 1964. Changes in the properties of cohesive soil lumps during belt transport. (In Polish), Zeszyty Nauk.Pol.Wrocł. Nr 105, Budownictwo XXIV, Wrocław.

Reclamation, Treatment and Utilization of Coal Mining Wastes, Rainbow (ed.) © 1990 Balkema, Rotterdam. ISBN 90 6191 154 0

The weathering of colliery spoil in the Ruhr – Problems and solutions

Michael Kerth
Geo-Infometric GmbH, Detmold, FR Germany

Hubert Wiggering
Hubert Wiggering, University of Essen, Essen, FR Germany

ABSTRACT: The physical weathering of colliery spoil from the Ruhr area causes a break-down of the coarse spoil particles. The effectiveness of weathering mechanisms is related to the petrography of the spoils. Within a few years of being heaped up the spoils acidify in the upper metre or so due to irondisulfide oxidation. The pH drops from values of about 8 to values of about 3. At these low pH values clay minerals are partly dissolved, leading to high amounts of potentially phytotoxic free and exchangeable alumina. Potential acid/base accounts of fresh spoils show a marked base deficit. A solution for reclamation problems which arise from strong acidification is the addition of calcareous materials, eg. wastes from the limestone industry in the southern reaches of the Ruhr area.

1 INTRODUCTION

In the Ruhr area of West Germany Upper Carboniferous coal is mined as deep as 1200 m. Because of the high automatisation in the coal mining industry about 1 metric ton of crushed rock is brought to the surface for every metric ton of coal. In 1989 about 50 million tons of colliery spoil had to be disposed of, most of which is heaped up. In the year 2000 about 40 km² or 1 % of the Ruhr area will be covered with colliery spoil.

Since the Ruhr area is densely populated, the coal mining industry and the local authorities have put strong efforts into reclaiming colliery spoil heaps for recreation and forestry purposes. However, adverse soil conditions endanger these reclamation schemes.

This paper presents the results of studies on spoil weathering and mine soil formation partly financed by the Kommunalverband Ruhrgebiet and the Ruhrkohle AG.

2 COMPOSITION OF COLLIERY SPOIL

Petrographically, the spoil consists of sand- and clay/siltstones (with the latter being predominant in the spoils of the Ruhr area) together with small amounts of coal, pyrite and siderite concretions. Minor amounts of underground building materials (concrete, slag, steel parts, polyurethane foam etc.) are also present.

Fresh spoil consists of the following minerals: the clay minerals chlorite, illite and kaolinite; quartz (Fig. 1) and small amounts of feldspar; the carbonate minerals siderite, ankerite, calcite; and the sulfides pyrite and marcasite. Furthermore, the spoil contains small amounts of clorides, mainly halite.

Additives used for coal processing such as magnetite are frequently present.

The composition of the clay minerals of the colliery spoils is dependent on the depositional lithofacies of the Carboniferous clay/siltstones and sandstones. Of the detrital clays kaolinite is concentrated in the finer grained sediments relative to the amounts of illite. The chlorite and interstratified illite-smectite undergoes diagenetic modification (Teichmüller 1962, Esch 1962).

The fresh, unweathered colliery spoils investigated in this study are dominated by illite, chlorite, and kaolinite (Wiggering 1987). Only a few vermiculites and illite-smectites have been retained by the rock material.

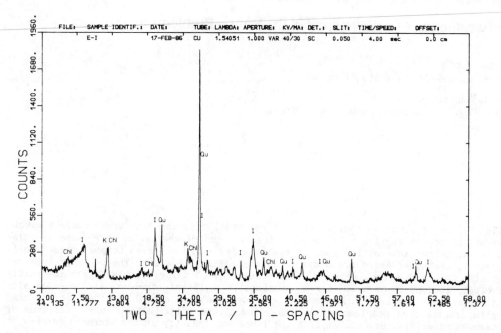

Fig. 1: X-ray diffraction analysis of fresh spoil from the Ewald colliery in Herten. The main components of spoils from the Ruhr area can be seen: Quartz (Qu) and the clay minerals illite (I), chlorite (Chl) and kaolinite (K).

3 PHYSICAL WEATHERING

The colliery spoil initially disintegrates due to the pressure release caused by extraction and the contact of the spoil with water in coal processing. Furthermore, some crushing of rock material is sometimes carried out in the coal processing.

The initial grain size distribution of the colliery spoil that is heaped up is thus affected both by natural characteristics of the spoil and the technical processes used in mining and processing operations.

After the spoil has been heaped up, physical weathering starts due to contact with air and water. Physical weathering mechanisms strongly depend on the climatic conditions of the area under concern. For the maritime, temperate climate of middle latitudes present in the Ruhr area abundant moisture, the frequency of wetting/drying (and to a lesser extent freezing/thawing) have to be taken into consideration.

Grain size analyses of spoils profiles of differing age show a decrease in spoil particle size with time (Fig. 2). In the upper few cm of spoil profiles the stone and gravel grain sizes show a significant

decline within 70 years. The fine fractions increase from 2-3 % to nearly 15 %. At a depth of only 10 cm breakdown rates are already low, while at a depth of more than 20 cm no breakdown occurs at all (Wiggering 1984).

The grain size distribution of the fraction < 63 µm shows that the amounts of the fraction < 2 µm are still very low in heaps 70 years old (Fig. 3).

Standard engineering tests under controlled conditions have been carried out on fresh colliery spoil to evaluate the effectiveness of different weathering mechanisms (Wiggering 1984). The tests simulate the response of rock material to frost action.

After several freeze-thaw cycles nearly 60 % of the clay/siltstones and 40 % of the sandstones broke down into smaller fragments (Wiggering 1984).

Comparison of experiments with different colliery spoil materials indicate that the larger the pore size of the rock material the more water is adsorbed and a strong breakdown of the rock material occurs. The smaller the pore size the less water is

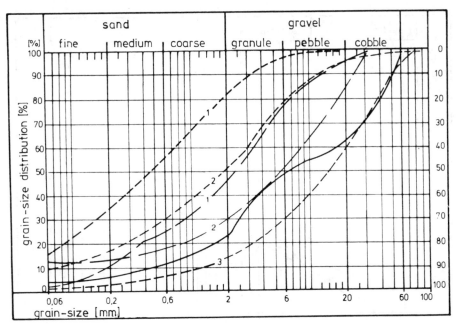

Fig. 2: Grain size distribution of fresh colliery spoil from the Ewald colliery in Herten (——), from the 4 year old Hoppenbruch heap (-- --) and the 70 year old Maximilian heap (- - - -). The numbers in the figure indicate sampling depth: 1 = 0 - 10 cm; 2 = 10 - 40 cm; 3 = 50 - 100 cm.

frozen and breakdown due to freezing is less efficient. Therefore, the degree of water saturation of the rock material must be considered as an important parameter in frost failure (Hudec 1977).

In comparison with sandstones the clay/siltstones generally have smaller pore sizes but significantly larger internal surface areas (Wiggering 1984). Accordingly the degree of saturation is higher. Therefore, the clay-/siltstones show a typical sorptive-sensitivity.

Wetting and drying have an effect similar to freezing and thawing in that the rock is alternately stressed by adsorbed water and unstressed when the water evaporates. During wetting-drying tests (Wiggering 1984) nearly 70 % of the clay/siltstones were broken down into smaller fragments as compared with only 20 % of the sandstones.

The experiments show that physical breakdown dynamics progress very rapidly. Grain size analyses, however, show that this breakdown mainly produced particles of gravel and sand grain size while the amounts o silt and clay sized grains are negligible (Wiggering 1984).

The weathering experiments and the field survey carried out by Wiggering (1984) show that the mechanisms for colliery spoil breakdown are as follows:

(i) The water, upon entering the rock pores, structures itself in an orderly manner on the walls of pores. The degree of structuring and the number of molecular layers of water structured in that way are functions of the internal surface and the polarity of the liquid. The structuring of water is akin to crystallisation.

(ii) If the pores are small enough, the water fills their entire volume and can begin to exert pressure. The action is similar to that of ice formation and explains the volume and pressure increase of colliery spoil upon contact with water.

(iii) Wetting and drying of the sorptive-sensitive rocks have an effect similar to freezing and thawing in that the rock is alternately stressed by adsorbed water and unstressed when water evaporates.

(iv) Temperature changes affect the number of layers of water structured by the sur-. face forces and produce an effect similar

419

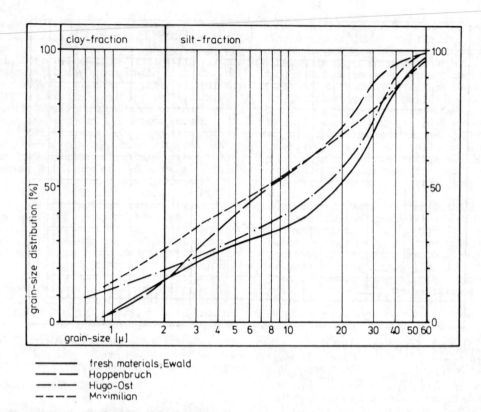

fresh materials; Ewald
Hoppenbruch
Hugo-Ost
Maximilian

Fig 3: Grain size distribution of the fraction < 63 µm of fresh spoil from the Ewald colliery, and from samples of the upper few cm of the Hoppenbruch heap (4 years old), the Hugo-Ost heap (40 years old) and the Maximilian heap (70 years old).

to wetting and drying. The greatest degree of structuring appears to take place in the 20°C to 30°C range (Wiggering 1984). In fact, water adsorbed at 20°C becomes in part freezable water at -16°C (Hudec 1977).

(v) Because of the stresses set up by the adsorbed water, the rock breaks down due to a combination of sorption-desorption and temperature changes.

Although freeze-thaw cycles are an effective physical breakdown mechanism for colliery spoil in weathering experiments, in the temperate climate of the Ruhr area it must be regarded as an less important breakdown mechanism in the field.

In the upper few cm of spoil profiles physical weathering is very intensive. The resulting rock fragments are transported to deeper levels by gravity and rainwater and accumulate in depths of 10-15 cm below the surface. As a result of the physical

weathering a 'consolidation-horizon' is formed (Wiggering 1984). This 'consolidation-horizon' slows down the weathering processes in the deeper parts of spoil profiles. Thus, efficient physical weathering only occurs in the upper few cm of spoil profiles.

4 CHEMICAL WEATHERING

For studying chemical weathering of colliery spoils of the Ruhr area spoil profiles of differing age have been sampled and analysed on two heaps (Kerth 1988). Furthermore, weathering experiments under controlled laboratory conditions have been carried out. In these experiments spoils from a number of collieries were subjected to an atmosphere with high water content and leached every two weeks. The leachate was analysed for pH and a number of ions, including iron and alumina (Kerth 1988).

The initial chemical process is the disso-

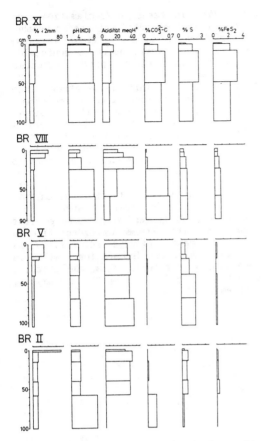

BR XI

BR VIII

BR V

BR II

Fig. 4: Compilation of analytical results of four sections on the Brassert heap. (Sections Br XI: 1 year old, Br VIII: 2 years old, Br V: 7 years old, Br II 9 years old). Acidity in mmol(eq) H⁺/100g; Carbonate carbon, total sulfur and pyrite content in weight-%.

lution and leaching out of the chlorides with rain water. In the chemical weathering experiments an almost instantaneous leaching out of chlorides can be observed (Kerth 1988)

Physical weathering produces small particles and so enhances the chemical weathering processes by exposing the spoils to oxygene and water. The dominant chemical weathering process is the bacterially catalysed oxidation of pyrite and marcasite. Within a few years time of being heaped up the irondisulfides are completely oxidised in the upper metre or so of the spoil profiles investigated (Fig. 4).

In a first chemical weathering stage with

pH values buffered by carbonates at about 7 the acid generated by irondisulfide oxidation is neutralised by the carbonates present in the spoil. By acid attack, even the less acid soluble siderite is dissolved (Fig. 4). Ironhydroxides, gypsum and thenardite are formed (Wiggering 1984; Kerth 1988). The latter results from exchange of calcium against natrium at the Na-saturated cation exchange sites of the spoil (Kerth, 1988). Concrete particles in the spoil are attacked and ettringite is formed. Through the action of rain water gypsum and thenardite are partly leached out of the spoils.

A second weathering stage starts when all carbonates are dissolved by the acid (Fig. 4). The pH rapidly drops to values of about 3. The presence of jarosite, together with kaolinite and silica, is responsible for buffering pH at this value (Miller 1979). Jarosite is a product of irondisulfide weathering at acid conditions (Brown 1971).

Buffering mechanisms can clearly be shown if pH-values of colliery spoil of differing age are plotted according to their abundance (Fig. 5). Distinct maxima occur at pH-values of about 7 and 3. Between 6.5 and 3.5 no buffering systems exist in colliery spoil thus leading to a fast drop of pH from near neutral to strongly acid conditions when all carbonates are dissolved.

In the weathering experiments, some spoils show a rapid drop of pH of leachates from 7 to about 3 within one year of investigation, a result which also illustrates the missing buffering system between pH 6.5 and 3.5 (Kerth 1988).

If neutralising materials are added to weathered spoil, jarosite becomes unstable and dissolves. The jarosite iron is precipitated as ironoxyhydroxide. Since the precipitation of ironoxyhydroxides leads to the formation of free acidity, jarosite stores the acidity formed by irondisulfide weathering (Van Breemen 1973, Kerth 1988).

Thus although the irondisulfides are oxidised within a few years, the acidity of spoil stays high due to the presence of jarosite.

The strong acidification of the spoils is generally followed by a chemical degradation leading to a progressive decomposition of clay minerals (Wiggering 1984).

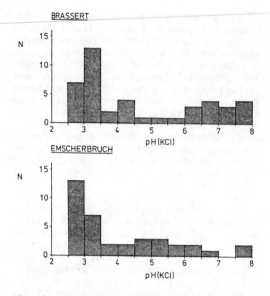

Fig. 5: Abundance of pH-values in 1 to 12 year old spoils from the Brassert and Em-scherbruch heaps.

5 CLAY MINERAL DEGRADATION

X-ray diffraction analyses of the < 2 μm fraction (see Wiggering 1987 for methodical procedure) indicate a `colliery spoil specific' course of weathering of the clay minerals.

During initial weathering (0 – 5 year old spoil), with near neutral pH and semi-saline conditions due to the presence of chlorides in the very beginning, small amounts of expansible layers occur in the illite structure. This causes an availability of plant nutrients such as K, Mg and Ca. The expansible interlayers already correspond to an open or degraded illite (Thorez 1976).

In moderately acidified spoils from 30 – 40 year old heaps (pH 4 – 5) the weathered illites show similarities of behaviour to a montmorillonite. According to Lucas (1963), the observed illites might be "open illites with montmorillonite-beha-viour" (expanded illite according to van der Marel 1962). During the weathering of colliery spoil there is a continuous tran-sition between well-crystallized illite, mixed layers built on illite layers, and montmorillonite interlayers. The varieties of the open illites may be considered as intermediate stages during weathering of a dioctahedral 10 A mica-like material (Tho-

rez 1976). In terms of reclamation pro-blems, it must be noted that these weathe-red montmorillonitic illites are ex-pandable and a very important source of, and an exchange site for, plant nutrients.

In fresh and only initially weathered col-liery spoils the chlorites foremost are "well-crystallized" ("normal" chlorite, Thorez 1976). Depending on the strength of interlayer bonding and chemical compo-sition during initial weathering, X-ray analyses show the same characteristics as the normal chlorite. When pH drops to va-lues between 6 and 5, the chlorites are hydrated. The instability of these chlori-tes is indicated by their collapse on heat treatment during analyses. These less stable chlorites are always identified in acidified materials (pH ≤4). With in-creasing chlorite degradation, Al^{3+} is more and more released from the chlorite structure (Fig. 6) and poses a toxicity problem for plants.

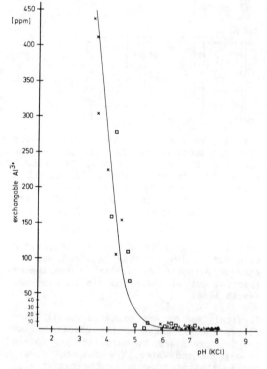

Fig. 6: Concentration of exchangeable alu-minum plotted versus pH of spoil. (Fresh spoils from the Ewald colliery, weathered spoils from the Hoppenbruch, Hugo-Ost and Maximilian heaps)

Tab. 1: Acid/base account of weakly weathered colliery spoil from two heaps (A and B) from the Ruhr area. Pyrite content determined with the method of Dacey and Colbourn (1979). H^+_{pot}: Potential acidity calculated from moles of pyrite x 3. H^+_{act}: Actual acidity already present in the spoil. Except first column all values in mmol/100g.

Sample	FeS₂ [M.-%]	FeS₂	H^+_{pot}	H^+_{act}	Base	Pred. Acidity
A1	3.07	26	78	11	35	54
A2	2.13	18	54	9	30	33
B1	3.31	28	84	20	29	72
B2	3.73	31	93	16	32	77

In weathered colliery spoil materials smectites, or randomly mixed layers composed of montmorillonite and illite layers, with X-ray patterns of poor quality occur. The poor quality of X-ray patterns is probably due to the degree of hydration or it might correspond to some randomly mixed layering involving smectite layers. Preliminary investigations on K-saturated preparations (for methods see Weaver 1958, Harward et al. 1969) showed that the smectites analyzed herein collapse after K-saturation and thus can be interpreted as illite-derived smectites. The smectites occur in such small amounts that they cannot play a role for the availability of plant nutrients.

Generally the clay minerals show a typical transformation during colliery spoil weathering:

(i) After a fast physical splitting of the colliery spoil during initial weathering these processes slow down and chemical attack on the clay minerals predominates. During initial weathering "open" illites with montmorillonite-behaviour occur (3-4 year old heaps).

(ii) During weathering there is a continuous transition between well crystallized illite, mixed layers built on illite layers, and montmorillonitic interlayers. Heat labile, unstable chlorites occur in colliery spoil with low pH (5-30 year old heaps).

(iii) In strongly weathered, old colliery spoils (40-70 year old heaps) smectites and/or randomly mixed layers composed of montmorillonite and illite layers occur. The smectites are interpreted to be illite-derived smectites.

6 POTENTIAL ACID/BASE ACCOUNT

In order to predict the acidification of spoils potential acid/base accounts can be calculated. A measure for the potential acid is the irondisulfide content. Depending on the reaction of irondisulfide to jarosite or ironhydroxides, 3 or 4 moles of acidity are generated for each mole of irondisulfide (Van Breemen 1973). In fresh, unweathered spoils determination of the total sulfur content is a simple method to evaluate the potential acid.

Neutralisation of acidity in spoils mainly takes place at pH values of about 7. Since at these pH values neutralisation reaction with earthalkali-carbonates only leads to formation of the hydrogene-carbonate anion, one mole of carbonate can only neutralise one mole of acidity. The hydrogene-carbonate is leached out.

Determination of the base content of fresh unweathered spoils can be carried out by stirring the spoil with acid followed by back titration with a base. The value determined by this method has to be divided by two due to the leaching out of hydrogene carbonate.

For weakly weathered colliery spoil an acid/base account was calculated. In addition to the potential acid the acidity already present has to be regarded (Tab. 1).

Comparison of the calculated acidity of the spoils with the acidities observed in strongly weathered spoil of the same heaps shows that the method seems to overestimate acidity. For heap A the observed acidities range from 25 to 38 mmoleq. H⁺/100 g spoil, for the heap B from 24 to 48 mmoleq. H⁺/100 g spoil. In comparing the predicted and observed acidities inhomogenities of the spoils have to be taken into account. For practical purposes potential acid/base accounts can be a useful tool for estimation of the acidity that will occur in the future.

7 CONCLUSIONS

The main problems of colliery spoil as a substrate for plant growth are:

(i) the small amounts of silt and clay grain sizes which are important for water storage even after physical weathering,

(ii) initial salinity problems,

(iii) acidification due to irondisulfide oxidation,

(iv) low levels of native plant nutrients in colliery spoil,

(v) degradation of clay minerals such as chlorites which leads to high amounts of free and exchangeable alumina being potentially phytotoxic.

In order to increase the formation of silt and clay sized particles physical weathering should be enhanced. This can be done by avoiding top soiling of the spoils and by intensive morphological structuring of the surface (Fohrmann et al. 1989).

To avoid strong acidification of colliery spoil the base deficit in the spoils should be decreased. This can be done by adding calcareous materials to the spoils. Near neutral to slightly acid conditions favor the formation of clay minerals with a high exchange capacity. Furthermore, phytotoxic substances such as alumina are not mobilised under these conditions. The problematical groundwater contamination by acid leachate of colliery spoil heaps (van Berk 1987) can be reduced.

REFERENCES

Brown, J.B. 1971. Jarosite-Goethite stabilities at 25 °C, 1 atm. Mineral. Deposita 6: 245-252.
Dacey, P.W. & P. Colbourn 1979. An assessment of methods for the determination of pyrite in colliery spoil. Reclamation Rev. 2: 113-121.
Esch, H.P. 1962. Zur Sedimentologie und Diagenese der Sandsteine und Schiefertone im Hangenden des Flözes P2 (oberes Westfal B) in der Emschermulde des Ruhr-

karbons. Fortschr. Geol. Rheinld. Westf. 3/2: 647-666.
Fohrmann, R.; N. Henkel; M. Kerth; F.B. Ludescher; J. Schwarz; H. Wiggering & P. Zimmermann 1989. Sekundärbiotope auf Bergehalden. Verhandl. Ges. Ökol. XVIII: 79-83.
Harward, M.E.; D.D. Carstea & A.H. Sayegh 1968. Properties of vermiculites and smectites: expansion and collapse. Clays Clay Min. 16: 437-447.
Hudec, P.P. 1977. Standard engineering tests for aggregate: What do they actually measure? Geol. Soc. Am. Eng. Geol. Case Hist. 11: 3-6.
Kerth, M. 1988: Die Pyritverwitterung im Steinkohlenbergematerial und ihre umweltgeologischen Folgen. Diss. Univ.-GHS Essen, 182 pp..
Lucas, J. 1963. La transformation des minéraux argileux dans la sédimentation. Etudes sur les argiles du Trias. Mém. Carte Géol. Als. Lorr. 23. 202 pp..
Miller, 1979. Chemistry of a pyritic coal mine spoil. PhD.-thesis Yale Univ., 201 pp..
Teichmüller, R. 1962. Zusammenfassende Bemerkungen über die Diagenese des Ruhrkarbons und ihre Ursachen. Fortschr. Geol. Rheinld. Westf. 3/2: 725-734.
Thorez, J. 1976. Practical identification of clay minerals. Dison: G. Lelotte.
Van Berk, W. 1987. Hydrogeochemische Stoffumsetzungen in einem Grundwasserleiter - beeinflußt durch eine Steinkohlenbergehalde. Bes. Mitt. Gewässerkundl. Jb. 49: 175 pp..
Van Breemen, N. 1973. Soil forming processes in acid sulphate soils. Proc. Intern. Symp. Acid Sulphate Soils, Wageningen, Netherlands 1972, Vol. I: 66-129.
Van der Marel, H.W. 1962. Quantitative analysis of the clay seperate of soils. Acta Univ. Carolinae Tchecosl. Geol. Suppl. 1: 23-82.
Weaver, C.E. 1958. The effects and geological significance of potassium "fixation" by expandable clay minerals derived from muscovite, biotite, chlorite and volcanic material. Amer. Min. 43: 839-861.
Wiggering, H. 1984. Mechanismen bei der Verwitterung aufgehaldeter Sedimente (Berge) des Oberkarbons. Diss. Univ.-GHS Essen, 228 pp..
Wiggering, H. 1987. Weathering of clay minerals in waste dumps of upper Carboniferous coal bearing strata, the Ruhr area, West Germany. Appl. Clay Sci. 2: 353-361

Reclamation, Treatment and Utilization of Coal Mining Wastes, Rainbow (ed.) © 1990 Balkema, Rotterdam. ISBN 90 6191 154 0

The C.B.R. index of coal mining wastes

J.González Cañibano & J.A. F.Valcarce
HUNOSA, Dirección Técnica, Oviedo, Spain

J. M. Rodríguez Ortiz
E.A.T., S.A., Madrid, Spain

J.A. Hinojosa
MOPU, Dirección General de Carreteras, Madrid, Spain

A. Falcón & J. L. Ibarzábal
MOPU, Demarcación de Carreteras del Estado, Oviedo, Spain

ABSTRACT: This article presents the results of laboratory tests carried out on washery and spoil heap coal wastes to determine their C.B.R. index (California Bearing Ratio). Swelling percentages of the coal wastes during these tests are also given. The results obtained are analysed according to type, particle size distribution of the coal waste, compacting energy applied, etc..
8

1 INTRODUCTION

Given that one of the most interesting applications of coal mining wastes is that of earthworks, as large amounts of material can be employed in this way, it is vital to understand the characteristics, properties and behaviour of such wastes from the point of view of soil mechanics.

While there is considerable bibliography regarding particle size distribution, plasticity, compaction, permeability, friction angle, etc. (1) (2) (3) (4), no data are available on the resistance of coal wastes used as soils.

It is for this reason that we herewith give the results obtained in laboratory tests on washery and spoil heap wastes designed to determine their C.B.R. index (California Bearing Ratio), from which, using abaci or formulae, their bearing capacity may be calculated.

2. TESTING PROCEDURE

C.B.R. resistance index is the name given to the percentage pressure exerted by a piston on a soil surface for a certain penetration with respect to the same penetration in a standard sample. The characteristics of the standard sample are in Table I.

C.B.R. index determination was carried out in accordance with Spanish Norm NLT-111/78 "Indice C.B.R. en el laboratorio" (C.B.R. index in laboratory) corresponding to norms A.S.T.M.:1883 and B.S.:1377.

For this test 3 moulds of 2320 m^3 capacity were taken and filled with material of optimum humidity obtained by a Proctor test. The material of one mould was compacted with 20%, the following with 50% and the last with 120% of the Proctor energy employed.

Where Proctor Normal was applied, the material, which has acquired optimum humidity in the Proctor Normal compaction test, was spread in each mould in three layers, each of which was compacted with a 2.5 kg rammer with a volume unit compaction energy of 1.18 – 2.97 and 7.10 kg.cm/cm^3 in the first, second and third mould respectively, equivalent to 12 – 30 and 72 blows.

On each mould a load greater than 4.54 kg was placed, depending on the weight the soil was to support, together with a device to determine whether or not swelling occurred. The moulds were submerged in water for 48 hours, after which time each was subjected to piston penetration, the resistance opposed to this giving the C.B.R. index for the density of material involved.

Representation of the density obtained for each compacting energy applied against the C.B.R. index allows the C.B.R. index for maximum density to be obtained (Graphs 1 and 2). The index thus obtained will be called Normal C.B.R. index.

If Modified Proctor energy is applied, the material, which will have optimum humidity obtained in the Modified Proctor compaction test, is spread in each of the moulds in 5 layers, each layer being compacted with 4.535 kg rammer, with 5.36 13.40 and 32.16 $kg.cm/cm^3$ energy in the first, second and third mould respectively, equivalent to 12 - 30 and 72 blows. Applying the same procedure as before, the C.B.R. index obtained will be the Modified C.B.R. index.

3 TEST RESULTS

North of Spain

Test were conducted on fresh washery wastes and minestones (only the fraction smaller than 40 mm), the particle size distribution of which is set out in Fig. 1.

The Normal C.B.R. indices corresponding to materials compacted with different energies are shown in Graph 1, from which it may be deduced that, for each material, the C.B.R. index rises with increased compacting energy, or equally, with density.

Normal C.B.R. indices, that is to say those corresponding to the maximum energy obtained in the Proctor Normal test, are set out en Table II, and from these it can be deduced that the C.B.R. index in coal washery wastes increases from sample 1 to sample 3, in which it reaches its highest value, diminishing then as the proportion of sizes greater than 10 mm increases, or that of those smaller than 10 mm decreases, as can be observed in Graph 2. This fall in C.B.R. index coincides with the fall in maximum density obtained in the Proctor Normal test, as can be appreciated in Table II.

Graph 3 shows the C.B.R. indices corresponding to materials compacted with different energies and Proctor Modified optimum humidity, from which it may be deduced that, in coal washery wastes, the C.B.R. index increases with density (or compacting energy) up to a certain maximum, and then decreases, that is, the soil resistance is inverted, thus as compacting energy continues to rise, the C.B.R. index disminishes.

The reason for this inversion can be found in the fact that as compacting energy rises, so does degradation of thicker material, and thus resistance is lower. This inversion does not appear in the case of Normal C.B.R., due to the fact that the energy applied is less. However, there are certain cases, as in Sample 4, where inversion begins to appear on the application of higher energy.

Modified C.B.R. indices, that is, those corresponding to maximum densities, are set out in Table II, from which it may be deduced that, in coal washery wastes, the Modified C.B.R. index increases with the rise in the percentage of particle sizes greater than 10 mm, that is, from sample 1 to sample 4.

At the same time, it can also be seen from Table II that in all cases, the Modified C.B.R. index is greater than the Normal one.

Determination of swelling gave values ranging from −0.34 to +0.1% for Normal and from −0.3 to +0.4% for Modified.

Central and Southern Spain

Tests were also carried out on bituminous coal and anthracite wastes from the Centre and South of Spain, the results of which appear in Graphs 4 and 5 and in Table III and from which it can be seen that, as in the previous case, the Modified C.B.R. indices are greater than the Normal ones in the case of Central bituminous coal wastes, the same for anthracite wastes and lower for coal waste from the South, due to the fact that as compacting energy increases so does degradation and resistance is lower. (The wastes from the South are highly degradable siltstones).

Determination of swelling gave similar values to those obtained in the case of washery waste samples, Proctor Normal densities varying between −0.1 and +0.7% and Proctor Modified densities varying from −0.3 to +0.3%.

4 CONCLUSIONS

It is deduced from the above that coal mining wastes have interesting C.B.R. indices which would suggest they have considerable resistance and only very slight swelling.

ACKNOWLEDGEMENTS

The authors wish to express their thanks to the Management of HUNOSA for the facilities afforded them in the preparation of this article, to Doña Josefina Grela Vàzquez for the typing and to D. Ceferino Fdez. Cuetos for the drawings.

REFERENCES

(1) Barnett, S.A. 1984. Recent case histories using minestone in construction of railway embankments in United Kingdom. Symposium on the Reclamation, Treatment and Utilisation of Coal Mining Wastes, Durham, England, September.

(2) Michalski, P.; Skarzynska, K.M. 1984. Compactability of coal mining wastes as a fill material. Symposium on the Reclamation, Treatment and Utilisation of Coal Mining Wastes, Durham, England, September.

(3) Berthe, M. 1986. La valorisation des schistes houillers. Industrie Minèrale, Aout-Septembre, 317.

(4) Gonzàlez Cañibano, J.; Rguez Ortiz, J.M. 1987. La utilizaciòn de los estèriles de la minerìa del carbòn en obras pùblicas. Carreteras, Noviembre-Diciembre, 47.

Table I. Characteristics of standard sample

Penetration		Pressure		
mm	inches	MN/m^2	kpf/cm^2	lb/in^2
2.54	0.1	6.90	70.31	1000
5.08	0.2	10.35	105.46	1500

Table II. Normal and Modified C.B.R. indices. North of Spain

	C.B.R. index		dry maximum density	
Sample	Normal	Modified	Normal Proctor	Modified Proctor
Washery wastes 1	8	12	1.98	2.02
Washery wastes 2	12	16	2.06	2.09
Washery wastes 3	17	17	2.09	2.11
Washery wastes 4	16	21	2.03	2.12
Minestone	9	11	1.92	1.96

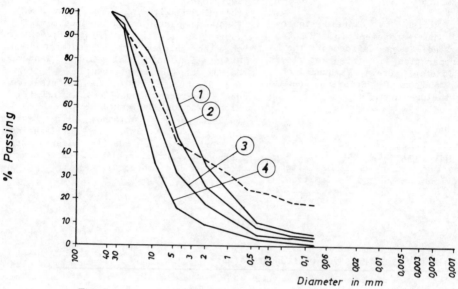

Fig. 1 – Particle size distribution of tested coal wastes

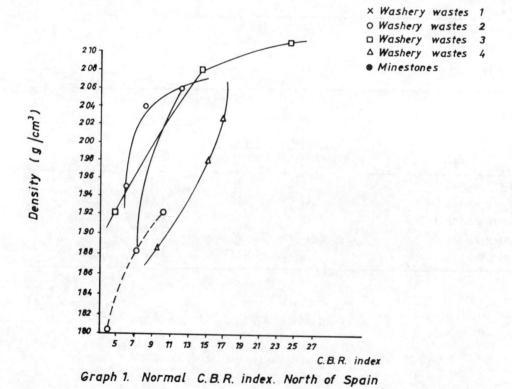

Graph 1. Normal C.B.R. index. North of Spain

Table III. Normal and Modified C.B.R. indices. Centre and South of Spain

| Coal wastes | C.B.R. indices | |
	Normal	Modified
Centre. Washery wastes	16	29
Centre. Washery wastes	11	22
South. Washery wastes	15	12
Centre. Washery wastes. Anthracite	22	22

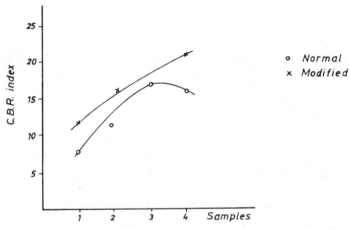

Graph 2. Normal and Modified C.B.R. index. Fresh washery coal wastes. North of Spain

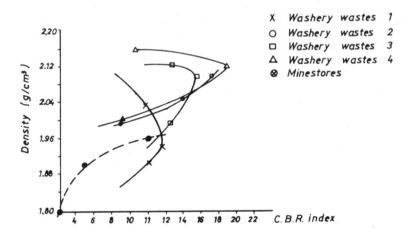

Graph 3. Modified C.B.R. index North of Spain

429

Centre. washery wastes   ×
Centre. washery wastes   o
South.   washery wastes   □
Centre. washery wastes. Anthracite ●

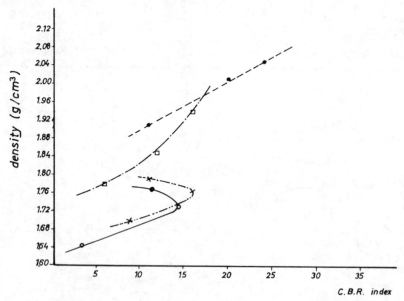

Graph 4.   Normal C.B.R. index. Centre and South of Spain

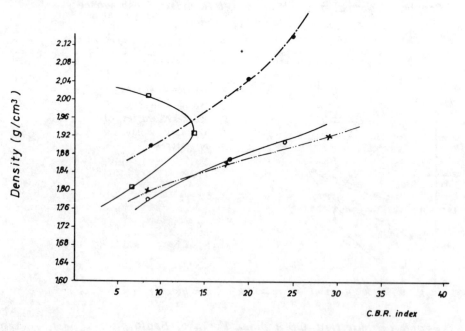

Graph 5. Modified C.B.R. index. Centre and South of Spain

430

Reclamation, Treatment and Utilization of Coal Mining Wastes, Rainbow (ed.) © 1990 Balkema, Rotterdam. ISBN 90 6191 154 0

Influence of sodium chloride on the freezing of minestone

K. M. Skarżyńska & M. Porębska
Department of Soil Mechanics and Earth Structures, University of Agriculture, Kraków, Poland

ABSTRACT: Paper presents the experiments on the influence of sodium chloride (NaCl) on the depression of freezing point of minestone. The analysis of the results confirmed such a depression following even very low NaCl concentration. It is possible, therefore, to use sodium chloride in earth works performed in winter conditions.

1 INTRODUCTION

Minestone has been more widely utilized in earth structures in consideration of the lack of conventional soils. The utilization of minestone demands, however, to test each time its basic geotechnical properties, and also, in some cases, to analyse its behaviour in different atmospheric and soil-water conditions. This is especially important when there appears a necessity to carry on the earth works in winter time. Such a case took place in Piekary Śląskie (Skarżyńska et al. 1989) where the construction of minestone embankment had to be continued in the temperature below 0°C.

Polish specifications (Instruction 1979) provide that earth works in negative temperature may be carried out only in a case of non-cohesive soils and only if the compaction methods guarantee the crushing of frozen lumps and compaction to the required state. It is generally known that application of sodium chloride (NaCl) causes the depression of freezing point of water and keeps moisture in the material. This phenomenon is advantageous when the earth works have to be conducted in winter. The problem is the amount of sodium chloride to be applied, the way of its application, and its possible negative effect to the environment.

It was decided, however, that the application of NaCl would allow to continue the construction in such critical situation. The recommended procedure implied the compaction of minestone in layers of 20-30 cm, sprayed with water-sodium chloride solution up to its optimum moisture content and using a heavy vibratory roller.

There were started the investigations on the application of sodium chloride to maintain the workability of minestone in assumed temperature -0.5 - -10°C. The experiments aimed at determining the freezing temperature of minestone resulting from the different concentrations of NaCl and at determining the optimum addition of sodium chloride to run the works in winter time.

2 INFLUENCE OF SODIUM CHLORIDE ON THE SOIL MEDIUM

The chemical and physical properties of sodium chloride and its water solutions determine its possible impact on the soil medium. The impact of NaCl is mainly of physical character and resolves itself into destruction of water structure resulting from the decreasing of its freezing point (proportionally to the salt concentration) and reducing the pressure of water vapour which increases the amount of water absorb-

ed by the material structure
(Kucharska et al. 1989).

The presence of sodium chloride in
the cooled soil medium brings about
the intensive migration of the solution in pores and microcrevices.
This will, especially during cyclic
freezing, influence the disintegration of the particles and aggregates.
In consequence the soil may be push-
ed forward towards the more frost
susceptible soils. On the other hand,
the maintenance of unfrozen water is
advantegous as it enables the respective soil compaction in the temperature below 0°C.

3 MATERIAL AND METHODS

Prior to the basic investigations of
minestone there were carried out the
experiments on the freezing point of
water beyond the tested soil medium.
Water originated from four sorces:
from the river and the reservoir of
mine industrial water along which
the earth works were conducted, tap
and distilled water. There were also
conducted the investigations of
freezing point of tap water of different NaCl concentration: 0%, 0.5%,
1%, 3%, 5%, 15%. Assumed temperature
was -11°C and freezing time 24 hours.

The minestone from Brzeszcze Coal
Mine was investigated. Its main component was claystone (97%). This rock
has some percentage of salinity by
chlorides which in carboniferous
formations amounts to 113-800 g, i.e.
0.113-0.8%. The material of grain
diameter $\phi < 10$ mm was used for the
experiments on the decreasing of
freezing point of minestone resulting from the application of sodium
chloride. The material was compacted
in plastic cylinders $\phi = 75$ mm and
100 mm high. The samples had initial
moisture content ca. 14% and bulk
density $\rho = 1.98$ g/cm³. Seven samples
were prepared, each of different
concentration of NaCl: 0%, 0.5%, 1%,
3%, 5%, 10%, 15%. The samples were
side-insulated so the temperature
could penetrate in one-way from the
surface towards the bottom. The
investigations were performed in
closed soil-water system, in the
temperature range +20 to -5°C. When
assumed temperature had been reached
and the freezing conditions stabilized, the temperature was lowered to
ca -10°C and kept such until it

stabilized again.

4 FREEZING POINT OF WATER BEYOND THE SOIL MEDIUM

The freezing temperatures of four
kinds of water are presented in
Table 1. The distilled water freezed
first in temperature close to 0°C
(-0.17°C), the others respectively
later and in lower temperatures.
The freezing points of water from
the river and from the reservoir
were similar: -1.2°C and -1.1°C,
and resulted from their salinity
(0.03%) and other chemical components, e.g. sulphur which was also
found in tested water.

Table 1. Freezing of water samples.

Type of water	Beginning of freezing (h)	Freezing point (°C)
From river	6.2	-1.23
From reservoir	5.4	-1.1
Tap water	3.2	-0.23
Distilled water	2.5	-0.17

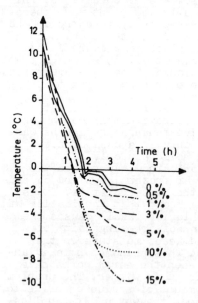

Fig.1. Course of temperatures in
the solution of various NaCl concentrations freezed in -11°C.

432

Freezing temperature of tested water-NaCl solutions is consistent with classical description of solutions from the references (Nalesiński 1963). Distinct bendings of the temperature curves at the beginning of crystallization make possible the reading of the freezing point for corresponding NaCl concentration (Fig.1). These results, given in Table 2, confirm that with the increase of the concentration of solution the freezing point decreases proportionally, and general tendency that the higher the concentration the more intensive process of heat absorption is maintained.

Table 2. Freezing of the solutions

Concentration of NaCl (%)	Freezing point ($^\circ$C)
0	-0.2
0.5	-0.4
1	-0.9
3	-2
5	-3.6
10	-7.1
15	-9.6

4.1 Depression of freezing point of minestone

The course of temperatures obtained for minestone with no admixture is classical, with clearly distinguished segment in which the crystallization of water occurs. On the curves of the sodium chloride addition from 0.5% to 3%, the segment corresponding to the crystallization of the mixture is not distinctly marked (Fig.2). The temperature curves of minestone with an admixture of NaCl from 5 to 15% are typical solidification curves of the solutions in which one component occurs in excess in proportion to another and its crystallization appears sooner (Nalesiński 1963).

Table 3 gives, for a comparison, the time of reaching chosen temperatures in the samples in a depth of 4.5 cm. The time of reaching 0°C is similar for all the samples. The sample with no admixture of NaCl differs distinctly from the others. After the temperature of -0.5°C is reached in this sample takes place a process of ice crystallization which delays the attaining of assumed minus temperature. In the samples containing the sodium chlo-

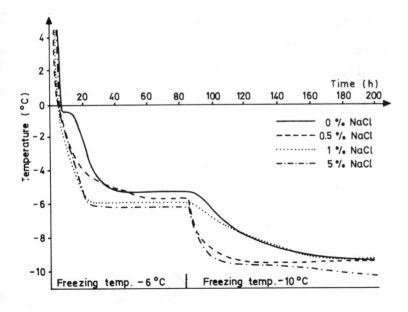

Fig.2. Course of temperatures in minestone samples with various sodium chloride concentration freezed in -10°C

433

Table 3. Time of reaching the chosen temperatures in minestone samples freezed with an admixture of NaCl

NaCl content (%)	Time of reaching the temperature on the depth of 4.5 cm (h)			
	$0^{o}C$	$-2^{o}C$	$-5^{o}C$	$-9^{o}C$
0	7	20	36	175
0.5	6	11	40	101
1	7	10	17	148
3	7.5	10.5	20	131
5	6	11	19.5	99
10	6.5	10	21	99
15	8	13	21	105

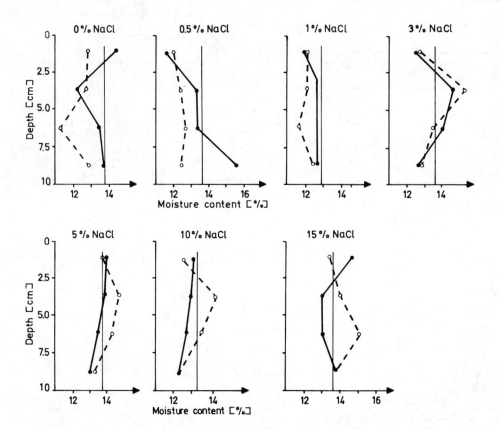

Fig.3. Moisture distribution in minestone samples of $\phi < 10$ mm with NaCl admixture (frozen in $-10^{o}C$). —— central zone, – – –side zone, —— initial moisture content.

ride from 1-15% the time of reaching the temperature $-2^{o}C$ and $-5^{o}C$ is very much alike. Greater differentiation can be observed only during reaching the temperature $-9^{o}C$.

When checking the state of samples with sodium chloride content it appeared that only some of them were entirely frozen. The samples with 0% and 0.5% NaCl freezed completely in -6°C. After decreasing the temperature of freezing to ca. -10°C the sample with 1% NaCl got also frozen. The samples with higher concentration of NaCl did not freeze at all, even in the temperature -10°C. The recorded depression of freezing point of minestone, proportional to NaCl concentration, is consistent with the decreasing of freezing point of the solutions tested beyond the soil medium.

The distribution of moisture in the samples is a consequence of the course of temperature. After completing the experiments there was calculated the distribution of moisture content on four depths in the samples (Fig.3). The configurations of moisture distribution obtained in these investigations are similar to such configurations acquired during freezing in closed system for the conventional soils and minestone with no admixtures (Porębska and Skarżyńska 1988). In the samples containing 0-1% NaCl can be observed, however, a decrease of moisture content in the whole sample (in relation to initial moisture content), on the average 0.7%. It can be especially seen that the central zone has higher water concentration that the side zone of which external edges crush. In the samples with higher content of NaCl (5-15%) the loss of moisture is inconsiderable and amounts to ca. 0.1% and, in the contrary to previous cases, less moist is the central zone of the samples. The sample with 3% NaCl has the characteristics of both groups.

The observations of the samples after the end of freezing show that in case of 3% and 5% of NaCl content the external surfaces are moist and glitter with water. However, in the sample with the highest NaCl concentration (15%), in spite of similar moisture distribution, no separated water was observed on the surface of the sample. The sample is more dense and hard, as if stronger bonds of material particles took place.

The decrement of moisture content in the minestone samples with 0-1% NaCl and practically not changed initial moisture of material with higher NaCl concentrations is consistent with general ascertainment that an admixture of sodium chloride decreases proportionally the amount of water returned by the material (Fig.4).

Analysing the distribution of temperatures and moisture content, checking the state of samples during the tests one can note the consistence and mutual complement of the results.

5 DISCUSSION

The sodium chloride influenced the depression of freezing point and keeping water in minestone. This was especially apparent in the samples with 3-15% NaCl concentration which, till the end of the experiments did not freeze (-10°C). between 4-6%. For proper compaction the optimal moisture content is 8-12% what causes a necessity of watering the minestone. When using water from industrial mine water reservoirs (which contains usually some quantities of sodium chloride) the earth works can be carried out in the temperatures below 0°C with no need of any addition of NaCl.

Small NaCl doses should not influence negatively the environment, the more so that in properly compacted minestone covered with top soil layer the process of washing the chlorides can be neglected.

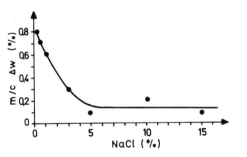

Fig.4. Sodium chloride concentration v. moisture content decrement.

REFERENCES

Instrukcja - Warunki techniczne wykonania i odbioru robót ziemnych. 1979. Budownictwo w dziedzinie gospodarki wodnej, Min.Roln. Warszawa, 40.

Kucharska, L., Maczko M., Tatarek J.
1989. Przyczyny niszczenia podbu-
dowy nawierzchni solami odladzają-
cymi. Drogownictwo, 3: 59-61.
Nalesiński, W. 1963. Układy wielo-
składnikowe, wielofazowe. W:
Chemia fizyczna, praca zbiorowa.
PWN Warszawa, 435-476.
Porębska, M., Skarżyńska, K.M. 1988.
Investigation of the frost heave
of colliery spoil. 5th Int.Symp.
on Ground Freezing, Nottingham,
England, 107-114.
Skarżyńska, K.M. 1980. Effect of
freezing on the selected properties of frost susceptible soils.
2nd Int.Symp. on Ground Freezing,
Sapporo, Japan, 2: 213-218.
Skarżyńska, K.M., Burda, H., Michal-
ski, P., 1989. Studium naukowo-
badawcze odnośnie możliwości wyko-
rzystania materiałów odpadowych
KWK "Powstańców Śląskich" do
budowy obwałowań rzeki Szarlejki.
Maszynopis.

Reclamation, Treatment and Utilization of Coal Mining Wastes, Rainbow (ed.) © 1990 Balkema, Rotterdam. ISBN 90 6191 154 0

Swelling of minestone in relation to its petrographic composition

K. M. Skarżyńska & E. Kozielska-Sroka
Department of Soil Mechanics and Earth Structures, University of Agriculture, Kraków, Poland

A. K. M. Rainbow
Minestone Services, British Coal, Hebburn, UK

ABSTRACT: The paper presents the results of an investigation of the swelling of minestone from Brzeszcze Coal Mine of different relative compaction and different saturation of the model samples. The material of full grain size distribution and selected fractions showed a constant, although slight, increase of swelling values. No stabilization was reached after three years of the experimentation. The results of swelling of minestone are discussed basing on the petrographic, mineralogical and chemical composition.

1 INTRODUCTION

Swelling of clay rocks is the subject of many papers. This problem is considered mainly in relation to the research on the properties and petrographic composition of rocks occurring in the roofs and floors of coal beds /e.g. Kozłowski 1966, Chmura and Kępa 1973/ and also in connection with the investigations on engineering characteristics of minestone and its weathering /NCB 1968, Taylor and Cripps 1984, Skarżyńska et al. 1987, Rainbow 1987, Taylor 1988/. Quite numerous are the papers devoted to the heaving of building subsoils built of clay rocks. The papers besides analysing these processes explain their causes and also give recommended preventive measures /Grattan-Bellew and McRostie 1982, Caldwell et al. 1984, Hawkins and Pinches 1987a, 1987b, Wilson 1987/.

Swelling of clay rocks and of minestone is a complex phenomenon and can result from many causes appearing in relation to the course of weathering process, both physical and chemical. Petrographic and mineralogical composition, water conditions, presence of air /conditions of oxidation/, and temperature play an important role. The intensity of weathering processes is also due to graining

of minestone and its compaction. The basic causes, described in literature, of swelling of clay rocks are:
- presence of swelling minerals, first of all of montmorillonite,
- alteration of clay minerals, i. e. transformation of illite into montmorillonite, and also forming of secondary minerals /jarosite and gypsum/ in the process of sulfide oxidation. These minerals, in a form of incrustation, may develop along the surface of interbeddings and discontinuities,
- imbibition of water by surface absorption on the edges and surfaces of the crystals of clay minerals and by sorption in micropores of coal matter.

The present paper tries to determine the effect of graining and compaction on the swelling of minestone, as well as to state the causes of its swelling in assumed conditions of the experiments.

2 INVESTIGATIONS ON SWELLING

The tests on the ability of swelling of minestone from Brzeszcze Coal Mine were carried out on the material of natural graining ϕ <100 mm and of selected fractions ϕ <40 mm and ϕ <10 mm /Fig.

1/. The samples, each weighing ca. 60 kg, were placed in big cylinders /ϕ 40 cm, h = 30 cm/ with perforated bottoms and then put in the tanks of tap water. The samples had moisture content of 9% and two different relative compactions: I_s = 0.85 and 0.90. There were also two variants of water conditions. In the first one water table in the tanks was kept at the level of upper surface of samples, ensuring the conditions of full saturation corresponding to the conditions of the deposits below water table. In the second variant water table was at the half of sample height modelling the conditions existing at the contact of aeration and saturation zones.

The investigations of the ability of minestone swelling were performed for over three years on the material of natural graining ϕ <100 mm, and for about 1.5 year on the material of graining ϕ <40 mm and ϕ <10 mm. The results show generally constant, although slight, increase of the sample height with marked tendency to stabilization under full saturation conditions /Table 1, Fig.2/. The values of swelling after 20 months are in the range 1.7–4.8 mm, the lowest being of the samples of graining ϕ <100 mm for both full and partly saturated, independently of the material compaction. The highest values correspond to partly saturated samples of graining ϕ <10 mm. After 40 months the samples of ϕ <100 mm showed further increase of swelling amounting to 2.91 mm for the fully saturated and to 4.17 mm for water level kept at the half of sample height /Fig.3/.

Basing on the analysis of the test variants one can note that the swelling value of minestone is influenced by:
- graining of the material. Distinct effect of minestone graining on the value of swelling was observed in all samples partly saturated when the compaction was higher /I_s = 0.90/. The highest swelling value was noted in fine material of ϕ <10 mm /Fig.2a/. When relative compaction was I_s = 0.85 and the samples fully saturated this regularity was not, however, entirely confirmed as the swelling of minestone of graining ϕ <40 mm was bigger than of the material of ϕ <10 mm /Fig.2b/.

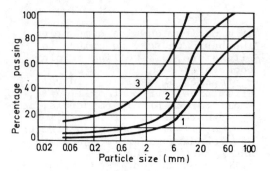

Fig.1. Particle size distribution of tested minestone. 1. ϕ <100 mm, 2. ϕ <40 mm, 3. ϕ <10 mm.

Table 1. Swelling of the minestone from Brzeszcze Coal Mine

Particle size	Relative compaction	Water level	Swelling /mm/ Duration of tests /months/			
			10	20	30	40
< 10	0.85	Up to half of	2.06	3.14		
< 10	0.90	sample height	2.52	4.80		
<40	0.90		2.12	3.77		
<100	0.90		1.53	2.28	3.40	4.17
< 10	0.85	Up to upper	1.78	2.11		
< 40	0.85	surface of	2.18	3.08		
<100	0.85	the sample	1.10	1.71	2.30	2.91
< 40	0.90		1.38	1.88		

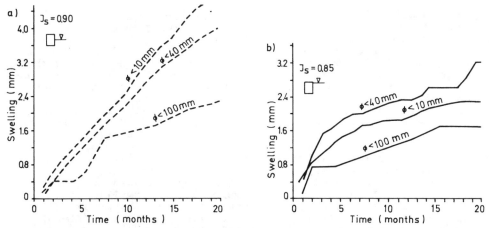

Fig.2. Effect of graining on minestone swelling. a/ partly saturated
samples, b/ saturated samples

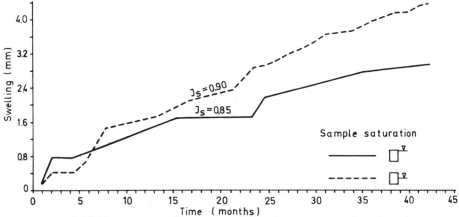

Fig.3. Course of minestone swelling of $\phi < 100$ mm for three years
of observations

- compaction of the material.
Distinct increment of swelling
was observed in more compacted
/I_s = 0.98/, partly saturated fine
material of $\phi < 10$ mm. This dependence is reciprocal for the fully
saturated material of $\phi < 40$ mm,
where lower swelling values were
obtained for highly compacted
minestone /Fig.4/.
- water conditions. Minestone
samples fully saturated /water
table kept at the level of their
upper surface/ were swelling less
than the samples in which water
level reached the half of their
height, the swelling of first ones
showing a tendency to stabilization /Fig.5a and 5b/. The effect

of water conditions on the swelling
ability of minestone seems to be
much bigger than of the compaction.

3 PETROGRAPHIC COMPOSITION OF
MINESTONE

Minestone from Brzeszcze Coal Mine
is almost entirely built of clay
rocks. Sandstone, carbonaceous
shales and coal make ca.10%. Claystones and clay shales in their
typical forms occur in 60%, the
remaining   are sandy claystones,
mudstones and clayey mudstones.
    Microscopic analysis of the
initial sample and of sample after

439

three year swelling, taken at random, showed that the tested minestone is represented by claystone. The texture of claystone is parallel, accentuated by directtional arrangement of foliate minerals. The analysis of mineralogical composition showed that the clay minerals are quite abundant and make ca.50% of the volume /Table 2/. Quarth is dominating breccial component giving 32%. The grains of this mineral are slightly rounded, in majority sharp-edged. There were also distinguished: muscovite, numerous lamellas of hydromuscovite, illite and opaque minerals, magnetite, ilmenite, limonite and fine crystalline

pyrite. Transparent heavy minerals appear in small quantities /Ratajczak 1989/.

In order to find clay minerals containing swelling packets there were performed X-ray radiographic examinations of sedimented and oriented preparations saturated with ethylene glycol which confirmed the presence of muscovite and micas of muscovite type. The indication tests showed the presence of chlorites and mixedpacket minerals of mica/smectite type. They characterize with prevailing not swelling packets connected with the presence of muscovite interlayers /Fig.6, Table 3/.

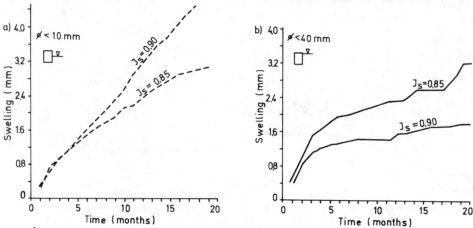

Fig.4. Effect of compaction on swelling of minestone. a/ partly saturated samples, b/ fully saturated samples

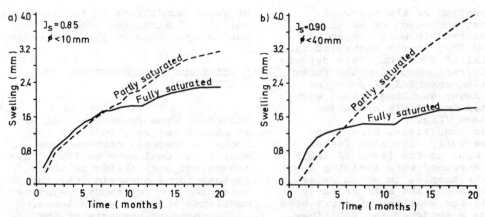

Fig.5. Ability of minestone swelling in relation to the sample saturation.

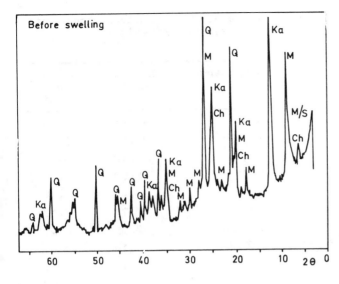

Fig.6. XRD traces of claystone. Ch - chlorite, M -
mica of muscovite type,
M/S - mixed-packet mineral
of mica/smectic type, K -
kaolinite, Q - quartz.

Table 2. Mineral composition of
claystone determined by microscopic tests

Components	Contents /% vol./
Quartz	32
Muscovite	6
Fragments of matrix rocks	3
Feldspars	2
Opaque minerals	4
Clay minerals	53

Thermic effects, observed on the
graphs from thermic differential
analysis, indicate the occurrence
of kaolinite in tested samples
/endothermic effects in 560-570°C
and exothermic ones in 950-960°C/.
Slight endothermic inflexion in
the temperature above 100°C and

corresponding inconsiderable mass
decrement proved the occurrence of
illite in traces. The tests proved
also the presence of organic matter
/Fig.7/.
Chemical analyses of minestone
indicated that the main mineralogical
phases are silicates and alumosili-
cates, the values of SiO_2 and Al_2O_3
for initial sample being 48% and
22.5% /Table 4/. The part of carbonates is small /1.1-1.4%/, as well
of manganese and phosphorus /0.05-
0.95%/. Pyrite occurs in relatively
big amounts. This assumption results
from quite high concentration of
sulfide S /2.39%/ and of Fe^{2+}. The
organic matter occurs in tested
minestone also in quite big amount
of 7% /Ratajczak 1989/.
Performed analyses showed that
the claystone from Brzeszcze Mine
is built of quart, kaolinite,
chlorite, mixed-packet minerals of

Table 3. Mineralogical composition of tested claystone determined by
diffractometric analysis /XRD/

Type of material	Main components /% of weight/			
	Quartz	Kaolinite	Mica	Chlorite + mixed-packet mineral mica/smectite/
Initial sample	30	20	10	40
Sample after swelling	25	25	10	40

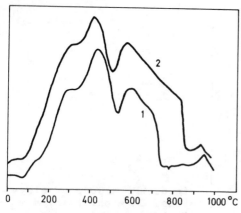

Fig.7. Derivatographic curves of claystone. 1. initial sample, 2. sample after swelling.

illite/smectite type and muscovite. The dominating component of clay substance is kaolinite.

4 DISCUSSION

The process of minestone swelling did not complete within the three year period of investigations. Under partial saturation there was observed further distinct increase of swelling values. Under the conditions of full saturation, however, the swelling tended to stabilization. Swelling values of the finest fractions $\phi < 10$ mm are the highest what should be attributed to the content of fraction $\phi < 0.06$ mm, which in the material $\phi < 10$ mm was 16%, in the samples $\phi < 40$ mm it was 6%, and in the material $\phi < 100$ mm only 2%. By analogy to mineralogical investigations of mudstones /Wilk 1973/ it may be stated that the considerable volumetric changes of minestone $\phi < 10$ mm are connected with higher content of clay minerals of kaolinite and illite type in the fractions $\phi < 0.06$ mm than in the coarser fractions where quartz is dominating.

The main cause of swelling of minestone from Brzeszcze Coal Mine can be the uptake of water by clay minerals into interlayer and inter-packet spaces and also its binding on boundary surfaces of material particles. The saturation of the

Table 4. Main chemical components of claystone /% weight/

Components	Initial sample	Sample after swelling
SiO_2	47.95	49.75
Al_2O_3	22.50	23.65
Fe_2O_3	1.10	0.95
K_2O	2.70	2.35
Na_2O	1.25	1.10
SO_3	0.71	0.53
sulfide S	2.39	2.07
H_2O^+	5.10	4.25
organic matter	7.15	6.23
pH	7.97	8.80

material decreases the rock density and increases the porosity of mine-stone. This manifests itself in the increment of specific surface of rock fragments and sorptive capacity. The process of material disintegration influenced by water, favourable parallel texture of rocks, the presence of clayey minerals of illite/smectite type, and significant amount of organic matter create good conditions for water absorbtion by tested mine-stone.

The changes of rock volume resulting from water absorbtion were also confirmed by the tests performed on various rock types by Caldwell et al. /1984/. These investigations showed that absorbtion of water occurred in clays, marls, shales and mudstones although all these materials were free of clay minerals of swelling structure. The investigations of mudstones from Chwałowice Coal Mine showed that their swelling was connected, first of all, with water absorbtion as these rocks had only 5% of montmorillonite /Chmura and Kępa 1973/. Earlier tests, carried out by Kozłowski /1966/ confirmed the influence of the type of clay minerals on the proportions of swelling of Carboniferous rocks originating from three coal mines in Upper Silesian Coal Fields /Fig.8/. He also proved a relation between swelling and soaking. The much soaked minestone weathered quickly, its frost-resistance was poor, and it was characterized by increased swelling.

Swelling of minestone can also result from the alterations of clay minerals initiated in the material. The analysis of petrographic composition shows that although the chemical reaction of minestone from Brzeszcze Coal Mine before and after swelling /pH = 7.97 and 8.80/ is slightly alkaline it can, however, demonstrate initiated in the sediments processes of forming the hydromicas /illite/ as well as the montmorillonization /pH ca. 8/. The latter process can explain the presence in both samples, before and after swelling, of degraded hydromicas, and first of all of minerals of hydromica mixed-packet structures. The obtained pH values rather exclude the processes of kaolinization of the sediments which take place when pH value is about 5 /Ratajczak 1989/.

The process of sulfide oxidation rather did not occur in the tested minestone as there was not observed any forming of gypsum crystals, as well as increase of SO_3 and decrease of pH. The tests performed for the hospital in Cardiff, where the swelling of shale subsoil resulting from gypsum-crystallization was noted, showed an increasing growth of SO_3 from 0.29% to 1.6% and a decrement of pH value from 7.1 to 5.3 in relatively short period of 17 months /Hawkins and Pinches 1987a, 1987b/. Minestone from Brzeszcze Mine is under conditions of flooding where, according to Twardowska et al. /1988/, the processes of sulfide oxidation are restrained and practically occur only in dry deposits /Table 5/.

Table 5. Constant velocities of sulphate production for Carboniferous rocks /Twardowska et al. 1988/.

Sample	Dry deposit /mval val G_s . d/	Flooded deposit
I	0.035	0.002
II	0.210	0.028
III	0.077	0.005
IV	0.075	0.005

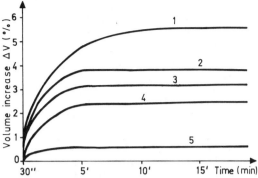

Fig.8. Swelling of Carboniferous rocks /Kozłowski 1966/. 1. Claystone microlaminated with montmorillonite layers - Chwałowice Coal Mine, 2. Mudstone /main clay components: illite and montmorillonite/ - Janina Coal Mine, 3. Claystone /main clay components: kaolinite and halloysite/ - Janina Coal Mine, 4. Illite-kaolinite claystone with 5-6% of organic matter - Jaworzno Coal Mine, 5. Sandstone with clay-siliceous binder - Jaworzno Coal Mine.

The complex investigations of minestone from Brzeszcze Coal Mine allow to formulate the final conclusion. The minestone swelling in assumed conditions should be mainly attributed to water imbibition by surface adsorbtion of clay minerals and to sorption of water in the macropores of coal substance. There were also observed no substancial changes of mineral and chemical composition in claystones after three year investigations of swelling compared with the initial material. No distinct process of forming or alterations of clay minerals was also noted. One can only suppose that these processes were initiated, what can be connected with relatively short period of investigations and with the conditions of saturation of minestone limiting very much the intensity of the processes of the alterations.

REFERENCES

Caldwell, J.A., A.Smith and J. Wagner 1984. Heave of coal shale fill. Can.Geotech.J. 21:

379-383.

Chmura, K., S.Kempa 1973. Budowa petrograficzna i charakter pęcznienia iłowców tworzących spąg i strop pokładów węgla 360/1 i 364/1 kopalni Chwałowice. Przegl.Górn. 9:344-349.

Grattan-Bellew, P.E., G.C.McRostie 1982. Evaluation of heave prevention for floors founded on shale in the Ottawa Region. Can.Geotech.J. 19:108-111.

Hawkins, A.B., G.M.Pinches 1987a. Sulphate analysis on black mudstones. Geotechnique 37,2: 191-196.

Hawkins, A.B., G.M.Pinches 1987b. Cause and significance of heave at Llandough Hospital, Cardiff - a case history of ground floor heave due to gypsum growth. Quart.J. of Eng.Geol.20:41-57.

Kozłowski, C. 1966. Rozmywalność skał karbońskich w zależności od pozycji stratygraficznej i warunków tektonicznych. Praca doktorska. Ph.D. Thesis.

National Coal Board 1968. Weathering of sedimentary rocks and mining spoil: A literature survey. Res. and Develop.Dept. 10.

Rainbow, A.K.M. 1987. An investigation of some factors influencing the suitability of minestone as the fill in reinforced earth structures. London: British Coal.

Ratajczak, T. 1989. Ustalenie składu mineralnego i chemicznego próbek odpadów powęglowych. Ms.

Skarżyńska, K.M., E.Kozielska-Sroka, H.Burda and P.Michalski 1987. Laboratory and site investigations on weathering of coal mining wastes as a fill material in earth structures. 2nd Int.Conf. on Reclam., Treatm.and Utiliz. of Coal Mining Wastes, Nottingham: 179-195.

Twardowska, I., J.Szczepańska and S.Witczak 1988. Wpływ odpadów górnictwa węgla kamiennego na środowisko wodne. Katowice: PAN.

Taylor, R.K., J.C.Cripps 1984. Mineralogical controls on volume change. In: Attwell, P.B., R.K. Taylor /eds/ Ground movements and their effects on structures. Surrey Univ.Press: 268-297.

Taylor, R.K. 1988. Coal measures mudrocks: composition, classification and weathering processes.

Quart.J. of Eng.Geol.21:85-89.

Wilk, A. 1979. Badania petrograficzne i wartość przemysłowa iłowców z górnych warstw załęskich kopalń Chwałowice, Staszic, Wieczorek. Prace Geol.: 117.

Wilson, E.J. 1987. Pyritic shale heave in Lower Lias at Barry, Glamorgan. Quart.J. of Eng. Geol. 20:251-253.

Reclamation, Treatment and Utilization of Coal Mining Wastes, Rainbow (ed.) © 1990 Balkema, Rotterdam. ISBN 90 6191 154 0

The effect of particle morphology on coal waste behaviour

M.R.Gonzalez Moradas
Departamento de Explotación y Prospección de Minas, Universidad de Oviedo, Spain

J.Gonzalez Cañibano
Hulleras del Norte, S.A. Hunosa, Oviedo, Spain

M.Torres Alonso
Departamento de Explotación y Prospección de Minas, Universidad de Oviedo & Consejo Superior de Investigaciones Científicas, Spain

ABSTRACT: The physical characteristics of coal wastes are studied, with a detailed analysis of particle morphology. Tailings (1mm), middle wastes (1-10mm) and coarse wastes (10-150mm) from the Candín washery of HUNOSA (Asturias, Spain) were studied together, independently of the nature of the particles, and, in parallel, for each of the basic components (shales, coal, quartz, calcite, etc.).

Morphological analysis was verified visually and by comparison with the Tables of POWERS (1953) and KRUMBEIN and SLOSS (1955) by means of minor element preparations and with the use of a binocular magnifier and petrographic microscope, thus determining percentage component content, sphericity and roundness, which then permitted statistical treatment and classification. The evolution of sphericity and roundness as particle size increased, dependent on component structure (shaleyness, microfracturing, etc.) was also studied.

The "coarse" fraction was studied by determining the major, intermediate and minor axes; then sphericity, flattening and roundness indices, and classifications according to ZINGG (1935), BREWER (1964) and SNEED-FOLK (1958) could be determined.

On the basis of the morphological study carried out, the direct influence on internal friction, degree of alterability, compacting processes, erosive phenomena, etc., is analysed.

1 INTRODUCTION

Coal exploitation involves the generation of large quantities of wastes, which pose a wide range of problems. Perhaps the most appropriate solution is to use them as new "geological resources", which makes study of their characteristics essential. This article centres on the specific aspect of particle morphology.

The first thing to bear in mind when undertaking a study of this kind is that the particles do not have a natural origin. Thus, for example, the original form of a particle -in this case a waste particle- will depend on the method of exploitation and the kind of tools used in extracting the coal. Size depends on technology. Neither the means nor distance of transport of these materials need be considered, as they are not parameters which differentiate one set of particles from another; in general, all are subject to the same conditions. It follows from this that since the technological conditions are the same for all the particles, the existence of intrinsic properties determining differences in particle form may be observed.

The main characteristics studied are sphericity, roundness and grain size. The results obtained permit the correlation of different parameters, and a morphological classification of the materials. All of this is directed towards a study of their subsequent behaviour with respect to internal friction, compacting, alterability, tube erosion, etc.

2 WORK METHODOLOGY

This present study was carried out on wastes from the Candín washery, belonging to HUNOSA (Asturias, Spain).

Washery wastes are divided into:
- -"tailings", size under 1 mm.
- -"middle", from 1 to 10 mm.
- -"coarse", from 10 to 150 mm.
- -"large", more than 150 mm.

in accordance with the classification of GONZALEZ CAÑIBANO and GARCIA GARCIA (1983), although size intervals have been varied for this study as follows: "tailing" less than 1 mm.; "middle wastes" 1-20 mm.; "coarse" 20-150 mm., so as to correlate more closely with the M.I.T., D.I.N., and A.S.T.M. systems. The other "large wastes" are not considered in this study.

The tools and methods employed were essentially different depending on whether we were dealing with "tailings", "middle wastes" or "coarse wastes".

Observations using the magnifier and microscope centred on sphericity, roundness and the major axis.

The difficulties inherent in three-dimensional (and even two-dimensional) measurement require determination by comparison with visual graphics. In these, sphericity is related to the proportion between the length and width of the images. Roundness was determined in an identical fashion. In this present work we have used the graphics of POWERS (1953), and KRUMBEIN and SLOSS (1955). The major axis has been assimilated arbitrarily to grain size.

The measurements made of coarse were those of the major, intermediate and minor axes. A set of parameters calculated from these allowed classification of the material in function of its form.

Once sphericity and roundness in each particle, were determined as described above, we proceeded to a statistical treatment of a representative sample of a population, always more than 50 elements (GRIFFITHS and ROSENFELD, 1954; ROSENFELD and GRIFFITHS, 1953)

The first task was to try to establish the general sphericity and roundness of the population by relative frequency histograms. Sphericity and roundness values which corresponded to the peaks of the histograms were taken as the values for that population, irrespective of the angularity of the deduced curve.

Given that a sample comprises particles of different sizes, the population represented being previously defined as to size intervals, all were, theoretically, represented in the sample; it is thus of interest to reduce grain size intervals and draw up histograms for sphericity and roundness for each of them. In other cases, the opposite procedure is more useful, that is, to set a fixed sphericity or roundness value and to observe the histograms according to different grain size intervals.

It is also important to determine the relationship between grain size and roundness. The method employed was determination of average grain size corresponding to each value for sphericity and roundness from which the correlation could be established. The aforementioned correlation could normally be established from five or six marker values (at most), depending on whether roundness or sphericity was the factor under consideration.

This approach has not been applied to "coarse". The study of these particles was much simpler: the principal axes were measured directly, and then pertinent indices were calculated. The results are given as averages, means and typical deviations.

Various kinds of classification have been applied, as the one employed depends on grain size, under consideration. Thus, the form of "tailings" and "middle wastes" was established according to the classification of POWERS (1953) and KRUMBEIN and SLOSS (1955), while that used for "coarse wastes" was obtained from two different classifications: that of ZINGG (1935) and BREWER (1964) and that of SNEED and FOLK (1958).

3 RESULTS

Results are presented individually for each of the various fractions.

"Tailings" .- General sphericity

for this fraction is 0.7 (see Fig. 1). For small size intervals, distribution is very similar to that observed for the general run of cases, the mode having practically the same values, but the greater the fractions, the more the percentage of lesser value sphericity increases.

Differentiating components, both shales and coal give a sphericity of 0.7 (see Figs. 2 and 3), but in the case of coal 0.5 sphericity , and to a lesser extent 0.3 are also important. Quartz and calcite have 0.7 sphericity (see Fig. 4).

Observing the relationship between particle size and sphericity, it can clearly be seen that sphericity decreases with increasing grain size. But the gradient for each component is different. That of coal is less than that for shales. Calcite too presents a steeper slope than quartz, this latter being practically the same as coal (see Fig. 5).

As for roundness, the general value for all "tailings" is 0.3 (see Fig. 6).

Shales have a roundness factor of 0.7 (see Fig. 7).

That of coal is 0.5 (see Fig. 8). Roundness for quartz and calcite is 0.3 (see Fig. 9).

The relationship between roundness and grain size for all components is clear: roundness increases with particle size. This relationship is more marked in shales than in coal (see Fig. 10).

"Middle wastes".- These were analysed before and after alterability tests, which meant they were separated into two groups (larger than 3mm. and smaller than 3mm.).

Sphericity for the fraction smaller than 3mm. is 0.7. If shales and coal are considered separately, shales have a value of 0.7 while that for coal is 0.5 (see Figs. 11 and 12). However, the difference between the frequencies of 0.5 and 0.7-0.9 is minimal (see Fig. 12).

In the graphs which relate grain size to sphericity, correlations follow the general pattern for this characteristic: sphericity diminishes as grain size increases (see Fig. 13). But contrary to what occurs with other fractions, the correlation line for coal is stripper than that for shales.

As for roundness, 0.5 is the major

frequency. Both shales and coal have the same roundness, although it is worthy of note that there is in shales a considerable percentage of particles with 0.7 roundness, while that of 0.3 is much lower. The opposite is the case with coal (see Figs. 14 and 15).

The graphs relating grain size to roundness behave similarly to that of the "tailings" fraction (see Fig. 16), although they are essentially different from "middle" greater than 3mm. In these materials, roundness increases with an increase in grain size, being greater for shales than for coal (see Fig. 16).

"Middle" smaller than 3mm. present a general sphericity of 0.7. Considered separately, both shales and coal meet this figure. The difference between them is that shales tend to have smaller sphericity, approaching 0.5, while coal has rather larger sphericity, about 0.9 (see Figs. 17 and 18).

The relationship between grain size and sphericity shows a clear correlation for both types of material: as size increases, sphericity diminishes. This reduction is smaller in coal than in shales (see Fig. 19).

General sphericity after the alterability test is 0.5 for shales and 0.7-0.9 for coal (see Figs. 20 and 21). There is a greater difference in behaviour between shales and coal when both are subjected to alterability; while in one (shales) sphericity diminishes, in the other (coal) it increases. This tendency is present in all sizes of the fraction under consideration and for both components.

There is no variation in the tendency remarked above in untested materials as regards the relationship between grain size and sphericity. The same thing happens: sphericity diminishes as particle size increases. As before, this reduction is more marked in shales than in coal. It is worth noting when comparing lines obtained before and after the tests that these latter are steeper, that is the reduction in sphericity is more marked as grain size increases in both materials. (See Fig. 19).

General roundness for "middle"

447

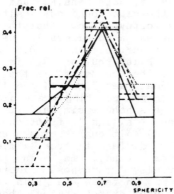

Fig. 1 – Histogram of relative frecuences, for different sphericitys.(Grain size < 1mm.)

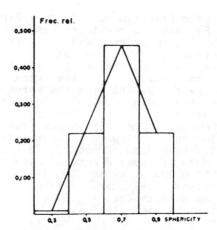

Fig. 2 – Histogram of relative frecuences, for different sphericitys. (Shale. Grain size < 1mm.)

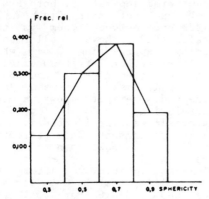

Fig. 3 – Histogram of relative frecuences, for different sphericitys. (Coal, grain size < 1mm.)

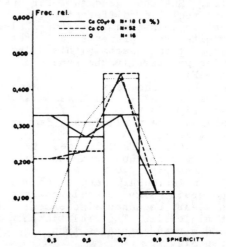

Fig. 4 – Histogran of relative frecuences, for different sphericitys. (Quarz and Calcite. Grain size < 1mm.)

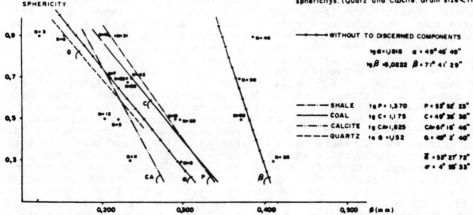

Fig. 5 – If grain size increases then sphericity decreases. All components are the same tendency.

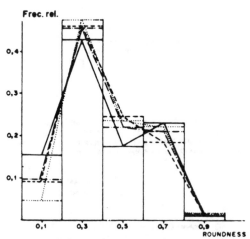

Fig. 6 - Histogram of relative frecuences, for different values of roundness.(Grain size < 1 mm.)

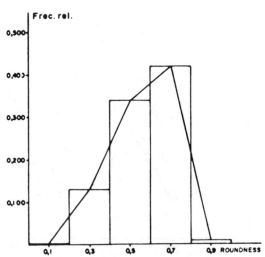

Fig. 7 - Histogram of relative frecuences, for different values of roundness.(Shale. Grain size < 1 mm.)

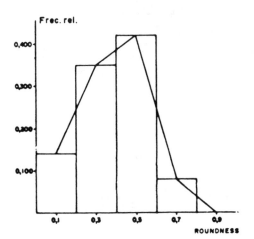

Fig. 8 - Histogram of relative frecuences, for different values of roundness.(Coal. Grain size < 1 mm.)

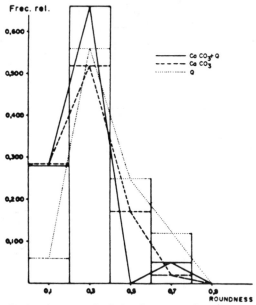

Fig. 9 - Histogram of relative frecuences, for different values of roundness.(Quarz and Calcite. Grain size < 1 mm.)

449

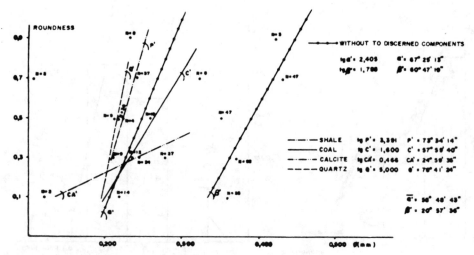

Fig. 10 – If grain size increases then roundness augments. All components are the same tendency. (Grain size < 1 mm.)

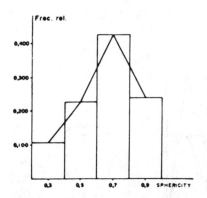

Fig. 11 – Histogram of relative frecuences, for different sphericitys. (Shale. Grain size < 3mm)

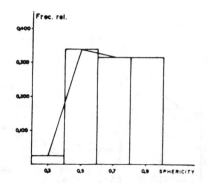

Fig. 12 – Histogram of relative frecuences, for different sphericitys. (Coal. Grain size < 3mm.)

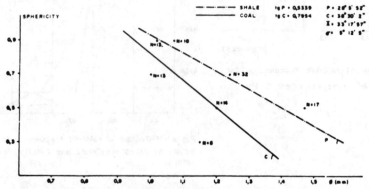

Fig. 13 – If grain size increases then sphericity decreases. (Grain size < 3mm.)

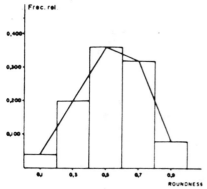

Fig. 14 – Histogram of relative frecuences, for different values of roundness. (Shale. Grain size < 3mm.)

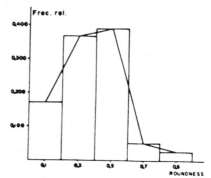

Fig. 15 – Histogram of relative frecuences, for different values of roundness. (Coal. Grain size < 3mm.)

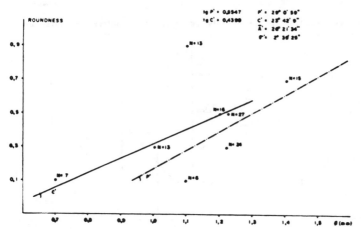

Fig. 16 – If grain size increases then roundness augments. (Grain size < 3mm.)

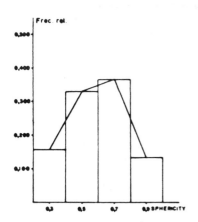

Fig. 17 – Histogram of relative frecuences, for different sphericitys. (Shale. Grain size between 3 and 20mm.)

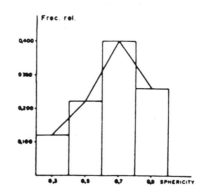

Fig. 18 – Histogram of relative frecuences, for different sphericity. (Coal. Grain size between 3 and 20mm.)

451

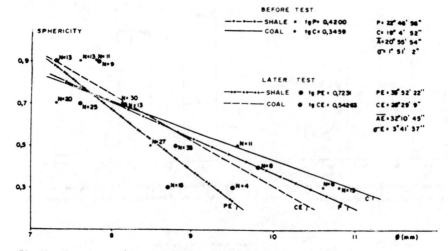

Fig. 19 - If grain size increases then sphericity decreases, both before and later of the alterability test.

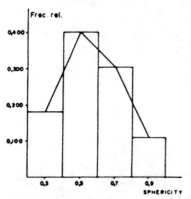

Fig. 20 — Histogram of relative frecuences, for different sphericitys.(Shale. Grain size between 3 and 20 mm. Later alterability test.)

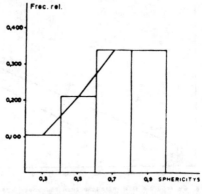

Fig. 21- Histogram of relative frecuences, for different sphericitys.(Coal. Grain size between 3 and 20 mm. Later alterability test.)

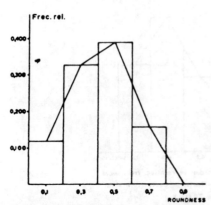

Fig. 22- Histogram of relative frecuences, for different values of roundness.(Shale. Grain size between 3 and 20 mm.)

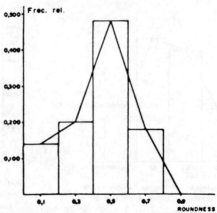

Fig. 23- Histogram of relative frecuences, for different values of roundness.(Coal. Grain size between 3 and 20 mm.)

452

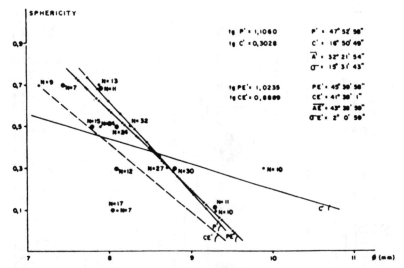

Fig. 24 – If grain size increases then sphericity decreases, both before and later of the alterability test.

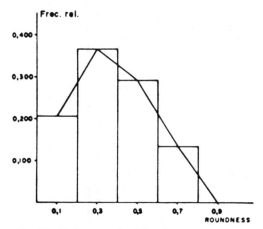

Fig. 25 – Histogram of relative frecuences, for different roundness. (Shale. Grain size between 3 and 20mm. Later alterability test.)

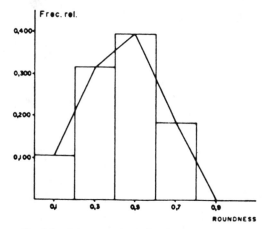

Fig. 26 – Histogram of relative frecuences, for different roundness. (Coal. Grain size between 3 and 20mm. Later alterability test.)

453

SUMMARY OF MORPHOMETRIC CHARACTERISTICS

WASTES OF CANDIN WASHING (HUNOSA)

	FRACTION: Ø(mm)<1					FRACTION: 1<Ø(mm)<20					
						>3mm				<3mm	
		DISCERNED COMPONENTS				BEFORE ALTERABILITY TEST		AFTER ALTERABILITY TEST			
	WITHOUT TO DISCERNED COMPONE					DISCERNED COMPONENTS		DISCERNED COMPONENTS		DISCERNED COMPONENTS	
		SHALE	COAL	CALCITE	QUARTZ	SHALE	COAL	SHALE	COAL	SHALE	COAL
COMPOSITION (%)	—	44(%)	46(%)	7(%)	1(%)	66(%)	38(%)	67(%)	31(%)	60(%)	33(%)
SPHERICITY	0,7	0,7	0,7	0,7	0,7	0,7	0,7	0,8	0,7-0,8	0,7	0,8
ROUNDNESS	0,3	0,7	0,8	0,3	0,3	0,6	0,6	0,6	0,6	0,7	0,3
MORPHOLOGY	Subangulous	Subrounded	Subangulous Subrounded	Subangulous	Subangulous	Subangulous Subrounded	Subangulous Subrounded	Subangulous Subrounded	Subrounded	Subangulous Subrounded	Subangulous Subrounded
IF GRAIN SIZE INCREASES — THE SPHERICITY	DECREASES ESF.	DECREASES ESF.	DECREASES ESF.	DECREASES ESF.	DECREASES ESF.	DECREASES ESF.	DECREASES ESF.	DECREASES ESF.	DECREASES ESF.	DECREASES ESF.	DECREASES ESF.
IF GRAIN SIZE INCREASES — THE ROUNDNESS	INCREASES RED.	INCREASES RED.	INCREASES RED.	INCREASES RED.	INCREASES RED.	DECREASES RED.	DECREASES RED.	DECREASES RED.	DECREASES RED.	INCREASES RED.	INCREASES RED.

454

greater than 3mm. before the alterability test is 0.5; this value is the same as that obtained for shales and coal, although shales already have a high roundness index, 0.3 (see Figs. 22 and 23).

The greater grain size, the lesser the roundness. Such reduction occurs more rapidly in shales than in coal (see Fig. 24).

After the alterability test, a reduction in roundness is observed, attributable to the test. (See Figs. 25 and 26).

Observation of the graphs relating particle size to roundness shows that both materials behave alike: roundness lessens as particle size increases. The difference between the two lies in the fact that shales vary rather more quickly than coal, although this phenomenon is not so marked as before the test. Material behaviour is subsequently more homogeneous. However, average slope before and after testing clearly reveals an increase in these gradients in materials subjected to testing (see Fig. 24).

It is important to note that, for these size, "middle" grains greater than 3mm., the relationship between grain size and roundness is just the opposite to that found in "tailings" and "middle" grains smaller than 3mm.; for these latter the correlation gradients are positive, and for the former negative.

"Coarse wastes".- In this fraction, once the principal axes had been determined, sphericity was calculated according to the authors cited in the section on methodology. The results were as follows:
- Krumbein sphericity:
* average   0.593
* standard deviation 0.098
- Sphericity according to Sneed and Folk:
* average   0.565
* standard deviation 0.127
The roundness index gave the following results:
* average   2.69
* standard deviation 1.01
* median   2.5
The flattening index, given by the average of the three values for the axes, depends to a large extent on the size of these axes. For these samples the results were as follows:
* average   2.70
* standard deviation 1.95

* median   1.5
In the light of the results thus far obtained with regard to sphericity and roundness materials smaller than 20mm. have been classified as shown in the summary chart below. They are in general subangulous materials.

"Coarse wastes" have been classified on the basis of two diagrams (ZINGG and BREWER), giving as a result DISCOIDAL particles, which are, however, notably dispersed.

This classification also relates the form of the particles to their Krumbein roundness. It can be seen that they are distributed between the 0.5 and 0.7 sphericity bands. The most frequent sphericity is that of 0.5-0.6. These results agree with those obtained when Krumbein´s sphericity formula is applied directly.

The results obtained from Sneed and Folk´s classification give a wide dispersion; one might say that there are coarse wastes of every kind of shape, although the most frequent ones are tabular (20%), flattened (18.9%) very tabular (17.4%) and very flat (12%). One definition of the form of the set would be TABULAR-VERY TABULAR (37.4%), rather than flattened-very flat (31.7%).

To sum up, it may be concluded that in these particles, one of the dimensions differs considerably from the other two.

4 EVALUATION OF THE RESULTS OBTAINED

In the "tailings" fraction, all the components present a sphericity of 0.7, while roundness varies. Components such as calcite and quartz are those which have least, perhaps due to the development of the exfoliation planes of calcite, which makes it break easily in these directions, giving it a more geometrical aspect with consequently less roundness.

For this same component, as grain size increases sphericity is less and roundness greater. This too can be explained by marked exfoliation. As their size increases, it is relatively difficult for the grains to fracture throughout their length, because of the exfoliation planes, and so the energy is applied in smoothing the surfaces by means of the development of small fractures which delimit the peaks. In the case of smaller grains, although the fractures which develop are of smaller dimensions, they are sufficient to divide the grain and give it a less roundned appearance. Dissolution phenomena also mean less roundness, although this may not be noticeable over short periods of time, except when the water used in the washeries is acidulous.

The explanation in the case of quartz is perhaps less obvious, although one must remember that it is by nature a resistant material without predominating fracture directions. As a resistant material, it has to be subjected to erosive action over a longer period of time before acquiring a more rounded appearance, and it must be borne in mind that the process to which it has been subjected does not extend over a long time, nor is it sufficiently brusque for the grain to break and create splinters in many cases rather than smoothe the surface. As with calcite, the larger the grain, the less is its sphericity and the greater the roundness. Although it is relatively difficult to produce fractures in quartz, as only those of small dimensions will be effective in breaking the grain, in those of smaller size less roundness will be observed as splinters are created by the breaking. With larger grains, only fractures of very small dimensions yield concrete results; it is precisely these which, along with others, delimit the splinters, resulting in a softening of the outlines, and so in an increase in roundness and sphericity. Dissolution phenomena can not be adduced for these materials.

Shales are subrounded, as they are less resistant than quartz or calcite. Alteration begins to advance along the edges, making them more vulnerable and leading to greater roundness. Schistocity too must be borne in mind, however. When the shale grains are large, it is more difficult for fractures to traverse them in all directions; they break along the planes of weakness and thus become longer, less spherical and more rounded; the energy is largely spent in eliminating small protrusions. In small sized grains, fracturing along the weakness planes is greater than for the others, and the effects of outline smoothing are lesser, resulting in less roundness.

Coal owes its morphological aspect to its easily nature.

Shales and coal with a grain size of between 1 and 3mm. behave in a similar way to those set out herewith, albeit they are more subrounded.

Among "middle wastes" larger than 3mm., an essentially different as regards roundness begins to appear with respect to material of smaller size. While in the former cases roundness increases with grain size, in these the process is the reverse. Behaviour with regard to sphericity is analogous. It follows from all this that greater or lesser roundness in governed by grain size.

The alterability test affects shales and coal differently owing to their different natures.

As regards "coarse wastes", both groups present considerably less sphericity than do "tailings" and "middle wastes", which goes to reaffirm what was noted earlier regarding the reduction in sphericity which accompanies increased grain size, not only in the intervals of small fractions, but also considering the whole set of size.

To explain the low sphericity of such materials, in addition to the reasons already adduced, mention should be made of a certain schistocity in shales, the short time of erosive action and large size, among others.

Particle geometry has an influence on the characteristics of the material, the most significant examples of which are:
- particle size distribution
- coefficient of friction
- internal friction angle
- abrasiveness

Particle size distribution in "tailings" is normally determined by methods the theoretical basis of which is the consideration of spherical particles falling into a fluid. Reality, though, is very different, as the particles have a sphericity substantially removed from that of theory, leading to errors of appreciation. It is, however, difficult to quantify this discrepancy. Roundness has not influence on this aspect.

Roundness does, however, considerably affect the coefficient of friction of the material. The less round and spherical the grains are, the more points of contact there are between the particles, and thus the greater is the coefficient of friction.

Roundness itself is the reason for the influence of particle form on the internal friction angle. As the number of points of contact between the particles is greater, greater effort is required to separate them than if they were balls. For this, a force opposed to the normal pressure is required, which must come from tangential pressure, which is greater than that necessary in the case of spheres, giving a greater internal friction angle.

In the case under consideration, the materials are angular in form, with relatively high internal friction angle, justified by the argument set out above. But with the passage of time, and owing to a variety of causes, the material becomes more rounded and thus a reduction in this angle is to be expected, although this has not been evaluated. There is another phenomenon to be borne in mind: the increase in cohesion, due precisely to the increase in "tailings", the majority of which will be of a clayey nature.

The geometry of the particles may cause greater or lesser erosion in the tubes of, for example, the heat interchangers of fluidised bed heaters as they strike them from outside, or in the transportation tubes for wastes to be used as infilling or to be poured into settling ponds. This erosion is greater when the particles are angular. Though coals and shales are not very abrasive; abrasion may be greater when the particles are more or less angular, as is the case with the materials here treated, such as quartz for example, give the intrinsically abrasive character of the same.

Particles rounded at certain temperatures, and with certain angles of inclination, run along the rubber, rubbing against it more than if they were slaking, simply falling into the belt and not moving during transport; angular particles are thus more suitable and in this respect the wastes are less abrasive.

To sum up, it may be established that angular materials will produce more wear than normal in those immobile conduction elements through the interior of which they circulate, while in the case of mobile conduction elements, where the particles should remain static, angular materials will cause less wear.

5 CONCLUSIONS

- Coal wastes, on leaving the washery, present a morphology from subangulous to subangulous - subrounded.
- The sphericity of particles decreases as grain size increases.
- Particle roundness varies as a function of grain size, increasing along with diameter up to 3mm. With larger sizes the tendency is reversed, and roundness diminishes.
- As well as by grain size, particle morphology is also conditioned by mineralogical and petrographical properties, such as exfoliation, shaliness, dissolution, etc.
- Particle morphology is decisive in the behaviour of the material, having a direct influence on particle size distribution, coefficient of friction, internal friction angle, erosionability, etc.
- The configuration of the particles conditions the processes of consolidation, compressibility, compactation and stability of the materials.

6 BIBLIOGRAPHY

GARCIA GARCIA, M.; GONZALEZ CAÑIBANO, J.: "Valoración de los estériles de carbón. La materia prima y sus utilizaciones". Energía, Enero-Febrero, 65-72, 1983.

GRIFFITHS, J.C.; ROSENFELD, M.A. (1954): "Operator variation in experimental research". Jour. Geol. Vol-62, pp 74-91.

KRUMBEIN, W.C.; SLOSS, F.A. (1955): "Stratigraphy and Sedimentation". Feedman & Co. 497 pp. San Francisco.

POWERS, M.C. (1953): "A new roundness scale for sedimentary particles" Jour. Sed. Soc. Geol. Petrology, vol. 23, 117-119.

ROSENFELD, M.A.; GRIFFITHS, J.C. (1953): "An experimental test of visual comparation technique in estimating two dimensional sphericity and roundness of quartz grains". Am. Jour. Sci. Vol-251, pp 553-585.

SNEED, E.D.; FOLK, R. (1958): "Pebbles in the Lower Colorado River, Texas a study in particle morphogenesis". Jour. Geol. Vol-66, pp 114-150.

ZINGG, T. (1935): "Beitrage zur Schotteranalyse". Schwiz. min. pet. Mitt. Vol 15, pp 34-140.

Reclamation, Treatment and Utilization of Coal Mining Wastes, Rainbow (ed.) © 1990 Balkema, Rotterdam. ISBN 90 6191 154 0

Development and utilization of coal washery refuse for strata control in underground coal mining operations

J.K.Hii, N.I.Aziz, S.Zhang & Y.H.Wu
Department of Civil and Mining Engineering, The University of Wollongong, N.S.W., Australia

ABSTRACT: This study is an integral part of a broad research programme on the development and utilization of coal washery refuse for strata control in Australian underground coal mining operations. A comprehensive laboratory test programme has been devised to investigate the mechanical properties and behaviour of cement and coal washery refuse (CCWR) material. An attempt is made to critically assess the present state of knowledge in the development and utilization of coal washery refuse (CWR) as a pump packing material in Australian underground coal mines, and understand the importance of the research in relation to future field application. It is anticipated that the contribution made with regard to the treatment and utilization of wastes culminating from coal washeries will play a major role in future mine planning as a result of the increasing importance of environmental issues; such as the disposal of CWR and surface subsidence.

1 INTRODUCTION

In New South Wales, Australia, the raw coal production in the year 1987-1988 alone was 76.3 million tonnes of which 12.3 million tonnes was extraneous dirt. According to NSW Government (1983), it is estimated that by the year 2000, the existing and planned coal mining activities could generate some 104 million tonnes of CWR in the Southern Coalfields alone, while the existing and planned surface emplacements have a capacity for only 30 million tonnes of CWR. There is a need for a full study of the problems associated with the generation, utilization and disposal of CWR in this state. The disposal of CWR in particular was considered to be a significant economic and environmental issue which needed to be resolved as a matter of urgency in the Illawarra region. It is envisioned that similar problems may be faced in other coalfields (NSW Government, 1983).

The main purpose of this investigation is to utilize CWR as a raw material for pump packing. Apart from the expensive commercial materials for pump packing, the utilization of CWR would have a two-fold advantage in costs reduction. There would be a reduction in both the dirt disposal costs and commercial materials costs. Much work however needs to be done in this field since many problems are being experienced in obtaining a consistent mix owing to the variation in products from coal washeries.

In order to substitute the expensive commercial materials for pump packing, it is necessary to improve its geotechnical characteristics. A comprehensive laboratory testing programme was therefore devised to investigate the geotechnical

properties of CWR as a function of various cement mixes over a 365 day period. The research programme aims at producing the most cost effective and economical monolithic pump packing system for strata control and increased recovery of coal. The investigation also aims: to optimize particular mechanical properties of CCWR materials, to develop a cheaper binding material, to increase the initial strength of the mixture, and to develop a material which is pumpable but can also give sufficient strength for use as a pack material in Australian underground coal mining operations. Test results on the effect of water content on the mechanical properties of CCWR material containing calcium chloride admixture has been reported elsewhere (Hii and Aziz, 1989).

2 OUTLINE OF EXPERIMENTAL PROGRAMME

The experimental programme was conducted in six series of tests which are listed chronologically as series I through VI in Table 1. All pouring and batching details of CCWR material specimens are presented in Tables 2 and 3. Test procedures for all test series and materials used in making test specimens have been reported in details (Hii and Aziz, 1986b, Hii et al, 1990). Tests in series V were carried out generally to AS 1012.13/Amdt 1/1986-12-05, Method for the Determination of Drying Shrinkage of Concrete.

The unconfined compressive strength, modulus of elasticity, triaxial strength, indirect tensile strength, sonic velocity and drying shrinkage of CCWR material were the primary properties investigated. The measurements of physical properties such as

Table 1. Summary of tests

Test Series	Description of Tests	Experimental Measurements	Total No. of Tests
I	Unconfined compressive strength tests. (Specimens with both lengths and diameters of 102.5mm).	Force, axial and lateral displacements.	365
II	Indirect tensile strength tests. (The discs were 56mm in diameters and 29mm thick).	Force and displacement.	42
III	Triaxial tests. (Specimens with lengths and diameters of 113mm and 56mm).	Force and displacement.	44
IV	Unconfined compressive strength tests. (Specimens with lengths and diameters of 113mm and 56mm respectively).	Force, axial and lateral displacements.	30
V	Drying Shrinkage tests. (Prisms 75mm x 75mm x approximately 285mm long).	Length.	24
VI	Unconfined compression strength tests. (Specimens with a height of 150mm and diameters of 55.5mm, 75.5mm, 102.8mm, 149.5mm, 206mm, 242mm and 300mm)	Force, axial and lateral displacement.	19

porosity, void ratio, specific gravity, bulk density and particle density were used to characterize the mechanical properties and behaviour of CCWR material.

3 TEST RESULTS AND DISCUSSION

3.1 Unconfined compressive strength tests

The results of unconfined compressive strength tests are summarised in Tables 4 and 5. Unconfined compressive strength and curing time relationships are shown in Figures 1 and 2. For brevity the results for mixes 1 through 5 only will be discussed in this paper. Mixes 2 through 5 exhibit good early strength followed by rapid strength development from 1.5 to 5 hours. Figure 3 graphically illustrates this property. These increases in maximum compressive strength for mixes 2 to 5 are quite pronounced. Increases in the compressive strength from 1 to 7 day curing time are still quite rapid although the rate is slightly lower than that described for up to 5 hour curing time. Thereafter, the increases in strength are not significant for mixes 3 through 5. Mix 1 displays a moderate strength increase from 7 to 90 day curing time while mix 2 displays a high strength increase. The results indicate that the specimens exhibit low residual strengths.

3.2 Triaxial strength tests

Results of triaxial strength tests carried out on

Table 2. Batching and pouring details of mixes 1 through 7.

		COAL WASHERY REFUSE				PORTLAND CEMENT		CALCIUM CHLORIDE			Total	Water			
Mix	Date	Moist Mass (g)	Dry Mass (g)	Water Content (%)	Mix Composition (%)	Dry Mass (g)	Mix Composition (%)	Moist Mass (g)	Solid Mass (g)	Liquid Contained (g)	Water Added (g)	Content (Wet basis) (%)	W/C Ratio	Air Temp. (°C)	Slump Test (mm)
1	31-7-86	60000	57278	4.54	84.4	10588	15.6	317.7	270.0	47.65	11764	14.72	1.1	15.0	110
2	6-8-86	45000	43709	2.87	87.5	6244	12.5	220.4	187.3	33.05	7521	13.0	1.2	15.0	---
3	19-8-86	47000	45247	3.73	90.0	4525	10.0	159.7	135.7	23.96	7985	13.8	1.8	17.0	15
4	21-8-86	47799	46000	3.91	92.5	3730	7.5	131.6	111.9	19.75	7476	12.6	2.0	16.0	10
5	25-8-86	48000	45692	4.81	95.0	2285	5.0	80.6	68.5	12.10	6967	12.0	3.0	16.5	10
6	8-10-87	50000	47717	4.78	85.0	9706	15.0	297.2	252.6	44.60	7690	12.0	0.9	20.0	---
7	14-10-87	48000	46937	2.26	85.0	8283	15.0	292.3	248.5	43.9	13638	21.0	1.6	17.0	C

C denotes collapsed slump.

Table 3. Batching and pouring details of mixes 8 through 14.

| Mix | Date | COAL WASHERY REFUSE | | | | PORTLAND CEMENT | | CALCIUM CHLORIDE | | | Total | Water | | | |
		Moist Mass (g)	Dry Mass (g)	Water Content (%)	Mix Composition (%)	Dry Mass (g)	Mix Composition (%)	Moist Mass (g)	Solid Mass (g)	Liquid Contained (g)	Water Added (g)	Content (Wet basis) (%)	W/C Ratio	Air Temp. (°C)	Slump Test (mm)
8	21-10-87	50000	48876	2.30	90.0	5431	10.0	215.8	162.9	52.9	12877	19.1	2.4	20.0	C
9	28-10-87	50000	48957	2.13	90.0	5440	10.0	216.2	163.2	53.0	7440	12.0	1.4	23.0	0
10	9-11-87	57500	54797	4.93	87.5	7827	12.5	463.6	234.8	228.7	12421	16.5	1.6	22.0	85
11	3-11-87	45000	43555	3.32	87.5	6222	12.5	247.3	186.7	60.6	11811	19.1	1.9	30.0	C
12	5-11-87	50000	47388	5.51	92.5	3842	7.5	152.7	115.3	37.4	12044	19.0	3.1	22.5	C
13	24-11-87	50000	47801	4.60	92.5	3876	7.5	229.5	116.3	113.3	10234	16.5	2.6	24.5	75
14*	27-9-88	125171	120000	2.70	85.5	21176	14.5	1254.1	618.8	635.3	24987	15.0	1.2	24.0	65

C denotes collapsed slump.

*For series VI experiments.

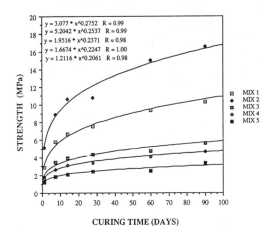

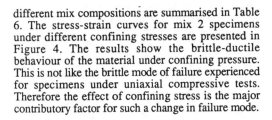

Figure 1. Strength-curing time relationships of mixes 1 through 5 of CCWR specimens.

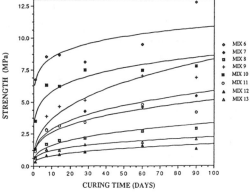

Figure 2. Strength-curing time relationships of mixes 6 through 13 of CCWR specimens.

different mix compositions are summarised in Table 6. The stress-strain curves for mix 2 specimens under different confining stresses are presented in Figure 4. The results show the brittle-ductile behaviour of the material under confining pressure. This is not like the brittle mode of failure experienced for specimens under uniaxial compressive tests. Therefore the effect of confining stress is the major contributory factor for such a change in failure mode.

Under triaxial stress conditions the post-failure stress-strain curves show that measured residual strength values constitute a high proportion of the peak strength values. In field application, the CCWR pack will initially be confined by the forms. As the pack is loaded it may expand laterally and confinement will be offered by adjacent packs. This means that a higher degree of resistance to closure will still be offered by the packs. Figures 5 and 6 show the peak strength envelope of CCWR specimens. Mixes 3 to 5 display an expected increase

461

Table 4. Effect of water content on the unconfined compressive strength of CCWR materials at different curing times.

Portland Cement (%)	Water Content (%)	Slump (mm)	Compressive Strength (MPa)													
			0.25H	0.5H	0.75H	1H	1.5H	2H	4H	1 Day	7 Days	14 Days	28 Days	60 Days	90 Days	365 Days
10	19.1	C								0.659	1.703	2.011	2.183	2.724	2.886	
10*	16.2				0.032			0.048			2.650	3.070	3.410			
10	13.8	15					0.039	0.042		1.799	3.449	3.882	4.344	4.699	5.631	8.875
10	12.0	10			0.041	0.036	0.047	0.078	0.271	3.495	6.305	6.232	7.474	7.433	7.795	
7.5	19.0	C								0.321	0.823	0.924	1.112	1.616	1.617	
7.5	16.5	75								0.594	1.289	1.213	1.680	1.864	2.097	
7.5	12.6	10					0.038	0.042	0.166	1.629	2.664	3.152	3.379	4.105	4.625	6.356
5.0	12.0	10							0.255	1.208	1.876	2.041	2.432	2.493	3.388	3.439

* (after Thomas, 1986).
C denotes collapsed slump.

Table 5. Effect of water content on the unconfined compressive strength of CCWR materials at different curing times.

Portland Cement (%)	Water Content (%)	Slump (mm)	Compressive Strength (MPa)													
			0.25H	0.5H	0.75H	1H	1.5H	2H	4H	1 Day	7 Days	14 Days	28 Days	60 Days	90 Days	365 Days
25*	16.3	·				0.104		0.551			14.690	18.500	21.80			
20*	15.96					0.051		0.229			11.70	14.10	15.19			
15	21.0	C								1.289	2.870	3.111	4.248	4.623	5.443	
15*	17.6				0.025			0.046			5.170	6.530	6.780			
15	14.7	110								2.589	5.774	6.620	7.539	9.325	10.282	13.559
15	12.0	0	0.037	0.033	0.042	0.068	0.111	0.231	0.892	6.714	8.552	8.682	8.126	9.493	12.773	
12.5	19.1	C								1.152	2.800	3.117	3.394	4.803	4.157	
12.5	16.5	85				0.009			0.013	1.365	3.895	4.670	5.134	7.029	6.875	
12.5	13.0	0					0.082	0.151	1.021	5.138	8.859	10.612	10.838	15.073	16.599	

* (after Thomas, 1986).
C denotes collapsed slump.

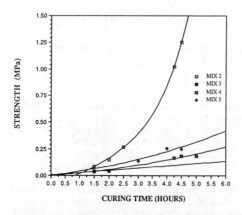

Figure 3. Stress-curing time curves of mixes 2 through 5 of CCWR specimens for up to 5 hours.

in the friction angle φ and a decrease in the cohesive strength C for a subsequent decrease in the mix cement percentage.

3.3 Indirect tensile strength test

Indirect tensile strength test results of different mix compositions are summarised in Table 7. The indirect tensile strength of CCWR specimens reflects that of the compressive strength. As would be expected the CCWR material is weaker in tension than in compression. A rule of thumb is that the tensile strength of the material is usually one tenth of its compressive strength.

The tensile properties of CCWR specimens are therefore slightly better than might be expected. After 28 day curing time the tensile strengths for mixes 3

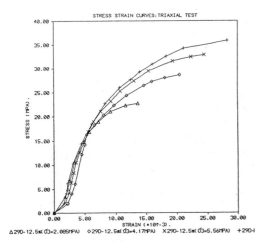

Figure 4. Triaxial strength test results for CCWR specimens at different confining pressure (12.5% cement, mix 2).

Table 6. Summary of triaxial test results for CCWR specimens.

Mix	Material friction angle (o)	Cohesive strength (MPa)	Unconfined compressive (MPa)
15.0%C-21.0%W	28	0.625	3.123
15.6%C-14.7%W	14	1.400	9.058
12.5%C-19.1%W	22	0.740	3.262
12.5%C-16.5%W	20	1.100	5.395
12.5%C-13.0%W	16	2.600	13.090
10.0%C-19.1%W	30	0.420	1.709
10.0%C-16.5%W	29	0.400	1.871
7.5%C-19.1%W	32	0.225	0.982
7.5%C-16.5%W	29	0.400	1.871
7.5%C-12.6%W	36	0.600	3.275
5.0%C-13.0%W	41	0.350	1.900

Table 7. Indirect tensile strength test results of different mix compositions

Mix	Tensile Strength (MPa)	Wet Density (t/m^3)	Water Content (%)
3	0.628	1.947	11.00
4	0.422	1.934	10.53
5	0.191	1.974	10.37
7	0.608		21.30
8	0.394		17.57
10	0.840		13.75
11	0.629		16.80
12	0.170		16.98
13	0.403		15.15

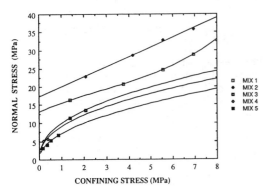

Figure 5. Peak strength envelope of CCWR specimens for mixes 1 to 5.

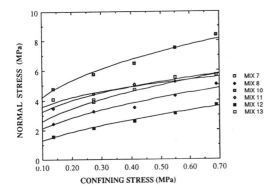

Figure 6. Peak strength envelope of CCWR specimens for mixes 7, 8, 10, 11, 12 and 13.

and 4 are approximately 15% and 13% of the compressive strength respectively. In practice, a large portion of a pack will be in compression, particularly the centre which will be confined by the outer material. The performance of the material therefore will not be significantly affected by its tensile property.

3.4 Sonic velocity values measured during curing

All velocities show a marked increase from 1 day curing time to 90 day curing time. This indicates stiffness increases in the specimens. Calculated

463

values of stiffness, Young's modulus and Poisson's ratio for the tested specimens confirm this proposal. The results show a slight but consistent increase with increased cement content. The results also indicate that the degree of compaction of the specimen may influence its sonic velocity value.

Scatter plots of sonic velocity against unconfined compressive strength of CCWR specimens for different mix formulations are illustrated in Figures 7 and 8. As expected the sonic velocity value of the specimen increases with a corresponding increase in its strength. This trend is consistent for all CCWR mixes. The results generated in this study may be used to predict the insitu strength of CCWR pack.

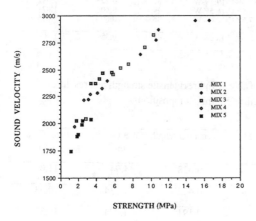

Figure 7. Sonic velocity-strength graph of CCWR specimens for mixes 1 through 5.

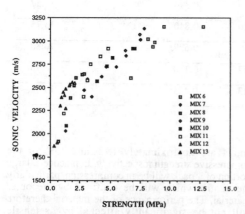

Figure 8. Sonic velocity-strength graph of CCWR specimens for mixes 6 through 13.

3.5 Physical properties of CCWR specimens

The relative density of coal washery refuse solids was found to be 2.31. The physical properties of representative CCWR specimens are summarised in Table 8. The results indicate that mix 1 has a

Table 8. Physical properties of CCWR specimens for different mixes.

Mix	Porosity	Void Ratio	Specific Gravity	ρ_{wet} (t/m³)	ρ_{dry} (t/m³)	ρ_s (t/m³)
15%C-12%W	11.516	0.114	2.054	2.049	1.941	2.193
15.6%C-14.7%W	25.990	0.351	1.930	1.928	1.681	2.272
15%C-21%W	24.728	0.247	1.960	1.955	1.709	2.271
12.5%C-13%W	18.580	0.228	2.033	2.031	1.853	2.276
12.5%C-16.5%W	20.303	0.202	2.013	2.007	1.807	2.268
12.5%C-19.1%W	23.030	0.229	1.952	1.947	1.729	2.246
10%C-12%W	14.494	0.145	2.083	2.077	1.935	2.264
10%C-13.8%W	23.990	0.316	1.966	1.963	1.733	2.280
10%C-19%W	25.898	0.259	1.961	1.956	1.698	2.291
7.5%C-12.6%W	23.600	0.309	1.979	1.977	1.752	2.293
7.5%C-16.5%W	24.995	0.248	2.011	2.006	1.757	2.343
7.5%C-19%W	28.311	0.282	1.976	1.971	1.690	2.358
5%C-12%W	21.480	0.274	1.972	1.969	1.776	2.261

relatively higher porosity (26% for mix 1 and 18.6% for mix 2) and void ratio (0.35 for mix 1 and 0.23% for mix 2) than mix 2.

3.6 Effect of water content on the drying shrinkage of CCWR material

The drying shrinkage of each specimen was measured at each drying period, namely after air drying of 2, 3, 4, 8 and 16 weeks. Drying shrinkage of CCWR material varies from 684 microstrain for 12.5% Portland cement mix (with moisture content of 19.1% dry basis) at two weeks to 2740 microstrain for 7.5% Portland cement mix (with moisture content of 19.0% dry basis) at 16 weeks. Figure 9 illustrates the effect of water content on the drying shrinkage of CCWR specimens.

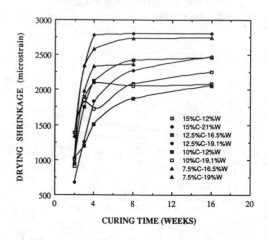

Figure 9. Effect of water content on the drying shrinkage of CCWR specimens.

It is evident from the tests that the drying shrinkage of wet mixes is larger than that of dry mixes. In general, moisture content affects the drying shrinkage of CCWR material as it reduces the volume of the restraining CWR. However, results indicate that drying shrinkage is insignificant in all mixes. Generally, small shrinkage means small microcracking and small crack connectivity, therefore, lower permeability and moisture absorption are expected.

4 CCWR PACK CHARACTERISTICS IN RELATION TO STRATA BEHAVIOUR

An attempt is made to understand the implications of the mechanical properties of CCWR material in the underground environment. The Longwall strata behaviour and the generation of pack load has been reported in some details in Hii and Aziz (1986a). However, an accurate assessment of the initial and ultimate strength requirements of the pack in the underground environment would be complex. Nevertheless, the initial strengths of 0.082 MPa in 1.5 hours, 0.151 MPa in 2 hours, 0.267 MPa in 2.5 hours, 1.251 MPa in 4.5 hours, 5.138 MPa in 24 hours and the long term strength of 16.599 MPa in 90 days for mix 2 indicate that the mechanical properties of CCWR material are close to those required. It is indicative that this material can offer a good early resistance, however, its strength should continue to develop in spite of accommodating the irresistible movement of the strata and possibly a degree of failure.

If the major displacement of the strata takes place within the first few days of the installation of CCWR pack, it may remain substantially unfailed. Contrary to this, the pack may fail considerably, however, the initial confinement of the pack by the retaining bags or forms and later by other packs can ensure a good residual strength for the pack. However, at this present period of research accurate prediction of the effect on strata behaviour, in general, by use of the pack material properties can not be quantified.

A Young's modulus of 2.847 GPa after 90 days (for mix 2 specimens) gives a satisfactory elastic modulus property of the material. Results indicate that specimens of mix 2 have on average a superior cohesive strength of 2.6 MPa (1.86 times the cohesive strength of mix 1) and this reflects their comparatively higher strength development pattern. The failure behaviour of the specimens indicates that they have a comparatively good cohesive property.

The high fluidity of mix 1 on pouring did not permit early strength testing of the specimens (1 and 2 hour curing time). The results for 1 day curing time specimens reveal a strength value of 2.589 MPa which is nearly twice the pack strength requirement as prescribed for UK mining situation (Whittaker and Woodrow, 1977, Smart et al, 1982, Buddery, 1984, Clark and Newson, 1985). Specimen strengths of 5.774 MPa after 7 days curing, 6.620 MPa after 90 days curing indicate very favourable pack properties. Young's modulus values of 0.386 GPa after 1 day

increasing to 1.658 GPa after 90 days indicate a satisfactory stiffness property of the material. For bulk transport of pack material a value of 1.95 t/m^3 may be used for design purposes. The failure behaviour of the specimens indicates that the cohesive strength of 1. MPa for this mix formulation is satisfactory.

The compressive strength test results show that the early strengths of the specimens of mix 3 are relatively low. This could extend deshuttering time in practice. The strength values after 1 day curing are favourable with the design criteria in the United Kingdom but the final strength of only 5.631 MPa after 90 days curing would appear to be insufficient for Australian pack requirements. The failure surfaces of the specimens appeared to be powdery indicating a low cohesive property of the material.

Mixes 4 and 5 exhibit relatively low compressive strength values which are lower than those required for the pack design. The final maximum strengths of 4.625 MPa for mix 4 and 3.388 MPa for mix 5 after 90 days curing are much lower than that needed for effective strata control. The mode of failure for both materials manifests as slabs crumbling and breaking away from the walls of the specimens. Again, the failure surfaces of the specimens appeared to be powdery and this reflects the low cohesive strengths (0.6 MPa for mix 4 and 0.35 MPa for mix 5) of the materials.

5 APPLICATION OF CCWR MATERIAL

In Australia, interests are now being shown in monolithic pump packing system utilising wastes culminating from power plants (flyash) and coal washeries (CWR). A pneumatic bulk handling system has been installed at West Cliff Colliery with the aim of constructing continuous lengths of monolithic pump pack support (using flyash and cement mixture) in Longwall gateroads. However, this has not been successful in terms of cost-effectiveness. The authors and other researchers have been committed to the evaluation of CWR for disposal and strata control in underground coal mines. Results obtained thus far have been very encouraging.

Due to the need to continually maximise productivity levels in the Australian underground coal mining industry, several alternative Longwall panel layouts utilising pump packing technology have been proposed by various researchers (Richmond, 1981, Hebblewhite, 1983, Richmond et al, 1985, Marshall and Lama, 1986, Lama, 1988). Some have the potential for increased development rate and/or improved recovery. The utilization of CWR as a pump packing material is an attractive proposition. Also, it is well known that the filling of underground coal mine openings utilizing CWR with or without cement and various additives may reduce surface subsidence (Hughson et al, 1987, Sinha, 1989).

It is envisaged that the experimental data generated from this study may become useful for future applications of CCWR materials in both engineering construction and its associated disciplines such as environmental engineering and geotechnical engineering.

6 CONCLUSIONS

Given the existing environmental constraints imposed on the coal mine operators, the utilization of CWR as a packing material is an attractive proposition. It has been demonstrated that the use of CCWR pack is a major step in the development of a cost-effective pack for strata control in Australian underground coal mining operations. This study has greatly improved understanding of the mechanical properties and behaviour of CCWR material produced.

As expected an increase in the strength of CCWR materials can be attained either by decreasing the water content or increasing the quantity of ordinary Portland cement used in the mixtures. The importance of water content with regard to its effect on strength, workability and other related mechanical properties of CCWR material is emphasised.

It is envisaged that the results generated from this study may become useful for future applications of CCWR materials in both engineering construction and its associated disciplines.

Suggestions for further work comprise the following:

1. Permeabilities are among the most important characteristics of materials as, for example, ventilation applications. Furthermore, permeabilities are, to some degree, a function of water/cement ratio and strength. The measurements of permeabilities are therefore suggested.

2. Quantitative rheological measurements for pumpability is recommended.

7 ACKNOWLEDGEMENTS

The support for this study in part by Kembla Coal and Coke Pty. Ltd. and in part by grants from The University of Wollongong is gratefully acknowledged. The authors are indebted to Dr. R.D. Lama, Manager, Technology and Development, Kembla Coal and Coke Pty. Ltd. for his helpful suggestions in the course of this work. Any opinions expressed in the paper however do not necessarily reflect those of others.

8 REFERENCES

AS 1012.13/Amdt 1/1986-12-05, Method for the Determination of Drying Shrinkage of Concrete, Australia.

Buddery, P.S. 1984. Grout Pack Strength Characteristics in Longwall Gate Roadways, Mining Science and Technology, Vol. 1, pp. 165-172.

Clark, C.A. and Newson, S.R. 1985. A Review of Monolithic Pumped Packing Systems, Mining Engineer, Vol. 144, No. 282, March, pp. 491-495.

Hebblewhite, B. 1983. Alternative Longwall Panel Layouts for Use in Australian Coal Mines, Paper 3, Proceedings of Longwall Mining Design, Development and Extraction Seminar, Sydney, NSW, Australia.

Hii, J.K. and Aziz, N.I. 1986a. The State of the Art of Pump Packing in Coal Mining, Research Report No.1, Department of Civil and Mining Engineering, The University of Wollongong, N.S.W., Australia, May, 76 Pages.

Hii, J.K. and Aziz, N.I. 1986b. An Evaluation of the Mechanical Properties of Cemented West Cliff Washery Refuse, Research Report No.2, Department of Civil and Mining Engineering, The University of Wollongong, N.S.W., Australia, September, 30 Pages.

Hii, J.K. and Aziz, N.I. 1989. Influence of Water Content on the Mechanical Properties of CCWR Concrete, First Conference on Concrete and Structures, Kuala Lumpur, Malaysia, 3rd and 4th October, pp.47-50.

Hii, J.K., Aziz, N.I., Zhang, S. and Wu, Y.H. 1990. Influence of Geometry on Strength and Elasticity of Cement and Coal Washery Refuse Material Models, 3rd International Symposium on the Reclamation, Treatment and Utilization of Coal Mining Wastes, The Kelvin Conference Centre, Wolfson Hall, The University of Glasgow, UK, 3rd-7th September, in Press.

Hughson, R., Tyminski, A. and Holla, L. 1987. A Review of Stowing and Packing Practices in Coal Mining, The AusIMM Bulletin and Proceedings, Australia, Vol. 292, No.9, December, pp. 79-86.

Lama, R.D. 1988. Developments in Underground Coal Mining Technology and Their Implications, The AusIMM Illawarra Branch, 21st Century Higher Production Coal Mining Systems, Their Implications, Wollongong, N.S.W., Australia, April, pp. 7-17.

Marshall, P. and Lama, R.D. 1986. Changes in Underground Coal Mining Technology - An Australian Outlook, 13th Congress of the Council of Mining and Metallurgical Institutions, Singapore, 11th-16th May, pp. 91-101.

NSW Government. 1983. Coal Reject Disposal in the Southern Coalfields, Coal Reject Disposal Sub-Committee, Report to the Coal Resource Development Committee (NSW Department of Mineral Resources: Sydney), May, pp. 1-47.

Richmond, A.J. 1981. Investigation into Monolithic Pump Packing as a Means of Underground Roof Support, Part A: Laboratory Studies to Determine Pack Properties, P.R. 81-8, ACIRL, August, 34 Pages.

Richmond, A.J., Skybey, G., Ross, A., Wypych, P. and Lama, R.D. 1985. Evaluation of Cementitious Support for Increased Recovery of Coal Mines, Progress Report, NERDDP Project Contract No. 84/4149.

Sinha, K.M. 1989. Hydraulic Stowing - A Solution

for Subsidence Due to Underground Mining in the USA, 30th U.S. Symposium on Rock Mechanics, West Virginia University, Morgantown, West Virginia, U.S.A. 19th-22nd June, pp. 827-834.

Smart, B.D.G., Isaac, A.K. and Roberts, D. 1982. Pack Design Criteria at Betws Colliery, Mining Engineer, Vol. 141, No. 230, July, pp. 15-21.

Whittaker, B.N. and Woodrow, G.J.M. 1977. Design Loads for Gateside Packs and Support Systems, The Mining Engineer, Vol. 136, No. 189, February, pp. 263-275.

Thomas, E.G. 1986. Report on Laboratory Strength Testing of West Cliff Washplant Reject, School of Mining Engineering, the University of New South Wales, N.S.W., Australia, pp. 1-40.

Reclamation, Treatment and Utilization of Coal Mining Wastes, Rainbow (ed.) © 1990 Balkema, Rotterdam. ISBN 90 6191 154 0

Study of the possibility of application of Haldex-Plant discards in civil engineering

P. Michalski & K. M. Skarżyńska
Department of Soil Mechanics and Earth Structures, University of Agriculture, Kraków, Poland

ABSTRACT: The aim of presented investigations and studies was to consider the possibility of the use of Haldex wastes for civil engineering purposes. The full range of geotechnical properties was analysed in the reference to particular uses being considered. The applicability of the material for river embankments, road construction, and as a subsoil was revealed.

1 INTRODUCTION

The activity of the Polish-Hungarian Mine Company "Haldex", having six plants within the Upper Silesian Coal Field, which is based on the extraction of coal from mine heaps, gives rise to large quantities of secondary wastes in the form of decarbonized shales /Herniczek and Laszlo 1989/. Since the area for depositing this material is strongly limited and because of necessary environment protection in the highly urbanized and industrialized Upper Silesian Industrial Region, the problem of the utilization of this material requires a prompt solution.

One of the main possible uses of large quantities of waste materials without any need for using additional preparation process is earthwork construction in a broad sense. Large earthworks are being at present conducted in the Upper Silesian Industrial Area and will continue to be constructed in future. Because of mining subsidences there is a need for comprehensive regulation of surface running waters, reconstruction of entire transportation system, and for filling sink holes and borrowpits to level the land surface. These works produce a demand for considerable amounts of soil materials which cannot be met by natural mineral soils because of serious deficit of this material in the area. Therefore, the use of industrial wastes, including those from Haldex plants, may bring major technical and economical benefits.

2 GENERAL PRINCIPLES FOR THE APPLICATION OF WASTE MATERIAL FOR EARTHWORK CONSTRUCTION

The application of materials of this type for earthwork construction requires comprehensive research aimed at determining the properties of these materials which are important from the earthwork construction need aspect and the development of principles for the erection of specific engineering structures.

The current standards and recommendations relating to earthwork construction do not take into account the possibility of using mining industry waste for this purpose. Thus the use of Haldex waste for engineering earthworks requires determination of the geotechnical properties of this material and any forecast changes in these features occurring after the material has been built into a structure as a result of external factors.

The Haldex wastes to be used for civil engineering purposes can

469

be generally considered as:
- fill material for construction of all types of hydraulic embankments,
- fill material for construction of road and railway embankments,
- material for subgrade and base course of roads, without or with admixtures of fuel ash or cement,
- subsoil for foundation purposes /when used for land-level works/.

For all types of considered applications /mentioned above/ the basic geotechnical properties should be determined, as listed below:
- grain size distribution
- natural moisture content
- optimum moisture content and maximum dry density
- permeability
- shear strength
- changes of these properties resulting from weathering

When the material is to be used for hydraulic embankment construction, the following extra tests ought to be carried out:
- slaking test
- freezing test

For the use of this material as a fill in road and railway embankment construction all mentioned above properties and additional features should be defined as follows:
- sand index
- passive capillary rise
- plasticity index
- compressibility

When material is to be used for the base course of roads, special properties should be determined, as below:
- californian bearing ratio /CBR/
- compression strength of mixtures with cement or fuel ash.

For foundation purposes the basic geotechnical properties and compressibility of the material considered as a subsoil should be defined.

Another important consideration when waste is used for earthwork construction purposes is specification of the optimum process for constructing a structure from the given material. This process should incorporate the principles for erection of embankments and a method of compacting the material. Correct compaction of an embankment, which governs the subsequent safe use of the structure, is the most important but at the same time the most difficult task in earthwork construction.

3 GEOTECHNICAL PROPERTIES OF HALDEX WASTES

The main aim of this paper is to present the detailed values of geotechnical properties of Haldex wastes obtained from the test carried out by the Department of Soil Mechanics and Earth Structures of the Agriculture University in Kraków /Skarżyńska and Michalski 1989, Skarżyńska et al. 1987, 1989, Michalski and Zawisza 1990/ and by the Road and Bridge Research Institute in Warsaw /Wileński 1987/.

Four kinds of material coming from three Haldex Works /"Makoszowy" fresh and weathered, "Brzezinka" and "Szombierki"/ were tested and obtained results were analized in the aspect of the use of this material for civil engineering purposes.

3.1 Basic geotechnical properties

Results of the tests conducted on the material from three Haldex plants, according to methods given by Polish Standards with small adaptations, are shown in Table 1 /Skarżyńska and Michalski 1989, Wileński 1987, Michalski and Zawisza 1990/.

The properties given in Table 1 are dependent moreover on the weathering and compaction of the material.

Weathering influences the grain-size distribution, presented on the example of wastes from Makoszowy Plant. It is visible that weathering results in an increase of fine fraction content and a decrease of d_{10} and d_{60} diameters, what gives the rise of uniformity coefficient. The change of grading causes an increase of natural moisture content, optimum moisture content and cohesion, however a decrease of maximum dry density value, the coefficient of permeability and the angle of internal froction. It is very similar to the material coming directly from coal mine /Skarżyńska et al. 1987/.

Table 1. Basic geotechnical properties of the material from Haldex Works.

Geotechnical property	"Makoszowy"		"Brzezinka"	"Szombierki"
	fresh	weathered	fresh	fresh
Grading: fraction content: $[\%]$				
- cobbles $/\phi > 40$ mm$/$	5	3	13	-
- gravel	73	75	72	82
- sand	10	10.5	10	15
- silt + clay	8	11.5	5	3
d_{10} $[$mm$]$	0.5	0.15	0.60	0.90
d_{60} $[$mm$]$	13	11	26	20
U.C. $[-]$	26	73	43	22
Natural moisture content Wn $[\%]$	6.3	7.3	8.3	-
Opt.moist.content W_{opt} $[\%]$ Max.dry density ρ_{ds} $[$Mg/m$^3]$ $/2.5$ kg rammer method$/$	9.15 1.995	11.10 1.880	10.0 1.99	8.4 2.03
Coefficient of permeability at R.C. = 95% k_{10} $[$m/s$]$	2×10^{-4}	6×10^{-5}	-	1×10^{-5}
Shear strength parameters at R.C. = 95%				
ϕ $[°]$	39.5	37.5	-	45
C $[$kPa$]$	30	50	-	20

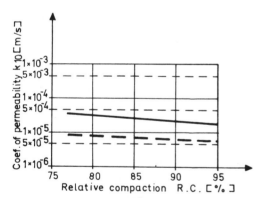

Fig.1 Relationship between coefficient of permeability k_{10} and compaction of material R.C. /material from Makoszowy Plant/. ——— fresh material, ———— weathered material

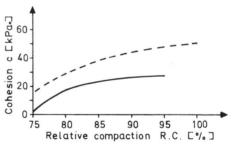

Fig.2 Relationship between cohesion c and compaction of material R.C. /material from Makoszowy Plant/. ——— fresh material, ———— weathered material

Compaction influences the permeability and cohesion as it can be seen in Fig.1 and Fig.2, respectively. Higher compaction causes the decrease of permeability coefficient and the increase of the cohesion.

3.2 Susceptibility to slaking and freezing

Slaking test was performed on the material from Makoszowy Plant. The samples of tested material were immersed in tap water and kept in it for 30 days. The grain size distribution was obtained before and after the test /Skarżyńska and Michalski 1989/. The results of this test are displayed in Table 2.

Table 2. Results of slaking test

Grading	Before test	After test
Fraction content *[%]*		
- cobbles	44	26
- gravel	45	60
- sand	7	9
- silt + clay	4	5
d_{10} *[mm]*	1.50	0.70
d_{60} *[mm]*	45	28
U.C. *[-]*	30	40

It can be seen from Table 2 that slaking causes mainly a disintegration of coarse fraction, diminishing d_{10} and d_{60} diameters and increasing the uniformity coefficient.

Freezing test was performed on the material coming from Szombierki Plant according to the methods given in Polish Standards /Wileński 1987/. Changes of grading after 25 cycles of freezing and thawing are presented in Table 3. It can be seen very clearly that freezing process causes considerable increase of fine fraction content, the decrease of d_{10} and d_{60}, and the rise of uniformity coefficient.

Table 3. Results of frost susceptibility tests

Grading	Before	After 25 freezing cycles
Fraction content *[%]*		
- cobbles	-	-
- gravel	85	65
- sand	13	27
- silt + clay	2	8
d_{10} *[mm]*	22	7
d_{60} *[mm]*	1	0.2
U.C. *[-]*	22	35

3.3 Sand index, plasticity index and passive capillary rise

The tests for obtaining the sand index, plasticity index and passive capillary rise were performed on the material from Szombierki Plant according to Polish Standards /Wileński 1987/. Results obtained are as follows:
- sand index $W_p > 30$
- plasticity index $I_p < 1\%$
- passive capillary rise $H_{kb} < 0.8$ m
These data allow to classify waste material as frost non-heave soil.

3.4 Californian bearing ration, compression strength

Californian bearing ratio and compression strength tests were performed on the material from Szombierki Plant according to Polish Standards /Wileński 1987/.

CBR tests were carried out on the material compacted up to R.C.= 95-96% at moisture content of W = 6-7%. The CBR values obtained immediately after compaction and after six day water saturation were in a range of 25-30% and no swelling of the material when saturated was observed.

The mixture of Haldex waste with 4% addition of fuel ash from brown coal was also tested. The CBR values, obtained immediately after compaction, were in a range of 39-45% and after 14 days of maturing /including six last days of the full saturation/ rose considerably exceeding 100%. These values meet the required specification for subgrade and base courses.

Compression strength tests were carried out on the material from Szombierki Plant mixed with fuel ash from brown coal or with cement /Wileński 1987/. Samples ⌀ 8 cm prepared from material with admixtures of 4, 8, 12% of fuel ash and 4, 6, 8% of cement were tested. Satisfactory compression strength for lower and upper layers of base courses was found to be for mixtures with cement only at sufficient cement percentage of 6%.

3.5 Compressibility tests

Compressibility tests were carried out on the material from Makoszowy

and Brzezinka Plants at natural
grain size distribution in a large
model box. Since these tests were
performed with the use of non-
standard methods and apparatus the
more explanations are needed
regarding the equipment and test
procedure applied /Skarżyńska et
al. 1989, Michalski and Zawisza
1990/.

Test equipment

Test equipment consisted of the
model box of inside dimensions:
width 50 cm, length 100 cm and
depth 90 cm, water control facility,
dial gauges to record surface displacements, and loading plates
with mechanical lever loading
system. Two kinds of loading
plates were applied; a big one
covering the whole upper surface
of the material in the box and a
smaller one of the dimensions 15x50
cm being placed in the centre of
the box perpendicularly to its
length. The loading system enabled
maximum stress of 50 kPa with the
use of the big plate and 300 kPa
with the use of the small one to be
applied. The time of maintaining
the load was unlimited and the
loaded material could be saturated
at each time with the use of
special sand filters.

Material tested and test procedure

The compressibility tests were
performed on two materials differently compacted. Material from
Brzezinka plant had grain size
distribution as given in Table 1
and relative compaction of 98%.
Material from Makoszowy Plant
tested in the model box was
compacted up to R.C. = 91% and had
the grading slightly different from
that given in Table 1, as below:
fraction content: /%/
- cobbles   27
- gravel   59
- sand   8
- silt and clay   6
d_{10} /mm/   0.7
d_{60} /mm/   20.0
U.C. /-/   28.6

Compressibility tests were
performed in two stages. In the
first stage all upper surface of

the material tested was loaded with
the big plate and the load was
increased step by step up to the
value σ_v of 50 kPa with the loading stress and displacements being
permanently recorded. In this
stage of the test the model box
worked as if a large-scale oedometer, therefore on the base of
the results obtained from the test
the modulus of compressibility M_o
was computed according to the
formula:

$$M_o = \frac{\Delta\sigma \cdot H}{\Delta h} \quad /kPa/$$

where:
$\Delta\sigma$ - increase of load /50 kPa/
H - active thickness of the
    material tested /48 cm/
Δh - average value of settlement
    of the plate /cm/
In the second stage the material
tested was unloaded and then
reloaded with the small plate. The
load was increased from σ_v = 0 to
σ_v = 300 kPa, and at σ_v = 50 kPa
and σ_v = 300 kPa the full stabilization of settlement was achieved
with stabilization time being
measured. As the result of this
test stage, the secondary modulus
of deformation E in the stress
range of σ_v = 0-50 kPa was computed
and for the stress range of σ_v =
50-300 kPa the primary modulus of
deformation E_o could be obtained,
both according to the below formula:

$$E /E_o/ = \frac{\Delta\sigma \cdot H}{\Delta h} \quad /kPa/$$

where:
$\Delta\sigma$ - adequate increase of load
    /kPa/
H - as above
Δh - adequate average settlement
    of the plate /cm/
The moduli of deformation E_{0-50}
kPa and $E_{050-300}$ kPa were recalcu-
lated to the moduli of compressibility M and M_o, respectively, by the
formulas as below:

$$M = \frac{E}{\delta} \qquad M_o = \frac{E_o}{\delta}$$

where: δ - correction factor = 0.9
/given by Polish Standards/
In the first stage of loading
/σ_v = 0-50 kPa/ in both materials
the moisture content of w = 7% was
maintained, then in the second
stage both materials were saturated
up to w = 10% /equal to W_{opt}/.
The material from Makoszowy
Plant was saturated at σ_v = 50 kPa

473

Table 4. Results of compressibility tests

Source of material tested	Compaction R.C. [%]	Moist. content W [%]	Modulus of compressibility for σ_v=0-50 kPa Mo [MPa]	Time of settlement stabilization at σ_v = 50 kPa t [days]	Modulus of deformation for σ_v=50-300 kPa E [MPa]	Modulus of compressibility for σ_v=50-300 kPa Mo [MPa]	Time of settlement stabilization at σ_v = 300 kPa t [days]	Secondary modulus of deformation for σ_v = 0-50 kPa E [MPa]	Secondary modulus of compress. for σ_v = 0-50 kPa M [MPa]
Haldex Brzezinka Plant	98	7	6.07	6	44.49	49.43	8	40.68	45.20
		10	-	-	27.54	30.60	70	-	-
Haldex Makoszowy Plant	91	7.6	5.86	46	-	-	-	12.01	13.34
		10	-	-	6.68	7.42	12 months	-	-

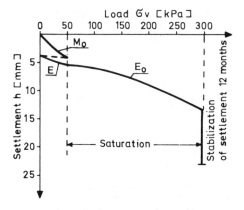

Fig.3 Compressibility curves of
the material from Makoszowy Plant

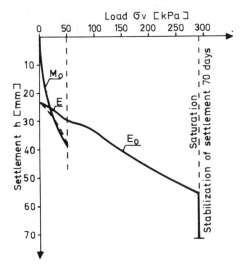

Fig.4 Compressibility curves of
the material from Brzezinka Plant

and this load was maintained for a
number of days - no collapsible
effect or an extra settlement was
observed. Then the load was
increased up to the value of $\mathfrak{S}_v$ =
300 kPa and this stress was kept
up to the full stabilization of
settlement at w = 10%.

The material from Brzezinka
Plant was saturated after the
stabilization of settlement at
$\mathfrak{S}_v$ = 300 kPa. After saturation the
vertical stress was continued when
the full stabilization of an extra
settlement has been achieved.

Test results and analysis

The compressibility curves are
displayed in Figs 3 and 4 and all
results of compressibility measurements and calculations are
presented in Table 4. Since the
origin and grading of these two
materials tested are very close
the results obtained are comparable and can be analysed. From the
data obtained it is very clearly
visible that compaction and
moisture content have considerable
influence on the compressibility
of the material however non-uniform
within the whole range of stresses
applied.

Within the stress range of $\mathfrak{S}_v$ =
0-50 kPa at the moisture content
of w = 7% the primary modulus of
compressibility M_o is the same for
both materials.

In the same stress range the
secondary modulus of compressibility M is 3.4 times higher for more
compacted /R.C. = 98%/ material
than for less /R.C. = 91%/ one.

Within the stress range of $\mathfrak{S}_v$ =
50-300 kPa and at the moisture
content of w = 10% the primary
modulus M_o is 4.1 times higher for
the material better compacted. The
settlement stabilization time at
$\mathfrak{S}_v$ = 50 kPa is 8 times and at $\mathfrak{S}_v$ =
300 kPa 6 times shorter for the
better compacted material than for
the looser one.

Influence of moisture content
on the compressibility of Haldex
wastes depends on the values of
vertical stresses applied. Thus at
$\mathfrak{S}_v$ = 50 kPa after extra saturation
any collapsible or extra settlement
was not observed in the material
less compacted, however, at $\mathfrak{S}_v$ =
300 kPa the increase of moisture
content up to w = 10% caused a
decrease of M_o value about 38%
with quasi-collapsible effect in
the material more compacted.

4. SUMMARY

Results of investigations presented
above can be summarized as follows:

The wastes from Haldex plants
constitute decarbonized shales with
very close grading of the material
coming from particular plants.

From the geotechnical point of
view the waste material has inter-

mediate properties between cohesive and non-cohesive soils with the predominance of gravel fraction /72-82%/ and low content of clay fraction /about 2%/.

Shear strength is relatively high; the angle of internal friction was found to be of $\phi = 40\text{-}45^{\circ}$ and the small cohesion of $c = 20\text{-}30$ kPa at R.C. = 95%.

The Haldex wastes are medium water permeable; coefficient of permeability was found to be of $k_{10} = 10^{-5}$ to 10^{-4} m/s at R.C. = 95%. Increase of compaction can diminish k_{10} about a range of values.

The material from Haldex plants is suceptible to slaking, freezing and weathering processes what results in the disintegration of material. The effects of this disintegration are the increase of cohesion and decrease of the angle of internal friction and permeability.

Values of the parameters obtained as sand index, passive capillary rise and plasticity index allow to qualify Haldex wastes for frost non-heave soil group.

CBR values obtained for the material compacted up to R.C. = 95 -96% were of 25-30% and for the material with 4% admixture of fuel ash from brown coal were considerably higher /39-45% immediately after compaction and over 100% after 14 days of maturing/.

Mixtures with cement /4-6%/ produce quite satisfactory compresion strength what allows to use them for base course.

Compressibility of Haldex wastes is realatively great and depends strongly on compaction and moisture content, however at the stress range exceeding $\sigma_v = 50$ kPa.
- Below $\sigma_v = 50$ kPa no influence of these factors on compressibility was found and any collapsible effect when material extra saturated was not observed. Modulus of compressibility M_0 is very low thus compressibility of the material is very high.
- Within the stress range of $\sigma_v = 50\text{-}300$ kPa values of the modulus M_0 were found to be of 50 MPa for the material more compacted /R.C. = 98%/ and they were much higher than for the material less compacted /R.C. = 91%/. These values are much

lower than in natural gravely soils.
- Extra saturation of the material at $\sigma_v = 300$ kPa produce the quasi-collapsible effect and reduction of M_0 value about 38%.

Haldex wastes do not contain pure coal therefore the risk of spontaneous ignition and combustion is very small.

5. CONCLUSIONS

From the test results and analyses displayed above the following general conclusions can be drawn:
- Waste materials from Haldex plants can be used for any hydraulic embankment construction but some extra technical protection against the seepage should be applied. Slaking, freezing and weathering processes reduce the permeability but do not affect the slope stability of embankments.
- Haldex wastes can be applied for road embankment construction with light and medium traffic under the condition that compaction of R.C. = 100% will be achieved and the embankment will be prevented an extra saturation.
- This waste material is applicable for road subgrades and with admixtures of cement can be used for base courses.
- When the material is to be used as a subsoil, light and non-susceptible to differential settlements buildings only can be founded and any extra saturation should be prevented.

REFERENCES

Herniczek, B. and M.Laszlo 1989. Działalność Polsko-Węgierskiej Górniczej Spółki Akcyjnej HALDEX w latach 1959-1989 /Activity of Polish-Hungarian Mine Company HALDEX in the years 1959-1989/. Mat.Symp. nt. Ekologiczne aspekty działalności Haldexu, Katowice.
Michalski, P. and E.Zawisza 1990. Interpretacja badań modelowych osiadań odpadów z zakładów prze-przeróbczych "Haldex" /Interpretation of compressibility model test of wastes from Haldex plants/. Dep. of Soil Mechanics

and Earthworks, Agricultural
Univ., Kraków, Poland, Ms.

Skarżyńska, K.M., H.Burda, E.
Kozielska-Sroka, P.Michalski
1987. Laboratory and site investigations on weathering of coal
mining wastes as a fill material
in earth structures. Proc. of 2nd
Int.Conf. on Reclam., Treatm. and
Utiliz. of Coal Mining Wastes,
Nottingham, England: 179-195.

Skarżyńska, K.M. and P.Michalski
1989. Zastosowanie odpadów popro-
dukcyjnych zakładów Polsko-
Węgierskiej Spółki Akcyjnej
HALDEX do ziemnych robót inży-
nierskich /The use of post-
production wastes from Polish-
Hungarian Mine Coop. "HALDEX"
works for earthworks/. Budown.
Węgl. Projekt-Problemy 11:20-24.

Skarżyńska, K.M., E.Zawisza. J.
Kurleto 1989. Badania osiadań
nawodnionych odpadów kopalni
węgla kamiennego pod wpływem
obciążenia /Study of the settlements of saturated coal mining
wastes under load/. Przegl.Górn.
6:17-19.

Wileński, P. 1987. Przydatność
łupka odwęglonego z zakładów
przeróbczych Haldex-Szombierki
do budowy nasypów komunikacyjnych
oraz warstw nośnych nawierzchni
drogowych /Applicability of
decarbonized shales from Haldex-
Szombierki works for road
embankment construction and for
base course of road pavement/.
Prace Inst.Badaw. Dróg i Mostów
3:95-109.

Reclamation, Treatment and Utilization of Coal Mining Wastes, Rainbow (ed.) © 1990 Balkema, Rotterdam. ISBN 90 6191 154 0

The obtainment of aluminiun compounds from coal mining wastes

Rubén G.Castaño, Ricardo Llavona & Julio Rodríguez
Area de Química Inorgánica, Universidad de Oviedo, Spain

José G.Cañibano
HUNOSA, Dirección Técnica, Oviedo, Spain

ABSTRACT: The bituminous coal mining wastes of the North of Spain are composed principally of philosilicates, with an aluminiun content of between 20 and 30%. This paper describes trials carried out to recover the aluminiun content of these wastes by leaching with sulphuric acid. The effect of calcination of the coal wastes on aluminiun extraction is studied as the influence of the variables involved in the leaching process. Using tailing fly ash as raw material in a fluidised bed, a recovery rate of more than 90% has been achieved.

1 INTRODUCTION

Extracting industries in general, and mining industries in particular, are among those generating the greatest quantities of wastes. In the bituminous coal mines of the North of Spain, as much waste as coal is extracted (3.8 million tons) (García 1983).

These masses of wastes, except for those used as infill, are stored in spoil heaps, giving rise to serious problems which may be divided into two main groups:

1. Storage: High cost, difficulty in finding suitable sites, large amounts to be stored

2. Environmental: Impact on the landscape, occupation of space suitable for other uses, contamination, etc.

The problems posed by the accumulation of these wastes, together with the fact that they have many useful applications, have prompted numerous studies to determine such uses.

Since the alumina content of coal wastes varies between 20 and 30%, it can be assumed that this alumina may be recovered.

Many studies have been conducted in recent years on the extraction of aluminium from non-bauxitic materials (Hamer 1981a, Crussard 1979, Sohn 1986, Padilla 1985a, Mahi 1985, Nath 1985, Livingstone 1983).

Fruit of this research has been the development of a series of processes, wich may be divided into two main groups:

1. Alkaline processes:

Used with materials having a high aluminium content but less reactive to acids. Among these are such minerals as anorthosite and nepheline (Crussard 1979), as well as the ashes proceeding from the burning of coal at power stations "fly ashes" (Hsieh 1981, Hamer 1981b, Padilla 1985b). They have also been applied to carbonous wastes.

In general, these processes consist in an alkaline fusion which transforms the aluminosilicates into soluble sodium aluminate. The aluminate is then selectively leached, although part of the silex is frequently also used.

2. Acid processes:

These are applicable to aluminosilicates with a low alkaline and alcalinoearth content. Schists and clays fall into this category. The aluminium is leached as Al^{3+} using one of the more important industrial acids, such as sulphuric (Nath 1985), hydrochloric (Bailey 1987) or nitric acid (Livingston 1983).

All the acid processes are essentially analogous, the only differences being the kind of acid used and the technique employed in subsequently purifying of the licor obtained.

In this work, a study on the influence of the variables involved in obtaining licors of aluminium sulphate by leaching tailing wastes using sulphuric acid as the extractant has been made.

2 EXPERIMENTAL METHODS

Although there are various kinds of wastes, this paper deals with tailings, beacuse they have the highest alumina content. Their size is under 1 mm and they have a calorific value of approximately 6,000 kJ/kg.

2.1 Chemical analysis

To verify the chemical determinations of the tailings a preliminary firing at 950°C was performed to destroy the organic matter present, and subsequently an alkaline fusion was conducted using lithium metaborate, analysing the resulting solutions (into Al^{3+}, Fe^{2+}, K^+, Ca^{2+}..) using a Perkin Elmer atomic absorption spectrometer model 200.

2.2 Mineralogical analysis

The semiquantitative mineralogical analysis was carried out using a Philips X-ray diffractometer model PV/1050, following the Shultz method (1960, 1964). Amounts of quartz and clay minerals were determined quantitatively by a study of the relative intensities of the characteristic reflections of the mineral species which make up tailings. The "factors of intensity" determined by Schultz for this kind of materials were used for this purpose.

2.3 Calcination of tailings

A fluidised bed burner was employed to achieve thermic activation of the wastes, that means using thermal treatment to increase their intrinsic reactivity.

This equipment proved suitable for this kind of process, since as well as allowing precise temperature control it permits the energy value contained in the wastes to be recovered.

Work has been conducted on the ashes proceeding from calcination of tailing wastes at three temperatures: 740, 800 and 860 °C.

To determine the degree of activation in the different materials we used a Philips X-ray diffractometer model PV/1050 and a Perkin Elmer infrared absorption spectrometer.

2.4 Leaching of tailing wastes or their ashes.

To determine the reactivity of the different materials: uncalcinated tailing wastes and ashes of tailing wastes burnt at temperatures of 740, 800 and 860 °C in a fluidised bed burner, standardised leaching was performed. Contact time was set at 90 minutes and the sulphuric acid/alumina relationship at four times the theorical value. To allow sufficient data to permit reliable conclusions various experiments were carried out, using extracting solutions of different sulphuric acid concentrations. The four materials were processed simultaneously, each extracting dissolution being at boiling point (20% sulphuric acid at 100°C, 30% at 105°C, 40% at 110°C and 50% at 120%). Lixiviation temperature was stabilised using a thermostatised bath.

To determine the influence of the variables involved in the lixiviation stage: concentration of acid in the extractant dissolution, contact time and the sulphuric acid/alumina relation, ashes burnt in a fluidised bed burner at 740°C were processed, proving to be the most reactive, in the following fashion:

1. Three contact times were used: 60, 120 and 180 minutes.
2. Sulphuric acid/alumina ratios two, four and six times the theoretical value were employed.
3. Extractant dissolutions of different sulphuric acid concentrations were used, operating at temperatures very close to the boiling point (20% at 100°C, 30% at 105°C, 40% at 110°C, 50% at 120°C and 60% at 138°C). The results obtained correspond to the combinations of these values fixed for the variables.

Aluminium was determined after solid/liquid separation using a Perkin Elmer atomic absorptiom spectrometer model 200.

3 RESULTS AND DISCUSSION

3.1 Chemical analysis

The chemical analysis of the coal tailing wastes, and of the calcination product, is detailed in Table I. It can be seen that in the calcination material the alumina content has increased considerably, and that there are significant amounts of iron and potasium.

Table I. Chemical analysis.

Component	%
SiO_2	43,1 (56,0)
Al_2O_3	23,2 (30,2)
Fe_2O_3	4,6 (6,0)
K_2O	2,9 (3,8)
Na_2O	0,3 (0,4)
CaO	2,0 (2,5)
MgO	1,2 (1,5)
TiO_2	1,1 (1,6)
V_2O_5	0,2 (0,3)
F	0,01
$S_{piritico}$	1,0
C	11,4

( ) refered to uncalcinated sample

3.2 Mineralogical analysis

The mineralogical analysis show the material to have an important clayey fraction make up of illite, kaolinite, chlorite, montmorillonite (in very small proportions), interstratificates, etc. The illite/quartz ratio is 8.2 and the clay/quartz and kaolinite/quartz fractions are 15.6 and 3.1 respectively.

3.3 Thermic activation of the material

The purpose of thermic activation is to increase the reactivity of a material, in this case in connection with a process the aim of which is to extract one of this components: aluminium.

Figure 1 shows the decline in reflection intensities characteristic of the different mineral species of which tailings are made up, under the effects of calcination.

This phenomenon is easily explained, as during the calcination process, beside the decomposition of oxosalts (carbonates, nitrates,...) there also occurs the elimination of the water content of the clays in the wastes, generating activated mineral phases (metakaolinite, illitemonohydrate, etc.).

A fluidised bed burner was used for thermal activation as this equipment allows recovery of the energy content of the wastes and gives ashes of high porosity very suitable for the extraction process.

To quantify the degree of activation of the various materials, they were submitted to a standarised extraction procedure so as to determine the reactivity of each (See EXPERIMENTAL METHODS). Figure 2 shows

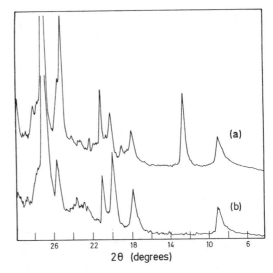

Fig.1 X-ray patterns of the uncalcinated tailing waste(a) and of the ashes of calcinated tailing waste (b).

the results obtained, from which it may be seen that:

1. In unactivated tailing, aluminium yield is low, rising markedly when the row material is calcinated.

2. The ashes obtained at 740°C are those presenting greatest reactivity, and thus those selected to study procedure variables.

The fact that reactivity diminishes in the sequence 740, 800 and 860°C would seem to be due to the fact that, during the calcination process, from about 800°C inert aluminosilicates begin to be formed.

3.4 Lixiviation of tailings ashes calcinated at 740°C

Of the different lixiviation temperatures tried in coal waste combustion, it is at 740°C that the most reactive ashes for the extraction process are obtained, wich is why this type of material has been used in the lixiviation stage.

In the present work, a study has been carried out of the different variables intervening in the lixiviation process, using sulphuric acid as the extracting agent: lixiviation time, lixiviation temperature, acid concentration, the solid/liquid ratio and sulphuric acid/alumina ratio calculated according to the following:

$$Al_2O_3 + 3 H_2SO_4 \longrightarrow Al_2(SO_4)_3 + 3H_2O$$

481

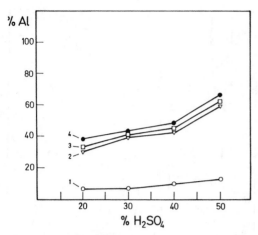

Fig. 2 Reactivity of the different raw materials against the aluminium extraction. Uncalcinated tailing wastes (1), calcinated tailing ashes at 860°C (2) at 800°C (3) and at 760°C (4).

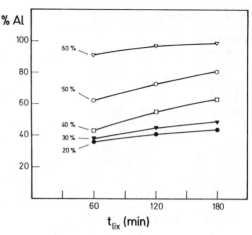

Fig. 3 Influence of the lixiviation time in the aluminium extraction by using tailing ashes calcinated at 760°C as raw materials. The molar ratio sulphuric acid/alumina used is four times the stoichiometric ratio. Sulphuric acid concentrations are specified in the diagram.

Acid concentration and solid/liquid and sulphuric acid/alumina ratios are interrelated variables.

Lixiviation was carried out, for thermodynamic and kinetic reasons, at temperature corresponding to the boiling points of the various acid solutions employed, there being a proportionality between the temperature at which these boiled and the concentration in sulphuric acid of the extractant solution.

As in acid lixiviation, together with the aluminium contained the greater part of the metallic oxides in the wastes are dissolved (Fe_2O_3, K_2O, CaO,..), it was decided, in the anticipation of a higher consumption of sulphuric acid than that which would be required were aluminium alone to be extractec, that stechiometric ratios 2, 4 and 6 times more than the theoretical value should be employed.

Figures 3 and 4 show the results obtained in the experiments carried out, it being observed that:

1. Aluminium recovery increases with acid concentration in the extractant solution, or, equally, with lixiviation temperature, as temperatures close to boiling point were used. These two variables were thus the most important factors in extraction yield.

2. The influence of the ratio sulphuric acid/alumina is moderate, increasing with sulphuric acid concentration in the extractant solution.

3. As regards the influence of lixiviation time, yield increases

significantly as this varies.

4 CONCLUSIONS

1. The use of tailing wastes as a source of aluminium would seem to be promising in view of the high yield obtained in lixiviation and the relative simplicity of the process.

2. The calorific potential of coal tailing wastes in Asturias is 6000 kJ/kg, and this energy can be used to develop the process at an industrial level (lixiviation and pyrohydrolysis of Al_2O_3).

3. The thermal activation phase is essential to obtain acceptable yield in subsequent lixiviation. Of the three temperatures tested 740, 800 and 860°C, it is the ashes burnt at 740°C which show greatest reactivity. The most suitable equipment for carrying out calcination of coal tailings is a fluidised bed burner, as it gives the most appropriate ashes for this process.

4. In lixiviation with sulphuric acid of tailing ashes burnt in a fluidised bed burner at 740°C, the most decisive variables are temperature and the concentration of acid in the extractant solution.

5. Working at boiling point, yields increase according to the concentration of acid in the extractant solution, that is,

482

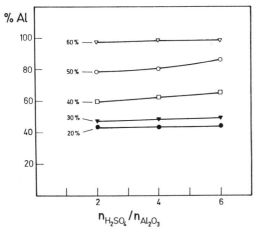

Fig. 4 Influence of the molar ratio sulphuric acid/alumina in the aluminium extraction by using tailing wastes ashes calcinated at 740°C. Lixiviation time was 180 minutes. Sulphuric acid concentrations are specified in the diagram.

they rise with lixiviation temperature, as the boiling point increases with the concentration in sulphuric acid, attaining yields of over 90% when working with sulphuric acid (60%) at 138°C.

6. The influence of contact time is normal for this type of process, while the effect of the sulphuric acid/alumina ratio is slight.

ACKNOWLEDGEMENTS

This work was supported by the FICYT for which grateful acknowledgement is made.

REFERENCES

Bailey, N.T. and Chapman, R.J. 1987. The use of coal spoils as feed materials for aluminium rocovery by acid leaching routes. 5. Hydrometallurgy Vol. 18: 337-350.

Crussard Ch. 1979. La valorisation des minerais alumineux. Annales des Mines 6: 81-92.

García, M.G. and Cañibano J.G. 1983. Valorización de estériles de carbón. La materia prima y sus utilizaciones. Energía 1: 65-72.

Hamer, C. 1981a. Acid extraction of alumina from Canadian nonbauxite sources at canmet. CANMET Rep. 81-2E.

Hamer, C. 1981b. Recovery of alumina from coal rejects. Proc. CIC Coal Symp.,

64tH, Can. Chem. Eng., Otawa, Ont: 546-549.

Hsieh, H.P. 1981. Extracting alumina from anthracite coal waste. Chem. Eng. Prog., Vol 77: 80-82.

Livingston, W.R., Rogers, D.A., Chapman, R.J. and Bailey, N.T. 1983. The use of coal spoils as feed materials for alumina recovery by acid-leaching routes. 1. Hydrometallurgy Vol. 10: 79-96.

Livingston, W.R., Rogers, D.A., Chapman, R.J. and Bailey, N.T. 1985. The use of coal spoils as feed materials for alumina recovery by acid-leaching routes. 3. Hydrometallurgy Vol. 13: 283-291.

Mahi, P. and Bailey, N.T. 1985. The use of coal spoils as feed materials for alumina recovery by acid-leaching routes. 4. Hydrometallurgy Vol. 13: 293-304.

Nath, K.C., Raja, K. and Karmarkar, G.H. 1985. Chemicals from coal wastes. Chem. Eng. World Vol. 20: 80-82.

Padilla, R. and Sohn, H.Y. 1985a. Sodium aluminate leaching and desilication in lime-soda sinter process for alumina from coal wastes. Metall. Trans. B, Vol. 16B: 707-713.

Padilla, R. and Sohn, H.Y. 1985b. Sintering kinetics and alumina yield in lime-soda sinter process for alumina from coal wastes. Metall. Trans. B, Vol. 16B: 385-395.

Sohn, H.Y. and Harbuck, D.D. 1986. Alumina from coal wastes through the formation of aluminium nitride by carbothermal reduction under nitrogen. Ind. Eng. Chem., Prod. Res. Dev., Vol. 25: 367-372.

Reclamation, Treatment and Utilization of Coal Mining Wastes, Rainbow (ed.) © 1990 Balkema, Rotterdam. ISBN 90 6191 154 0

Experimental research on reinforced fly-ash ceramsite concrete members

D.J.Ding
Nanjing Institute of Technology, People's Republic of China

ABSTRACT: In this paper, the author introduces briefly the utilization of coal waste in civil engineering in China, including the publication of the Design Specification. He presents mainly the experimental research on reinforced fly-ash ceramsite concrete members more than 100 tested at NIT. From tests some important conclusions were drawn. On this basis, the corresponding calculations were suggested.

1 GENERAL SITUATION

In China the solid or hollow blocks for wall are sometimes made of cinder as aggregate,or of fine fly-ash as cement.The gangue from coal mine is sometimes used for sintering bricks.The gangue spontaneously combusted in outdoor pile-up is also used as aggregate to produce fully-lightweight concrete. The ceramsite made of clay sintered with fly-ash is used as coarse aggregate with sand to produce light-weight aggregate concrete and reinforced light-weight aggregate concrete for structural members, of which the unit weight is equal to 17-19kN/m^3.

At the beginning of the eighties, a multi-story reinforced concrete frame building in Nanjing, built about 45 years ago, but stopped at the 2nd floor due to war, should be added to 7 stories above ground. If the conventional concrete would be used, then the reinforcements in a half of original columns should be strengthened through the examination and calculation with consideration of aseismic requirements. The author suggested to use fly-ash ceramsite concrete (unit weight=18 kN/m^3),finally,the total added dead load was reduced by 1/4 and only four corner columns should be strengthened slightly. It greatly saved material and quickened the pace of construction. This engineering has been completed several years ago (Ding 1986a).

Since the later stage of the seventies,some universities and research institutes in China have conducted a series of researches on the behaviours of material and structural members of lightweight aggregate concrete. NIT has also carried out the experimental research on 62 reinforced fly-ash ceramsite concrete beams with rectangular, T and inverted T section under short-term loading and that on 6 beams with rect. section made of the same material under long-term loading (Ding 1986b). Based on these research series, the "Design Specification for Reinforced Lightweight Aggregate Concrete Structures" (JGJ12-82) had been published in China in 1982 as the guidance of design.

Later, the author and his colleagues have conducted too the tests on 3 uncracked and 5 cracked reinforced fly-ash ceramsite concrete beams under long-term loading sustained respectively for $2\frac{1}{2}$ years (Ding 1985b) and $2\frac{2}{3}$ years.

At the middle stage of the eighties, 7 master students at NIT under the guidance of the author had finished their thesis tests on studying fly-ash ceramsite concrete members.

2 EXPERIMENTAL RESEARCH ON STIFFNESS AND NORMAL CRACK AND DESIGN PROPOSALS

2.1 Under short-term loading

Measurements showed that the distribution of mean strains along beam depth is in good accordance of plane hypothesis up to the failure moment as it behaves in conventional concrete members (Ding 1989a). Besides, another two conclusions drawn from conventional concrete members (Ding 1989a) hold also, i.e., at the service stage after concrete cracks and before steel yields, the curves of $\varepsilon_s - \varepsilon_{s,mean}$ are parallel to one another and the elastoplastic section modulus coefficient $\zeta = M/\varepsilon_{c,mean} b_w d^2 E_c$ for calculating mean strain $\varepsilon_{c,mean}$ along compressive concrete edge can be taken as a constant disregarding steel ratios, but concerning with section shapes, where ε_s, $\varepsilon_{s,mean}$ are steel strain calculated at cracked section and mean strain measured respectively; b_w, d — web width and effective depth, E_c — elasticity modulus of concrete.

1. Stiffness   According to plane hypothesis of mean strains along depth, the short-term stiffness B_s can be found as follows:

$$B_s = \frac{E_s A_s d^2}{\frac{\psi}{\eta} + \frac{\alpha_E \rho}{\zeta}} = \frac{E_s A_s d^2}{1.15\psi + \frac{0.15+5\alpha_E \rho}{1+2\gamma'}} \quad (1)$$

the formula of nonuniformity coefficient ψ of steel strain can be derived following the parallelism of curves $\varepsilon_s - \varepsilon_{s,mean}$ as follows:

$$\psi = 1.1[1-0.235(1+1.5\gamma_1 +0.4\gamma_1')b_w d^2 f_{ct}/M] \quad (2)$$

and

$$\frac{\alpha_E \rho}{\zeta} = \frac{0.15+5\alpha_E \rho}{1+2\gamma'} \quad (3)$$

where A_s — section area of tensile steel; η — coeff. of internal force arm, $\eta \cong 0.87$; $\alpha_E = E_s/E_c$, E_s — elasticity modulus of steel; ρ — steel

ratio, $\rho = A_s/b_w d$; γ', γ_1', γ_1 — strengthening coeffs.: $\gamma' = (b-b_w)t/b_w d$, $\gamma_1' = (b-b_w)t/b_w h$, $\gamma_1 = (b_t - b_w)t_t/b_w h$, b, t, or h_t, t_t — width and thickness of flange compressive or tensile; h — overall depth of section; f_{ct} — tensile strength of concrete.

2. Crack   Mean width w_{mean} of crack can be determined as follows:

$$w_{mean} = \psi \sigma_s s_{cr}/E_s, \quad \sigma_s = M/A_s 0.87d \quad (4)$$

$$s_{cr} = [7.5+0.03(1+1.5\gamma_1 +0.4\gamma_1')d_b/\rho]\nu \quad (5)$$

where constant 75 in mm; d_b — diameter of bars, in mm; $\nu = 1.0$ for plain bars and $\nu = 0.8$ for deformed bars.

2.2 Under long-term loading

1. For nonprestressed concrete beams, the test showed that the long-term deflection (δ_1) increase of fly-ash ceramsite concrete beams drawn on semi-logarithmic paper of time doesn't appear linear (Fig.1) as conventional concrete beams, and isn't also greater than that of the latter, because the shrinkage of fly-ash ceramsite concrete is larger but the creep is smaller. The mean increase coeff. θ of long-term de-

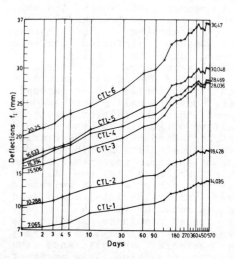

Fig.1 Curves of f_1-lnt

flections from tests = δ_l/δ_s=1.855. Therefore it is proposed to take θ=2.0 as for conventional concrete beams (TJ10-74 1974). Referring the long-term tests of the latter (Ding 1985b), θ should be increased by 20% for inverted T beams and should also be reduced for doubly reinforced beams (Ding 1985b).

For prestressed fly-ash ceramsite concrete beams, the calculation can be done following the author's proposals (Ding 1985a).

2. Crack   From tests, the mean ratio of the max. crack width under long-term loading to that under short-term loading was equal to 1.76, so it is proposed to take 2.0 as for conventional concrete beams (TJ10-74 1974).

3 DIAGONAL CRACK AND SHEARING STRENGTH OF THIN-WEBBED I-BEAMS

Master students Mr. Qiu Hongxing & Mr. Mai Weian have conducted their thesis tests for two topics on 20 thin-webbed I-beams, of which the section is shown in Fig.2.

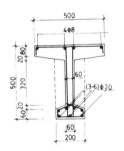

Fig.2 Section of I-beams

In order to get the more precise comparison materials of shearing carrying capacity, one halves of 11 beams were reinforced with 45° inclined stirrups (the tests by another master student Mr. Diao Aiguo under the guidance of the author showed the effects of 45° inclined stirrups were the best among 30°, 45° and 60° inclined ones) (Diao 1985) and the another halves reinforced with vertical stirrups; 6 beams were all reinforced with 45° inclined stirrups and the other 3 beams with vertical stirrups. After

one end was failed under shear, the shear-span ratio of the unfailed end kept unchanged and the support of the failed end moved toward span center, the beam was reloaded so as to obtain the shearing strength under the same shear-span ratio. For this purpose the parts of 1/4 span in the middle of beams were thickened as done at supports and shown in Fig.2 by dot line so as that new supports could be placed there.

3.1 Diagonal Crack

The diagonal cracks were approximately parallel to one another and distributed near uniformly (Fig.3)

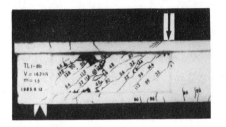

Fig.3 Photo of cracks

and symmetrically between two halves of beams. The widths opened more in the middle and less nearby two ends. The measurements of the stress in stirrups showed this phenomenon (Fig.4).

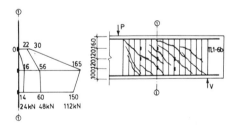

Fig.4 Stress of stirrups in MPa

The most of diagonal cracks extended across the tensile flanges and the most of the critical diagonal cracks didn't extended into the compressive flanges(Fig.3). The widths of diagonal cracks were measured nearby their middle parts with 40-fold reading magnifying glass.

The max. width of diagonal crack is proposed to be calculated as follows:

$$w_{max}^{90}=(1.3s+0.2d_b/\rho_{sti})\,\sigma_{sti,max}/E_{sti} \quad (6)$$

$$w_{max}^{45}=(0.36s+0.06d_b/\rho_{sti})\,\sigma_{sti,max}/E_{sti} \quad (7)$$

where $\sigma_{sti,max}$ — max. stress in stirrups:

$$\sigma_{sti,max}^{90}=0.5\frac{f_{ct}}{\rho_{sti}}+\frac{1.06+0.21m}{\sqrt{m}\,\rho_{sti}b_wd}(V-V_{cr}) \quad (8)$$

$$\sigma_{sti,max}^{45}=0.5\frac{f_{ct}}{\rho_{sti}}+\frac{\sqrt{m}(2.12+0.42m)}{(m+1.2)\rho_{sti}b_wd}\times \\ \times (V-V_{cr}) \quad (9)$$

Shear against diagonally cracking

$$V_{cr}=[1+\sqrt{2}\alpha_E\rho_{sti}(\sin\alpha+\cos\alpha)]\,b_wh_wf_{ct} \quad (10)$$

ρ_{sti} — stirrups ratio$=A_{sti}/b_w\sin\alpha$,
s — horizontal spacing of stirrups,
α — inclination angle of stirrups,
m = a/d — shear-span ratio, h_w — web depth of I-beam.

From the above it can be seen that the max. width of diagonal crack in beams with inclined stirrups is reduced greatly.

Checking the results measured with those calculated shows: mean V_{cr}^t/V_{cr} = 1.00, deviation coeff. C_v=0.272; mean $\sigma_{sti,max}^t/\sigma_{sti,max}$ = 0.99, C_v = 0.197 and mean w_{max}^t/w_{max}=1.02, C_v = 0.220.

3.2 Shearing strength

Fig.5 shows the measured result of longitudinal steel strain in shear span of beam (without stirrups) by means of continuously sticking electronic gauges in two keyways planned along two ribs of steel bar. it can be seen that the strains of longitudinal steel are inconsistent with those calculated according to the

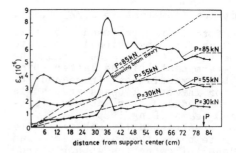

Fig.5 Distribution of strain of steel along its length

normal bending of beam theory without consideration of diagonal bending after cracking diagonally. There also occurs tensile strain of steel with certain magnitude at the support, so a sufficient anchorage length should be required for the steel bars in support.

The tests showed that the shearing strength (expressed in terms of nominal shearing stress τ=V/b_wd) of beam reinforced with 45° stirrups is higher than that of beams with vertical stirrups (Fig.6, where f_c

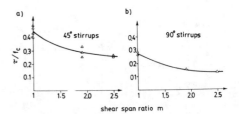

Fig.6 Relationship of $\frac{\tau}{f_c}$ - m

— compressive strength of prism with section of 10x10cm and length equal to 30cm, f_c = 0.85f_{cu} can be taken, f_{cu} — compressive strength of 20cm cube).

According to the analysis, the shearing strength formula can be given as follows:

$$\tau/f_c=\lambda(\cot\alpha+\cot\theta)/(1+\cot^2\theta) \quad (11)$$

where λ — coeff., indicating the ratio of diagonal compression strength f_{cd} as the web fails to the prism strength f_c; θ — inclination angle of diagonally failed section. Cotθ should be determined by trial and error.

Following regression analysis, Mr. Qiu proposed to take 1/λ as follows

$$\frac{1}{\lambda} = 3.44 + \frac{0.7}{\xi} - \frac{2}{m} \qquad (12)$$

where $\xi = \rho_{sti} f_y / f_c$, f_y — yield point of steel.

Comparing with the former materials of NIT (Diao 1985) shows that the shearing strength of fly-ash ceramsite concrete beams is lower than that of conventional concrete beams, because the inclination of crack is smaller so as that its capacity for transferring shear is reduced. The Specification (JGJ12-82) pricribes to reduce by 17%.

For checking the results of 27 ends failed under shear, the mean ratio of the calculated value τ/f_c to the tested one is equal to 0.999, $C_v = 0.236$, among which 13 data of beam halves with vertical stirrups, the mean is equal to 1.045, $C_v = 0.28$, but following the Specification (JGJ12-82), the mean is equal to 1.324, $C_v = 0.262$.

4 CRACKS IN PARTIALLY PRESTRESSED BEAMS

Master student Mr. Tian Zhuping has conducted his thesis test on 10 partically prestressed beams.

In the distance $s_{cr,min}$ from the crack occurred, the bond makes the tensile stress in concrete to reach f_{ct}, where a new crack will form, then the distance $s_{cr,min}$ is the min. spacing of cracks. When the tensile concrete stress between two cracks is equal to σ_{ct}, if σ_{ct} can reach f_{ct} during the increase of loading, then the spacing between these two cracks is the max. one, i.e., $s_{cr,max}$. The mean crack spacing s_{cr} is equal to $(s_{cr,min} + s_{cr,max})/2$ and the mean crack width $w_{mean} = 2s$ (mm), s in the relative slip between steel and concrete in the spacing s_{cr}.

For fly-ash ceramsite concrete, the relationship of bond-slip can be expressed as follows:

$$\tau = 7.11 \times 10^3 s - 3.75 \times 10^6 s^2 + 4.62 \times 10^8 s^3 \qquad (13)$$

Calculation of crack width was finished with computer. The ratio of results measured to those calculated is equal to 1.015. $C_v = 0.170$. For crack width.

At different stages of loading the distribution of crack widths approximates to normal one (Fig.7). According to that as the agreed probability of being over is equal to 5%, the characteristic crack width

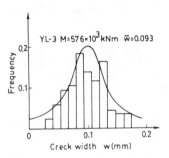

Fig.7 Diagram of crack width distribution

w_k can be taken as follows:

$$w_k = 1.61 w_{mean} \qquad (14)$$

5 REDISTRIBUTION OF INTERNAL FORCES AT SERVICE STAGE AND SHEARING STRENGTH OF CONTINUOUS BEAMS

Master students Mr. Wu Zhishen and Mr. Lu Yong have conducted their thesis tests on four continuous beams with total length equal to 4.0m and each span length — 1.8m, and two simply supported beams for constrast.

5.1 Redistribution of moments

Fig.8 shows the measured results of strain along top steel in the field of 80cm on each side of the internal support in a beam (Ding 1986b). From this figure, it can be seen that the positions of inflection point in

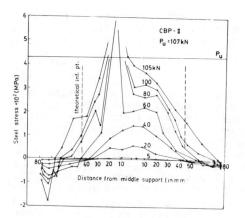

Fig.8 Change of position of inflection points

beam change continuously with the increase of loading.This phenomenon shows that the redistribution of internal forces in beam occurs continuously.

Following the stiffness formula of fly-ash ceramsite concrete beams under service loads and the min. stiffness principle (Ding 1985b), i.e., the stiffness values in the regions with positive and negative moments are assumed to be constant and those at the sections with max. moment in span and min. moment at support are taken respectively,considering the continuity of strains at support it can find out the corresponding support and span moments with micro-computer. The calculated values show good consistence with tested results.

5.2 Shearing strength

Tests showed that the shearing strength of reinforced lightweight aggregate concrete continuous beams, as shown in conventional reinforced concrete continuous beams, is lower than that of simply supported beams with the same generalized shear-span ratio.

According to the method of plastic theory,the calculation formulas of ultimate shearing strength of continuous beams with rectangular section were derived and gave satisfactory calculated results as comparison with tests. These formulas are omitted here due to being too long.

6 SHEAR-WALL

Master students Mr. Ren Zhenhua and Mr. Li Yanfei have finished their thesis tests on 6 specimens of shear wall with edge frame. These walls had height-width ratio equal to 3.14 and were tested with "shear-span" equal to 3.

6.1 Top displacement

Measurements showed that the distribution of mean strains along section depth in shear-walls (deep beams) doesn't appear linear (Fig. 9). In service stage after cracks

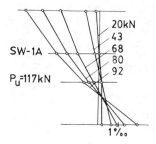

Fig.9 Distribution of mean strains along section depth in shear-wall

occur,the top displacement consists of 3 parts:
1. δ_1 due to the bending deformation in upper region uncracked;
2. δ_2 due to the deformations of the region cracked;
3. δ_3 due to anchorage slip of steel at bottom of shear-wall. Δ can be determined as follows:

$$\Delta = \frac{H^3}{[A_s + 0.3(\frac{\rho_{sti}}{\rho} + \frac{A_{sd}}{A_s})]E_s d^2} \times$$

$$\times [(0.55 + \frac{2\alpha_E \rho}{1+2\gamma'})P - 0.55P_{cr}] +$$

$$+ \frac{19.2d_b H(PH - 0.43Nd)}{(A_s + 0.3A_{sd})E_s d^2} \quad (15)$$

$$P_{cr} = (\frac{N}{A} + 1.75f_{ct})\frac{W}{H} \quad (16)$$

where H — height of wall; A_s, d_b —
section area of concentrated reinforcements in flange (edge frame)
and their diameter; P,N — horizontal and vertical loads; P_{cr} —
horizontal load against cracking;
ρ_{sti} — ratio of horizontal steel
in the field of a diagonal crack;
A_{sd} — section area of vertical distributed steel; W — elastic section modulus, the other notations
have been given formerly. In Eq.(15),
the first term is obtained through
regression after finishing the integration operation with computer,
and the second one is δ_3.

6.2 Ductility coefficient of top displacement

Following the detailed analysis of
the influence factors on the ducti-
lity coefficient μ_Δ, such as vertical load, steel ratios, section
area of flanges, etc., the calculation formula can be given as follows:

$$\mu_\Delta = \frac{1}{0.42 - \dfrac{0.02}{\bar{n}}} \qquad (17)$$

as $n \geqslant \gamma_1'$, $\bar{n} = n$ $\qquad (18)$

as $n < \gamma_1'$, $\bar{n} = (n + 0.25\rho\dfrac{f_y}{f_c} + 0.5\rho_d\dfrac{f_y'}{f_c})/$

$$/(1 + 2\gamma_1') \qquad (19)$$

where $n = N/f_c A$, A — section area of
wall, $\rho = A_s/b_w h$, $\rho_1 = A_{sd}/b_w h$, A_s, A_{sd}
— section areas of steel in flange
and in web, respectively.

Checking gives the mean $\mu_\Delta^t/\mu_\Delta = 0.982$, $C_v = 0.075$.

REFERENCES

TJ10-74 1974. Design code for rein-
forced concrete structures (in
Chinese), p.55-60. Beijing: China
Building Industry Press (CBIP).
JGJ12-82 1982. Design specification
for reinforced lightweight aggregate concrete structure (in
Chinese), 115pp. Beijing: CBIP.
Ding, D.J. et al. 1984. Experimental
research on reinforced and prestressed concrete beams under long
term. loading. Journal of Danish
Society for Structural Science and
Engineering. No.4, 55:91-117.
Diao, A.G., Ding, D.J. et al. 1985.
Behaviours against shear of thin-
web reinforced concrete beams with
inclined stirrups(in Chinese). Industrial Construction. No.1:28-33.
Ding, D.J. et al. 1985. Experimental study and calculation proposal for stiffness of prestressed fly-ash ceramsite concrete
beams under long-term loading (in
Chinese). Industrial Construction.
No.6:8-13.
Ding, D.J. et al. 1985. Stiffness
and crack width of reinforced
concrete members, Journal of Society for Structural Science and
Engineering. No.4, 56:81-125.
Ding, D.J. 1986. Application and
research of lightweight concrete
structures in China. Proceedings
of 1st International Conference
on Lightweight Architectural
Structures (NSW), 711-718.
Ding, D.J. et al. 1986. A study for
stiffness and crack of reinforced
fly-ash ceramsite concrete beams
under short- and long-term loading, Proceedings of International
Symposium on Fundamental Theory
of Reinforced and Prestressed
Concrete (Nanjing). 3:1296-1303.
Ding, D.J. et al.1989. Experimental
research on stiffness and crack
width of reinforced concrete members and proposals for the calculation. Selected Papers from Chinese Journal of Structural Engineering, approved for publication
by the Structural Division of the
ASCE (New York). 216-226.
Ding, D.J. 1989. Calculation for
deflection of continuous reinforced concrete beams in consideration of moment redistribution. Testimonials, 75° compleanno
di Franco Levi (Politecnico di
Torino). 279-283.

Reclamation, Treatment and Utilization of Coal Mining Wastes, Rainbow (ed.) © 1990 Balkema, Rotterdam. ISBN 90 6191 154 0

Fly ash utilization in environmental engineering: The case of Greece

A. I. Zouboulis
Laboratory of Gen. and Inorg. Chemical Technology, Department of Chemistry, Faculty of Natural Sciences, Aristotelian University, Thessaloniki, Greece

R. Tzimou-Tsitouridou
Laboratory of Anal. Chem., Department of Chemical Engineeering, Faculty of Engineering Sciences, Aristotelian University, Thessaloniki, Greece

ABSTRACT: Applications of fly ash, a cheap by-product produced during the combustion of coal mainly for electricity production, in wastewater engineering, were examined for the case of Greece. Particularly, fly ash has been used for the removal of some toxic metals (Pb,Cr,Mn), from their dilute solutions with very good results (removals were near to 100%). Fly ash has been also examined as a possible additive in filtration experiments, for (domestic and industrial) sludge dewatering using a filter press in pilot-scale experiments. Also in this case good results, in terms of specific resistance lowering, moisture of the filter cake lowering, etc, were obtained.

1 INTRODUCTION

Fly ash is obtained as a fine particulate mineral residue, left behind after the majority of the combustibles in the coal are burnt out. In thermal power stations pulverised coal is used and fly ash is obtained as a waste product in large quantities. The ash is recovered with the help of cyclones and/or electrostatic precipitators from the flue gases and collected in hoppers. Fly ash properties result primarily from the type of coal burned, the type of combustion equipment used and the fly ash collection mechanism employed.

With regard to Greece, lignite cambustion in power stations has been grown steadily the last 30 years to reach today a consumption of about 50 million tons per year and projections through 2000, based on the gonerments energy programme, foresee a further increase up to about 60 million tons. Around 15% of this amount is produced as fly ash (Liatis, 1984). This imply the big environmental problem of safe disposal of huge amounts of this waste product. To stress more this point it is sufficient to note, that most power stations have to deal with thousands of tons of fly ash per day.

From the other hand, the increasing demand of raw materials and the limited availability of natural resources gave rise to investigate industrial by-products for their possible reuse. Particular attention was given to the possible reuse of several waste solid materials (like red mud, mineral fines, fly ash, etc.), for environmental engineering purposes.

Especially for the case of fly ash, leaving out the long term economic perspective of recovering valuable metals contained in it, for the present, the most promising ways for large-scale utilization of fly ash, are seen in the civil and industrial engineering, taking advantage of the pozzolanic properties of fly ash and using it as a constituent in concrete and concrete-related products or employing fly ash as a builder in soil or roadway conditioning. But the use of fly ash for all the above mentioned cases does not consume the corresponding production (Torrey, 1978). It has been calculated for the case of Greece, that only about 20% is reused as cement additive, while the rest quantity is disposed (Liatis, 1984).

A wide range of fly ash applications in the field of wastewater treatment has been already proposed, e.g. for treatment of polluted water (Tenney, 1970), for removal of toxic metals, like Cd(Yadava, 1987), Cu (Chu, 1978), B (Hollis, 1988), phenols (Pankajavalli, 1987), for treatment of dyed wastewaters (Gupta, 1988), for abatement of flue gas (Peloso, 1983), for sludge dewatering (Moehle, 1967, Nelson, 1979, Papachristou, 1988, Tenney, 1968), etc.

Being of clay origin and heat treated (heated to 1,500°C and cooled) and aided by properties like porosity, high specific surface and relative chemical inertness to

493

water and dilute solids, fly ash is well-suited for adsorption purposes in aqueous or gas media.

As the composition of fly ash is largely dependent from the origin of the produced coal, generalizations are difficult to be applied, and preliminary experiments must be always carried out to determine the optimum conditions for the specific uses of fly ash.

This paper, which is our contribution to the increased interest, that has appeared the last few years among the greek scientists, taken advantage of the specific properties of fly ash, discusses the use of a waste by-product:

1. To treat other wastes, namely to remove toxic metals (Pb, Cr, Mn), from industrial wastewaters and

2. To dewater sludges, domestic or industrial origin, acting as a filtering aid using a filter press.

Other related published works from greek scientists, in the field of environmental engineering, include the use of fly ash for lead removal from battery-shop effluents (Papachristou, 1983), for phosphate removal (Tzennini, 1986) and for destruction of some pesticides in mixtures with soil (Albanis, 1988).

2 EXPERIMENTAL

Two different fly ash samples, collected according to conventional methods, from power plants at two different regions of Greece (Ptolemais, Megalopolis), were used for our studies.

The minerals in the lignite, which are present as hydrated silica, calcite, quartz, etc, in varying proportions, determine the chemical composition of the fly ash. These minerals during combustion transform to mullite, magnetite, tridymite etc., forming the composites of the fly ash. Although fly ash is a complex, heterogeneous material exhibiting wide variations in physical and chemical properties, average ranges of certain specific properties can nevertheless be cited. Their average composition is given in Table 1 from where the main difference between them seems to be the silica and calcium oxide content.

The color of fly ash is a shade of gray, with the degree of gray coloring depending on the amount and size of the associated particles of carbon. Most fly ash particles approach spherical shapes. Table 2 indicates typical values for some selected physical properties of fly ash.

Table 1. Average composition of main fly ash constituents (% w/w).

Component	Origin	
	Ptolemais	Megalopolis
SiO_2	30	42.5
Al_2O_3	17.5	17.5
Fe_2O_3	6	6
CaO	33	13.5
MgO	3	2
SO_3	6	2
K_2O	1	1.7
Na_2O	0.7	0.5
TiO_2	0.5	0.8
CO_2	2.5	0.3
Cu	0.01	0.04
Cr	0.06	0.04
Mn	0.03	0.05
Ni	0.04	0.04
Zn	0.02	0.01
Ba	0.07	0.12
Loss on ignition	5	2

Table 2. Typical physical properties of fly ash (Ftikos, 1985).

Physical property	Value
Average percent of particle size passing 34 μm sieve	50%
Range of particle size (90%)	0.1-126 μm
Specific gravity	1.9-2.6 g/cm^3
Specific area-Ptolemais (Blaine)	500-600 m^2/kg
Specific area-Megalopolis (Blaine)	250-300 m^2/kg
Alcalinity	773 meq/l

As someone can notice there is a wide range of particle sizes. In Fig. 1 it is given the particle size distribution of Ptolemais fly ash.

The treatment of fly ash samples, in order to find out the sorption efficiency was done according to the following way. Different quantities of fly ash (0.1-2.0 g)

494

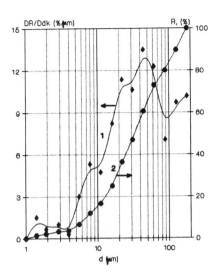

Fig.1. Particle size distribution of Ptolemais fly ash (mean value of four repli-
cates), where (1) additive, (2) differential distribution.

and the metal ion solution, which was prepared from the corresponding sulfate salt,
were placed in centrifuged glass tubes
(total volume 20 ml) and shaken for 30
min. After equilibrium was achieved, the
mixtures were centrifuged for 10 min at
3,000 rpm and the clear uper part of the
mixture was analyzed for the residual metal
ion concentration. The metal analysis was
performed by Atomic Absorption Spectropho-
tometry (A.A.S.). The temperature in all
the experiments was kept constant (25° C).

Desorption experiments were also carried
out for the metals adsorbed. Tap water
was used as washing solution. The mixtures of the washing solution and the obtained sludge were placed in centrifuged
glass tubes (volume 20 ml) and shaken for
60 min. Then the mixtures were centrifuged for 10 min at 3,000 rpm and the possible metal ion desorbed was determined
in the supernatant, using A.A.S.

The filtration experiments, in order to
determine the effectiveness of sludge dewatering using fly ash as a physical additive, were performed using a pilot-scale,
batch operated filter press (Fig.2), mainly consisted from plates and frames placed
alternatively, with a filter cloth between
them. The sludge is pumped between the
plates, filling up the empty space. The
applied pressure was kept in most experiments constant at 220 kN/m^2 (around 30
psi) unless other stated, and the duration
of the experiment was 20-180 min.

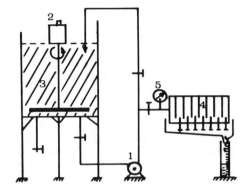

Fig.2. Pilot-scale filter press, used in
filtration experiments, where (1) centrifugal pump, (2) mechanical mixer, (3) feed
tank (volume 100 l), (4) filter press section and (5) manometer.

The used filter cloth, was consisted of
natural cotton fibers, with a total filtration area of 19.3×10^{-2} m^2. The solids
content of the filter cake, was measured
by heating at 105°C up to constant weight
and expressed %. This normally varies in
the range 10-50%. The volatile solids content was then measured by incineration of
the dry cake to 550°C for 2 h.

Two types of sludges, from different o-
rigins were also examined:
1. Thickened secondary industrial sludge,
aerobically digested, from a treatment effluent pland working with the activated
sludge process.
2. Thickened secondary sewage sludge from
trichling filter.

It has to be pointed out, that differences in daily quality of the used sludges,
make necessary the characterization and e-
xamination of all the used sludge samples.

The efficiency of the filter press was
measured in terms of specific resistance
(r), calculated from the basic filtration
equation, which for constant pressure and
after integration, gives:

$$\frac{t}{V} = \frac{\mu W}{2PA^2} V + \frac{\mu R_m}{PA}$$

where   t,V: time (s) and volume (m^3) of
            filtrate respectively
        P : pressure (N/m^2)
        A : filter area (m^2)
        μ : viscosity (poise)
        r : specific resistance of filter
            cake (m/kg)

495

W : mass of dry solids per unit volume of filtrate (kg/m^3)

R$_m$: resistance of filter medium (m^{-1})

By plotting t/V versus V (laboratory data) and if the slope of this line is m, then

$$r = \frac{2PA^2m}{\mu W}$$

Specific resistance can be viewed as the reciprocal of filterability (which in turn can be defined as the ease at which the sludge gives up water), where a high value of resistance means a poor value for filtration and vice versa.

3 RESULTS AND DISCUSSION

Concerning the utilization of Greek fly ash on heavy metals removal from their solutions, the adsorption technique has been found to be a useful mean in our laboratory experiments. The case of Mn and Cr was examined, using fly ash from Ptolemais and the obtained data are given in Figure 3 and Table 3 and 4.

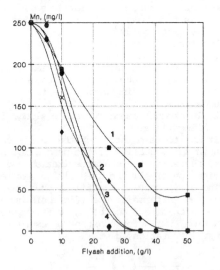

Fig.3. Mn removal using different initial Ptolemais fly ash concentrations at 4 different pH values, with constant initial [Mn] = 250 mg/l, where (1) pH 6, (2) pH 7, (3) pH 8, (4) pH 9.

For the case of Mn, using initial concentrations 250 mg/l, the maximum sorption of the metal was found by using 30 g/l fly ash at pH values 8-9; the given explanation was, that this is due:

1. To the interaction of adsorbate with the insoluble constituents of fly ash like silica, alumina or other products formed after the reaction of fly ash with water and especially those formed because of the considerable amount of existed calcium oxide (up to 40%).

2. To precipitation process of the corresponding low solubility metal hydroxide.

Table 3. Cr sorption results, using different initial Cr concentrations, but keeping constant the used amount of Ptolemais fly ash (75 g/l). The final pH value was regulated with NaOH, to be around 8 while the initial pH was 4-5.

Initial [Cr^{3+}] mg/l	Final [Cr^{3+}] mg/l
500	0
600	0
700	0
800	0
1000	0

Table 4. Cr sorption results using different initial Ptolemais fly ash concentrations, but keeping constant the [Cr] and equal to 500 mg/l; final pH value was keeping constant at 9.

Fly ash addition (g/l)	Final [Cr^{3+}] mg/l
5	0
10	0
25	0
35	0
40	0
50	0
60	0
75	0

For the case of Cr using initial concentration 500 mg/l and for the pH value 9, only 25 g/l was found to be sufficient for a 100% Cr removal. Changing the initial concentrations (500-1000 mg/l) and keeping constant the used fly ash at 7.5 g/l, also 100% Cr removal was achieved for all the examined cases.

In comparison with the above mentioned experiments, relating to lead sorption, it has been found, that using initial concentration of 500 mg/l, the maximum removal of Pb ions was obtained at pH around 8, using 35 g/l (Papachristou, 1983).

Leaching experiments of the used fly ash and the produced sludge with tap water (the results are presented in Table 5), showed that the percentage of the above ions leached was between zero and 0.2%, which means that the examined metal ions have been not-reversibly sorbed an the fly ash, i.e. the disposal of these sludges is environmentally quite safe.

Table 5. Desorption results for Pb, Mn, and Cr samples treated with P-fly ash, using tap water as desorbent. The quantity of fly ash used was 35 g/l for Pb, variate for Mn and 75 g/l for Cr.

Pb

Initial [Pb] mg/l	pH	[Pb] sorbed mg/l	[Pb] desorbed mg/l
250	10.2	248	0
500	10.0	496	0
750	9.0	744	0
1000	8.0	993	0

Mn

Initial [Mn] mg/l	Fly ash used g/l	pH	[Mn] sorbed mg/l	[Mn] desorbed mg/l
250	25	8.0	234	0.22
500	40	7.4	270	0.14
750	40	6.5	290	0.06
1000	50	6.5	350	0.11

Cr

Initial [Cr] mg/l	pH	[Cr] sorbed mg/l	[Cr] desorbed mg/l
500	9.0	500	0.65
600	8.0	600	0.84
700	8.0	700	1.59
800	9.0	800	1.40
1000	8.0	1000	2.05

Further on, the obtained data were used to design a pilot plant, with the aim to evaluate the laboratory-scale obtained results for their possible utilization in industrial scale for wastewater treatment. Comparative studies were also performed using instead of fly ash, only $Ca(OH)_2$ (20 g/l). The results are shown in Table 6 for two different pH, using an actual industrial effluent of initial [Mn] 150 mg/l (pH around 1 to 2). It becomes quite clear, that fly ash seems to act more active than $Ca(OH)_2$ alone, i.e. better results, which means less [Mn] left in su-

pernatant, were obtained even though fly ash was used in proportions 6 times less than $Ca(OH)_2$.

Table 6. Mn removal using fly ash (3.5 g/l) or $Ca(OH)_2$ (20 g/l).

pH	[Mn] left in supernatant (mg/l)	
	Fly ash	$Ca(OH)_2$
8.5	1.5	30.5
10.0	0	1.5

The removal of the anions, fluoride and phosphates, from their solutions was also examined in the same way as described above, using the adsorption technique. For this case the retention capacity was found to be at the level of 80 g of F^-/kg of fly ash and 0.064-0.099 mmol PO_4^{3-}/g of fly ash respectively. For more informations see also the literature (Tsitouridou, 1985 and 1987).

The final conclusion which was drawn about the mechanism of sorption for the last two cases was again that adsorption and precipitation processes are operating concurrently, while the latter one seems to be more probable.

The application of fly ash as an additive, to help sludge dewatering using a filter press, is presented with the following some typical results in Figures 4 and 5.

As it could be easily concluded from Fig. 4, (1,3) the fly ash addition decreases substancially the necessary filtration time to collect a certain amount of filtrate (e.g. 1 l). It has been calculated, that the specific resistance of filter cake (r) is greatly reduced from 1762.1 m/kg to 199.8 after the fly ash addition, thus providing an easier filtration, the resistance of filter cloth (R_m) is also reduced from 88.3 m^{-1} to 43.3 m^{-1}, the solids content of the filter cake is increased from 11.6% to 35.3%, while the volatile solids from 70.9%, increased to 92.6%; the appearance of the cake is dense, dry and textured.

Increasing the applied pressure the specific resistance is reduced, while the filtration rate (kg/m^2h) is increased, as it could be concluded from Fig. 5.

From the relults presended in Fig.6, it could be noticed, that increasing the fly ash content the specific resistance of the filter cake is substancially reduced, while the solids content of the filter cake is steadily increased.

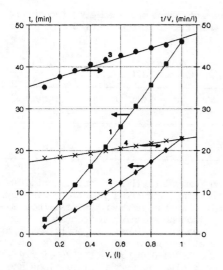

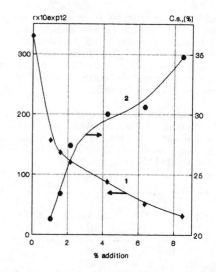

Fig.4. Example of a filter press test run, whithout (1,3) and with 2% w/v Megalopolis fly ash (2,4) using domestic sewage sludge of initially 3% w/v solids content, from which 68.6% were volatile solids.

Fig.6. Effect of increasing Ptolemais fly ash concentration to the specific resistance of filter cake (1) and to the solids content of the filter cake (2). Sewage sludge was used, with 4% w/v initial solids cake content.

The filtrate from the pressure filter is nearly free of suspended solids. At the beginning of each experiment, the filtrate contains a lot of suspended solids (very dirty), but rapidly as the filter cake is formed, the filtrate becomes clear without observable suspended solids; but it has to contain soluble salts from the fly ash (sulphates, etc.), as well as soluble organic matter not sorbed to the fly ash particles (see also Tenney (1968). Because of the low suspended solids content of the filtrate and of his relative small volumes, it can be discharged directly to the entrance of the biological treatment unit.

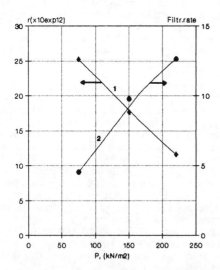

Fig.5. Influence of applied pressure to the specific resistance (1) of filter cake and to the filtration rate (2) using 10% w/v Ptolemais fly ash and industrial sludge of initially 3.65% w/v solids content from which 71.7% were volatile solids.

The sludge pH, as well as the filtrate pH, was increased from around 7 (before fly ash addition) to more alkaline values, which are depending on the fly ash percentage used.

Some general concluding remarks could be drawn down from the filtration experience:

1. Fly ash acts as an effective filter aid. Filtration is improved because fly ash bild up a porous, permeable, and rigid lattice structure which retains solid particles and allows the liquid to pass through. This is true especially for our case where a compressible sludge, filtered without additives, plug the filter cloth.

2. Fly ash could be, at least partially, decrease the necessary chemical conditioning requirements, thus decreasing substancially the sludge dewatering costs.

3. The dewatered sludge has no discremible odor, as the bacterial sectivity of the digested sludge is discoureged, due to

the addition of fly ash oxides, especially of the CaO and to the subsequent pH increase, usually in values more then 10.

4 LITERATURE

Albanis, T.A., Pomonis, P.J.and Sdoukos, A.T. 1988. The influence of fly ash in aqueous soil suspensions on hydrolysis, degradation and adsorption of methyl parathion. Toxicol. and Environm. Chem. 17: 351-62.

Chu, T., Steiner, C. and McEnture, C.L. 1978. Removal of complex copper-ammonia ions from aqueous wastes with fly ash. J. Wat. Poll. Contr. Feder., 50: 2157-74.

Ftikos, Ch., et al 1985. Feasibility studies for the beneficiation of Greek fly ashes. Report, Athens, Techn. Chamber Greece. (Gr.)

Gupta, G.S., Prased G., Panday, K.K. and Singh, V.N. 1988. Removal of chrome dye from aqueous solutions by fly ash. Water, Air and Soil Poll., 37: 13-24.

Hollis, J.F., Keren, R. and Gal, M. 1988. Boron release and sorption by fly ash as affected by pH and particle size. J. Environm. Qual., 17: 181-4.

Liatis, T.S. 1984. Greek fly ashes and the possibilities of their further use. 10th Nat. Chem. Cont., Athens, 5-10 Nov., 751-70, (Gr.).

Moehle, F.W. 1967. Fly ash aids in sludge disposal. Environm. Sci. and Techn., 1: 374-379.

Nelson, R.F. and Brattlof, B.D. 1979. Sludge pressure filtration with fly ash addition. J. Wat. Pol. Contr. Fed., 51: 1024-31.

Pankajavalli, R., Balachandran, T.R. and Shanmugan, T. 1987. Fly ash in effluent treatment. Ind. J. of Environm. Prot., 7: 209-11.

Papachristou, E., Guitonas, A. and Zouboulis, A. 1988. Fly ash employment in conditioning biological sludges for dewatering. 2nd Symp. on Wat. Poll. Contr., IAWPRC, Bangkok, 9-11 Nov., 405-9.

Papachristou, E., Vasilikiotis,G. and Alexiadis, C. 1983. Fly ash use in the treatment of lead containing industrial effluents. 2nd Intern. Symp. on Environm. Poll. and its Impact on Life in the Medit. Reg., MESAEP, Iraklion, Greece, Sept. 6-9, 437-47.

Peloso, A., Rovatti, M. and Ferraiolo, G. 1983. Fly ash as adsorbent material for toluene vapors. Resources and Conserration, 10: 211-20.

Tenney, M.W. and Cole, T.G. 1968. The use of fly ash in conditioning biological sludges for vacumn filtration. J. Wat.

Pol. Contr. Fed., 40: R281-301 (Part 2).

Tenney, M.W. and Echelberger, W.F. 1970.Coal ash utilization in the treatment of pollutedwater. Proc. 2nd Ash Util.Symp., March 10-11, 237-68.

Torrey, S. (ed.). 1978. Fly ash Utilization, Noyes Data Corpor., New Jersey, USA.

Tsitouridou, R. and Georgiou, J. 1988. A contribution to the study of phosphate sorption by three greek fly ashes. Toxicol. and Environm. Chem., 17: 129-38.

Tsitouridou, R., Papachristou, E., Alexiadis C. 1985. Fly ash for defluoridation. Water and Waste Treatm., May: 40, 48-9.

Tzannini, A., Vordonis, L., Koutsoukos, P. and Lykourgiotis, A. 1986. Removing orthophosphates with fly ash. 11th Nat. Chem. Conf., Athens, 2-5 Dec., 103-7 (Gr.).

Yadava, K.P., Tyagi, B.S., Panday, K.K. and Singh, V.N. 1987. Fly ash for the treatment of Cd(II) reach effluents. Environm. Technol. Lett., 8: 225-34.

Reclamation, Treatment and Utilization of Coal Mining Wastes, Rainbow (ed.) © 1990 Balkema, Rotterdam. ISBN 90 6191 154 0

Development of flyash composites for building elements

D. Desai Mahesh
Applied Mechanics Department of SVR College, Surat, India

J. M. Patel
Broach

R. C. P. G. Verma
SVRCET

ABSTRACT: India's major source for power supply is thermal power plants which has produced millions of tonnes of Ash and by 1990, the annual production might reach 12 million tonnes per year. The disposal without pollution of land and air is critical problem.

The need for housing in India is enormous. This would mean demand for building materials like bricks, blocks, panel walls etc has been summarised. The extension work was carried out to evolve a polymer flyash composite (PFC) which can be used for casting building elements and sanitary fittings (substituting Chinawares). The literature is reviewed. Flyash is used as filler phase of composite. The matrix material thermoplastic polyester (a product of petrochemical industries, now growing fast in India) was selected as it has cross links between long chain molecules. As theoretical approach is yet to be evolved, the trial mix empirical approach was adopted and specimens were casted with different proportions for preliminary estimates of engineering properties. Isophthalic grade polyester resin made locally with methyl - ethyl ketone as catalyst, cobalt naphthenate solution as accelerator and styreve as monomer has been selected for study.

The composite PFC specimens were casted for the four different proportions of ingredient and the specimens have been tested for compressive, split tensile, flexural, hardness and durability tests.

The properties of composites are compared with Indian code requirements for floor materials, structural components, repairs jobs, sanitary equipments and pipes.

The work has established the technical feasibility of PFC as an alternative material for housing. Economics will change from time to time and place to place.

1 PRESENT SCENE

India, a fast developing country, has 60 major thermal power stations to generate 734×10^3 MWh (1982) electricity. The need for power is rapidly increasing and present production of 19 million tons of flyash is expected to reach 33 million tons. This fine dust produced is a waste creating land and air pollution.

Most of the cases partial disposal to lagoons and keeping it wet is estimated to cost Rs.300 million every year at current rate of Rs.4/-per MWh power produced.(2) The agricultural lands required to dump ash, estimated at 3.15×10^{-6} hectares/1 MWh/ year, is 2×10^6 hectares/year. The pollution consciousness, enforcement of act, inflation in economy and non availability of waste land for disposal of ash has brought out need for finding alternative systems. The use of ash must be such that instead of spending millions in partial disposal, it must create products which could generate wealth.(8)

2 WASTE MATERIAL

There are two major waste products: ash from coal lignite (Neyveli) in India. The ash is collected and disposed of in most of the plants by wet process. Dry process is rare. The bottom ash and ash from precipitators are mixed with water to form a slurry which is pumped to Lagoons.

The ash collected from electro static precipitators is finer than bottom ash.

First hopper collects coarser fraction with higher carbon content. The fineness of ash in each succeeding stage is increasing with carbon content decreasing. The lagoon sample of ash is a mixture of all ashes, sedimented in irregular stratifications. The material is fine grained non plastic soil for all practical purposes.

3 PHYSICAL & CHEMICAL CHARACTERISTICS

The samples from lagoons and different stations exhibit very wide range of properties but in all the cases of major plants the ash confirms to IS 3812.(4)

The major fraction passing is 45 micron with 20% retained on 75 micron sieve. It has bulk density of 1.2 t/m^3, specific gravity 2.25, Lime reactivity 42 kg/cm^2 (dry ash from height shows 60 kg/cm^2) and 30 days drying shrinkage 0.07%.

Chemically ash excluding from legnite has $S10_2$:45 to 60%, $A1_20_3$: 10-20%, Fe_20_3(4 to 6%); LOI: 0.3 to 15% and small percentage of Calcium, Magnesium, Alkalies, and sulphates. The glass content of Indian flyash is reported to be only 20 to 30% as against 80% in ashes from USA and France. Hence lesser reactivity is expected.

4 BULK USAGES

Desai M.D.(3) has reviewed the extensive work done starting with Comar ('74), the contributions of Gardine M.D.(E'78), Shastri V.R.(8), N.B.O.('85). Recently central fuel research institute ('86), Sharma C.L.(6), Alluwalia S.C.(1) and Sinha H.P.(7) have evaluated bulk use of ash for marking housing elements.

The uses cover technology for producing bricks, blocks, hollow blocks, roofing and flooring tiles, mortar admixture to Cement, replacement of Cement in RCC, fill material for Embankments – highways and dams. The industry has yet to overcome inertia to adopt new technological options.

The main thrust has been to develop bricks-blocks with cement (5-10%) or lime (10-12%) or cement and lime as binder with 0.2% accelerator. The blocks are air dried for 2 hrs and then either cured under water for 10 days or steam cured for 5-8 hrs.

The main justification for the development is estimated shortage of 25,000 million bricks as per the findings of All India Bricks & Tiles Federation. The present market prices for bricks are Rs.700/- per thousand in most of urban growth centres. The poor shape, 15% breakages, high water absorbtion are added disadvantages. The cost of rendering is also high.

Thus industry was forced to search alternatives. The present production of 20 million tons of ash is enough to cover shortage of bricks. The industrial trials by Central Fuel Research Institute, TISCO, and Neyveli plant have produced bricks at marketable rates of 700/- Rs., 200/- Rs. per 1000(2). These blocks do not have most of the disadvantages of bricks cited earlier.

Neyveli project has succeeded in producing insulation bricks for intermediate lining using flyash lime and foaming agent (Alluminium powder) which can stand 800°C. The cost is half of conventional alternatives.

Though uses in small quantities of flyash are reported in canal linings, dams cellular concrete, light weight aggregates, R.C.C. Still percentage utilisation is insignificant. (1)

5 IMPACT ON ECONOMY

Housing sector in India has vast potential for growth.(3) The demand is ever increasing and hence cost escalation is logical consequence. The targets for low law cost rural and urban housing could be attained by flyash bricks and blocks. In continuation of work (Desai 1981-87) further work on blocks and hollow blocks with present cost is shown in (TABLE : 1) photo plate shows solid and hollow flyash blocks.

The use of total flyash will provide savings of Rs.300 millions on disposal cost and save another Rs.300 million or more by cutting down cost of bricks from Rs.700 to Rs.450, at least, per 1000 bricks. These flyash hollow bricks with low moisture absorbtion and better finish, now designed, can save on plaster as well. Economics, at cost of living today, may jusify brick plant as a part of thermal power project.

6 POLYMER-FLYASH COMPOSITE

Technically applications of ceramics and polymers almost unknown until 1900AD, have increased several folds. The advancement in polymer technology is tremendous. The new polymers are suitable to form composite materials. The flyash waste material is studied as filler phase of a composite. To develop compatible phase with flyash the matrix material must have link between long chain molecules of epoxy, polyester. These two are selected under availability and curing at room temperature.

Design Mix for flyash blocks and costing

Sr. No.	Proportion by weight Cement:Flyash:Sand:Grit	Average Com.Stress after 35 days (kg/cm²)	% Water absorbtion	Total Cost Rs./per block*	Remarks
A 1	1 : 2 : 3 : 2	44.2	19.5	5.80	Ukai flyash from pond
A 2	1 : 3 : 2 : 3	37.1	18.9	5.55	"
A 3	1 : 3 : 3 : 2	40.0	13.5	5.42	"
C 1	1 : 3 : 4 : 4	154.6	7.3	4.44	Ukai flyash from precipitator.
C 2	1 : 3 : 4 : 5	130.8	6.9	4.22	"
C 3	1 : 3 : 5 : 5	140.5	10.0	4.00	"

(IS class 50)   (IS limit 20%)
& above.

*   Solid Blocks of 30 cm x 20 cm x 15 cm
(Note standard bricks 19 cm x 9 cm x 4 cm (IS 1077-1975)

TABLE : 2

Curing Time for Polyester Resin (in minutes)

Catalyst	Accelerator			Remark
	1%	2%	3%	
1%	93	51	28	* Methylethyl ketone peroxide
2%	40	23	13	** Cobalt napthenate solution
2.5%	28	20	11	

Trail Mixes by Weight

Composite	Binder System Resin	Monomer by wt of resin	Filler Systems (by wt) F.A. sand, Marble, grit			Binder filler
PFC 1	1	-	4	-	-	1:4
PFC 2	1	0.25	6	-	-	1.4.8
PFC 3	1	0.4	5	5	-	1.7.13
PFC 4	1	0.25	2.5	3	3.5	1.7.2

7 BINDER PHASE

The thermosetting plastic (polyester & epoxy unsaterated resins) which looses reversibility to soften on reheating, isoth-elmic resin has been tested.

8 FILLER

The fillers could be fibres, laminates or powder. The use of Quartz, Chalk, Barites, sawdust have been reported to obtain heat and acid resistant products, plasticizers, dyes, lubricants, catalysers are introduced with filler (cement Research Institute of India, New Delhi (1988)). Use of polyester resin with Benzol peroxide catalyst and styrene as copolymer agent has been studied. The properties achieved are strength, density, corrosion and alkaly resistance, poor heat conductivity.

9 LITERATURE STUDY

The polymer composite have been developed to overcome deficiency of concrete in special applications. Literature survey (9) indicates work done on polymer impregnated concrete (PIC), polymer concrete (PC), Polymer cement concrete (PCC), we will limit to PC only.

Kukacka L.E. et at ('71) has used polymer (MMA) for high strength concrete. Valore R.C. '&Naus D.J. ('75) studied polyester bound aggregate and epoxy bound aggregate systems. Compressive strength of both decrease rapidly with increaing temperature. Kobayashi (775) found strength decreases in PC is linear for range of 20°-100°C. Inoue S ('75) discussed PC using synthetic resins for structural elements. Klocker W ('75) reported about light weight PC using

polyester resins to form new building material. Use of acrl-concrete (MMA) for kerbstone, sanatory articles, staircases etc. has been discussed by Koblischeck ('75). Browne R.D. et..al ('75) have brought out advantages and disadvantages of PC for building industry. Nagaraj ('87) used sugar industries waste furfurol acetone monomers for PC.

10 POLYMER FLYASH COMPOSITE

The size, shape and particle size distribution mechanical interaction with matrix. The surface area, wetting, distribution of aggregates effect adhesion between two phases. Volume changes on curing induces transient internal stresses at particle interface.

The variables for filler phase are sand and flyash. The binding is attained by resin system which is fluid enough for penetration and which polymersies without heat or pressure. Unsaterated polyester (Isothalric resin) provided principal molecular chain and styrene used provides cross link ie. rigidity in three dimensions. For workability of few hours organic catalyst is added. The accelerators are used to initiate polymansation.

For experimental work Isophthelic polyester resin Glefstar India Ltd. having following properties were viscocity at 25[C, poise, sp. gravity = 1.12, setting time could be adjusted by propertioning catalyst and accelerator (TABLE : 2). Cast polyester laminates shows hardness 110 (Rockwell M Scale), heat distortion 70°C, impact (Izod) 0.54 cmkg, sp. heat 0.55 cal g^{-1}, Thermal Conductivity 5 x 10^{-4} Cal Sec^{-1}, Cm^{-1} deg c^{-1} cm^{-2}.

Filler system is flyash from pond and dry collection from hopper from Ukai power plant and fine sand from sea at Dumas. The ash had particle size 0.002 mm to 0.075 mm whereas sand had range 0.06 to 0.2 mm. proportioned amount (TABLE : 3) by weight of resin, accelerator and styrene were mixed in a container. The filler material of calculated proportions is then added until stiff constancy is attained. Catalyst is then added. After thorough blending sand was added gradually until the mixture remained workable and compatible. This blended PFC was used to cast specimens for flexure (40x15x200 mm), compression (30x30mm) 70 mm diameter and 140 mm cylinders were casted for splite tensile strength. The compaction was done by rodding. The test results for four trials are shown in Table 3 and Table 4.

10 APPLICATION OF PFC

As per the composites performance depends on characteristics and proportions of fillers and binders and adhesive forces binding the two phases. The PFC could be designed to obtain required properties for a product.

PFC could be used for floor material, structural components, repair of structures, senatory equipments pipes etc. IS 774 (1970) specifications of absorbtion, impact strength and suitability against chemical attack could be met with by mix 'PFC 4' using marble chips as aggregate with modifications. It can also be used for a seamless flooring.

The structural components subjected to high wear and tear, high bending tensile and compressive stresses and resistance to weathering could be made from PFC without any reservation except costing. Products like washbasins, baths, kitchen platforms, stair cases, doors and window frames, architectural effects could be well planned using PFC.

The sewage or industrial waste drains, sanatory work and pipes (IS 651-1971) could be fabricated using PFC. They will be thin in section and light in weight. The material having also better bond for strengthening old concrete structures.

11 ACKNOWLEDGEMENT

The authors gratefully acknowledge the encouragement given by College authorities in this project. Also thanks to Miss P.N. Mehta who has actively assisted in preparation of paper and Shri R.T. Patel (Lecturer) has been associated with experimental work.

12 CONCLUSION

In addition to introduction of flyash as a basic raw material for hollow or solid building blocks or bricks, ample scope exists in development of Polymer Flyash Concrete (PFC) composite for use in Civil Engineering. This technical feasibility, though may not prove economical for each element at present has signified potentiality to meet challenging assignments. The technology is based on waste to wealth and can be considered as a option available for exploitation.

TABLE : 4

Properties of PFC

Composite	Strength (KG/cm²)			Water absorbtion	Acid resistances loss % by wt.	Hardness (Rockwell Scale)	Unit weight KN/m³
	Compressive	Tensile	Flexural				
P F C 1	728	527	578	0.15	1.6	H 60	16
P F C 2	682	419	550	0.30	1.9	H 55	16.2
P F C 3	852	266	311	0.60	1.6	H 75	19
P F C 4	652	–	207	0.55	2.0	H 82	17.2

REFERENCES

1) Alluwalia S.C.(1988) "Utilisation of
Flyash for manufacture of building
materials - developments prospects and
problems." Proceedings of National
Seminar in Building Material Industries
in Conversion of Waste into Wealth,
Cement Research Institute New Delhi,
Rookee (II : 67)

2) Bose M.C. (1988) "Utilisation of ash
from thermal power station in Building
Industries". Proceedings of National
Seminar on role of Building Material
Industries in conversion of waste into
wealth New Delhi Roorkee (II - 39-80)

3) Desai M.D. and Raijiwala D.B. (1987)
"Building materials from industrial
waste of Coal based power plants".
Proceedings of Reclamation, Treatment
and Utilisation of Coal Mining Waste,
Elsevier science publishers, Amsterdam
(p.p 661-667)

4) IS code of practice 3812 (1981)
"Specification for flyash for use as
pozzolana and mixture" Indian Standard,
new Delhi.

5) Ramkrishna G. (1988) "Utilisation of
Flyash for Building materials".
Proceedings National Seminar on role of
building material industries in
conversion of wastes into wealth New
Delhi Roorkee (I -22-26)

7) Sinha H.P. Ct.. al (1988) "Manufacture
of building bricks from tisco flyash".
Proceedings of National Seminar on role
of building materials New Delhi Rookree
(III 33-37).

8) Shastri V.R. (1978) "Feasibility of
bricks manufacture from Ikai flyash". ME
thesis South Gujarat University, Surat.

9) Varma R.C. (1979) "Polymer flyash
Composite" ME thesis South Gujarat
University, Surat.

Reclamation, Treatment and Utilization of Coal Mining Wastes, Rainbow (ed.) © 1990 Balkema, Rotterdam. ISBN 90 6191 154 0

Extraction of saleable components from Latrobe Valley fly ash/char mixtures

K.A. Laws-Herd
L.H.S. Research Services Pty. Ltd, Mitcham, Vic., Australia

C. Black & D.J. Brockway
State Electricity Commission of Victoria, Australia

ABSTRACT: About 500,000t of solid ash waste is generated from power station combustion of low rank brown coal in Victoria each year. The amount is small compared to other countries, but environmental issues and disposal costs have led to active research into the potential for its large scale utilisation.

Due to its unique composition most of the ash is unacceptable for the more conventional usages of bituminous fly ashes and other similar solid wastes. Recent research, however, has shown that there are opportunities unique to the properties of the material which could yield substantial commercial returns in addition to addressing the disposal issues.

Opportunities have been identified to supply Australian markets with high purity sodium sulphate, which makes up about 11% by weight of the ash as thenardite and with carbonaceous char which comprises about 10% by weight of the total annual solid waste output. Both these components are easily removable by flotation and water leaching. A major opportunity exists for the extraction of high purity magnesium metal. After removal of the char and sodium salts, magnesium oxide comprises about 25% by weight of the remaining residues. The potential for production of precipitated calcium carbonate is also discussed.

The paper presents a resume of research into the extraction of commercially valuable components from Latrobe Valley brown coal solid waste. The economic and technical feasibility of the most prospective process is considered.

1 INTRODUCTION

The Latrobe Valley, situated approximately 150km to the east of Melbourne in Victoria, contains large deposits of low rank brown coal dating from the Tertiary period. The brown coal seams extend continuously for some 70km, from Yallourn to near Sale, and range in width from 10 to 30km. There are a number of major areas where thick coal seams underlie thin layers of overburden. Latrobe Valley brown coals (LVBC) are strongly banded due to variations in the organic matter deposited, resulting in varying coal lithotypes.

Latrobe Valley brown coal is characterised by a high moisture content and low ash production (Brockway and Chalmers 1986). Average moisture content of the coals range between 52% and 67%.

The high moisture content of brown coal is reflected in its low net specific energy values which range from 6.5 to 11.5MJ/kg.

The average ash content ranges between 1% and 4%, on a dry basis.

About 5000MW of the electricity generated in Victoria is derived from the combustion of the LVBC in three main power stations at Yallourn, Hazelwood and Loy Yang. Each is situated next to an open cut mine based on a different seam. Figure 1 shows the locations of the open cuts and power stations in Latrobe Valley and the location of Latrobe Valley in Victoria.

Total solid combustion waste (fly ash, char and bottom ash) amounts to about 500,000 tonnes per annum with increases expected to bring this amount to over 550,000 tonnes/annum by 1994. Disposal of the solid waste using a wet

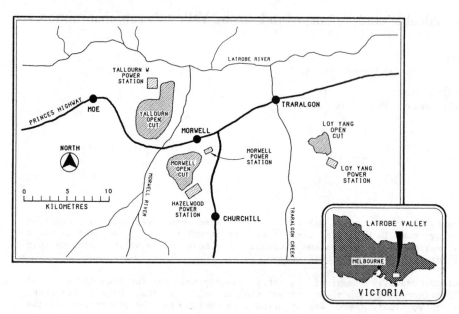

Figure 1 Location of brown coal open cut mines and power stations in the Latrobe Valley

sluicing system in Latrobe Valley power stations is an operation with significant costs and inherent on-going technical and environmental problems related to the unique nature of the fly ash/char mixture. Coupled to these are severe limitations on the sites for placement of future waste dumps and the move toward backdumping in older open cuts. Large scale utilisation of this waste is one option under consideration to resolve many of the current issues.

The composition and distribution of the ash forming constituents in the coal gives rise to a unique fly ash/char mixture during combustion which is different in most respects to all other fly ashes. It has virtually no potential for utilisation in the large volume usage categories common to bituminous and other lignite coal ashes.

Recent studies, however, have indicated that there is an opportunity to develop an industrial project which produces marketable chemicals and high purity magnesium metal from the ash. When the abundance of power and the well developed industrial infrastructure existing in the Latrobe Valley, the significant environmental gains and reduction of the other ash disposal problems are considered the opportunity looks even more attractive.

A preliminary cost study based on a conceptual process flow sheet has demonstrated that the process could be developed in two stages. The first stage produces sodium sulphate and low ash char

of a quality suitable for sale. The remaining washed ash slurry is suitable for either safe, lower cost disposal, or further processing by second stage higher technology treatment to produce calcium carbonate and magnesium metal. The final remaining leached residue may have application as a low grade agricultural fertiliser or could be disposed of with confidence.

2 FORMATION OF LVBC FLY ASHES AND CHAR

The main reason for the formation of this unique ash is the way in which the inorganic constituents are distributed in the LVBC. In black coals the major inorganic ash forming constituents are usually present as discrete mineral particles sporadically distributed throughout the coal i.e. quartz, siderite and gypsum. They undergo little chemical change as a result of combustion. By contrast the major ash forming constituents in LVBC occur as inorganic cations bound to carboxylic functional groups. The cations present in highest concentration are Ca^{++}, Na^{+}, Mg^{++} and Fe^{++}. Aluminium is also sometimes present, as a complex cation, particularly near the base of seams and adjacent to interseam sediments. Sodium chloride is observed in some of the coal. Only small amounts of discrete mineral particles comprising of quartz, kaolinite, muscovite and pyrite/marcasite

Table 1: Comparison of LVBC fly ashes and black coal ashes (results expressed in terms of the oxide).

OXIDE %	LATROBE VALLEY BROWN COAL ASH			AUSTRALIAN BLACK COAL ASH		
	YALLOURN W	HAZELWOOD	LOY YANG	AWABA	TARONG	CURRAGH
SiO_2	1.4	6.6	60.4	64.6	59.5	63.5
Al_2O_3	2.1	1.8	13.3	26.4	36.0	21.7
Fe_2O_3	24.5	8.7	8.5	3.4	1.4	6.4
TiO_2	0.1	0.2	1.7	1.1	2.2	1.0
K_2O	0.4	0.4	1.2	2.6	0.1	2.2
MgO	18.0	18.8	2.2	0.6	0.1	1.3
Na_2O	11.0	4.5	2.1	0.2	0.1	1.0
CaO	12.3	28.4	1.0	0.5	0.1	0.9
Mn_3O_4	0.2	0.2	<0.1	0.1	<0.1	<0.1
P_2O_3	<0.1	<0.1	0.1	0.1	<0.1	0.4
SO_3	21.7	15.6	3.4	0.4	0.1	0.6
Cl	<0.1	3.4	<0.1	-	-	-
LOI	8.2	11.7	7.6	-	-	-

are present and usually occur in higher concentrations in the top and bottom few metres. Table 1 presents the comparative analysis between LVBC fly ashes and Australian Black Coal fly ashes (Brockway and Chalmers 1986).

During combustion the organically bound cations undergo a series of oxidation and sulphation reactions. In particular sodium and calcium after initial oxidation react with sulphur released from the coal as a gaseous oxide to form sodium sulphate and calcium sulphate. Magnesium undergoes oxidation to magnesium oxide. A significant proportion of the Fe^{++} cations after complete oxidation appear to react with magnesium and calcium oxides to form magnesium or calcium ferrites at temperatures above 1000°C.

Another feature of the LVBC fly ash is the presence of high levels of low-volatile-matter char which derives from the woody coal content in some of the coal. Its presence is due to difficulties in milling the fibrous woody coal fragments down to 100μm. Although the fragments do not fully combust, most of the volatile matter is lost in the process and small size char particles 1 to 2mm in diameter are carried out of the boiler.

The char content in the ash and its high relative volume leads to rapid filling in ash settling ponds.

3 THE CHARACTER OF LVBC FLY ASHES AND CHARS

Table 2 lists the compounds found to be present in the three fly ashes and char products, the % amount of each component and the total annual productions rates. The % amounts are expressed in terms of compounds in this table rather than the standard expression of results for ash analysis in terms of the oxide.

A large number of characterisation tests were carried out on a series of representative samples and has led to a significantly increased understanding of the waste products (Laws-Herd 1989).

3.1 Yallourn fly ash and char

Yallourn fly ash is typically high in sodium sulphate (over 20%), magnesium oxide (about 18%), and calcium sulphate (about 12%). Calcium oxide content is usually low but is sufficient to give rise to a pH of about 10 when in contact with water. The iron oxide content is most often very high. Haematite is the major iron oxide but magnetite and/or magnesium ferrite are also present (in total these oxides amount to 25% by weight). Acid extraction tests indicate that magnesium ferrite predominates and probably about 10% of the iron is bound in this way.

Table 2: Composition of LVBC fly ash/char mixtures
(results expressed in terms of compounds)

	YALLOURN W 1989		HAZELWOOD 1989	LOY YANG 1989		LOY YANG 1994*
	Fly Ash	Char	Fly ash/ Char	Fly Ash	Char	Fly Ash
QUANTITIES						
kt/y	86	38	135	104	23	40
COMPOUNDS						
Na_2SO_4	25.2	5.8	2.0	4.8	6.2	16
NaCl	–	–	5.6	–	–	13
CaO	7.0	9.4	15.0	–	–	–
MgO	14.4	13.0	18.4	1.8	2.0	20
$CaSO_4$	12.7	4.0	20.4	–	–	11
Fe_2O_3	11.9	10.9	–	8.5	8.4	6
$Ca_2Fe_2O_5$	–	–	13.7	–	–	–
$MgFe_2O_4$	15.4	14.1	–	–	–	–
SiO_2	1.4	6.0	6.0	59.3	56.9	16
%LOI	8.2	32.5	11.7	7.6	18.5	8
Other	3.8	4.3	7.2	18.0	8.0	10

* Composition predicted from coal bore analysis data.

The high woody content of Yallourn coal gives rise to the char production during combustion. Char collectors are necessary in the plant up-stream of the electrostatic precipitators. The product resembles the fly ash but has a much higher percentage loss on ignition (%LOI), as would be expected, at over 30%.

It features a significantly reduced sulphate content with a higher quartz content compared to the fly ash.

3.2 Loy Yang fly ash and char

Loy Yang Power Station and open cut commenced operation in 1985. The fly ash content is at present masked by high levels of quartz and clay minerals which are prevalent in the upper parts of the new open cut. Char content in the ash is significant. Bore samples indicate that when deeper coal is mined the fly ash will resemble a mixture of Yallourn and Hazelwood fly ash but with increased amounts of sodium chloride. An estimate of the future ash composition has been made from bore analysis data and is listed in Table 2.

3.3 Hazelwood fly ash/char

The fly ash/char mixture is typically high in calcium sulphate, calcium oxide (up to 35% in total), and magnesium oxide (up to 18%). The major iron compound is calcium ferrite ($Ca_2Fe_2O_5$ at 14%) with small amounts of haematite. Sodium chloride and silica are usually present (about 6% each) and the level of sodium sulphate is most often low (about 2%). The %LOI is usually less than 12% and due to low char production. The pH of aqueous solutions of this ash are typically about 12 arising from dissolution of sparingly soluble calcium oxide.

4 ASSESSMENT OF THE POTENTIAL FOR UTILISATION

4.1 Low and medium technology processing

There are stringent requirements for the use of fly ash in Portland cement concrete under the Draft Australian Standard (Fly Ash Use in Concrete) which provides for three grades of fly ash; fine, medium and coarse (DR87242). Fine and medium grades are intended for normal concrete whereas coarse grades are used for soil stabilisation and similar applications, lean mix concrete or intergrinding with Portland cement clinker to produce blended cements.

Of the fly ashes collected for these studies only the current Loy Yang fly ash may be suitable for the applications covered in the Draft Standard. In the raw state the ash meets the coarse grade requirements and it appears technically possible to upgrade this ash to meet the medium grade requirements by reduction of the % LOI. The composition, however, is not expected to be sustained in the medium to long term when deeper coal devoid of discrete mineral particles will predominate.

The suitability for intergrinding all three fly ashes with novel cement raw blends prior to kilning is under investigation and there may be potential to use some of the higher calcium content fly ashes as an additive in ceramic tiles. None of the fly ashes meet the Australian Standard for Mineral Fillers in Asphalt (AS2357 1980).

The total requirement for suitable material in Victoria in these categories amounts to no more than 100,000 to 150,000 t/a, thus even if the total market could be captured there would still be a substantial continuing volume requiring disposal.

4.2 High technology processing

The present research indicates that the logical approach to utilisation of LVBC fly ash on a large scale is the economic extraction of its commercially valuable components.

There have been studies in a number of countries to examine the possibilities of utilising fly ashes in this way. The most comprehensive to date is the study carried out by the Electric Power Research Institute (EPRI) in USA which ran from 1979 to 1986 (EPRI 1987). It was mainly concerned with the direct concentrated acid leaching (DAL) of bituminous and sub-bituminous fly ashes for removal and formation of a range of marketable minerals and leached residues.

Applying this approach directly to LVBC ashes and char would not be viable. For example the high calcium levels would produce prohibitively high amounts of gypsum. The leached ash residues would not be suitable for the applications investigated by EPRI and the economies of scale would be too low. The direction the present research has taken is to develop a process based on water and dilute acid leaching of the ash to produce a small range of products resulting from the unique properties of the LVBC fly and char. As seen from Table 2 the material has and will continue to contain relatively high levels of magnesium, sodium, calcium, sulphur, iron and char.

In addition to addressing the major issues, the approach has added appeal in that it is in line with two important State Government policies. The policies which encourage establishment of energy intensive industries in Victoria to utilise cheap energy (Victorian Government 1987) and the waste minimisation policy in which fly ash is listed as a prescribed waste (Environmental Protection Authority 1988).

5 MARKET SURVEY

About thirty compounds and metals could in principle be extracted from LVBC fly ash, however, the number is narrowed substantially when practical considerations are taken into account and of the thirty only ten products were subsequently surveyed for market potential; these were low ash char, sodium sulphate, calcium carbonate, silica, alumina, iron oxides, magnesia, magnesium metal, titanium dioxide and titanium metal.

Four products were ultimately selected as the most prospective with regard to ease of production and accessible markets. These were char, sodium sulphate, calcium carbonate and magnesium metal.

6 CONCEPTUAL PROCESS DESCRIPTION

The development of a conceptual process was based on an intensive research program directed at the short list of prospective products (McLennan and Magasanik (1) 1989 & Laws-Herd 1989).

Figure 2 is a simplified schematic showing inputs to and outputs from the conceptual process.

The combined findings have indicated that the project could be developed in two

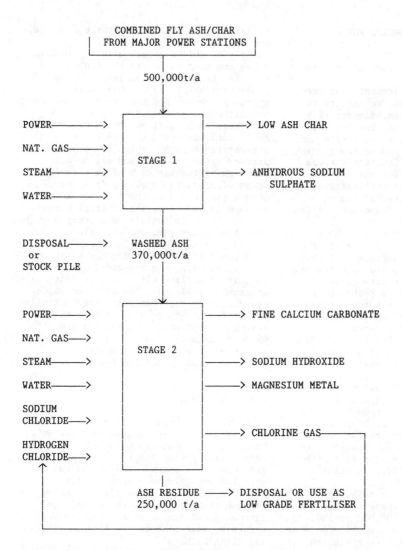

```
              COMBINED FLY ASH/CHAR
            | FROM MAJOR POWER STATIONS |


                    500,000t/a
                        │
                        ▼
POWER────────>  ┌──────────────┐  ───────> LOW ASH CHAR
                │              │
NAT. GAS───────>│              │
                │   STAGE 1    │
STEAM──────────>│              │  ───────> ANHYDROUS SODIUM
                │              │                    SULPHATE
WATER──────────>│              │
                └──────────────┘
                        │
DISPOSAL────────>       │
    or          WASHED ASH
STOCK PILE      370,000t/a
                        │
                        ▼
POWER────────>  ┌──────────────┐  ───────> FINE CALCIUM CARBONATE
                │              │
NAT. GAS──────>│              │
                │   STAGE 2    │
STEAM────────>  │              │  ───────> SODIUM HYDROXIDE
                │              │
WATER────────>  │              │  ───────> MAGNESIUM METAL
                │              │
SODIUM          │              │
CHLORIDE──────> │              │  ───────> CHLORINE GAS───┐
                │              │                          │
HYDROGEN        │              │                          │
CHLORIDE──────> │              │                          │
                └──────────────┘                          │
                        │                                 │
                ASH RESIDUE ──────> DISPOSAL OR USE AS     │
                250,000 t/a         LOW GRADE FERTILISER   │
```

FIGURE 2:   SIMPLIFIED CONCEPTUAL PROCESS SCHEMATIC

stages. The first stage involves the separation and removal of the two medium value products, low ash char and anhydrous sodium sulphate, using well known technology.

Wet separation of char from ash occurs readily. The char floats immediately on contact with water and can be simply skimmed from the surface and collected as a low ash content-low volatile matter char generally greater than 200μm in size and with a moisture content of 60% to 70%. It can be readily dried down to 15% moisture and in this state has identified potential as a fuel. The separation has been demonstrated in laboratory trials.

Sodium sulphate and sodium chloride salts are easily leached from the fly ash by using warm to hot water. The concentration of sodium sulphate must be raised sufficiently to enable selective crystallisation of Glaubers Salt ($Na_2SO_4.10H_2O$). The final concentration depends on the amount of sodium chloride present as this tends to beneficially increase the crystallisation temperature of sodium sulphate.

The second stage involves the high technology production of magnesium metal and calcium carbonate. The residue after water leaching is devoid of its high volume char and water soluble components and contains magnesium, iron, silica, other minor components and most of the calcium. About 90% of the magnesium (present as MgO) can be leached out with hot dilute hydrochloric acid along with a portion of the calcium oxide, some of the calcium sulphate and also that part of the iron present as haematite.

Calcium sulphate and ferric hydroxide can be precipitated from solution by neutralisation with sodium hydroxide to produce a concentrated solution containing chloride salts of magnesium, calcium and sodium.

From this point onwards the objective is to produce 99.9% anhydrous magnesium chloride and then to electrolyse a melt into chlorine and magnesium metal.

There are several methods, technologies and patents for production of magnesium metal but the question of technology availability requires resolution. Recycling of chlorine after conversion to hydrochloric acid is considered essential to the economic feasibility of the process. Production of calcium carbonate can occur by carbonation of the solution left after magnesium chloride has been removed.

7 ECONOMIC ASSESSMENT

A preliminary economic feasibility study and cost estimates were based on plant with proven technology to process 500,000 t/a of LVBC fly ash and char, situated in a centralised location in the Latrobe Valley (McLennan Magasanik (2) 1989). The assessment could only provide indicative figures (± 50%). Both capital and operating costs were estimated and compared with potential income from the four products, low ash char, anhydrous sodium sulphate, fine calcium carbonate, and magnesium metal. A fifth product, sodium hydroxide, resulting from operation of the chlor-alkali plant was also included in the assessment.

It was shown that the capital costs of Stage 1 was only a small proportion of that for Stage 2 (approximately 10%) and that the former is a very promising proposition. It also indicated that Stage 2 has the potential to be economically viable but requires a more detailed design and economic assessment be carried out based on more accurate details of plant design, the reliability of the resource and future plans for electricity production in the

Latrobe Valley. The cost sensitivity on options for several plant components, for instance, anhydrous magnesium chloride plant, is required before there is sufficient confidence for a large capital and operating investment to be made in a full scale pilot plant.

It is intended that this more detailed design and economic assessment will be made within the next year.

In the cost study to date no attempt has been made to assess the overall environmental cost/benefits of the project, however, these are expected to be significant and full credit to this aspect ought to be made in a more accurate cost/benefit study.

8 CONCLUSIONS AND DISCUSSION

On the basis of the laboratory research there is little doubt from the technical point of view that the process postulated can produce the four products at the required purity. They are all common commodities on the chemicals and metals market. The recent market survey has shown that there is an identified demand for all the products in the quantities which could be produced.

The indications are that Stage 1 is economically viable in its own right if it is justifiable to credit as little as 30% of the resulting savings in ash disposal costs due to Stage 1 proceeding. In addition to ash disposal costs, Stage 1 is sensitive to capital costs.

Stage 2 may be economically viable, in combination with Stage 1. The major uncertainties are the capital costs to which Stage 2 is sensitive. It is also sensitive to magnesium prices.

The state of knowledge in relation to the processing involved in Stage 2 is relatively less well developed than for Stage 1. In view of this it will be necessary to assess the technologies available for the production of anhydrous magnesium chloride and its electrolysis to produce the metal and chlorine by-product.

Further revision of the capital and operating costs and re-evaluations of the economics will be necessary as the project proceeds. Another important aspect will be to identify priorities for further Research and Development.

The environmental effects/benefits of utilising ash on this scale can fairly clearly be identified. However, the costs have not been quantified and are not included. For example elimination of the soluble and alkaline components from the disposal waste alone removes the

possibility of contamination of the
environment. Similarly additional control
over leachable trace elements such as
selenium would also be possible by this
processing.

9 REFERENCES

Australian Standard AS2357 1980. Mineral
    Fillers in Asphalt.

Brockway D.J. & Chalmers 1986. Chemical
    Assessment of Coal for Power Generation
    Chem. in Aust.

Draft Australian Standard DR87242 1987.
    Fly Ash for Use in Concrete.

Environment Protection Authority, 1988.
    Draft Industrial Waste Management Policy
    on Waste Minimisation.

EPRI 1987. Recovery of Metal Oxides from
    Ash including Beneficiation Products,
    Report CS 4384 Vol 1.

Laws-Herd K.A. 1989. Potential Utilisation
    of Latrobe Valley Fly Ashes and Char -
    Laboratory Research Program Vol. 1. State
    Electricity Commission of Victoria,
    Research & Development Department Report
    SC/89/162.

McLennan Magasanik Associates (1), 1989.
    Potential Utilisation of Latrobe Valley
    Fly Ashes and Char Confidential Report to
    State Electricity Commission of Victoria.

McLennan Magasanik Associates (2), 1989.
    CoalAsh Development Project Economic
    Study, Confidential Report to State
    Electricity Commission of Victoria.

Victorian Government 1987. Victoria - The
    Next Decade, Victorian Government
    Publishers.

Reclamation, Treatment and Utilization of Coal Mining Wastes, Rainbow (ed.) © 1990 Balkema, Rotterdam. ISBN 90 6191 154 0

Ash and slag to be used for making peripheral embankments around ash and slag disposals

D. Knežević & M. Grbović
Mining Institute, Belgrade, Yugoslavia

P. Djuknić
Heat Power Generation Plant 'Nikola Tesla'-Obrenovac, Yugoslavia

ABSTRACT: Paper presents technology process of thickening and clasifying the ash-and-slag hydromixture in thermal power plants fueled by coal in order to provide a suitable dilution to build up peripheral embankments. The process has been applied in a number of Yugoslav and Czechoslowakian thermal power stations with satisfactory results. By making peripheral embankments from ash, the available dumping capacity of the disposals is enlorged, establishing of more than one levels without consolidating and drying up of the disposal is made possible, investment capital and operating costs are decreased and the environment protection is improved.

1 INTRODUCTION

More than half of Yugoslav electric power is produced in thermal power plants fueled by coal. Low grade coals - lignite (some 80%) and brown coal (some 20%) are used for this purpose. In this process approximately 11.10^6 tons of ash and slag are produced as a by-product every year. These quantities rank Yugoslavia among ten major world ash "producers". After interpolating these parameters against the number of population, then Yugoslavia with 460 kg of ash yearly per inhabitant takes a second position, immediately after Czechoslowakia.

This situation has caused many problems associated with providing disposal area as well as with ecological requirements for a healthy and clean environment.

In Yugoslavia, hydraulic methods of transport and disposal in the specially prepared areas near the thermal power plants´ facilities are the ones most often applied. Mining Institute from Belgrade has developed and applied in serval Yugoslav and Czechoslowakian thermal power plants an original technology of making peripheral embankments using ash and slag as the building material.

2 BASIC FEATURES OF ASH AND SLAG

Yugoslav lignites are characterized by a high moisture content (40-50%), ash (15-20% on an average) and combustible matter (approximately 40%). Largest coal producers for the thermal power plants are Kolubara mines and Kostolac mines, some 100 km far from Belgrade. Yearly 14 thermal power plants fueled by this coal produce some 6×10^6 tons of ash and slag.

Chemical composition of the ash and slag is made mainly from silica oxide, iron and allumina (Table 1).

Table 1. Chemical Composition of Ash and Slag

	TPP "Nikola Tesla-A"	TPP "Kostolac-A"	TPP "Kostolac-B"
SiO_2	55,40	50,35	53,67
Al_2O_3	24,87	24,39	21,06
Fe_2O_3	8,87	9,67	9,93
CaO	4,55	8,56	9,23
MgO	2,07	4,20	3,62
SO_3	1,38	1,13	0,86
P_2O_5	0,09	0,22	0,20
Na_2O	0,19	0,22	0,20
K_2O	1,79	0,45	0,41
TiO_2	1,06	0,87	0,92

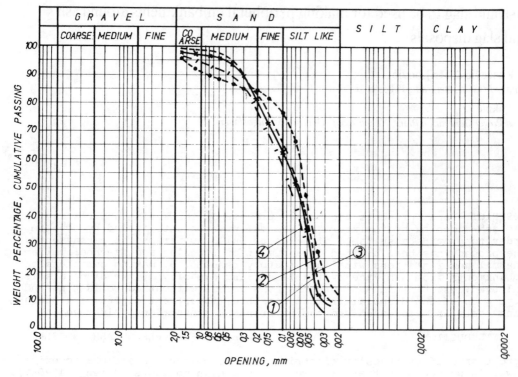

| GRAVEL | | | SAND | | | | SILT | CLAY |
| COARSE | MEDIUM | FINE | CO ARSE | MEDIUM | FINE | SILT LIKE | | |

① TPP ,,Nikola Tesla'' - A
② TPP ,,Nikola Tesla'' - B
③ TPP ,,Kostolac''- A
④ TPP ,,Kostolac''- B

Fig. 1. Granulometric composition of ash and slag in the
thermal power plants: Nikola Tesla A and B and
Kostolac A and B.

Mineralogically, glass makes the highest share (~54%), followed by mullite (~20%), magnetite (~ 6%), quartz (~ 4%), calcite, anhydrite, geothite, periclase, corundum etc. The soluble saults make only a slight share ranging from 0,5-1,5%.

By their granulometric composition the ash and slag make a mixture of the crushed anorganic components from the sand-and-dust group. Fig. 1 presents the granulometric composition of ash and slag from the thermal power plants: Nikola Tesla-A, Nikola Tesla-B, Kostolac-A and Kostolac B.

The ash from these thermal power plants is a hydrofobous material, dark in colour, easily susceptible to evapo-transpiration.

The ash is characterized by a good water permeability with a respective coefficient of $1,2-8,9 \times 10^{-6}$ m/s, a low compression coefficient (1,5-2,5%), module of compressibility between 150-200 daN/cm^2, angle of internal friction from 29-30 degrees and lack of cohesive power among the particles.

Comparing the features of these ashes to the features of the flotation wastes which have been used for deceades to make the peripheral embankments at the waste disposals, a high degree of similarity was noticed. In many elements, primarily

516

in size, chemical composition and water permeability even higher values have been noted. This ispired an aproach to the serious investigations concerning the possibility to use the ash and slag as a building material for peripheral embankments.

3 RESEARCH TESTS

The purpose of research tests was to determine the conditions to provide the class of ash and slag having the suitable geomechanical characteristics. But, contrary to the flotation wastes which leave the process at a density of 22-40% of solid by weight, ash and slag are being transported to the disposal at a density of hardly 3,2-7,1% of solid by weight. Change of density degree would require considerable amount of reconstruction in the system of collecting the ash and slag in the heat power generation plant, and therefore the problem of the hydromixture density had to be solved within the technological process of providing the suitable ash class.

First phase of research was focused to finding out the device which would increase this initial density of 3,2-7,1% of solid by weight to some 20% of solid. The tests were carried out with the thickeners, cone type classifiers (cones) and hydrocyclones. The thickener turned out to be inefficient (without suitable flocculants) because the overflow was impured so that it had to be additionally cleared in the pool. In order to achieve the adequate density of the underflow to meet the requirements of TPP "Nikola Tesla-A, it was necessary to make a thickener with a diameter of 200 m (or a number of smaller units). Basic reason of an inefficient operation of the thickener is an austandingly low thickening velocity of 57 mm/h.

Tests of thickening in a cone gave similar results. It became clear that the thickening process could not be performed exclusively through gravitation. The turn was therefore made to tests the option to use the hydrocyclones as thickeners.

The first problem encountered was the choking of hydrocyclones caused by the presence of rod-like, wooden pieces usually accompanying xylite coals such as Yugoslav lignites. The hydrocyclone was, therefore, designed having diameter of 1270 mm. The operation of this hydrocyclone was satisfactory on the ground of more than one reason:

- required density of the underflow (~20% of solid) was provided,
- satisfactory mass distribution to overflow and underflow (33%:67%) was provided,
- majority of the rod-like pieces was provided to be separated through the overflow thus reducing the danger of clogging the hydrocyclone,
- separation of the muddy portion throuhg the overflow was provided.

By a successful solution of the thickening problem the basic problem was solved and the classifying problem of the thickened product was solved by using the classical hydrocyclone-classifier. By using the hydrocyclone having diameter of 350 mm the dilution with satisfactory geomechanical features was provided.

4 TECHNOLOGY OF MAKING PERIPHERAL EMBANKMENTS

The flowsheet of the technological process for the preparation of the required ash class for making the peripheral embankments is plotted in the Fig.2.

The ash and slag being transported as the watery pulp to the disposal have to be firstly subjected to the thickening and afterwards to classifying. Both the operations are using hydrocyclones. Thickening is performed in the so called primary cyclone (D=1270 mm). Water and smallest particles are firstly separated through the overflow. The underflow free from slurry and water is having density of approximately 20% of solid by weight.
Distribution by volume in the primary hydrocyclone is 80% overflow: 20% underflow. The overflow is directly poured within the peripheral embankment into the disposal. The underflow from the primary hydrocyclone is passed through the protective grating (strainer) in

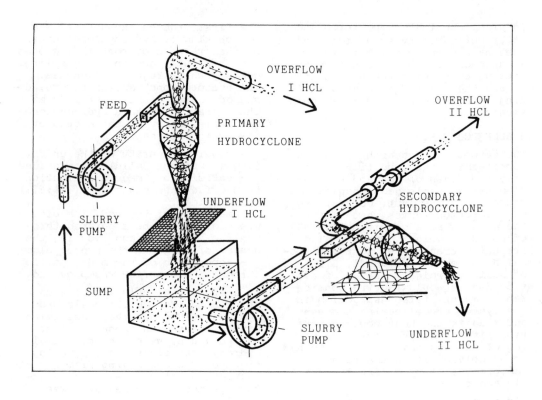

Fig. 2. Flowsheet of the technologic process

order to separate the largest pieces of the slag which have passed through the primary hydrocyclone. Large pieces are kept at the perimeter of the grating from where they are taken away from time to time. The strained product is passed through the feeding box into the slurry pump. The purpose of this slurry pump is to transport the underflow from the primary hydrocyclone to the secondary hydrocyclone. This pump provides also a required input pressure into a secondary hydrocyclone. The transport of the pulp is carried out through the plastic pipeline. The pipeline is extended depending on the advance of the embankment erection advance.

The fraction suitable for making the peripheral embankment is formed in the secondary hydrocyclone. The overflow of this hydrocyclone contains mainly water and majority of smallest grains not suitable to be deposited in the embankment. This class is directly poured within the embankment into the disposal.

In order to make the formation of the embankment possible it is necessary to provide the hydromixture density at the output for the sand of 40-50% of solid by weight.

In order to eliminate the negative effect of the input density variations a special bolt is mounted at the overflowing pipeline the regulation of which provides the required density and distribution.

The technology sketched in this way was firstly applied at the disposal of the thermal power plant "Nikola Tesla-A" in Obrenovac and it gave satisfactory results.

5 FEATURES OF THE PERIPHERAL EMBANKMENT

By a two-phase hydrocycloning the material is gained which not only by its consistence but also by its granulometric composition makes the formation of the peripheral embankments possible.

Fig. 3 presents the granulometric composition of ash at the input into the preparation process, in the course of preparation and at the output from the process as the material for making the embankment.

It is obvious that the grain size was improved because the portion of the smallest size classes was decreased and the portion of the largest grains considerably increased.

Portion of grains smaller than 74μm is below 10% which completely fits into the knowledge achieved trough studies about optimum participation of these grains in the material used for making the embankment.

Water permeability of the material prepared for making the embankment is closely associated with the granulometric composition. The permeability was increased more than ten times reaching the value of $1,2-7,8 \times 10^{-5}$ m/s .

This makes it possible to make the

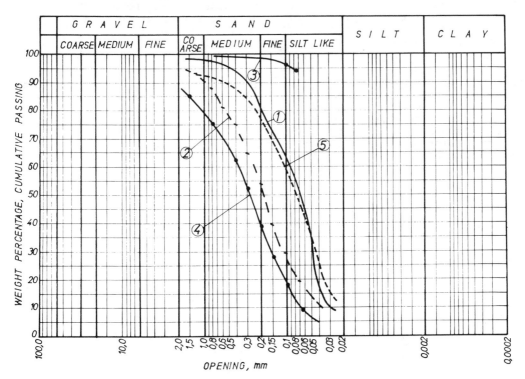

①   Feed in primary hidrocyclone
②   Underflow of primary hydrocyclone/feed in secondary hydrocyclone
③   Overflow of primary hydrocyclone
④   Underflow of secondary hydrocyclone
⑤   Overflow of secondary hydrocyclone

Fig. 3 Granulometric composition of ash and slag through the process flowsheet

519

embankment at the hydromixture density of only 40-50% of solid.

Relative compactness of the material used for making the embankment is 0,770. Maximum relative compactness of 0,840 with a decrease of the water permeability, coefficient to $2,7x10^{-6}$ m/s, was determined by the laboratory tests. This indicates that with the increase of the disposal height and subsequent increase of the disposal height and subsequent increase of load to the lower layers, water permeability might be expected to decrease. Water permeability will also decrease because of the introduction of the smallest particles into the embankment caused by the infiltration water. All this leads to a conclusion that the situation at the high disposals should be under continuous control and that more than one of the adequate drainage rings should be incorporated.

The angle of the internal friction of the classified ash was increased to approximately 32° while the classified as well as non-classified ash does not show the existence of the cohesion forces amont the particles. Such a high angle of the internal friction indicates that the embankments at the levels might be made at aquite understandable incline of 1:2,5-1:3.

The compressibility was also improved. Degree of compression is below 2% and the module of compressibility was increased to 200 daN/cm^2.

6 CONCLUSION

Ash and slag produced by the lignite combustion process were successfully used to make the peripheral embankments at the disposals. In this way the available accumulation area at the disposals was increased because the embankments were not built by the borrowed material or by the refulled river sand. The costs for making the peripheral embankments were therefore manifold decreased. At the disposals where the peripheral embankments are made from the"barrowed" material it is necessary before building up the next level, 3-5 m high, to provide the consolidation and drying up of the disposal to make it possible

for the building machines to move over the disposal. In case of disposals where the embankments are made by the hydrocyclones from the classified ash and slag it is possible to make 4-5 levels, each 3,0 high, without leaving the disposal. It is possible because the embankment making is continuous and without the building machines. Longer period of work at one disposal, or at a part of it, decreases the costs for the environmental protection and increases the degree of the surrounding area protection.

Geomechanical features of the classified ash are fully comparable to the material by which the peripheral embankments at the flotation disposals are made. The features of the classified ash are not inferior to those of the "borrowed" material either, but offer an advantage to "pack" the hydrologically treated material in a better and more uniform way compared to the mechanical compression.

In the last eight years the described technological process has been applied in four thermal power plants in Yugoslavia (TPP "Nikola Tesla" A&B, TPP"Kostolac"A&B) and in two in Czechoslowakia (TPP "Počerady" and TPP "Pruneřov")

REFERENCES

Yugoslav patent N° 28011/81 - P-3000/81
  Postupak za kontinuirano formiranje deponija pepela TE na ugalj"
Knežević D.,Košutić Lj. 1985 "Realizacija novog postupka deponovanja pepela i šljake u termoelektrani "Nikola Tesla-A", Obrenovac", Precedings of 10th Yugoslav Symp. on Mineral Ore Dressing, Struga.
1972,"Tailing Disposal łoday"proccedings of the First Inter. Tail. Symp., Tucson, USA.
Monz O.E.,1980: "Ash Production and Utilization in the World", Sem.on the Extraction, Removal and Use of Ash from Coal-fired Thermal Power Stations, Zakopane, Poland.
Douša, K.,1987. "Vystavba zvyšovacich hrazi odkališt popela primym plavenim pomoci hydrocyklonu", procedings of Hornicka Pribram,1987.

Reclamation, Treatment and Utilization of Coal Mining Wastes, Rainbow (ed.) © 1990 Balkema, Rotterdam. ISBN 90 6191 154 0

Coal preparation wastes of the USSR and their utilization

M.Ya.Shpirt & V.A.Ruban
Fossil Fuel Institute, Ministry for Coal Industry of the USSR

ABSTRACT: The investigations of composition, properties of coal preparation wastes (CPW) and technologies of their utilization have been carried out at our Institute for more than 20 years. The quantities of these wastes (CPW) utilized for the construction of the artificial ground structures and building materials production increase every year. The specifications for using CPW as the only or the main component of the blend for producing bricks, cement, aluminium compounds and other materials have been established.

The first shop for making high quality bricks from CPW without additives have been put into operation and now its capacity exceeds the design output.

The classification for the solid wastes from coal mining and preparation proposed earlier has been modified according to the recommendations of the group of experts of the United Nations Economic and Social Council. The computerized data bank on the composition, properties and the utilization trends of CPW from 50 coal preparation plants with CPW annual yield more than 56 mln.tonnes have been worked out.

In the USSR the relative and absolute yields of coal preparation wastes (CPW) increase from year to year due to geological characteristics of coal deposites technologies of mining and increased volumes of coal subjected to the preparation. The volumes of CPW increases by 2-4% annually, and in 1988 every 1 t of coal subjected to preparation yielded 0.38 t of CPW in the average. The chemical and minerological composition of CPW even of one and the same deposit for different plants can differ greately, though the change in time for every deposit is negligible. The conclusion is grounded on the comparison of 1979-1982 and 1983-1988 data summerized in table 1 and in our previous report /1/.

The CPW of one plant have rather stable composition due to blending in the process of coal prepa tion. Besides, the needed organic matter content can be established due to the variation of the regimes of the preparation process. Therefore, CPW are viewed as more promising for the utilization than the coal wastes of mining operations and dumps.

As it will be shown /1-4/, CPW application can be based on their physical, physico-chemical properties and composition according to the following main directions:

-as the base for artificial ground structures, highway constructions, reclamation, filling the holes, levelling the lay of the ground;

-as the additives to composters or for sulfur compounds production;

- as the organic-mineral raw feedstock, for building materials production (bricks, porous aggregate, cement and others), in the ferrous and non-ferrous metallurgy for different silicon-aluminium alloys, alumina, aluminium sulphate or oxychloride, silicon-carbide materials and other valuable products.

Table 1. Coal preparation wastes composition of the main USSR coalfields (method of preparation).

Content, %	Coalfields (method of preparation)				
	Donbass		Kuzbass		Karag.
	grav.	float.	grav.	float.	grav.
A^d	69-88	53-75	61-88	49-77	68-86
C_t^d	6-20	10-28	4-21	11-45	8-22
S_t^d	<1-4.2	<1-3.9	<1-1.4	<1-1.1	<1-2
SiO_2	57-63	51-59	56-78	51-66	52-64
Al_2O_3	14-31	17-32	14-28	15-31	23-35
Fe_2O_3	3-17	3-13	2-10	2-9	4-7
CaO^*	0.3-5	2-4	1-7	1-10	1-6
MgO^*	0.8-2	1-2	0.3-3	0.7-3	0.3-1

	Karaganda	Pechora	Mosbass
	float.	grav.	grav.
A^d	64-73	61-89	70-78
C_t^d	18-25	6-26	10-18
S_t^d	<1-2.4	<1-3.3	2.8-8.6
SiO_2	54-56	62-67	56-61
AL_2O_3	23-29	19-25	25-33

1	2	3	4
Fe_2O_3*	6-13	5-8	4-12
CaO*	1-5	1-2	1-5
MgO*	1-2	2-4	1-3

*In ash

The largest quantities of CPW are utilized in the USSR as the earth ground (12-14 mln.t/year) and for the production of bricks (more than 1 mln.t/year).

Table 2. Coefficients of the equation (1) for the calculation of carbon content (Co^d) as a function of ash.

Coal preparation wastes of coalfields	Coefficients	
	a_i	b_i
Donbass		
coal B.B.	71.4	0.76
" C.B.	77.8	0.83
" H.B.	85.7	0.91
antracite	92.1	0.97
Kuzbass		
coal B.B, C.B.	80.3	0.85
" H.B.	89,3	0.95
Karaganda	78.3	0.82
Pechora	78.3	0.79
Mosbass	67.9	0.75

Coal preparation wastes of coalfields	Coefficients correlation r	limits of application of equation(1) for Co
Donbass		
coal	0.94	5-19
"	0.95	6-20
"	0.95	7-29
antracite	0.97	7-30
Kuzbass		
coal	0.97	7-26
"	0.95	9-37
Karaganda	0.99	9-23
Pechora	0.97	4-21
Mosbass	0.98	7-25

The volumes of CPW utilized for artificial ground constructions is increased constantly and certain specifications are being worked out in some regions of the USSR. The content of the organic carbon (Co) is one of the main parametres for the evaluation of CPW as the organic-mineral raw feedstock. Summing up the experimental data we have obtained the linear correlation equation (1) showing the dependence of Co on ash (A^d):

$$Co = a_i - b_i A^d \quad (1)$$

This equation can be applied for
CPW of many coalfields though the
coefficients (a_i, b_i) differ (table 2). Now the energetic potential
of CPW is utilized mostly in producing bricks from the feedstock containing 8-12% of CPW. The special
technologies of producing bricks
from the feedstock containing CPW
as the main component are being
worked out in the USSR. The content
of CPW in the feedstock depends on
their chemical composition (table 3)

on CPW of different coalfields and
compositions (Table 4). The porosity is 27% or 12% by plastic formation or semi-dry pressing, respectively. The first shop (in Kuzbass)
for making the high quality bricks
(compressive and bending strength
15 and 8 MPa) from CPW without additives has been put into operation
and now its capacity exceeds the design output. Now some shops for
producing 60 mln. of bricks per year
from CPW without additives are designed by coal preparation plants
in different coalfields.

Table 3. The basic parameters of the
specification in using CPW for brick
production.

Parameter	Feedstock CPW content,%		
	100	>40	<40
Ash,%	$\geqslant$80	$\geqslant$75	<80
C_o^d	$\leq$8	$\leq$16	>10
S_t^d	$\leq$1,2	$\leq$2	$\leq$4
Al_2O_3*	$\geqslant$15	no limit	no limit
CaO+MgO*	$\leq$6	8	no limit
W_t	$\leq$10	$\leq$12	$\leq$25

*In ash

The methods of plastic formation(pf)
or semi-dry pressing (sdp) are used
in the technological processes. The
validity of these processes was shown

Table 4.The properties of bricks
(250x120x65mm) produced with CPW.

Coalfie- ld,Coal prepar. plant	Method of forma- tion	Strength		Water permea- bility, %
		Compres- sing	Bend- ing	
Donbass plant 1				
C_o 8%	pf	21	4 4	10
C_o 8%	sdp	17	2 6	8
plant 2				
C_o 9%	pf	34	2.8	8
plant 3				
C_o 7%	sdp	16	3	8
plant 4 C_o=6	sdp	15	4	8
Karaganda plant 5				

1	2	3	4	5
$C_o=9\%$	sdp	32	3	12
$C_o=9\%$	pf	33	4	7
Pechora				
plant 6				
$C_o=5\%$	pf	24	3.8	11
$C_o=5\%$	sdp	38	5.6	13

The CPW are also used for producing cement. Their content in the
feedstock constitutes 5-12%. The
technology of producing porous ag-
glomerates by the agglomeration of
the feedstock containing 70-100%
of CPW has been worked out. The composition of CPW applied in this pro-
cess has been established ($A^d=65-90\%$; $C_o=4-20\%$; $S_t^d\leq 2\%$; contents in
ash,%:$Al_2O_3=15-40$; $Fe_2O_3\leq 18$; CaO+
+MgO≤ 12; $K_2O+Na_2O\leq 5$; $SiO_2+Al_2O/$
$/Fe_2O_3+CaO+MgO+Na_2O+K_2O=4-20$).

The investigations were carried
out on a pilot or semi-industrial
scale to evaluate the CPW as a feedstock for producing different aluminium compounds. The following
parameters of CPW composition have
been established. For alumina - by
the method of sintering with $CaCO_3$:
Al_2O_3 (in ash)$\leq 32\%$; $Al_2O_3:SiO_2\geq$
0.6; $A^d\geq 65-70\%$); by the method of
acid leaching:Al_2O_3 (in ash)$\geq 28\%$;

$Al_2O_3:Fe_2O_3> 5$; $A^d\geq 65\%$.
For alloys of Al-Si-Fe:$S_t^d\leq 0.8\%$;
$P_t\leq 0.15\%$; $C_t^d\geq 17\%$; $Al_2O_3^*> 28\%$;
$Fe_2O_3\leq 2\%$ (for components of Al-
Si-alloys) or$\leq 7\%$(for using instead
of Al by the oxidization of steel).
The investigation of the CPW composition and properties are conducted constantly to estimate the possibility of the utilization of CPW
of every coal preparation plant.
Summing up all the data resulted in
the computerized data bank on the
composition, properties and the utilization trends of CPW of 50 coal
preparation plants with the annual
yield of CPW more than 56 mln.ton-
nes. The accomplished investigation
confirmed the applicability of the
proposed classification /1/ of coal
mining and preparation wastes as a
raw material for the utilization.
However, we see the increasing attention of the society to the problems of ecology. Therefore, this
classification would be added by
two more parameters estimating the
ecological danger which can arise
in utilizing coal wastes; as they
contain benza-oxypurines and radioactive elements (K,Ra and Th isotopes). The quantitative values of

these parameters are likely to differ in different countries depending on their state regulations. We
hope that the discussion of this
question will make these parameters
more precise and thus favour the
increase of the volumes of coal preparation wastes utilized.

REFERENCES

1. M.Ya.Shpirt, V.A.Ruban, "Utilization of mining operations and
   coal preparation processes wastes
   in the USSR and the principles of
   their classification". Report for
   the Second International Symposium on the Reclamation, Treatment
   and Utilization of Coal Mining
   Wastes, 7-11, 69.87, Amsterdam,
   Elsevier Science Publ., 1987.
2. M.Ya.Shpirt, "Bezotkhodnaya tekh-
   nologiya of Coal Mining and Treatment Wastes"(Russ.), Moscow, Nedra, 1986.
3. M.Ya.Shpirt, V.R.Kler, I.S.Per-
   sikov, "Inorganic Components of
   Solid Fuel"(Russ.), Moscow, Chemistry, 1990.
4. M.Ya.Shpirt, V.A.Ruban, Yu.V.It-
   kin, "Ispolzovanie toplivocoder-
   zhaschikh otkhodov dobychi i obo-
   gascheniya ugley"(russ.), Moskva,
   Nedra, 1990.

Reclamation, Treatment and Utilization of Coal Mining Wastes, Rainbow (ed.) © 1990 Balkema, Rotterdam. ISBN 90 6191 154 0

Author index